Methods in Enzymology

Volume 408
DNA Repair
Part A

METHODS IN ENZYMOLOGY

EDITORS-IN-CHIEF

John N. Abelson   Melvin I. Simon

DIVISION OF BIOLOGY
CALIFORNIA INSTITUTE OF TECHNOLOGY
PASADENA, CALIFORNIA

FOUNDING EDITORS

Sidney P. Colowick and Nathan O. Kaplan

Methods in Enzymology

Volume 408

DNA Repair
Part A

EDITED BY

Judith Campbell

BRAUN LABORATORIES
CALIFORNIA INSTITUTE OF TECHNOLOGY
PASADENA, CALIFORNIA

Paul Modrich

HOWARD HUGHES MEDICAL INSTITUTE
DUKE UNIVERSITY
DURHAM, NORTH CAROLINA

AMSTERDAM • BOSTON • HEIDELBERG • LONDON
NEW YORK • OXFORD • PARIS • SAN DIEGO
SAN FRANCISCO • SINGAPORE • SYDNEY • TOKYO
Academic Press is an imprint of Elsevier

Academic Press is an imprint of Elsevier
525 B Street, Suite 1900, San Diego, California 92101-4495, USA
84 Theobald's Road, London WC1X 8RR, UK

This book is printed on acid-free paper. ∞

For information on all Academic Press publications
visit our Web site at www.books.elsevier.com

ISBN-13: 978-0-12-182813-4
ISBN-10: 0-12-182813-1

PRINTED IN THE UNITED STATES OF AMERICA
06 07 08 09 9 8 7 6 5 4 3 2 1

Table of Contents

Contributors to Volume 408

Article numbers are in parentheses and following the name of contributors.
Affiliations listed are current.

DIRK K. ANDERSON (32), *Medical Microbiology & Immunology, Creighton University, Omaha, Nebraska*

KATJA BAERENFALLER (18), *Institute of Molecular Cancer Research, University of Zurich, Zurich, Switzerland*

VISWANATH BANDARU (2), *Department of Microbiology and Molecular Genetics, Markey Center for Molecular Genetics, University of Vermont, Burlington, Vermont*

WILLIAM A. BEARD (7), *Enzymology Section, Laboratory of Structural Biology, NIEHS-NIH, North Carolina*

SERGE BERGERON (32), *Creighton University, Medical Microbiology and Immunology, Omaha, Nebraska*

PENNY J. BEUNING (20), *Department of Biology, Massachusetts Institute of Technology, Cambridge, Massachusetts*

JEFFREY O. BLAISDELL (2), *Department of Microbiology and Molecular Genetics, Markey Center for Molecular Genetics, University of Vermont, Burlington, Vermont*

SERGE BOITEUX (6), *CEA, CNRS, Laboratory of Radiobiology DNA, Department Radiobiology and Radiopathology, Aus Roses, France*

JOE BUDMAN (27), *Departments of Medicine and Biochemistry, Stanford University School of Medicine, Stanford, California*

GILBERT CHU (27), *Departments of Medicine and Biochemistry, Stanford University School of Medicine, Stanford, California*

HEATHER A. COKER (10), *DNA Editing Laboratory, Cancer Research United Kingdom, Clare Hall Laboratories, South Mimms, Hertfordshire*

JAMES C. DELANEY (1), *Department of Chemistry and Biological Engineering Division, Massachusetts Institute of Technology, Department of Chemistry, Cambridge, Massachusetts*

BRUCE DEMPLE (4), *Harvard University, Department of Genetics and Complex Diseases, Boston, Massachusetts*

JEAN-MARC EGLY (15), *Institut de Génétique et de Biologie, Moléculaire et Cellulaire, Illkirch Cedex, France*

JEROEN ESSERS (29), *Department of Cell Biology & Genetics, Erasmus MC, Rotterdam, The Netherlands*

JOHN M. ESSIGMANN (1), *Department of Chemistry and Biological Engineering Division, Massachusetts Institute of Technology, Department of Chemistry, Cambridge, Massachusetts*

FRANZISKA FISCHER (18), *Institute of Molecular Cancer Research, University of Zurich, Zurich, Switzerland*

PAULA L. FISCHHABER (22), *Department of Pathology, University of Texas Southwestern Medical Center, Dallas, Texas*

ix

ERROL C. FRIEDBERG (22), *Department of Pathology, University of Texas Southwestern Medical Center, Dallas, Texas*

LIONEL GELLON (4), *Harvard University, Department of Genetics and Complex Diseases, Boston, Massachusetts*

JOCHEN GENSCHEL (17), *Department of Biochemistry, Duke University Medical Center, Durham, North Carolina*

VERONICA G. GODOY (20), *Department of Biology, Massachusetts Institute of Technology, Cambridge, Massachusetts*

MYRON F. GOODMAN (23), *Department of Biological Sciences and Department of Chemistry, University of Southern California, Los Angeles, California*

LIYA GU (19), *Departments of Toxicology and Pathology, University of Kentucky Medical Center, Lexington, Kentucky*

MARIE GUILLET (6), *Department of Pediatric Onocology, Dana-Farber Cancer Institute and Division of Hematology/ Onacology, Children's Hospital of Boston and Harvard Medical School, Boston, Massachusetts*

JAMES E. HABER (26), *Rosenstiel Center, Brandeis University, Waltham, Massachusetts*

PHILIP HANAWALT (14), *Department of Biological Sciences, Stanford University, Stanford, California*

TAPAS K. HAZRA (3), *Sealy Center for Molecular Science and Department of HBC&G, University of Texas Medical Branch, Galveston, Texas*

ADRIAAN B. HOUTSMULLER (29), *Department of Cell Biology and Genetics, Erasmus MC, Rotterdam, The Netherlands*

DANIEL F. JAROSZ (20), *Department of Biology, MIT, Cambridge, Massachusetts*

QINGFEI JIANG (23), *Departments of Biological Sciences and Chemistry, University of Southern California, Los Angeles, California*

JOSEF JIRICNY (18), *Institute of Molecular Cancer Research, University of Zurich, Zurich, Switzerland*

ROBERT E. JOHNSON (24), *Sealy Center for Molecular Science, University of Texas Medical Branch, Galveston, Texas*

ROLAND KANAAR (29), *Department of Cell Biology and Genetics, Erasmus MC, Rotterdam, The Netherlands*

PATRICIA KANNOUCHE (25), *Laboratory of Genetic Instability and Cancer, CNRS, Institut Gustav Roussy, Villejuif, France*

THOMAS A. KUNKEL (21), *Laboratory for Molecular Genetics, NIEHS-NIH, North Carolina*

ISAO KURAOKA (13), *Human Cell Biology Group, Laboratories for Organismal Biosystems, Graduate School of Frontier Biosciences, Osaka University, Yamadaoka, Suita, Osaka, Japan*

JEAN-PHILIPPE LAINE (15), *Institut de Génétique et de Biologie, Moléculaire et Cellulaire, Illkirch Cedex, France*

JI-HOON LEE (33), *Institute of Cellular and Molecular Biology, University of Texas at Austin, Austin, Texas*

ALAN LEHMANN (25), *Genome Damage and Stability Centre, University of Sussex, Falmer, Brighton, United Kingdom*

GUO-MIN LI (19), *Departments of Toxicology and Pathology, University of Kentucky Medical Center, Lexington, Kentucky*

MICHAEL R. LIEBER (31), *Departments of Pathology, Biochemistry and Molecular Biology, Microbiology, and Biology, Norris Comprehensive Cancer Center, University of Southern California Keck School of Medicine, Los Angeles, California*

TOMAS LINDAHL (8), *London Research Institute, Cancer Research United Kingdom, Clare Hall Laboratories, South Mimms, Hertfordshire, United Kingdom*

A-LIEN LU-CHANG (5), *University of Maryland, Department of Biochemistry, Baltimore, Maryland*

YUNMEI MA (31), *Departments of Pathology, Biochemistry and Molecular Biology, Microbiology, and Biology, Norris Comprehensive Cancer Center, University of Southern California Keck School of Medicine, Los Angeles, California*

MARGARET MACRIS (28), *Department of Molecular Biophysics and Biochemistry, Yale University School of Medicine, New Haven, Connecticut*

SCOTT D. MCCULLOCH (21), *Laboratory for Molecular Genetics, NIEHS-NIH, North Carolina*

LISA D. MCDANIEL (22), *Department of Pathology, University of Texas Southwestern Medical Center, Dallas, Texas*

SANKAR MITRA (3), *Sealy Center for Molecular Science and Department of HBC&G, University of Texas Medical Branch, Galveston, Texas*

VINCENT MOCQUET (15), *Institut de Génétique et de Biologie, Moléculaire et Cellulaire, Illkirch Cedex, France*

PAUL MODRICH (17), *Department of Biochemistry, Duke University Medical Center, Durham, North Carolina*

HUGH D. MORGAN (10), *Laboratory of Developmental Genetics and Imprinting, Developmental Genetics Programme, The Babraham Institute, Cambridge, United Kingdom*

TANYA T. PAULL (33), *Institute of Cellular and Molecular Biology, University of Texas at Austin, Austin, Texas*

SVEND K. PETERSEN-MAHRT (10), *DNA Editing Laboratory, Cancer Research United Kingdom, Clare Hall Laboratories, South Mimms, Hertfordshire*

GERD P. PFEIFER (14), *Division of Biology, City of Hope Cancer Centre, Duarte, California*

LOUISE PRAKASH (24), *Sealy Center for Molecular Science, University of Texas Medical Branch, Galveston, Texas*

SATYA PRAKASH (24), *Sealy Center for Molecular Science, University of Texas Medical Branch, Galveston, Texas*

RAJENDRA PRASAD (7), *Enzymology Section, Laboratory of Structural Biology, NIEHS-NIH, North Carolina*

ULRICH RASS (30), *Cancer Research United Kingdom, London Research Institute, Clare Hall Laboratories, South Mimms, Hertfordshire, United Kingdom*

JOYCE T. REARDON (12), *Department of Biochemistry and Biophysics, University of North Carolina School of Medicine, Chapel Hill, North Carolina*

JAMES REID (16), *Cancer Research United Kingdom, London Research Institute, Clare Hall Laboratories, South Mimms, Hertfordshire, United Kingdom*

PETER ROBINS (8), *London Research Institute, Cancer Research United Kingdom, Director of Clare Hall Laboratories, South Mimms, Hertfordshire, United Kingdom*

AZIZ SANCAR (9,12), *Department of Biochemistry and Biophysics, University of North Carona School of Medicine, North Carolina*

GWENDOLYN B. SANCAR (9), *Department of Biochemistry and Biophysics, University of North Carona School of Medicine, North Carolina*

KATHARINA SCHLACHER (23), *Departments of Biological Sciences and of Chemistry, University of Southern California, Los Angeles, California*

BARBARA SEDGWICK (8), *London Research Institute, Cancer Research United Kingdom, Clare Hall Laboratories, South Mimms, Hertfordshire, United Kingdom*

MICHAEL G. SEHORN (28), *Department of Molecular Biophysics and Biochemistry, Yale University School of Medicine, New Haven, Connecticut*

SHAROTKA M. SIMON (20), *Department of Biology, Massachusetts Institute of Technology, Cambridge, Massachusetts*

GRACIELA SPIVAK (14), *Department of Biological Sciences, Stanford University, Stanford, California*

KAORU SUGASAWA (11), *Cellular Physiology Laboratory, RIKEN Discovery Research Institute, Hirosawa, Wako, Saitama, Japan*

NEAL SUGAWARA (26), *Rosenstiel Center, Brandeis University, Waltham, Massachusetts*

PATRICK SUNG (28), *Department of Molecular Biophysics and Biochemistry, Yale University School of Medicine, New Haven, Connecticut*

JUNG-SUK SUNG (4), *Department of Biology, Dongguk University, Seoul, South Korea*

JESPER Q. SVEJSTRUP (16), *Cancer Research United Kingdom, London Research Institute, Clare Hall Laboratories, Blanche Lane, South Mimms, Hertfordshire, United Kingdom*

PATRICK C. SWANSON (32), *Creighton University, Medical Microbiology and Immunology, Omaha, Nebraska*

KIYOJI TANAKA (13), *Human Cell Biology Group, Laboratories for Organismal Biosystems, Graduate School of Frontier Biosciences, Osaka University, Yamadaoka, Suita, Osaka, Japan*

STEPHEN VAN KOMEN (28), *Department of Molecular Biophysics and Biochemistry, Yale University School of Medicine, New Haven, Connecticut*

GRAHAM C. WALKER (20), *Department of Biology, MIT, Cambridge, Massachusetts*

SUSAN S. WALLACE (2), *Department of Microbiology and Molecular Genetics, Markey Center for Molecular Genetics, University of Vermont, Burlington, Vermont*

STEPHEN C. WEST (30), *Cancer Research United Kingdom, London Research Institute, Clare Hall Laboratories, South Mimms, Hertfordshire, United Kingdom*

SAMUEL H. WILSON (7), *Enzymology Section, Laboratory of Structural Biology, NIEHS-NIH, North Carolina*

ROGER WOODGATE (23), *Section on DNA Replication, Repair and Mutagenesis, National Institutes of Health and Human Development, Bethesda, Maryland*

METHODS IN ENZYMOLOGY

VOLUME 391. Liposomes (Part E)
Edited by NEJAT DÜZGÜNEŞ

VOLUME 392. RNA Interference
Edited by ENGELKE ROSSI

VOLUME 393. Circadian Rhythms
Edited by MICHAEL W. YOUNG

VOLUME 394. Nuclear Magnetic Resonance of Biological Macromolecules
(Part C)
Edited by THOMAS L. JAMES

VOLUME 395. Producing the Biochemical Data (Part B)
Edited by ELIZABETH A. ZIMMER AND ERIC H. ROALSON

VOLUME 396. Nitric Oxide (Part E)
Edited by LESTER PACKER AND ENRIQUE CADENAS

VOLUME 397. Environmental Microbiology
Edited by JARED R. LEADBETTER

VOLUME 398. Ubiquitin and Protein Degradation (Part A)
Edited by RAYMOND J. DESHAIES

VOLUME 399. Ubiquitin and Protein Degradation (Part B)
Edited by RAYMOND J. DESHAIES

VOLUME 400. Phase II Conjugation Enzymes and Transport Systems
Edited by HELMUT SIES AND LESTER PACKER

VOLUME 401. Glutathione Transferases and Gamma Glutamyl Transpeptidases
Edited by HELMUT SIES AND LESTER PACKER

VOLUME 402. Biological Mass Spectrometry
Edited by A. L. BURLINGAME

VOLUME 403. GTPases Regulating Membrane Targeting and Fusion
Edited by WILLIAM E. BALCH, CHANNING J. DER, AND ALAN HALL

VOLUME 404. GTPases Regulating Membrane Dynamics
Edited by WILLIAM E. BALCH, CHANNING J. DER, AND ALAN HALL

VOLUME 405. Mass Spectrometry: Modified Proteins and Glycoconjugates
Edited by A. L. BURLINGAME

VOLUME 406. Regulators and Effectors of Small GTPases: Rho Family
Edited by WILLIAM E. BALCH, CHANNING J. DER, AND ALAN HALL

VOLUME 407. Regulators and Effectors of Small GTPases: Ras Family
Edited by WILLIAM E. BALCH, CHANNING J. DER, AND ALAN HALL

VOLUME 408. DNA Repair (Part A)
Edited by JUDITH L. CAMPBELL AND PAUL MODRICH

VOLUME 409. DNA Repair (Part B) (in preparation)
Edited by JUDITH L. CAMPBELL AND PAUL MODRICH

[1] Assays for Determining Lesion Bypass Efficiency and Mutagenicity of Site-Specific DNA Lesions *In Vivo*

By JAMES C. DELANEY and JOHN M. ESSIGMANN

Abstract

DNA damage, if left unrepaired, may hinder translesion synthesis, leading to cytotoxicity, and instruct a DNA polymerase to incorporate an incorrect incipient base opposite the damage, leading to mutagenicity. This chapter describes technology used to measure quantitatively the degree to which a specific type of DNA damage impedes DNA replication. The technology also quantifies the mutation frequency and specificity of such damage after replication within cells. If cells with defined defects in DNA repair are used as hosts for replication, one can pinpoint the specific enzymes or pathways of repair that are operative on specific types of DNA damage.

Introduction

Exposure of cells to DNA-damaging agents usually results in a large array of structurally heterogeneous chemical–DNA adducts. In order to make the task of defining which lesions are cytotoxic and mutatgenic tractable, a technology was developed whereby it is possible to situate defined lesions at known sites in replicable genomes. Such a genome can be passed through cells and analyzed for cytotoxicity and mutagenicity by the methods described in this chapter.

While it falls outside the scope of this chapter to discuss in detail the intricacies of all methods for synthesizing site-specifically modified DNA, the most common methods for placing a modified base within any sequence context employ the use of (1) a suitably protected phosphoramidite and a DNA synthesizer or (2) a modified dNTP, a DNA polymerase (Klenow exo$^-$), and a suitable primer/template to allow for a single base extension of the primer with the modified dNTP; the newly extended primer can then be ligated to another oligonucleotide to internalize the modified base. Because enzymatic incorporation and ligation may not be very efficient, the latter method is usually reserved for cases when a phosphoramidite is not commercially available, easily synthetically accessible, or the modification would not survive the rigors of current DNA synthesis and/or deprotection conditions. Once the modified oligonucleotide is synthesized

METHODS IN ENZYMOLOGY, VOL. 408   0076-6879/06 $35.00

and deprotected (and postsynthetically derivatized, if a "convertible nucleoside" is used), good chemical practice dictates that it be characterized after gel and HPLC purification. Matrix-assisted laser desorption ionization/time of flight mass spectrometry is a fast and convenient ways to verify the mass of the oligonucleotide, whereas HPLC analysis of an enzymatic or chemical hydrolysate will reveal the characteristic retention time and ultraviolet (UV), fluorescent, or electrochemical properties of the modified base which can be characterized further by mass spectrometry or other analytical techniques.

A main focus of our laboratory has been to elucidate how a particular DNA lesion interferes with DNA replication by hindering the polymerase and causing mutations, as well as to measure the repair of DNA lesions by proteins that work on single-stranded (ss) DNA by a direct reversal of base damage mechanism, such as Ada, Ogt, and AlkB. Site-specifically modified ss-M13 genomes are ideal for studying these phenomena, and the following sections describe their construction and use in determining the toxicity, mutagenicity, and repairability of site-specific DNA lesions in a high-throughput manner based on the entire population of progeny after passage through isogenic cells differing in DNA polymerase or repair status.

Genome Construction

A versatile method to construct ss-M13 genomes with DNA adducts in any desired sequence context is shown in Fig. 1. This method was originally used to study the effect of nearest neighbors on O^6-methylguanine mutagenesis in the presence of Ada and Ogt methyltransferases. This study required an efficient, high-throughput method to construct the 16 possible genomes containing the lesion flanked by G, A, T, and C, without ligation bias (Delaney, 1999; Delaney and Essigmann, 1999, 2001). Using a modification of the technique developed in the Lawrence laboratory (Banerjee et al., 1988; Gibbs and Lawrence, 1993), ss-M13mp7(L2) is treated with EcoRI, which cuts the stem of a hairpin, thus linearizing the vector. Instead of placing one large scaffold to span the new ss-vector ends, thus creating a nook for our site-specifically modified oligonucleotide to be nestled, we use two smaller scaffolds, each spanning one end of the cut vector and the modified oligonucleotide insert. The advantage of this approach is twofold in that no base is placed opposite the lesion so spurious signals in bypass and mutagenesis assays, which may have resulted from incomplete destruction of the scaffold, are avoided. Furthermore, one achieves the same ligation efficiency, regardless of the lesion, which becomes an issue when using bulky DNA adducts. After annealing of the 5'-phosphorylated

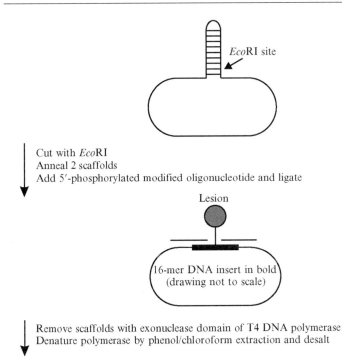

FIG. 1. Construction of site-specifically modified single-stranded viral genomes.

oligonucleotide, T4 DNA ligase cements the modified oligonucleotide into the M13 genome. The scaffolds are destroyed using the technique of Moriya (1993), taking advantage of the powerful 3' to 5' exonuclease of T4 DNA polymerase. Having done this, we perform a phenol/chloroform/ isoamyl alcohol extraction (to denature the T4 DNA polymerase, thus eliminating its possibility of being associated with the genome upon transfection and giving rise to the observed mutations) and desalt using a Sephadex G-50 Fine column or Centricon 100 so that electroporation can be used to introduce the viral DNA into the host.

Protocol for Preparing ss-M13 DNA Vector Starting Material

The preparation of M13 viral DNA starting material is based on the work of Delaney and Essigmann (1999). Briefly, a starter culture is made by plugging a well-isolated plaque with a sterile Pasteur pipette from a lawn of *Escherichia coli* containing an F' episome (GW5100 or NR9050) that has been infected with M13 phage. The plug is vortexed in 1 ml of LB media, and 200 μl of the supernatant is mixed with 10 μl of GW5100 cells (grown overnight, O/N) and grown in 10 ml 2× YT at 37° for 7 h to achieve a typical titer of $>10^{12}$ pfu/ml. Approximately 250 ml of 2× YT is innoculated with 500 μl of GW5100 O/N cells and grown with aeration in a 500-ml baffled flask at 37° for 2 h, followed by the addition of 1 ml of the supernatant from the phage starter culture. After continued growth for 8 h, cells are pelleted and discarded. Prior to workup, it may be prudent to titer the phage in the supernatant and use it to infect F' cells for a double-stranded M13 miniprep, which will reveal if undesired miniphage are present. The yield of purified ss-M13 genome is typically greater than half of what one would expect based on titer (using the conversion factors of 1 $A_{260} = 33$ μg/ml and 2.36 μg/pmol M13). Phage are precipitated with 4% PEG 8000 MW and 0.5 M NaCl at 4° O/N (after 2 h on ice water, a noticeable suspension of phage precipitate should be seen), resuspended in 5 ml TE (pH 8), and extracted 4 × 3 ml with phenol/chloroform/isoamyl alcohol (25:24:1) until the aqueous/organic interface does not change appearance. The aqueous phase is passed through a small hydroxyapatite column (0.5 g Bio-Rad resin, 12 mm D × 16 mm H), washed with 3 ml TE (pH 8), and eluted in ~8 ml of phosphate buffer, which is exchanged for TE (pH 8) using Centricon-100 spin-dialysis columns (Millipore). DNA should be stored at 0.7–1.4 μM at −20°. As an important caveat, if one avoids the hydroxyapatite step, the genomes become nonviable upon the heating conditions described (Delaney and Essigmann, 1999). The yield of purified genome is typically ≥1 pmol/ml 2× YT.

Protocol for Generating M13 Site-Specific Genomes

The construction of site-specifically modified ss-M13 genomes is outlined in Fig. 1. Because many lesions are sensitive to heat, the scaffolds are preannealed to the linearized vector. While the following protocol is based on a 20-pmol vector scale, it can be scaled proportionately; however, concentration of the components is discouraged due to the formation of a viscous mass after heating. A thermocycler with a heated and weighted lid to prevent evaporation is convenient for performing all incubations described in this chapter. Oil-free polymerase chain reaction (PCR) cycling is performed in low-profile plates using Microseal 'A' film, generally avoiding placing samples in the peripheral wells. Other incubations are generally performed in Hard-Shell plates using Microseal 'F' film (MJ Research).

The *Eco*RI digest is performed at 23° for 8 h in 40 μl using 20 pmol ss-M13mp7(L2), 1× *Eco*RI buffer, and 40 U *Eco*RI (New England Biolabs, NEB). Cut vector should not be stored for extended periods of time to reduce the possibility of termini degradation. In a minimal volume to the digestion mixture are added 25 pmol of each scaffold, which are annealed to the newly formed ends of the M13 genome using the profile 50° for 5 min and then 50° to 0° over 50 min. In the illustrated case, the scaffolds (5'-GGT CTT CCA CTG AAT CAT GGT CAT AGC-3' and 5'-AAA ACG ACG GCC AGT GAA TTG GAC GC-3') anneal to 20 bases of the vector, with 6- or 7-base overhangs to house a modified 16-mer (5'-GAA GAC CTX GGC GTC C-3'), leaving a 3-base central gap containing the lesion X. Meanwhile, 30 pmol of the 16-mer insert is 5'-phosphorylated in 30 μl containing 1× T4 polynucleotide kinase (PNK) buffer, 1 m*M* ATP (Amersham Biosciences), 5 m*M* dithiothreitol (DTT), and 15 U PNK (USB Corporation/Amersham Biosciences). Although lower temperatures and shorter incubation times can suffice, typical conditions are 37° for 1 h. Just prior to being added to the kinase-phosphorylated oligonucleotide, the cold annealing mixture is supplemented to achieve the following concentrations (calculated after combination with the oligonucleotide, but excluding the ATP and DTT contributions from the kinase reaction): 1 m*M* ATP, 10 m*M* DTT, 25 μg/ml bovine serum albumin, and 800 U T4 DNA ligase (NEB) in a target volume of 75 μl. After incubation at 16° for 8 h, T4 DNA polymerase (0.25 U/μl, NEB) is added and incubated at 37° for 4 h to remove the scaffolds. The mixture volume is increased to 110 μl with water, extracted once with 100 μl phenol/chloroform/isoamyl alcohol (25:24:1), and the aqueous phase passed through a home-made Sephadex-G50 Fine (Amersham Biosciences) spin column (28 mm packed height of spun-dried resin) to remove traces of phenol and salt (as verified by a 320 nm–220 nm UV scan). Vector recoveries can range from 15%–50% depending on lot, but are consistent when run concurrently. Alternatively, the aqueous phase is mixed with 2 ml of water and spun through a Centricon 100 followed by 3 × 2 ml water washes and final volume adjustment to ~130 μl with a vector recovery of ~30%. Empty columns (QS-GSEM) and frits (D-2262M) can be obtained through PerkinElmer.

Genome Normalization

This section, whereby the relative concentration of genome constructs are determined, is important for the lesion bypass assay, which is explained later. As outlined in Fig. 2, an excess of scaffolds is added to the genome construct, which is incubated with *Hin*FI to cleave the vector 5' to the insert, 5'-radiolabeled, and treated with *Hae*III to cleave the vector 3' to the insert. After polyacrylamide gel electrophoresis (PAGE), the 34-mer resulting from the insert (16-mer) being fully ligated to the M13 vector is quantified. We find that the competitor genome used for the bypass assay (see later) makes a convenient marker (which releases a 37-mer). While multiple PAGE bands are seen due to labeling of full-length and

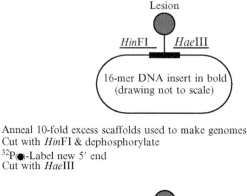

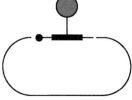

FIG. 2. Normalization of fully ligated genomes used in the CRAB lesion bypass assay (Delaney *et al.*, 2005).

enzyme-digested scaffolds, as well as cleavage of the vector at sites distal to the insert region [*Hin*FI and *Hae*III can cut ss-DNA (NEB technical literature)], the region used for quantification is easily seen. This method can also distinguish the genome construct from uncut M13mp7(L2) wt and competitor, unlike our previous method (Delaney and Essigmann, 2004).

Protocol for Genome Normalization

To desalted vector (~0.35 pmol in 5 μl) are added a 10-fold molar excess of scaffolds used to construct the genome, 1× buffer 2 (NEB), 10 U *Hin*FI (NEB), and 1 U shrimp alkaline phosphatase (Roche), which are incubated in ~8 μl at 37° for 1 h, 80° for 5 min, and then ramped to 20° at 0.2°/s. The mixture is supplemented with 1× buffer 2, 5 mM DTT, ATP (150 pmol cold, premixed with 1.66 pmol [γ-^{32}P]ATP at 6000 Ci/mmol), 5 U PNK, and 10 U *Hae*III (NEB, stock 250 U/μl), which are incubated in 11 μl at 37° for 1 h, after which 9 μl of formamide loading buffer is added and the products separated using 20% PAGE until the xylene cyanol dye has migrated ~12 cm (4 h). PhosphorImagery reveals the relative concentrations of fully ligated genomes, which can now be normalized to one another.

Lesion Bypass (CRAB) Assay

Traditional survival assays used to study lesion bypass rely on scoring the reduction in the number of plaques or colonies formed after transfection and immediate plating of a genome containing a lesion, as compared with its separately transfected nonlesion reference. The competitive replication of adduct bypass (CRAB) assay outlined in Fig. 3 was developed to address several issues (Delaney and Essigmann, 2004). Because transfection efficiencies can be somewhat variable, premixing the lesion-bearing genome with a nonlesion competitor genome prior to transfection allows one to score accurately blocks to replication as a decrease in the lesion-to-competitor output signal ratio, with the competitor acting as an internal standard. Because the transfected cells are allowed to grow in solution prior to analysis, lesions that hinder replication, but may have eventually allowed the formation of a plaque or colony, would be scored properly as replication blockades and their cytotoxicity would not be underestimated. The same global output population is used for both the CRAB and mutation assays to provide a convenient, statistically robust analysis of lesion bypass and mutagenesis.

The lesion/competitor output ratio is determined by the quantification of size-discriminated radiolabeled bands on a gel, after unbiased PCR amplification and *Bbs*I digestion of the mixture, followed by ^{32}P labeling the newly exposed 5' ends and trimming with *Hae*III, as outlined in Fig. 3. The competitor genome was designed to be identical to the lesion-bearing genome, with the exception of a three-base (5'-TAG-3') addition and change of sequence 3' to the lesion region, which allows one to PCR amplify selectively the output from genomes that contained the lesion for the mutagenesis assay described in the next section. The in-frame TAG codon in the competitor also allows one to perform a plaque color assay (Delaney and Essigmann, 1999) to test the lesion-to-competitor input genome ratio prior to doing a full-scale experiment. In this case, the model for a "lesion" is a constructed control genome containing 100% of either G, T, or C at the lesion site in the 16-mer shown in Fig. 3 (to avoid a stop codon), which will form dark blue plaques on X-gal indicator plates, while light blue plaques are formed from the competitor. Because all genome constructs have been carefully normalized to one another as described earlier, the proportion of dark-blue plaques from a G, T, or C control is

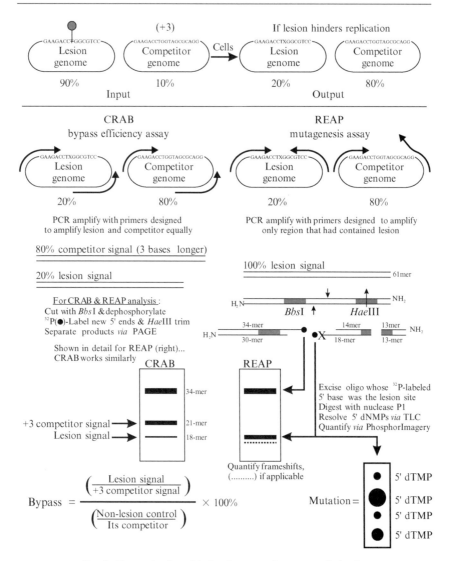

FIG. 3. Determination of lesion bypass and mutagenesis *in vivo*.

identical to the proportion of any bona fide lesion-bearing genome. The CRAB and mutation assays are essentially identical in technique, and the experimental design elements presented here and later should be viewed *in toto*.

Protocol for Lesion Bypass (CRAB) Assay

Desalted viable genomes containing a lesion ($\geq$150 fmol) are mixed with competitor genomes at a 9:1 ratio in $\sim$10 μl prior to transfection. *E. coli* are made electrocompetent by innoculation of 3 $\times$150 ml LB with 3 $\times$ 1.5 ml of cells from an O/N culture. The cells are grown to midlog phase (OD$_{600}$ $\sim$ 0.4–0.5), after which they are kept at 4°, centrifuged, thoroughly resuspended in a minimal volume of water, pooled, and desalted by 3 $\times$ 175 ml water washes and centrifugations, with the final pellet resuspended in a final volume of 6 ml with 10% glycerol. SOS-induced cells are prepared as described by Delaney and Essigmann (2004). A mixture of 100 μl (1 $\times$ 10^9) cells with 10 μl genome is exposed in a 2-mm electroporation cuvette to $\sim$2.5 kV and $\sim$125 Ω, delivering a typical time constant of >4.5 ms, and is immediately transferred to 10 ml LB, after which 1–10 μl is immediately plated onto a lawn of cells to verify that there are enough initial independent events to make a statistically sound conclusion of lesion cytotoxicity and mutagenicity. We typically obtain 10^5–10^6 initial events after electroporation of 150 fmol of viable genome bearing our most toxic lesion. Initial events are obtained by growing a 1:6 dilution of NR9050 O/N cells in 2$\times$ YT at 37° for 90 min and mixing 300 μl cells, 25 μl 1% thiamine, 10 μl isopropyl β-D-thiogalactopyranoside (24 mg/ml), and 40 μl 5-bromo-4-chloro-3-indolyl β-D-galactopyranoside (40 mg/ml X-gal in *N,N*-dimethylformamide) with the transfected cells that secrete M13 virus, which infects NR9050 cells, forming plaques. The components are mixed quickly in B-broth soft agar at 52° and plated on B-broth plates, which are allowed to set for 10 min prior to incubation at 37° for 16 h. This procedure is also used to test the lesion-to-competitor input genome ratio *vide supra*.

The 10-ml culture is grown with agitation at 37° for 6 h, centrifuged, and the supernatant containing progeny phage (as well as genome still bearing the lesion) is stored at 4°. Prior to PCR amplification, it is imperative to make sure that the amplified signal will come from events *within* the cell rather than from artifacts created by action of the PCR DNA polymerase on the lesion. Indeed, with minor adjustments, the mutation assay (see later) can be modified to look at the distribution of bases placed opposite a lesion *in vitro* by a variety of purified DNA polymerases. The progeny: lesion template ratio is increased several orders of magnitude by infection of an O/N culture of SCS110 *E. coli* (10 μl containing $\sim$3 $\times$ 10^6 cells) in 10 ml LB with a well-represented population of M13 progeny (100 μl containing >10^5 phage), which are grown at 37° for 6–7 h to yield $\sim$10^{11} pfu/ml, of which 700 μl is passed through a QIAPrep spin M13 kit (Qiagen) for isolation of PCR template DNA in 100 μl.

PCR for the CRAB assay is performed using the primers 5'-YCAG
CTA TGA CCA TGA TTC AGT GGA AGA C-3' and 5'-YCAG GGT
TTT CCC AGT CAC GAC GTT GTA A-3' (Y is an amino group, which
removes the 5' hydroxyl functionality of DNA). A master mix containing
25 pmol of each primer, 5 nmol each dNTP, 1× cloned *Pfu* DNA polymerase reaction buffer (for 25 μl), and 1.25 U *PfuTurbo* DNA polymerase
(Stratagene, added last) in 10 μl per sample is added to 15 μl of DNA
template and cycled using the following program: 94° for 1 min, followed
by 30 cycles of 94° for 30 s, 67° for 1 min, and 72° for 1 min, with a final
extension of 72° for 5 min. After the addition of 85 μl water, the mixture is
extracted once with phenol/chloroform/isoamyl alcohol (25:24:1), and the
aqueous portion is passed through a dry Sephadex G-50 Fine spin column
(prepared with 2 ml of slurry from 2.5 g washed resin in 38 ml water, which
may need adjustment and calibration depending on lot to verify, by a
320 nm–220 nm UV scan, the complete removal of dNTPs). Four microliters of eluant is treated in a 6-μl reaction volume with 1.5 U *BbsI* (NEB)
and 0.3 U shrimp alkaline phosphatase (Roche) in 1× buffer 2 at 37° for 4
h, followed by heating at 80° for 5 min and cooling to 20° at a rate of 0.2°/s.
This is treated in an 8-μl reaction volume with 1× buffer 2, 5 mM DTT,
ATP [20 pmol cold mixed with 1.66 pmol [γ-^{32}P]ATP (1 μl of NEG-002Z
from PerkinElmer at 6000 Ci/mmol and 10 μCi/μl)] and 5 U PNK at 37° for
15 min, followed by heating at 65° for 20 min and cooling to 23° at a rate of
0.1°/s. After supplementation with 10 U *Hae*III, the 10-μl reaction mixture
is incubated at 37° for 2 h, followed by quenching with 10 μl formamide
containing xylene cyanol FF and bromphenol blue dyes. The mixture is
loaded onto a small 20% denaturing gel prepared as described by Delaney
and Essigmann (1999) and is electrophoresed at 550 V for ~3.5 h until the
18-mer band has migrated through 12.2 cm of gel, as judged by a 10.5-cm
migration of xylene cyanol, thus resolving the 18-mer lesion signal from its
frameshift signals (if present) and the 21-mer competitor signal. The gel is
then exposed to a PhosphorImager screen and scanned with a Storm 840
for quantification of bands (Amersham Biosciences).

Lesion Mutagenesis (REAP) Assay

The restriction endonuclease and postlabeling analysis of mutation
frequency and specificity (REAP) assay was invented to determine the
mutagenicity of several DNA lesions in a variety of sequence contexts,
based on the entire output population of progeny. The base composition at
the lesion site after cellular processing is determined by radiolabeling of
the phosphate 5' to this site and separating the enzymatically hydrolyzed
5'-^{32}P dNMPs on a TLC plate. The assay is high throughput, as mutations

are not scored by sequencing or by plaque hybridization analysis of individual clones, and does not rely on phenotypic selection. The REAP assay can also be adapted easily to verify the integrity of a site-specifically incorporated DNA lesion after genome construction (Delaney and Essigmann, 2004).

As shown in Fig. 3, the sequence of the reverse PCR primer used in the REAP assay is different from that used in the CRAB assay, thus allowing for the selective amplification of progeny from the genome that originally contained the lesion. Because the other (forward) primer anneals to the 5′ end of the insert and PCR is performed until most of the primer has been converted to product, the signal intensity remains robust and nearly equal for all lesions processed in all cellular situations. Progeny containing genetic engineering artifacts such as large deletions that do not contain the insert or uncut wt M13mp7(L2) are not amplified. The 28-mer primers contain an amino group at the 5′ end and therefore do not become radiolabeled, allowing for clean signals on PAGE.

The oligonucleotide insert used to construct the genome also contains the recognition sequence for the type IIs restriction endonuclease *Bbs*I, which will cleave the double-stranded PCR product two and six bases away from its 5′-GAAGAC-3′ recognition site, yielding four-base 5′ overhangs, one of which contains the lesion site at its 5′ terminus. Using a type IIs enzyme grants unfettered access to the lesion site, which can be situated within any local surrounding sequence context. We chose *Bbs*I because it exhibits little exonuclease and nonspecific endonuclease cleavage of DNA, its buffer is compatible with the other enzymes of the assay, and the 5′ overhangs that it generates allows for the efficient removal of the cold 5′ phosphate with shrimp alkaline phosphatase, as well as the efficient addition of ^{32}P with PNK. None of the possible mutation outcomes of any lesion within the sequence context shown in Fig. 3 will generate restriction endonuclease sites for *Bbs*I or *Hae*III, ensuring the proper representation of all possible outcomes.

After PCR, the mixture is phenol extracted to destroy the polymerase and exonuclease activities of *Pfu* DNA polymerase and desalted prior to *Bbs*I cleavage, as dNTPs compete for and are inhibitors of the alkaline phosphatase and PNK used for radiolabeling. Shrimp alkaline phosphatase is conveniently heat inactivated, which is essential prior to phosphorylation with a kinase. It is important to ensure that an excess ratio of ATP to 5′-phosphorylatable DNA ends be maintained to eliminate the preferential ^{32}P labeling of one particular base over another, which PNK may exhibit (Lillehaug and Kleppe, 1975). Indeed, small amounts of a contaminating phosphatase in some commercial preparations of PNK will cause excess ATP to be consumed. The manifestation of any bias that an enzyme may have toward a particular base at the lesion site may be eliminated by making sure that all enzymatic steps have gone to completion for each possible base. One can be confident that the base composition at the lesion site is portrayed accurately by taking progeny that have a nearly equal percentage of G, A, T, and C at the lesion site through the entire REAP assay and verifying that the output percentages of each base remain unchanged.

The radiolabeled *Bbs*I cleavage products are of similar size, and *Hae*III cleavage of the fragment containing the lesion site as the 5′ base allows for their resolution on a small 20% denaturing gel. Indeed, the desired 18-mer generated after *Hae*III treatment is resolved from oligonucleotides differing in length by a single base. Because the PCR primers do not anneal to the five-base region containing the central lesion site, the most common types of frameshift mutations can be quantified directly on the gel, and it is useful to carry a genome through the assay that lacks the lesion site to act as a −1 marker. This control can also serve as a check on PCR contamination, as the ^{32}P-labeled 17-mer produced should not contain 18-mer signal; likewise, another control containing 100% of a particular base at the lesion site should not contain 17-mer signal. Substitution of *Hae*III with *Tsp*509 I can further increase the resolution for frameshift detection, as the band of interest is trimmed down to an 8-mer.

The 5′-^{32}P-labeled 18-mer migrates as a well-defined band on a small 20% denaturing polyacrylamide gel, as the complementary 14-mer cannot remain in equilibrium with the duplex form. During excision of the 18-mer from the gel, it is important not to include any frameshift (or non-18-mer) band, as this will contribute to the final base composition readout on the TLC plate, which is expected to be 100% from the lesion site. However, use of a longer gel is discouraged because the 18-mer band itself will become partially resolved as a function of the 5′ base, and exclusion of such signals would bias results. The small gels described in Delaney and Essigmann (1999) are convenient to handle and take into account these concerns.

After the band containing the ^{32}P-labeled lesion site has been isolated by "crush and soak," it must be thoroughly desalted on a Sephadex column prior to nuclease P1 treatment and TLC resolution. The Sephadex resin is washed several times with water to remove any loose carbohydrates, and the appearance of a white residue after lyophilization of eluted DNA may signify inefficient desalting. Nuclease P1 is favored over snake venom phosphodiesterase (SVPD), as commercial preparations of SVPD sometimes contain adenosine deaminase, which would convert the 5′-^{32}P-dAMP to 5′-^{32}P-deoxyinosine monophosphate. SVPD also exhibits more of a contaminating phosphatase, as evidenced by a more intense radioactive solvent front. Streaking of the radioactive 5′-dNMP spots on the polyethyleneimine (PEI) TLC plate is minimized by washing the plate briefly in water, which presumably removes loose polymer, and by using the liquid from a saturated salt solution for development. All 5'-dNMPs are resolved on PEI-TLC over a pH range of 6.4 to 5.4 for a saturated solution of ammonium phosphate dibasic, and pH 5.8 appears to be optimum.

Protocol for Lesion Mutagenesis (REAP) Assay

The methodology is identical to that used in the CRAB assay, but using the PCR primers 5'-YCAG CTA TGA CCA TGA TTC AGT GGA AGA C-3' and 5'-YTGT AAA ACG ACG GCC AGT GAA TTG GAC G-3'. The resolved 18-mer is excised from the gel as described (Delaney and Essigmann, 2004), crushed and soaked in 200 μl water, and desalted with a Sephadex G-50 Fine spin column prepared as described earlier. The lyophilized residue is resuspended in a 5-μl solution containing 30 mM sodium acetate (pH 5.3), 10 mM ZnCl$_2$, and 1 μg nuclease P1 (Roche) and incubated at 50° for 1 h. PEI-TLC plates (20 × 20 cm, J. T. Baker) that have been soaked for in water for 2 min and dried are spotted with 0.5 μl of digest using Drummond microcaps and are eluted with 200 ml of (NH$_4$)$_2$HPO$_4$ (the supernatant from a saturated solution is adjusted to pH 5.8 with concentrated H$_3$PO$_4$, and the buffer can be reused several times). After the solvent front has migrated to the top ($\sim$12 h), the plates are weighted down at the corners to prevent curling and air dried (>3–4 h) prior to PhosphorImagery.

Conclusion

This chapter described the latest versions of the lesion bypass (CRAB) and mutagenesis (REAP) assays developed in our laboratory. These high-throughput assays greatly simplify the study of how cells deal with a DNA lesion in a well-defined system and are efficient, as the output from a single biological experiment can be used for both assays. Data generated are a snapshot of the entire population of progeny and are therefore statistically robust. Because PCR is used to amplify the region that contained the lesion, one can envision the ability of the REAP assay to evaluate the mutagenicity of lesions integrated within human chromosomes. To date, the REAP mutagenesis assay has been used to evaluate the mutation frequency and specificity of nearly 50 lesions spanning alkylative and oxidative damage, as well as nonnatural atomically mutated base analogs (Delaney, 1999; Delaney and Essigmann, 1999, 2001, 2004; Henderson *et al.*, 2002, 2003, 2005; Delaney *et al.*, 2003, 2005; Kroeger *et al.*, 2004a, b, c; Neeley *et al.*, 2004; Kim *et al.*, 2005, and unpublished), in *E. coli* and cells of higher organisms of different repair and replication background.

References

Banerjee, S. K., Christensen, R. B., Lawrence, C. W., and LeClerc, J. E. (1988). Frequency and spectrum of mutations produced by a single cis-syn thymine-thymine cyclobutane dimer in a single-stranded vector. *Proc. Natl. Acad. Sci. USA* **85**, 8141–8145.

Delaney, J. C. (1999). "Influence of DNA Sequence Context on O^6-Methylguanine Mutagenicity in *Escherichia coli*." Ph.D. Dissertation, The Massachusetts Institute of Technology.

Delaney, J. C., and Essigmann, J. M. (1999). Context-dependent mutagenesis by DNA lesions. *Chem. Biol.* **6**, 743–753.

Delaney, J. C., and Essigmann, J. M. (2001). Effect of sequence context on O^6-methylguanine repair and replication *in vivo*. *Biochemistry* **40**, 14968–14975.

Delaney, J. C., and Essigmann, J. M. (2004). Mutagenesis, genotoxicity, and repair of 1-methyladenine, 3-alkylcytosines, 1-methylguanine, and 3-methylthymine in *alkB Escherichia coli*. *Proc. Natl. Acad. Sci. USA* **101**, 14051–14056.

Delaney, J. C., Henderson, P. T., Helquist, S. A., Morales, J. C., Essigmann, J. M., and Kool, E. T. (2003). High-fidelity *in vivo* replication of DNA base shape mimics without Watson-Crick hydrogen bonds. *Proc. Natl. Acad. Sci. USA* **100**, 4469–4473.

Delaney, J. C., Smeester, L., Wong, C., Frick, L. E., Taghizadeh, K., Wishnok, J. S., Drennan, C. L., Samson, L. D., and Essigmann, J. M. (2005). AlkB reverses etheno DNA lesions caused by lipid oxidation *in vitro* and *in vivo*. *Nature Struct. Mol. Biol.* **12**, 855–860.

Gibbs, P. E., and Lawrence, C. W. (1993). U-U and T-T cyclobutane dimers have different mutational properties. *Nucleic Acids Res.* **21**, 4059–4065.

Henderson, P. T., Delaney, J. C., Gu, F., Tannenbaum, S. R., and Essigmann, J. M. (2002). Oxidation of 7,8-dihydro-8-oxoguanine affords lesions that are potent sources of replication errors *in vivo*. *Biochemistry* **41**, 914–921.

Henderson, P. T., Delaney, J. C., Muller, J. G., Neeley, W. L., Tannenbaum, S. R., Burrows, C. J., and Essigmann, J. M. (2003). The hydantoin lesions formed from oxidation of 7,8-dihydro-8-oxoguanine are potent sources of replication errors *in vivo*. *Biochemistry* **42**, 9257–9262.

Henderson, P. T., Neeley, W. L., Delaney, J. C., Gu, F., Niles, J. C., Hah, S. S., Tannenbaum, S. R., and Essigmann, J. M. (2005). Urea lesion formation in DNA as a consequence of 7,8-dihydro-8-oxoguanine oxidation and hydrolysis provides a potent source of point mutations. *Chem. Res. Toxicol.* **18**, 12–18.

Kim, T. W., Delaney, J. C., Essigmann, J. M., and Kool, E. T. (2005). Probing the active site tightness of DNA polymerase in sub-angstrom increments. *Proc. Natl. Acad. Sci. USA* **102**, 15803–15808.

Kroeger, K. M., Goodman, M. F., and Greenberg, M. M. (2004a). A comprehensive comparison of DNA replication past 2-deoxyribose and its tetrahydrofuran analog in *Escherichia coli*. *Nucleic Acids Res.* **32**, 5480–5485.

Kroeger, K. M., Jiang, Y. L., Kow, Y. W., Goodman, M. F., and Greenberg, M. M. (2004b). Mutagenic effects of 2-deoxyribonolactone in *Escherichia coli*: An abasic lesion that disobeys the A-rule. *Biochemistry* **43**, 6723–6733.

Kroeger, K. M., Kim, J., Goodman, M. F., and Greenberg, M. M. (2004c). Effects of the C4′-oxidized abasic site on replication in *Escherichia coli*: An unusually large deletion is induced by a small lesion. *Biochemistry* **43**, 13621–13627.

Lillehaug, J. R., and Kleppe, K. (1975). Kinetics and specificity of T4 polynucleotide kinase. *Biochemistry* **14**, 1221–1225.

Moriya, M. (1993). Single-stranded shuttle phagemid for mutagenesis studies in mammalian cells: 8-Oxoguanine in DNA induces targeted G·C → T·A transversions in simian kidney cells. *Proc. Natl. Acad. Sci. USA* **90,** 1122–1126.

Neeley, W. L., Delaney, J. C., Henderson, P. T., and Essigmann, J. M. (2004). *In vivo* bypass efficiencies and mutational signatures of the guanine oxidation products 2-aminoimidazolone and 5-guanidino-4-nitroimidazole. *J. Biol. Chem.* **279,** 43568–43573.

[2] Oxidative DNA Glycosylases: Recipes from Cloning to Characterization

By VISWANATH BANDARU, JEFFREY O. BLAISDELL, and SUSAN S. WALLACE

Abstract

As new organisms are being sequenced on a daily basis, new DNA glycosylases that recognize DNA damage can be easily identified in an effort to understand both their phylogenetics and substrate specificities. As a practical matter, existing bacterial and human homologs need to be readily available as laboratory reagents in order to compare the activities of the novel enzymes to existing enzymes. This chapter attempts to provide a primer for cloning, expression, and assay procedures for bacterial and human DNA glycosylases that recognize oxidative DNA damages. These methodologies can be translated readily to novel DNA glycosylases or to DNA glycosylases that recognize other types of DNA damages.

Introduction

Reactive oxygen species (ROS) or free radicals are constantly being generated during electron transport, fatty acid metabolism, and inflammatory responses and can interact with DNA, causing base and sugar damages (for reviews, see Breen and Murphy, 1995; Cadet *et al.*, 2003). The radiolysis of water upon exposure to ionizing radiation also generates ROS and, when present in the vicinity of DNA, can cause singly or multiply damaged sites (for reviews, see Blaisdell *et al.*, 2001; Wallace, 1998). The first step in the base excision repair (BER) process that removes ROS-induced DNA damage is catalyzed by DNA glycosylases. A number of different DNA glycosylases, often with redundant substrate specificities, are present in cells serving as backup activities to remove the damages, thus maintaining genomic stability (for reviews, see Izumi *et al.*, 2003; Slupphaug *et al.*, 2003; Wilson *et al.*, 2003).

METHODS IN ENZYMOLOGY, VOL. 408
0076-6879/06 $35.00
DOI: 10.1016/S0076-6879(06)08002-5

In *Escherichia coli*, four different glycosylases that target oxidative DNA damage, namely formamidopyrimidine DNA glycosylase (EcoFpg), adenine DNA glycosylase, MutY (EcoMutY), endonuclease VIII (EcoNei), and endonuclease III (EcoNth), have been identified, whereas in humans, five oxidative DNA glycosylases, 8-oxoguanine (8-oxoG) DNA glycosylase (hOGG1) and the human homologs of *E. coli* endonuclease III (NTHL1), *E. coli* mutY (MUTYH), and *E. coli* endonuclease VIII (NEIL1 and NEIL2), have been isolated to date. Human OGG1 and *E. coli* Fpg recognize oxidized purine damages, whereas the *E. coli* and human homologs of endonucleases III and VIII recognize oxidized pyrimidine bases. The mismatch repair-specific human and *E. coli* mutY homologs recognize and remove adenine when mispaired with 8-oxoG (for reviews, see Izumi *et al.*, 2003; Slupphaug *et al.*, 2003; Wilson *et al.*, 2003).

Although various laboratories have expressed, purified, and extensively characterized the substrate specificities of the aforementioned DNA glycosylases extensively, this chapter describes laboratory procedures to study DNA glycosylases, including protocols for overexpression and purification. These enzymes serve as laboratory stocks and are used routinely as controls for further characterization of existing DNA glycosylases or when characterizing new DNA glycosylases.

Purification of DNA Glycosylases

Two different expression systems, the IMPACT system (New England Biolabs, Beverly, MA) and the pET System (Novagen, Madison WI), have been tested for overexpression of the DNA glycosylases. The IMPACT system is the preferred choice for expression when a BER protein is readily expressed in *E. coli*. In addition, using the IMPACT system, native protein without a tag can be obtained because the intein tag is cleaved during the purification. In the case of NEIL1 and hOGG1, where the IMPACT system did not yield any protein, we used the pET system to overexpress proteins with a hexa-his tag. For NEIL1, a C-terminal rather than an N-terminal hexa-his tag is required, as Fpg/Nei proteins require a processed N-terminal end to expose the catalytic proline residue (Rieger *et al.*, 2000; Zharkov *et al.*, 1997). In the case of hOGG1 expression, two different pET vectors, pET29a and pET30a, which yield C-terminal tagged or N-terminal tagged proteins, respectively, were tested for expression (Fig. 1A). As shown in Fig. 1B, hOGG1 was expressed only when the gene was spaced ~137 bp from the ribosome-binding site (RBS) in the pET30a vector and as a fusion protein with a ~43 amino acid N-terminal tag, which includes the hexa-his tag and a enterokinase recognition sequence. The optimized expression systems and induction and storage conditions for all the

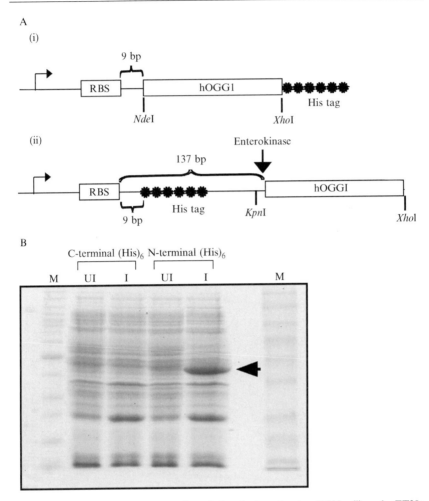

FIG. 1. (A) Schematic representation of the cloning sites in pET29a (i) and pET30a (ii) vectors used for hOGG1 expression. (B) Overexpression of hOGG1 using the pET system. pET29a-hOGG1 or pET30a-hOGG1 vectors were induced with 1 mM IPTG in BL21 (DE3) cells carrying the pLysS/RIR tRNA vector (constructed in our laboratory). Uninduced (UI) and induced (I) whole cell extracts of *E. coli* were loaded onto a 12% SDS–PAGE gel. M, marker.

oxidative DNA glycosylases, as well as for human APEX1, are summarized in Table I.

All the glycosylases except for hOGG1 in the pET system have been cloned between *Nde*I and *Xho*I sites of the expression vector to yield a

TABLE I
EXPRESSION SYSTEMS, INDUCTION AND STORAGE CONDITIONS FOR OXIDATIVE DNA GLYCOSYLASES

BER enzyme	Expression vector	Tag used for purification	Induction strain	Induction conditions	Column(s) for purification	Storage buffer in 50% glycerol[g]
EcoFpg[a]	pTYB2[a]	C-terminal intein	JM109 (DE3)[c]	1 mM IPTG; 16° for 16 h	Chitin-agarose[a] and HiTrap SP FF[f]	50 mM HEPES (pH 7.5), 150 mM NaCl, 1 mM DTT
EcoNei	pTYB2	C-terminal intein	ER2566 fpg[−d]	0.4 mM IPTG; 37° for 3 h	Chitin-agarose HiTrap SP FF	20 mM Tris (pH 7.5), 150 mM NaCl, 1 mM DTT
EcoNth	pET22b[b]	C-terminal intein	BL21(DE3)[b]	1 mM IPTG; 16° for 16 h	HiTrap chelating HP and HiTrap SP FF	50 mM HEPES (pH 7.6), 100 mM NaCl, 1 mM DTT
hOGG1[b]	pET30a[b]	N-terminal (His)$_6$	Rosetta (DE3) pLysS[b]	1 mM IPTG; 16° for 16 h	HiTrap chelating HP[f] and HiTrap SP FF	20 mM Tris (pH 7.5), 100 mM NaCl, 1 mM DTT
NEIL1	pET30a	C-terminal (His)$_6$	Rosetta (DE3) pLysS[b]	1 mM IPTG; 16° for 16 h	HiTrap chelating HP and HiTrap SP FF	20 mM HEPES (pH 7.6), 150 mM NaCl, 1 mM EDTA, 1 mM DTT
NTHL1[h,i]	pTYB2	C-terminal Intein	ER2566[e]	0.3 mM IPTG; 16° for 16 h	Chitin-agarose	50 mM Tris (pH 8.0), 150 mM NaCl, 0.01% Tween 20, 1 mM DTT
APEX1[h,i]	pTYB2	C-terminal Intein	ER2566	0.3 mM IPTG; 30° for 16 h	Chitin-agarose	50 mM Tris (pH 8.0), 100 mM NaCl, 1 mM DTT

[a] New England Biolabs Inc. (Beverly, MA).
[b] Novagen/EMD Biosciences Inc. (Madison, WI).
[c] Promega Corporation (Madison, WI).
[d] Constructed in our laboratory by transducing the fpg⁻ (tetracycline resistance) allele into ER2566.
[e] ER2566: F-lamda-fhuA2 [lon] ompT lacZ::T7 gene1 gal sulA11 D(mcrC-mrr)114::IS10R(mcr-73::miniTn10–TetS)2 R(zgb-210::Tn10) (TetS) endA1 [dcm]; New England Biolabs Inc.
[f] Amersham Biosciences (Piscataway, NJ).
[g] Tris and HEPES are buffered with HCl and NaOH or KOH, respectively.
[h] Human Genome Nomenclature Committee-approved gene symbols for human endonuclease III (hNTH1) and human apurinic/pyrimidinic nuclease (hAPE1).
[i] From Wain et al. (2002, 2004).

C-terminal His tag. For hOGG1, the gene was cloned between *Kpn*I–*Xho*I sites of the pET30a vector with an enterokinase site immediately before the first amino acid of the protein. Induction of pET vectors carrying BER proteins in bacterial strains is carried out essentially as described in the pET manual (Novagen, Madison, WI). The hexa-his-tagged proteins are purified using ÄKTAprime plus (Amersham Biosciences, Piscataway, NJ) according to the manufacturer's recommendations. The conditions for purification on a HiTrap chelating HP column followed by a HiTrap SP FF column are given in Table II. While the amount of protein induced varies with the particular BER enzyme, typically 5-ml columns are used to purify proteins from 2 to 6 liter of induced cultures. The lysate is loaded onto the chelating column with 100% buffer A, washed with 2–4 column volumes of buffer A before eluting the protein with a linear gradient of 0–100% B in 10–20 column volumes. Following elution from the chelating column, protein-containing fractions are pooled and dialyzed into HiTrap SP FF buffer A before loading onto the second column. The conditions for load, wash, and elution are same as those for the chelating column. After purification on the HiTrap SP FF column, protein-containing fractions are pooled, concentrated, and dialyzed against storage buffer in two steps: first into storage buffer with 25% glycerol and then into the storage buffer with 50% glycerol. All enzymes are quantified by the Bradford assay and stored at –20° until use.

All the BER genes in the IMPACT system have been cloned between *Nde*I and *Sma*I sites of the pTYB2 vector to allow expression of the protein fused to the N terminus of the intein. Conditions for purifying BER proteins using the IMPACT system are summarized in Tables III and IV. The protocols for induction and purification are essentially as described in the IMPACT manual (New England Biolabs, Beverly, MA). The low salt buffer containing 1 mM TCEP·HCl (reducing agent) (Pierce Biotechnology, Inc.) and 20 μM phenylmethylsulfonyl fluoride (PMSF) (a serine protease inhibitor) are used for cell lysis by a sonicator or French press. Following elution, the proteins except EcoFpg and EcoNei are concentrated and dialyzed into storage buffer containing 25% glycerol first and subsequently into storage buffer with 50% glycerol. A second round of purification on a HiTrap SP FF column is included for EcoFpg and EcoNei before dialyzing the protein into storage buffer (Table IV).

Preparation of Oligonucleotide Substrates

Double-stranded oligonucleotides carrying oxidative DNA damages such as 8-oxoguanine and thymine glycol are used routinely to test the activity of DNA glycosylases (for examples, see Bandaru *et al.*, 2002;

TABLE II
Buffer Conditions for Purification of Oxidative DNA Glycosylases in the pET System Using ÄKTAprime Plus

Column for purification	Buffer A	Buffer B	ÄKTA conditions
HiTrap chelating HP Column charged with 100 mM nickel sulfate	50 mM sodium phosphate (pH 8.0) 100 mM NaCl 10 mM imidazole 10% glycerol 5 mM β-mercaptoethanol	50 mM sodium phosphate (pH 8.0) 150 mM NaCl 500 mM imidazole (pH 8.0) 10% glycerol 5 mM β-mercaptoethanol	Flow rate: 2 ml/min Max pressure: 0.5 MPa Fraction: 1 ml Gradient: 0–100% B in 10–20 column volumes
HiTrap SP FF	20 mM HEPES (pH 7.6)[a] 150 mM NaCl 10% glycerol 5 mM β-mercaptoethanol	20 mM HEPES (pH 7.6)[a] 1 M NaCl 10% glycerol 5 mM β-mercaptoethanol	Flow rate: 2 ml/min Max pressure: 0.7 MPa Fraction: 1 ml Gradient: 0–100% B in 10–20 column volumes

[a] HEPES is buffered with either NaOH or KOH.

TABLE III

BUFFER CONDITIONS FOR PURIFICATION OF OXIDATIVE DNA GLYCOSYLASES IN THE IMPACT SYSTEM ON A CHITIN AGAROSE COLUMN

Column for purification	BER enzyme	Low salt buffer[a]	High salt buffer[a]	Cleavage buffer[a]	Elution buffer[a]
Chitin agarose	EcoFpg	50 mM HEPES (pH 7.5) 100 mM NaCl 0.1 mM EDTA	50 mM HEPES (pH 7.5) 1 M NaCl 0.1 mM EDTA 0.1% Tween 20	50 mM HEPES (pH 7.5) 1 M NaCl 0.1 mM EDTA	50 mM HEPES (pH 7.5) 100 mM NaCl 1 mM DTT
	EcoNei	50 mM Tris (pH 7.5) 100 mM NaCl 0.1 mM EDTA	50 mM Tris (pH 7.5) 500 mM NaCl 0.1 mM EDTA 0.1% Tween 20	50 mM Tris (pH 7.5) 200 mM NaCl 50 mM DTT	50 mM Tris (pH 7.5) 200 mM NaCl 1 mM DTT
	APEX1	50 mM Tris (pH 8.0) 50 mM NaCl 1 mM EDTA	50 mM Tris (pH 8.0) 500 mM NaCl 1 mM EDTA 0.1% Tween 20	50 mM Tris (pH 8.0) 50 mM NaCl 50 mM DTT	50 mM Tris (pH 8.0) 100 mM NaCl 1 mM DTT
	NTHL1	50 mM Tris (pH 8.0) 50 mM NaCl 1 mM EDTA	50 mM Tris (pH 8.0) 500 mM NaCl 1 mM EDTA 0.1% Tween 20	50 mM Tris (pH 8.0) 50 mM NaCl 50 mM DTT	50 mM Tris (pH 8.0) 100 mM NaCl 1 mM DTT 0.01% Tween 20

[a] Tris and HEPES are buffered with HCl and NaOH or KOH, respectively.

TABLE IV
BUFFER CONDITIONS FOR PURIFICATION OF OXIDATIVE DNA GLYCOSYLASES IN THE
IMPACT SYSTEM ON A CATION-EXCHANGE COLUMN

Column	BER enzyme	Buffer C[a]	Buffer D[a]
HiTrap SP FF	EcoFpg	50 mM HEPES (pH 7.5)	50 mM HEPES (pH 7.5)
		100 mM NaCl	1 M NaCl
		5% glycerol	5% glycerol
		1 mM DTT	1 mM DTT
	EcoNei	20 mM Tris (pH 7.5)	20 mM Tris (pH 7.5)
		100 mM NaCl	500 mM NaCl
		10% glycerol	10% glycerol
		1 mM DTT	1 mM DTT

[a] Tris and HEPES are buffered with HCl and NaOH or KOH, respectively.

Harrison et al., 1999; Hatahet et al., 1994; Jiang et al., 1997; Purmal et al., 1998). Although damage-containing oligonucleotides can be purchased from several companies, the quality of the product must be verified before use. For example, oligonucleotides containing 5,6-dihydrothymine (DHT) or 5,6-dihydrouracil (DHU) require the use of ultramild deprotection and hence the use of ultramild phosphoramidites during synthesis. In our past experience, we have received oligonucleotides containing DHU or DHT that were synthesized with regular phosphoramidites, which upon deprotection resulted in the loss of the damaged base, creating abasic sites in the oligonucleotides. Other problems included incomplete deprotection of oligonucleotides synthesized under milder conditions. Performing mass spectral analysis to verify the molecular weight of oligonucleotides can assure their quality. However, oligonucleotides longer than 50 bases do not yield reliable mass spectral data. Furthermore, oligonucleotides that are G rich also have problems with mass spectral analysis. Before phosphoramidites for damaged bases were available commercially, we incorporated modified nucleoside triphosphates enzymatically, such as 8-oxodGTP, with DNA polymerases to make 54-mer oligonucleotides (Hatahet et al., 1993). However, with the availability of a wide range of phosphoramidites, including damages such as thymine glycol and cis–syn thymine dimers from Glen Research (Sterling VA), we started to purchase damage-containing 35-mer oligonucleotides (5'-TGTCAATAGCAAGXGGAGAAGTCAA TCGTGAGTCT-3'; X = damaged base) from Midland Certified Reagent Company (Midland, TX) with the damage placed asymmetrically in the sequence.

Oligonucleotides synthesized on either 0.2 or 1.0 μM scales are purified as follows. Oligonucleotides are resuspended in water, mixed with an equal volume of formamide, and loaded onto a 15% polyacrylamide (PAGE) gel. Bromphenol blue is added to an additional lane to determine the migration distance of oligonucleotides if needed. Following the run, the oligonucleotide is visualized by ultraviolet (UV) shadowing with a hand-held longwave UV light and excised by cutting the top 75% of the band from the gel to guarantee removal of any incomplete synthesis products (n-1 moieties). Fluorescent thin-layer chromatography plates are not needed to visualize the band when preparing at or above the 0.2 μM scale; glass plates alone can be used. The gel fragments are transferred into sterile, disposable 15-ml conical tubes and "mashed" via agitation (vortexed and shaken by hand, repeatedly). A 15-ml solution consisting of 10 mM Tris (pH 8.0), 1 mM EDTA, and 0.5 M NaCl is added to the sample and allowed to soak with rocking overnight at room temperature. The solution is decanted and placed at 4° for storage while a second 15-ml soak is performed overnight. After removing the gel fragments by filtering through a 0.45-μm syringe filter, the sample is loaded onto a Sep-Pak C18 column (Waters Corporation, Milford, MA) as follows: 7 ml of 100% acetonitrile is passed through to prepare the column followed by a 20-ml sample of "binding buffer" (25 mM TEAB in water). Next, the sample is passed through the C18 column followed by a 4-ml wash with binding buffer. Finally, 4 ml of "elution buffer" (0.1 M TEAB, 75% acetonitrile in water) is used to elute the sample from the column and the sample is dried on low heat in a Speed-Vac or lyophilized.

In a laboratory setting with multiple researchers working with many different enzymes, the quality and supply of damaged oligonucleotides can easily become problematic. Although the wide range of phosphoramidites available these days has made obtaining damaged substrates trivial, how the end user handles the product should not be. In order to maximize both reliability and reproducibility (and, if necessary, regulation) our laboratory has been using a centralized storage approach. Specifically, after drying the purified oligonucleotide, the sample is resuspended in 1 ml of water. The concentration of the sample is determined via spectrophotometry, and the entire stock is diluted to a 2-pmol/μl solution. Fifty microliter volumes (100 pmol) are aliquoted to multiple capless microcentrifuge tubes (Fisher Scientific) and either lyophilized or placed on low heat in a Speed-Vac until dry. Samples are then capped (Fisher Scientific) and labeled with preprinted clear labels and stored at -20° until use. Depending on the synthesis and purification, a 0.2 μM scale can yield anywhere from 200 to 500 samples per oligonucleotide. We started this approach approximately 2 years ago and have not seen any degradation or loss of material to date. In preparing 8 to 10 different damages, including all complements, even in a single sequence context, one can easily see that there is a large initial investment in time and money. However, we feel the benefits of using this approach far outweigh the cost.

Damaged oligonucleotides are 5′ labeled by T4 polynucleotide kinase (T4 PNK) (New England Biolabs, Beverly MA) as follows: 1 μl of T4 PNK (1–5 units) is added to a 19-μl mixture containing 1 μl (1 pmol) of damaged oligonucleotide, 2 μl 10× T4 PNK buffer, 15 μl water, and 1 μl of a 10-μCi/μl stock of [γ -^{32}P]ATP (6000 ci/mmol) (NEN). After incubating at 37° for 30 min, the reaction is terminated by heating at 95° for 5 min, followed by ethanol precipitation to remove unincorporated radioactivity. Following a 70% ethanol wash, the oligonucleotides are dried under low heat in a Speed-Vac. The dried product is then resuspended in 20 μl of 10 mM Tris·HCl (pH 8.0), 1 mM EDTA, and 50 mM NaCl containing 9 pmol of cold damaged oligonucleotide and 10–15 pmol of the complementary strand oligonucleotide. To make a double-stranded substrate, the sample is heated to 95° for 3 min and allowed to slow cool for 30 min at room temperature.

Activity Assays for Oxidative DNA Glycosylases

Activity Assays with Purified Recombinant Enzymes

Assays with damage-containing double-stranded oligonucleotides are done to (a) test the activity of the purified recombinant protein both qualitatively and quantitatively, (b) verify knockout or knockdown mutants devoid of the BER enzyme of interest, and (c) obtain a qualitative activity in cells overexpressing BER enzymes.

The creation of a DNA substrate panel for purified glycosylases is a practical first experiment for any enzyme with previously unknown specificity. Although qualitative in nature, substrate panels are very informative when comparing multiple enzymes with various damaged oligonucleotides. When working with new enzymes, particularly mutant proteins, an important issue to take into consideration is the dissociation constant (K_d) of the enzyme/substrate complex. To elucidate this further, assume Michaelis–Menton kinetics given the following enzymatic mechanism:

$$E + S <==> ES : k_1, k_2 \tag{1}$$

$$ES \rightarrow E + P : k_3 \tag{2}$$

where E, S, and P are enzyme, substrate, and product, respectively; k_1 is the kinetic constant specific to association, k_2 to dissociation, and k_3 the catalytic constant; we can therefore formulate the following standard equations:

$$K_M = (k_2 + k_3)/k_1 \qquad (3)$$

$$K_d = k_2/k_1 \qquad (4)$$

Examining the aforementioned equations, it is clear that an enzyme displaying a decreased activity on a particular substrate may be doing so based on a decreased affinity for the substrate (greater K_d), a decreased catalytic activity (lower k_3) once bound, or a combination of both. Therefore, increasing the initial concentrations of enzyme, substrate, or both may be necessary to determine substrate specificity in some cases before a qualitative ruling of "activity" is made. A typical substrate panel for a DNA glycosylase, in this case human endonuclease VIII homolog endonucleus VIII homolog NEIL2, is shown in Fig. 2. It can be seen in Fig. 2 that like EcoNth, EcoNei, NTHL1, and NEIL1, the DNA glycosylase of interest (NEIL2) cleaves a duplex substrate containing thymine glycol (Tg) or 5,6-dihydrothymine (DHT). Unlike EcoFpg, 8-oxoguanine is a poor substrate for NEIL2 (see also Hazra *et al.*, 2002; Wallace, 1998). When designing a substrate panel, a number of controls should be run alongside experimental lanes. First, enzymes with known activity should be used to determine the end chemistry of the cleaved products, i.e., β elimination (Nth family) or β,δ elimination product (Fpg/Nei family) (Fig. 2). Second, a no enzyme control should always be used to verify that the substrate is not degraded during storage or handling (Fig. 2; lanes 1, 9, and 17). As shown in Fig. 2 (lane 9), deprotection of ring saturation damages such as DHT can sometimes produce abasic sites. Finally, because NEIL1 and NEIL2 have been shown to be active on single-stranded DNA (Dou *et al.*, 2003), it is important to include single-stranded DNA substrates when doing substrate panels with novel DNA glycosylases.

The assay buffers for various DNA glycosylases used in our laboratory are summarized in Table V. It is imperative that reaction conditions such as pH, bovine serum albumin (BSA), and salt concentration be optimized for a new DNA glycosylase by picking a good substrate for that enzyme based on a qualitative assay with different substrates. Once optimized, these reaction conditions can be used to generate a substrate panel for the glycosylase of interest with appropriate controls as shown in Fig. 2.

Determination of Kinetic Constants Using Purified Enzymes

Substrate panels, as described earlier, are a useful qualitative analysis of enzyme activity over a broad range of substrates. However, for quantitative comparisons of a particular enzyme activity on different substrates or for comparisons of different enzymes on common substrates, kinetic rate constants (k_{cat}, K_m, V_{max}) must be calculated. Traditionally, the rate constants

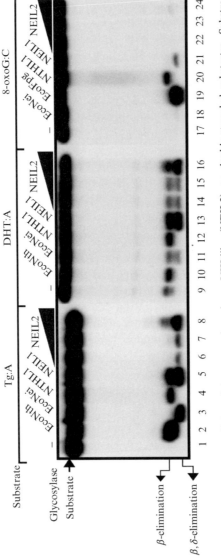

Fig. 2. Substrate specificity of human endonuclease VIII-like (NEIL2) on double-stranded substrates. Substrate (2.5 nM) containing thymine glycol [(Tg) lanes 1–8], 5,6-dihydrothymine [(DHT) lanes 9–16], or 8-oxoguanine [(8-oxoG) lanes 17–24] was incubated with 5 nM of EcoNth, EcoNei, and NTHL1, 10 nM NEIL1, and various concentrations of NEIL2: 10 nM (lanes 6, 14, and 22), 25 nM (lanes 7, 15, and 23), and 250 nM (lanes 8, 16, and 24).

TABLE V
REACTION BUFFERS FOR VARIOUS DNA GLYCOSYLASES

BER enzyme	Reaction buffer[a]
EcoFpg	10 mM Tris (pH 7.5), 50–100 mM NaCl, 1 mM EDTA
EcoNei	10 mM Tris (pH 7.5), 50–100 mM NaCl, 1 mM EDTA
EcoNth	10 mM Tris (pH 7.6), 50–100 mM NaCl, 1 mM EDTA, 0.05 mg/ml BSA
hOGG1	20 mM Tris (pH 7.4), 100 mM NaCl, 0.15 mg/ml BSA
NEIL1	20 mM CHES (pH 9.5), 100 mM NaCl, 1 mM EDTA
NTHL1	50 mM Tris (pH 8.0), 75 mM NaCl, 1 mM DTT
APEX1	10 mM HEPES (pH 6.5), 100 mM KCl, 10 mM MgCl$_2$ [b]

[a] Tris and HEPES/CHES are buffered with HCl and either NaOH or KOH, respectively.
[b] Same as 1× REC buffer 7 from Trevigen (Gaithersburg, MD).

are determined under multiple turnover conditions from a plot of initial velocity (v_0) vs substrate concentration [S] for a single enzyme concentration and multiple substrate concentrations. Data may then be transformed into a linear plot to determine kinetic constants, e.g., a Lineweaver–Burk or Eadie–Hofstee plot. The main issue with this approach is that linear regressions assume residuals that are distributed normally such that the standard deviation is the same for all values on the X axis, an assumption that cannot be guaranteed when data are linearized.

Fortunately, many software packages are now capable of performing nonlinear regression (Seber and Wild, 2003) on arbitrary models that define Y axis values as a function of X axis values. However, most require that data be in the form of initial velocity vs substrate concentration due to the difficulties associated with fitting progress curves (substrate cleaved vs time). The problem with this approach is that extracting the linear portion of a progress curve to determine the initial velocity is often difficult because the curve, by definition, is nonlinear. Additionally, due to the speed of the reaction, a degree of imprecision is often added by determining the initial velocity with a single time point. In order to fit progress curves, a program that fits data to models defined by differential equations must be used (Duggleby, 1995). The software used currently in our laboratory for the calculation of kinetic parameters is the program Dynafit (Kuzmic, 1996). The main benefit of Dynafit is the ability to define models in a mechanistic form similar to Eqs. (1) and (2) given earlier and the automatic generation of differential equations describing that model. The differential equations define dY/dX as a function of one or more variables and are used during nonlinear least-squares regression to find a best fit of experimental data to the defined model.

Independent of the approach or software chosen, there are two important considerations with regard to experimental design. First, determination of the active fraction of enzyme is necessary for obtaining accurate rate constants. This is typically done using a Schiff base assay as described later. However, if the enzyme–product complex dissociates at a much slower rate then the catalytic cleavage step, the active fraction can be determined via the burst phase (Porello *et al.*, 1998; Williams and David, 2000). Second, under multiple turnover conditions, the substrate concentration range chosen should vary from well below the estimated K_m of the enzyme to well above whenever possible. Accordingly, in some cases it may be necessary to increase both enzyme and substrate concentrations used, notably with enzyme mutants.

Schiff Base Assay

A key intermediate in the reaction mechanism for all bifunctional DNA glycosylases that retain both glycosylase activity (cleave the damaged base) and AP lyase activity (ability to cleave the phosphodiester backbone) is the formation of a Schiff base (Bailly and Verly, 1987; Dodson *et al.*, 1994; Kow and Wallace, 1987; Mazumder *et al.*, 1991). This intermediate can be trapped covalently by reducing the transient intermediate with sodium borohydride ($NaBH_4$) or sodium cyanoborohydride ($NaCNBH_3$) (Dodson *et al.*, 1994; Nash *et al.*, 1996; Sun *et al.*, 1995). While this trapping strategy is used routinely to demonstrate the bifunctionality of a DNA glycosylase, this approach has also been used successfully in determining crystal structures of DNA glycosylases bound to the substrate but missing the damaged base (for a review, see Verdine and Norman, 2003).

Although reducing agents $NaBH_4$ and $NaCNBH_3$ can covalently trap the Schiff base intermediate, the ability to efficiently trap the intermediates formed depends on the choice of the reducing agent. Specifically, when $NaCNBH_3$ is used, and because it is a weak reducing agent, a large percentage of Schiff base intermediates proceed to product formation rather than being trapped (Burgess *et al.*, 2002). However, using $NaBH_4$, which is a stronger reducing agent, the abasic site or the nonreducing sugar moiety is less stable with a half-life of 12 s compared to a half-life of 6 h in $NaCNBH_3$ (Manuel *et al.*, 2004; McCullough *et al.*, 2001). Hence the choice of reducing agent may depend on the DNA glycosylase and/or the substrate, e.g., an oligonucleotide with an abasic site versus an oligonucleotide with a damaged base. Regardless, the protocol for trapping the Schiff base intermediate with either reducing agent is the same as that used for DNA glycosylase activity. Briefly, the 5′ end-labeled substrate in the DNA glycosylase reaction buffer is incubated with 25–100 mM of the freshly prepared reducing agent. The reaction is initiated by adding the enzyme simultaneously with the addition of the reducing agent and allowed to incubate for 30–60 min. The concentration of the reducing agent may need to be optimized depending on the DNA glycosylase or the type of reducing agent used, but typically 50 mM of the reducing agent can be used as a starting point for optimization. The reaction is terminated by the addition of SDS–PAGE loading buffer [125 mM Tris (pH 7.6), 10% β-mercaptoethanol, 4% SDS, 10% glycerol, and 0.004% bromphenol blue), denatured by heating to 95° for 5 min, and loaded onto a 10% SDS–PAGE gel. We routinely use the Schiff base trapping assay to determine the percentage active enzyme for kinetic analysis.

Preparation and Activity of Bacterial Cell Extracts

The following procedure is used to make cell extracts from BER enzyme mutants or bacteria overexpressing a DNA glycosylase. *E. coli* cells are harvested from a 10-ml overnight culture of each strain and are resuspended in 1 ml 2× TE (10 mM Tris, 1 mM EDTA) followed by the addition of 1–2 μl of 10% lysozyme in water. The lysozyme-treated cell suspension is then incubated at 37° for 10 min followed by two freeze/thaw cycles: 10 min at –70° and 2 min at 37°. The thawed samples are then sonicated at 4° for 5 min to shear the genomic DNA. This is necessary to allow radiolabeled oligonucleotide susbtrates to migrate out of the wells when high percentage polyacrylamide gels are used. For the glycosylase assay, 5 fmol of the double-stranded 5′ end-labeled damage-containing substrate is incubated with 1–2 μg of crude cell extract in 10 mM Tris·HCl (pH 8.0), 50 mM NaCl, and 10 mM EDTA at 37° for 10 min. The addition of EDTA is particularly important so as to inhibit degradation of the substrate and/or the product by cellular exonucleases. The reactions are stopped by the addition of an equal volume of formamide and separated on a PAGE gel. A glycosylase assay using 8-oxoG and Tg substrates with cell extracts made from glycosylase mutants of the bacterial strain BW35 is shown in Fig. 3. As shown, wild-type extracts (lanes 11 and 12) and mutant extracts (lanes 1–10) when incubated with damage-containing Tg:A or 8oxoG:C substrates generate products with differing end chemistries: β (from Nth lyase activity) β,δ (from Fpg or Nei lyase activity), or 3′ hydroxyl (from AP endonuclease processing of the lyase products). Therefore, assays with cell extracts are particularly useful in verifying knockout mutants and, as mentioned earlier, provide a quick method to test the activity not only of wild-type enzymes, but also to study the effects of site-directed mutations on the activity of a specific DNA glycosylase.

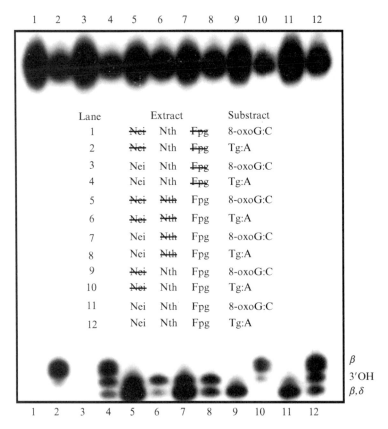

FIG. 3. Glycosylase assay using bacterial cell extracts. Six extracts of *E. coli* strain BW35—*nei fpg* double mutant (lanes 1 and 2), *fpg* mutant (lanes 3 and 4), *nei nth* double mutant (lanes 5 and 6), *nth* mutant (lanes 7 and 8), *nei* mutant (lanes 9 and 10), and wild-type (lanes 11 and 12) cells—were incubated with either a Tg:A substrate (even lanes) or an 8-oxoG:C substrate (odd lanes), as described in the text. Glycosylase enzymes (Nei, Nth, and Fpg) not present in the extract due to genomic mutations are shown using strike-through text, whereas functional enzymes present in the extract are shown with standard text (no strike through).

Preparation of Human Whole Cell Extracts

DNA glycosylase assays using whole cell extracts of human cells provide a method to test the incision capability of extracts devoid or overexpressing DNA glycosylases. We used this approach to measure incision activity in a variety of substrates in cell extracts from human B lymphoblastoid TK6 cells (Inoue *et al.*, 2004), from TK6 cells overexpressing hOGG1 or NTHL1 (Yang *et al.*, 2004), and from TK6 cells where hOGG1 or NTHL1 was knocked down using siRNA (Yang *et al.*, 2005). The protocol for making whole cell extracts is as follows. Cells ($\sim 10^8$) are pelleted by centrifugation at 3000 rpm (1000*g*) for 15 min and washed with ice-cold phosphate-buffered saline (PBS). The washed pellet is resuspended in 1 ml of lysis buffer [70 mM HEPES·NaOH (pH 7.6), 0.4 M NaCl, 1 mM EDTA, 1 mM dithiothreitol (DTT), and 10% glycerol] and flash frozen until use. To make the extract, the cell suspension is thawed on ice and treated with 0.2% Triton X-100 (v/v) to complete the cell lysis. Because the amount of extract used to detect activity depends on the expression levels of a given BER enzyme, one should optimize the amount of cell extract needed to detect activity. Furthermore, optimization of salt (NaCl/KCl), EDTA, and magnesium chloride concentrations for each DNA glycosylase is essential to obtain reproducible results with whole cell extracts. This approach has also worked well in our hands for HeLa cell nuclear extracts (Kathe *et al.*, 2004) and nuclear, mitochondrial, and whole cell extracts from IMR 90 fibroblasts (Shen *et al.*, 2003).

Acknowledgment

Research from this laboratory was supported by NIH PHS CA 33657 awarded to S.S.W. by the National Cancer Institute.

References

Bailly, V., and Verly, W. G. (1987). *Escherichia coli* endonuclease III is not an endonuclease but a beta-elimination catalyst. *Biochem. J.* **242,** 565–572.

Bandaru, V., Sunkara, S., Wallace, S. S., and Bond, J. P. (2002). A novel human DNA glycosylase that removes oxidative DNA damage and is homologous to *Escherichia coli* endonuclease VIII. *DNA Repair* **1,** 517–529.

Blaisdell, J. O., Harrison, L., and Wallace, S. S. (2001). Base excision repair processing of radiation-induced clustered DNA lesions. *Radiat. Protein Dosimetry* **97,** 25–31.

Breen, A. P., and Murphy, J. A. (1995). Reactions of oxyl radicals with DNA. *Free Radic. Biol. Med.* **18,** 1033–1077.

Burgess, S., Jaruga, P., Dodson, M. L., Dizdaroglu, M., and Lloyd, R. S. (2002). Determination of active site residues in *Escherichia coli* endonuclease VIII. *J. Biol. Chem.* **277,** 2938–2944.

Cadet, J., Douki, T., Gasparutto, D., and Ravanat, J. L. (2003). Oxidative damage to DNA: Formation, measurement and biochemical features. *Mutat. Res.* **531,** 5–23.

Dodson, M. L., Michaels, M. L., and Lloyd, R. S. (1994). Unified catalytic mechanism for DNA glycosylases. *J. Biol. Chem.* **269,** 32709–32712.

Dou, H., Mitra, S., and Hazra, T. K. (2003). Repair of oxidized bases in DNA bubble structures by human DNA glycosylases NEIL1 and NEIL2. *J. Biol. Chem.* **278,** 49679–49684.

Duggleby, R. G. (1995). Analysis of enzyme progress curves by nonlinear regression. *Methods Enzymol.* **249,** 61–90.

Harrison, L., Hatahet, Z., and Wallace, S. S. (1999). *In vitro* repair of synthetic ionizing radiation-induced multiply damaged DNA sites. *J. Mol. Biol.* **290,** 667–684.

Hatahet, Z., Purmal, A. A., and Wallace, S. S. (1993). A novel method for site specific introduction of single model oxidative DNA lesions into oligodeoxyribonucleotides. *Nucleic Acids Res.* **21,** 1563–1568.

Hatahet, Z., Kow, Y. W., Purmal, A. A., Cunningham, R. P., and Wallace, S. S. (1994). New substrates for old enzymes: 5-Hydroxy-2'-deoxycytidine and 5-hydroxy-2'-deoxyuridine are substrates for *Escherichia coli* endonuclease III and formamidopyrimidine DNA N-glycosylase, while 5-hydroxy-2'-deoxyuridine is a substrate for uracil DNA N-glycosylase. *J. Biol. Chem.* **269,** 18814–18820.

Inoue, M., Shen, G. P., Chaudhry, M. A., Galick, H., Blaisdell, J. O., and Wallace, S. S. (2004). Expression of the oxidative base excision repair enzymes is not induced in TK6 human lymphoblastoid cells after low doses of ionizing radiation. *Radiat. Res.* **161,** 409–417.

Izumi, T., Wiederhold, L. R., Roy, G., Roy, R., Jaiswal, A., Bhakat, K. K., Mitra, S., and Hazra, T. K. (2003). Mammalian DNA base excision repair proteins: Their interactions and role in repair of oxidative DNA damage. *Toxicology* **193,** 43–65.

Jiang, D., Hatahet, Z., Melamede, R. J., Kow, Y. W., and Wallace, S. S. (1997). Characterization of *Escherichia coli* endonuclease VIII. *J. Biol. Chem.* **272,** 32230–32239.

Kathe, S. D., Shen, G. P., and Wallace, S. S. (2004). Single-stranded breaks in DNA but not oxidative DNA base damages block transcriptional elongation by RNA polymerase II in HeLa cell nuclear extracts. *J. Biol. Chem.* **279,** 18511–18520.

Kow, Y. W., and Wallace, S. S. (1987). Mechanism of action of *Escherichia coli* endonuclease III. *Biochemistry* **26,** 8200–8206.

Kuzmic, P. (1996). Program DYNAFIT for the analysis of enzyme kinetic data: Application to HIV proteinase. *Anal. Biochem.* **237,** 260–273.

Manuel, R. C., Hitomi, K., Arvai, A. S., House, P. G., Kurtz, A. J., Dodson, M. L., McCullough, A. K., Tainer, J. A., and Lloyd, R. S. (2004). Reaction intermediates in the catalytic mechanism of *Escherichia coli* MutY DNA glycosylase. *J. Biol. Chem.* **279,** 46930–46939.

Mazumder, A., Gerlt, J. A., Absalon, M. J., Stubbe, J., Cunningham, R. P., Withka, J., and Bolton, P. H. (1991). Stereochemical studies of the beta-elimination reactions at aldehydic abasic sites in DNA: Endonuclease III from *Escherichia coli*, sodium hydroxide, and Lys-Trp-Lys. *Biochemistry* **30,** 1119–1126.

McCullough, A. K., Sanchez, A., Dodson, M. L., Marapaka, P., Taylor, J. S., and Lloyd, R. S. (2001). The reaction mechanism of DNA glycosylase/AP lyases at abasic sites. *Biochemistry* **40,** 561–568.

Nash, H. M., Bruner, S. D., Scharer, O. D., Kawate, T., Addona, T. A., Spooner, E., Lane, W. S., and Verdine, G. L. (1996). Cloning of a yeast 8-oxoguanine DNA glycosylase reveals the existence of a base-excision DNA-repair protein superfamily. *Curr. Biol.* **6,** 968–980.

Porello, S. L., Leyes, A. E., and David, S. S. (1998). Single-turnover and pre-steady-state kinetics of the reaction of the adenine glycosylase MutY with mismatch-containing DNA substrates. *Biochemistry* **37,** 14756–14764.

Purmal, A. A., Lampman, G. W., Bond, J. P., Hatahet, Z., and Wallace, S. S. (1998). Enzymatic processing of uracil glycol, a major oxidative product of DNA cytosine. *J. Biol. Chem.* **273,** 10026–10035.

Rieger, R. A., McTigue, M. M., Kycia, J. H., Gerchman, S. E., Grollman, A. P., and Iden, C. R. (2000). Characterization of a cross-linked DNA-endonuclease VIII repair complex by electrospray ionization mass spectrometry. *J. Am. Soc. Mass Spectrom.* **11,** 505–515.

Seber, G. A. F., and Wild, C. J. (2003). "Nonlinear Regression." Wiley, New York.

Shen, G. P., Galick, H., Inoue, M., and Wallace, S. S. (2003). Decline of nuclear and mitochondrial oxidative base excision repair activity in late passage human diploid fibroblasts. *DNA Repair* **2**, 673–693.

Slupphaug, G., Kavli, B., and Krokan, H. E. (2003). The interacting pathways for prevention and repair of oxidative DNA damage. *Mutat. Res.* **531**, 231–251.

Sun, B., Latham, K. A., Dodson, M. L., and Lloyd, R. S. (1995). Studies on the catalytic mechanism of five DNA glycosylases: Probing for enzyme-DNA amino intermediates. *J. Biol. Chem.* **270**, 19501–19508.

Verdine, G. L., and Norman, D. P. (2003). Covalent trapping of protein–DNA complexes. *Annu. Rev. Biochem.* **72**, 337–366.

Wain, H. M., Bruford, E. A., Lovering, R. C., Lush, M. J., Wright, M. W., and Povey, S. (2002). Guidelines for human gene nomenclature. *Genomics* **79**, 464–470.

Wain, H. M., Lush, M. J., Ducluzeau, F., Khodiyar, V. K., and Povey, S. (2004). Genew: The human gene nomenclature database, 2004 updates. *Nucleic Acids Res.* **32**, D255–D257.

Wallace, S. S. (1998). Enzymatic processing of radiation-induced free radical damage in DNA. *Radiat. Res.* **150**, S60–S79.

Williams, S. D., and David, S. S. (2000). A single engineered point mutation in the adenine glycosylase MutY confers bifunctional glycosylase/AP lyase activity. *Biochemistry* **39**, 10098–10109.

Wilson, D. M., 3rd, Sofinowski, T. M., and McNeill, D. R. (2003). Repair mechanisms for oxidative DNA damage. *Front. Biosci.* **8**, d963–d981.

Yang, N., Chaudhry, M. A., and Wallace, S. S. (2005). Base excision repair by hNTH1 and hOGG1: A two-edged sword in the processing of DNA damage in γ-irradiated human cells. *DNA Repair*.

Yang, N., Galick, H., and Wallace, S. S. (2004). Attempted base excision repair of ionizing radiation damage in human lymphoblastoid cells produces lethal and mutagenic double strand breaks. *DNA Repair* **3**, 1323–1334.

Zharkov, D. O., Rieger, R. A., Iden, C. R., and Grollman, A. P. (1997). NH2-terminal proline acts as a nucleophile in the glycosylase/AP-lyase reaction catalyzed by *Escherichia coli* formamidopyrimidine-DNA glycosylase (Fpg) protein. *J. Biol. Chem.* **272**, 5335–5341.

[3] Purification and Characterization of NEIL1 and NEIL2, Members of a Distinct Family of Mammalian DNA Glycosylases for Repair of Oxidized Bases

By Tapas K. Hazra and Sankar Mitra

Abstract

NEIL1 and NEIL2 were newly discovered as mammalian orthologs of *Escherichia coli* Nei and Fpg, oxidized base-specific DNA glycosylases. These are distinct from previously characterized OGG1 and NTH1, the other two glycosylases for repairing oxidatively damaged bases in mammalian cells, in regards to reaction mechanism. Recombinant human NEIL1 and NEIL2 were purified from *E. coli* and biochemically characterized.

METHODS IN ENZYMOLOGY, VOL. 408
Copyright 2006, Elsevier Inc. All rights reserved.

0076-6879/06 $35.00
DOI: 10.1016/S0076-6879(06)08003-7

Some damaged bases are common substrates for both groups of enzymes. However, in contrast to the lack of activity of NTH1 and OGG1 for substrate lesions in single-stranded DNA, the NEILs have unique preference for bubble or single-stranded DNA substrates, suggesting their preferential involvement in repairing transcribed or replicating DNA sequences.

Endogenous and Induced DNA Damage

Reactive oxygen species (ROS) are formed endogenously as by-products of oxidative metabolism (Gotz *et al.*, 1994; Grisham *et al.*, 1998; Parkins *et al.*, 1997; Ramana *et al.*, 1998) and are exogenously induced by a variety of environmental agents, including ionizing radiation (Breen *et al.*, 1995; Ward, 1994). ROS have profound impact on survival, evolution, and homoeostasis of all living organisms (Breen and Murphy, 1995; Breimer, 1990; Meneghini, 1997). ROS have been implicated in the etiology of many pathophysiological states, including aging, arthritis, and carcinogenesis, as well as atherosclerosis (Ames *et al.*, 1993; Breimer, 1990; Fraga *et al.*, 1990; Gryfe *et al.*, 1997; Lavrovsky *et al.*, 2000). The ROS reaction with DNA is complex, leading to a multitude of modifications to DNA bases, along with base loss and DNA strand cleavage. The induction of mutations is a known prerequisite for many diseases, particularly cancer (Ames *et al.*, 1995; Breen *et al.*, 1995; Cadet et al., 1997; Wallace, 1998). Many oxidized base lesions are mutagenic because of mispairing and replication by lesion bypass DNA polymerases. Such mutations may be oncogenic and also trigger other diseases.

5-Hydroxyuracil (5-OHU) and 8-oxoguanine (8-oxoG) are among the major ROS reaction products in the genome generated from C and G, respectively. Many point mutations in oncogenes and tumor suppressor genes have been identified in sporadic cancers, among which GC-to-TA and GC-to-AT mutations are predominant (Cheng *et al.*, 1992). Assuming that ROS-induced mutations are responsible for such cancers, both GC-to-TA and GC-to-AT mutations could arise from 5-OHU mispairing with T and A, respectively; pairing of 8-oxoG with A will also contribute to GC-to-TA mutations.

Repair of Oxidized Base Lesions in the Genome

Oxidized base lesions in DNA are repaired via a highly conserved base excision repair (BER) pathway (Friedberg *et al.*, 1995; Krokan *et al.*, 1997, 2000). It consists of five steps, involves at least four proteins, and is initiated with excision of the damaged base by a DNA glycosylase to form an abasic (AP) site. These glycosylases also possess intrinsic AP lyase activity and utilize an N-terminal Lys or an N-terminal Pro as the active site nucleophile, which is involved both in cleaving the N-glycosyl bond of the damaged base and in attacking the sugar–phosphate bond via concerted lyase reaction. The final product is a DNA strand break at the damage site with a blocked 3' terminus. The blocking group has to be removed before a DNA polymerase can fill in the gap in the next step of repair (Dodson et al., 1994; Hazra et al., 1998). Such 3' end cleaning is usually carried out by an AP-endonuclease (APE) (Demple and Harrison, 1994). Finally, after gap filling the nicked strand is sealed by a DNA ligase (Tomkinson and Mackey). Only two mammalian-oxidized, base-specific DNA glycosylases, NTH1 and OGG1, were characterized earlier (Ikeda et al., 1998; Lu et al., 1997). Although both enzymes have a broad substrate range, NTH1 (the human ortholog of E. coli Nth), prefers oxidatively damaged pyrimidines as substrate, whereas OGG1 (the functional counterpart of E. coli MutM/Fpg) is most active with purine-derived lesions. The homozygous null mouse mutants for OGG1 and NTH1 do not show any obvious disease phenotype (Minowa et al., 2000; Ocampo et al., 2002; Takao et al., 2002b), although damaged bases accumulate in their tissue genomes. Furthermore, despite the absence of the major repair activity for NTH1 substrates, e.g.,thymine glycol (Tg) or 5-OHU, NTH1 null mouse embryonic fibroblasts (MEF) are no more sensitive to ROS than those from NTH1 heterozygous mice (Takao et al., 2002b). Additionally, OGG1 null MEFs are proficient in repairing 8-oxoG, preferentially from the transcribed strand (LePage et al., 2000). Taken together, these results suggest the presence of additional DNA glycosylases and/or pathways in mammalian cells for the repair of oxidative DNA damage, particularly from the transcribed sequences.

Two Classes of Oxidized Base-Specific Glycosylases

Based on structure and mechanism, the oxidized base-specific glycosylase/AP lyases belong to two classes, whose prototypes are E. coli Nth and Nei/Fpg. Nth utilizes an internal Lys residue as the active site nucleophile and carries out a β elimination reaction to break the phosphodiester bond and produces $3'$-$\alpha\beta$-unsaturated aldehyde and 5'-phosphate. N-terminal Pro is the active site of Nei/Fpg (MutM) type enzymes, which carry out a $\beta\delta$ elimination reaction at the AP site, leaving behind 3'P and 5'P termini at the strand break (Zharkov et al., 1997). The 3'-blocked termini of both β and $\beta\delta$ elimination products are removed in E. coli by the APEs (Xth and Nfo) to provide the 3'-OH terminus required for DNA repair synthesis. However, in mammalian cells, APE1, the only APE, hydrolyzes efficiently only the β elimination product, but its 3' phosphatase activity, required for removal of the 3' P generated by the $\beta\delta$ elimination reaction, is very weak. At the same time, polynucleotide kinase (PNK), which is present in mammalian cells but absent in *E. coli*, is the primary contributor of DNA 3′ phosphatase activity in mammalian cells, in addition to its function as a 5′ DNA kinase (Caldecott, 2002; Habraken and Verly, 1988). PNK functions in the repair of DNA single-strand breaks generated by ionizing radiation (Whitehouse *et al.*, 2001). When NTH1 and OGG1 were the only known oxidized base-specific DNA glycosylases in mammalian cells, the 3′ phosphatase activity of APE1 appeared unnecessary for oxidized base repair because both NTH1 and OGG1 catalyze the β elimination reaction. $\beta\delta$ elimination activity in the HeLa extract was detected in the late 1990s (Hazra *et al.*, 1998), and two new enzymes based on *E. coli* Nei homology, which carry out ßδ elimination, were subsequently identified (Hazra *et al.*, 2002a,b). It has been shown that PNK rather than APE1 functions in BER initiated by these glycosylases (Wiederhold *et al.*, 2004.)

Human Genome Database Analysis and Discovery of MutM/Nei-Like Proteins

The NCBI and Celera Genomic Databases were searched using the BLASTP program to identify human cDNA clones with significant homology to *E. coli* MutM or Nei. Two putative clones of human genes along with their mouse homologs were identified that were originally named NEH (Nei homolog) and have since been renamed NEIL (*Nei-like*). The cDNA for human NEILs (NEIL1; Accession No. AAH10876 and NEIL2; Accession No. BC013964), obtained from Research Genetics (Huntsville, AL), were inserted between the *Nde*I and the *Xho*I sites of pRSETB plasmid, and their identities were confirmed by sequence analysis (Hazra *et al.*, 2002a, b).

Unlike the previously known oxidized base-specific DNA glycosylases OGG1 and NTH1, NEILs contain N-terminal Pro, in the sequence PE(L/G)P(E/L), which is also present in *E. coli* Fpg and Nei. NEILs also share a potential helix-two-turn-helix (H2TH) motif with Fpg and Nei instead of the classical helix-hairpin-helix motif present in Nth-type DNA glycosylases, including NTH1 and OGG1 (Hazra *et al.*, 2002a, b). Because of the active site of NEILs at the N terminus, we could not produce N-terminal fusion polypeptides of NEILs, which could have facilitated their purification. Instead, we purified unmodified, recombinant human NEIL1 and NEIL2 as follows.

Purification of NEIL1 from E. coli

Because plasmid-encoded expression of NEIL1 and NEIL2 is low in common *E. coli* strains, presumably because of the requirement of rare codon usage, we introduced the recombinant plasmid into *E. coli* BL21(DE3) codon plus (Stratagene) and confirmed significant expression of NEIL1 by SDS–PAGE of the soluble extract. We then used the following protocol for large-scale purification of wild-type NEIL1.

1. Transfer several NEIL1 plasmid-containing *E. coli* colonies to 100 ml Luria broth (LB) containing 150 μg/ml ampicillin and 40 μg/ml chloramphenicol. Grow the cells at $37°$ until OD_{600} reaches 0.5–0.6.
2. Transfer the culture to 900 ml LB containing ampicillin and chloramphenicol in a 2-liter flask. Grow the culture at $37°$ until the OD reaches 0.3–0.4.
3. Cool the culture down to $16°$ and add isopropyl-β-D-thiogalactoside to 0.2 mM for inducing expression of NEIL1. Continue to grow the culture on a rotary shaker (150 rpm) for 12–15 h at $16°$.
4. Harvest the cells at $4000g$ at $4°$ for 10 min. Discard the supernatant and suspend the bacterial pellet in 50 ml ice-cold buffer containing 50 mM Tris·HCl (pH 7.5), 400 mM NaCl, 1 mM dithiothreitol (DTT), 0.1 mM EDTA, and a protease inhibitor cocktail (Roche).
5. Place the suspension on an ice bath and sonicate five to six times, 45 s each using a Braunsonic sonicator, allowing a 1-min interval to cool in between sonication.
6. Remove cell debris by centrifugation at $14,000g$ for 30 min.
7. Add polymin P (final concentration 0.5%) and keep for 1 h at $0°$ to precipitate nucleic acids. Centrifuge ($14,000g$, 15 min) to remove the precipitate.
8. Add ammonium sulfate to the supernatant to 60% saturation (36.1 g/100 ml) and stir gently for 2 h at $0°$.

After dissolving the precipitate in a minimum volume (2–3 ml) of buffer A (25 mM Tris·HCl, pH 7.5, 10% glycerol, 1 mM DTT, 0.1 mM EDTA) containing 150 mM NaCl and chromatography in a 110-ml Superdex 75 gel filtration column, fractions containing NEIL1 (as indicated by the presence of a NEIL1 band in SDS–PAGE of various fractions) are pooled and adjusted to 50 mM NaCl before applying to 5-ml HiTrap Q and SP columns (Amersham Pharmacia) connected in tandem. After a wash with 25 ml of the starting buffer, the Q column is disconnected, and the proteins are eluted from the SP column at 500 mM NaCl with a 100-ml linear gradient of NaCl ranging from 50 mM to 1 M. The final step involves chromatography on a 1-ml Mono S (preequilibrated with buffer A containing 100 mM NaCl) column, and NEIL1 is eluted without detectable impurity at 500 mM NaCl (Fig. 1, lane 2). The final yield of the pure protein is about 5 mg/liter culture.

The sequence of purified NEIL1, N-Pro-Glu-Gly-Pro-Glu-Leu-His-Leu-Ala-Ser, as determined by automated Edman degradation, matches

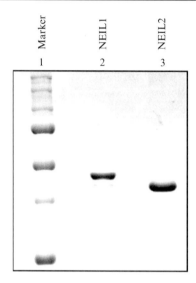

FIG. 1. SDS–PAGE (12%) of purified NEIL1 and NEIL2 stained with Coomassie blue.

perfectly with the predicted sequence from the Gene Bank database (after cleavage of the initiator Met residue). Its molecular weight of 43,633, as determined by MS analysis, is close to the predicted value of 43,582.

Purification of Recombinant NEIL2 from E. coli

Recombinant NEIL2 is also purified from the *E. coli* BL21(DE3) codon plus in which its expression level is less robust than of NEIL1. Wild-type NEIL2 is purified using the same sequence of steps as described for NEIL1 with some modifications (Hazra *et al.*, 2002a). Briefly, the sonicated extract is adjusted to 25 m*M* Tris·HCl, pH 7.5; 10% glycerol, 1 m*M* DTT, and 100 m*M* NaCl (buffer A) and is passed through tandemly attached Q and SP-Sepharose columns as before. After washing with 50 ml buffer A and disconnecting the Q column, NEIL2 is eluted from SP-Sepharose at 300 m*M* NaCl. NEIL2- containing fractions are chromatographed further on a 1-ml HiTrap heparin column (preequilibrated with buffer A) and the enzyme is eluted at 0.5 *M* NaCl. The final step involves gel filtration in a 25-ml Superdex 75 column equilibrated in 25 m*M* Tris·HCl (pH 7.5) and 10% glycerol containing 200 m*M* NaCl. NEIL2 is extremely unstable and loses activity nearly completely after storage at −80° or after a 15-min incubation at 37°. Thus the purified enzyme is never frozen. The active fractions (Fig 1, lane 3) are stored in phosphate-buffered saline (PBS) containing 1 mM DTT and 50% glycerol at $-20°$. The final yield of the pure enzyme is about 4 mg/liter culture.

Enzymatic Activity of NEIL1 and NEIL2 with Duplex or Bubble Oligonucleotide Substrates

Oligonucleotide Substrates

Two 51-mer oligonucleotides containing either 5-OHU or 8-oxoG at position 26 (Table I) are from Midland Certified Reagent. The undamaged control oligonucleotide contains C at position 26. Sequences of complementary oligonucleotides, purchased from Invitrogen, contain G or C opposite the lesion or contain sequences for producing bubble structures with 5, 11, and 19 unpaired bases. The bubble-containing oligonucleotides are named B5, B11, and B19, respectively (Table I). Two hundred fifty picomoles each of the lesion strand and its complementary strand are heated together at 94° for 2 min in 50 μl PBS and then slowly cooled to room temperature. To produce ^{32}P-labeled substrates, the single-stranded oligonucleotides are labeled at the 5′ terminus with [γ-^{32}P] ATP and T4 polynucleotide kinase before annealing with various complementary oligonucleotides to generate duplex or bubble oligonucleotides. The melting temperatures (T_m) of the annealed duplexes in the presence of 50 mM NaCl, calculated according to Oligo Analyzer 3.0 (Integrated DNA

TABLE I

SEQUENCES OF LESION-CONTAINING AND COMPLEMENTARY OLIGONUCLEOTIDES D, COMPLEMENTARY STRAND FOR GENERATING DUPLEX (N:G OPPOSITE 5-OHU), OR BUBBLES OF 5 (B5), 11 (B11), AND 19 (B19) NUCLEOTIDE UNPAIRED REGIONS[a]

Lesion-containing strand (X : 5-OHU)
5′-GCT TAG CTT GGA ATC GTA TCA TGT AXA CTC GTG TGC CGT GTA GAC CGT GCC-3′

Complementary strand
D, 5′-GGC ACG GTC TAC ACG GCA CAC GAG TNT ACA TGA TAC GAT TCC AAG CTA AGC-3′

B5, 5′-GGC ACG GTC TAC ACG GCA CAC GA<u>A GCC C</u>CA TGA TAC GAT TCC AAG CTA AGC-3′

B11, 5′-GGC ACG GTC TAC ACG GCA CA<u>A ACA GCC CAC G</u>GA TAC GAT TCC AAG CTA AGC-3′

B19, 5′-GGC ACG GTC TAC ACG G<u>AC ACA ACA CCC CAC CAC CCC</u> GAT TCC AAG CTA AGC-3′

[a] Noncomplementary sequences to generate bubbles are underlined.

Technologies, Inc.), range from 50° to 74°, depending on the bubble size. Thus the duplexes and bubble structures are stable in subsequent assays.

Detailed analysis of the structure of the duplex and B11 oligonucleotides containing a lesion or a normal base is carried out (Dou *et al.*, 2003). Both strands in annealed DNAs are labeled with [32]P. A single band of expected mobility after electrophoresis in nondenaturing gels for both 5-OHU- and normal base-containing oligos is observed (Fig. 2). The mobility of the duplex is in between that of ssDNA and the bubble oligonucleotide. These results indicate that the duplex and bubble structures are stable and homogeneous without contamination of single-stranded oligonucleotides (Fig. 2A). We confirmed the presence of a single-stranded region in the control B11 oligonucleotide containing C at position 26 by examining its susceptibility to the single-stranded DNA-specific mung bean endonuclease (Fig. 2B). Duplex sequences, 20-mer long on either side of the bubble in the B11 oligonucleotide, should be resistant to this endonuclease, whereas the single-stranded sequences in these oligonucleotides

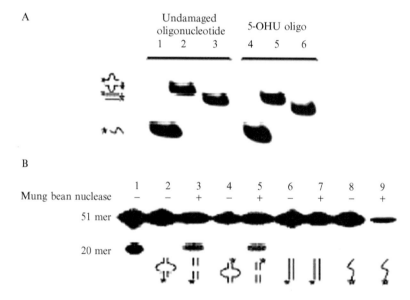

FIG. 2. Characterization of single-stranded, bubble-containing, and duplex oligonucleotides. (A) PAGE of undamaged (lanes 1–3) and 5-OHU-containing oligonucleotides (lanes 4–6) annealed to generate duplex and bubble structures as described in the text. Their secondary structures are diagrammed on the left. Asterisks indicate ^{32}P label. (B) Sensitivity of nondamaged oligonucleotides to mung bean nuclease. Lane 1, 51-mer substrate and predicted 20-mer nuclease product as markers. Even-numbered lanes, control; odd-numbered lanes, after nuclease treatment. Diagrams below the lanes depict oligonucleotide structures.

would be degraded. Figure 2B shows that the duplex oligonucleotide with limited nuclease digestion remains unchanged (lanes 6 and 7), whereas a significant fraction of the B11 DNA in which both strands were radiolabeled (lanes 2 and 4) is converted into 20- to 22-mer fragments from either strand (lanes 3 and 5) after denaturing gel electrophoresis. As expected, the ssDNA is degraded extensively by nuclease treatment (lane 9). Thus, the bubble oligonucleotide contains stable 20-mer duplex sequences flanking an ssDNA region. It should also be noted that no significant hairpin duplex region in the oligonucleotides is predicted from the sequence analysis under our assay condition (Oligo Analyzer 3.0, Integrated DNA Technologies, Inc.).

Incision Assay of DNA Glycosylases

The enzymatic activity of NEIL1 and NEIL2 is analyzed by quantitating the cleavage product of the damage-containing strand generated due to the lyase activity of the enzyme. Both duplex and bubble oligonucleotides containing 5-OHU or dihydrouracil (DHU) in the ^{32}P-labeled strand are incubated at 37° for various times in a 10-μl reaction mixture containing 40 mM HEPES-KOH (pH 7.5), 50 mM KCl, 5% glycerol, and 100 μg/ml bovine serum albumin. After terminating the reactions with 70% formamide and 30 mM NaOH, the cleaved oligonucleotide products are separated by denaturing gel electrophoresis in 20% polyacrylamide gel containing 7 M urea, 90 mM Tris borate (pH 8.3), and 2 mM EDTA.

Both NEILs prefer ROS-derived lesions of pyrimidines. NEIL1 acts like Fpg in excising both Fapy-G and Fapy-A from natural DNA. It is also active with several other pyrimidine lesions, such as thymine glycol and 5-OHU. NEIL2 is active in excising several cytosine-derived lesions with robust activity for 5-OHU and weaker activity for DHU with the same oligonucleotide substrate. NEIL2 has low activity for 5-hydroxycytosine, very weak activity for thymine glycol and 8-oxoguanine, and is inactive for uracil, 2-hydroxyadenine, hypoxanthine, and xanthine. As expected, both NEIL1 and NEIL2 have strong lyase activity for the AP site (data not shown).

MutM and Nei generate 3' phosphate termini, whereas Nth generates 3' phosphoe-α,β-unsaturated deoxyribose by cleaving the AP site after damaged base excision. The 3' P-containing product has higher electrophoretic mobility than that with 3' phosphosugar terminus (Hazra *et al.*, 1998). A comparison of the mobility of NEIL1 digestion products with those of *E. coli*, Nei, and Nth from the same DHU-containing oligonucleotide substrate shows that NEIL1 carries out $\beta\delta$ elimination like Nei/MutM and not β elimination like Nth (Fig. 3A). A similar comparison of the

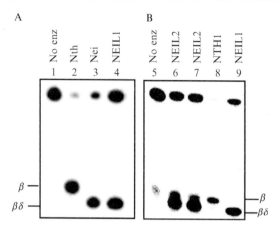

FIG. 3. (A) Substrate specificity and AP lyase activity of purified hNEIL1. Lane 4, purified NEIL1 (0.5 pmol) was incubated with 5′ ^{32}P-labeled DHU-containing duplex oligonucleotide (4 pmol). Lanes 2 and 3, incision activity of Nth and Nei, respectively, with DHU·A oligonucleotide. (B) Incision activity of purified NEIL2. NEIL2 (0.25 pmol) was incubated with 0.1 pmol of 5′ ^{32}P-labeled duplex oligonucleotides containing 5-OHU·A (lane 6) and 5-OHU·G (lane 7). Lanes 8 and 9, incision products of NTH1 and NEIL1 with 5-OHU·A-containing oligonucleotide as marker. Lane 5, no protein. Positions of β and $\beta\delta$ elimination products are indicated.

mobility of incision products of NEIL2 (lanes 6 and 7), NTH1 (lane 8), and NEIL1 (lane 9) indicates that NEIL2 also carries out $\beta\delta$ elimination, while all three enzymes excise 5′-OHU. Thus, based on the structural analysis and reaction mechanism, both NEIL1 and NEIL2 are orthologs of *E. coli* MutM/Nei.

Unique activity of NEIL1 and NEIL2 with DNA Bubble Substrates

We compared the DNA glycosylase activity of NEIL1, NEIL2, and NTH1 in excising 5-OHU from duplex, single-stranded, or bubble oligonucleotides B5, B11, and B19. The lesion base was positioned at the middle of the bubble in all cases. We confirmed an earlier observation (Takao *et al.*, 2002a) that NEIL1 is active with ssDNA (Fig. 4A, lane 6). We further showed that NEIL2 (Fig. 4A, lane 11) but not NTH1 (Fig. 4B, lane 5) was similarly active with ssDNA. More remarkably, both NEIL1 and NEIL2 were active in excising 5-OHU when it was located inside the bubble of B5, B11, and B19 (Fig. 4A, lanes 3–5 and 8–10). In contrast, NTH1 was completely inactive with the bubble substrates (Fig. 4B, lanes 3 and 4). This was expected because NTH1 was also inactive with the ssDNA substrate

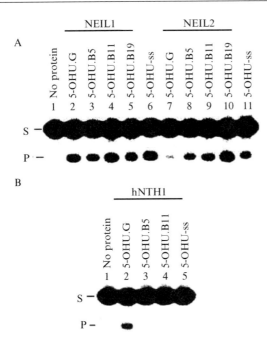

FIG. 4. Activity assay of NEIL1, NEIL2, and NTH1 with 5-OHU-containing substrates in different structures. Identical 5-OHU-containing oligonucleotide strands were used as is (5-OHU-ss) or annealed with a complementary strand containing G opposite 5-OHU (5-OHU·G) or with a noncomplementary strand to generate B5, B11, or B19 bubbles flanked by duplex sequences, as described in the text. Activity of NEIL1 and NEIL2 (A) and of NTH1 (B). S, substrate; P, product.

(Fig. 4B, lane 5). NEIL1 is generally more active than NEIL2 in excision of bases from duplex DNA. However, while both NEIL1 and NEIL2 have comparable activity with the bubble substrates, NEIL2 has significantly higher excision activity with 5-OHU when present in B11 or B19 relative to the duplex oligonucleotide.

We determined the kinetic parameters of NEIL1 and NEIL2 with oligonucleotides containing 5-OHU either paired with G or present in the B11 oligonucleotide. The G•5-OHU pair is generated in the genome *in situ* after oxidative deamination of C. For analysis of enzyme kinetics, we incubated the 5-OHU•G-containing duplex oligonucleotide and 5-OHU•B11 oligonucleotide substrates (7.8–125 nM) with either NEIL1 (at 17.5 nM for 3.5 min) or NEIL2 (at 20 nM for 5 min) at 37°. The rate of product formation was linear for both enzymes under these conditions. The radioactivity in the DNA bands was quantitated by analysis in a

TABLE II

Kinetic Parameters of NEIL1 and NEIL2 with Duplex (5-OHU•G) and Bubble (5-OHU•B11) Substrates

	K_m (nM)		k_{cat} ($\times 10^3$/min)		k_{cat}/K_m (min^{-1} nM^{-1})	
	5-OHU•G	5-OHU•B11	5-OHU•G	5-OHU•B11	5-OHU•G	5-OHU•B11
NEIL1	12.8 ± 1.1	9.0 ± 1.6	65.8 ± 0.1	124.5 ± 11.9	5.2 ± 0.62	14.0 ± 1.7
NEIL2	21.9 ± 2.5	15.6 ± 3.5	8.2 ± 2.1	37.8 ± 1.9	0.37 ± 0.08	2.5 ± 0.6

PhosphorImager (Amersham Biosciences) using Image-Quant software. Both NEIL1 and NEIL2 have higher catalytic specificity in excising 5-OHU from an 11 nucleotide bubble than from a G•5-OHU pair in the duplex (Table II). Although both NEIL1 and NEIL2 prefer 5-OHU located in the bubble, NEIL1 has a higher turnover and catalytic specificity than NEIL2. However, the preference of NEIL2 for the 11 nucleotide bubble substrate relative to duplex DNA is higher (about sevenfold) than that of NEIL1 (about threefold).

Detection of DNA Glycosylase Activity of Recombinant NEILs by Trapping Analysis

All DNA glycosylase/AP lyases, regardless of preference for the substrate base, form a transient Schiff base with the free AP site in DNA, which is stabilized after reduction with NaCNBH$_3$ (or NaBH$_4$); the stable structure is called a "trapped complex" (Dodson et al., 1994). The DNA-trapping reaction is performed by incubating a ^{32}P-labeled oligonucleotide substrate with the purified enzyme in a 20-μl reaction mixture in the presence of 25 mM NaCNBH$_3$ at 37° for 30 min. The reaction is stopped by heating at 95° for 5 min with 2× gel loading buffer, and the trapped complexes are separated by SDS–PAGE (12% polyacrylamide [Hazra et al., 1998]).

Figure 5 shows SDS–PAGE of ^{32}P-labeled trapped complexes of NEIL2 (lane 2) and NEIL1 (lane 4) along with the N terminus-tagged NEIL2 mutant (lane 3) with an AP site-containing oligonucleotide. Wild-type NEILs formed a trapped complex, indicating that recombinant NEILs are active as an AP lyase. However, the lack of formation of a trapped complex with the N-terminal His-tag fusion of NEIL2, which also shows no strand incision activity, is consistent with Pro1 being the active site because it is no longer available in the N-terminal His-tagged NEIL2.

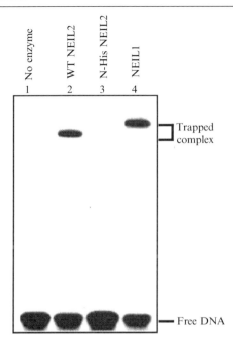

FIG. 5. Trapping analysis of purified (0.15 pmol) enzyme with 0.1 pmol of AP site-containing duplex oligonucleotide. Lane 1, no protein; lanes 2–4, wild type, N-terminally-tagged NEIL2 and NEIL1, respectively. The positions of free DNA and trapped complexes are indicated.

Perspective/Concluding Remarks

All organisms ranging from bacteria to mammals have evolved with expression of multiple DNA glycosylases, which have broad substrate range, although with distinct preference for specific damaged bases. Thus the redundancy of DNA glycosylases, particularly for oxidatively damaged bases, may have a teleological basis in ensuring protection of the genome from ROS, the most insidious genotoxic agents. However, NEILs are distinct from other mammalian DNA glycosylases because of the reaction mechanism and preference for bubble DNA. Because such bubble structures in DNA are transiently formed during both transcription and replication, it is tempting to speculate that NEILs are preferentially involved in repair of oxidative base damage during replication and/or transcription. Finally, the discovery of the NEIL family led to identification of an APE-independent subpathway of base excision repair. We have shown that the NEIL-generated 3′ P at the DNA strand break could be removed efficiently by polynucleotide kinase but not by APE1. Although *E. coli*, unlike mammalian cells, does not express PNK, this enzyme is not essential because the *E. coli* APEs efficiently remove all 3' blocking groups. Thus this alternative route of BER may provide a critical safeguard against oxidative DNA damage in mammalian genomes.

Acknowledgments

Research in the authors' laboratories is supported by USPHS Grants R01 CA81063, CA102271, and NIEHS Center Grant ES06676.

References

Ames, B. N., Gold, L. S., and Willett, W. C. (1995). The causes and prevention of cancer. *Proc. Natl. Acad. Sci. USA* **92,** 5258–5265.

Ames, B. N., Shigenaga, M. K., and Hagan, T. M. (1993). Oxidants, antioxidants, and degenerative diseases of aging. *Proc. Natl. Acad. Sci. USA* **90,** 7915–7922.

Breen, A. P., and Murphy, J. A. (1995). Reactions of oxyl radicals with DNA. *Free Radical Biol. Med.* **18,** 1033–1077.

Breimer, L. H. (1990). Molecular mechanisms of oxygen radical carcinogenesis and mutagenesis: The role of DNA base damage. *Mol. Carcinogen.* **3,** 188–197.

Cadet, J., Berger, M., Douki, T., and Ravanat, J. L. (1997). Oxidative damage to DNA: Formation, measurement, and biological significance. *Rev. Physiol. Biochem. Pharmacol.* **131,** 1–87.

Caldecott, K. W. (2002). Polynucleotide kinase: A versatile molecule makes a clean break. *Structure (Camb.)* **10,** 1151–1152.

Cheng, K. C., Cahill, D. S., Kasai, H., Nishimura, S., and Loeb, L. A. (1992). 8-Hydroxyguanine, an abundant form of oxidative DNA damage, causes G>T and A>C substitutions. *J. Biol. Chem.* **267,** 166–172.

Demple, B., and Harrison, L. (1994). Repair of oxidative damage to DNA. *Annu. Rev. Biochem.* **63,** 915–948.

Dodson, M. L., Michaels, M. L., and Lloyd, R. S. (1994). Unified catalytic mechanisms for DNA glycosylases. *J. Biol. Chem.* **269,** 32709–32712.

Dou, H., Mitra, S., and Hazra, T. K. (2003). Repair of oxidized bases in DNA bubble structures by human DNA glycosylases NEIL1 and NEIL2. *J. Biol. Chem.* **278,** 49679–49684.

Fraga, C. G., Shigenaga, M. K., Park, J. W., Degan, P., and Ames, B. N. (1990). Oxidative damage to DNA during aging: 8-Hydroxy-2'-deoxyguanosine in rat organ DNA and urine. *Proc. Natl. Acad. Sci. USA* **87,** 4533–4537.

Friedberg, E. C., Walker, G. C., and Siede, W. (1995). "DNA Repair and Mutagenesis." ASM Press, Washington, DC.

Gotz, M. E., Kunig, G., Riederer, P., and Youdim, M. B. (1994). Oxidative stress: Free radical production in neural degeneration. *Pharm. Therap.* **63,** 37–122.

Grisham, M. B., Granger, D. N., and Lefer, D. J. (1998). Modulation of leukocyte-endothelial interactions by reactive metabolites of oxygen and nitrogen: Relevance to ischemic heart disease. *Free Radical Biol. Med.* **25,** 404–433.

Gryfe, R., Swallow, C., Bapat, B., Redston, M., Gallinger, S., and Couture, J. (1997). Molecular biology of colorectal cancer. *Curr. Prob. Can.* **21,** 233–300.

Habraken, Y., and Verly, W. G. (1988). Further purification and characterization of the DNA 3'-phosphatase from rat-liver chromatin which is also a polynucleotide 5'-hydroxyl kinase. *Eur. J. Biochem.* **171,** 59–66.

Hazra, T. K., Izumi, T., Boldogh, I., Imhoff, B., Kow, Y. W., Jaruga, P., Dizdaroglu, M., and Mitra, S. (2002a). Identification and characterization of a human DNA glycosylase for repair of modified bases in oxidatively damaged DNA. *Proc. Natl. Acad. Sci. USA* **99,** 3523–3528.

Hazra, T. K., Izumi, T., Maidt, L., Floyd, R. A., and Mitra, S. (1998). The presence of two distinct 8-oxoguanine repair enzymes in human cells: Their potential complementary roles in preventing mutation. *Nucleic Acids Res.* **26,** 5116–5122.

Hazra, T. K., Kow, Y. W., Hatahet, Z., Imhoff, B., Boldogh, I., Mokkapati, S. K., Mitra, S., and Izumi, T. (2002b). Identification and characterization of a novel human DNA glycosylase for repair of cytosine-derived lesions. *J. Biol. Chem.* **277,** 30417–30420.

Ikeda, S., Biswas, T., Roy, R., Izumi, T., Boldogh, I., Kurosky, A., Sarker, A. H., Seki, S., and Mitra, S. (1998). Purification and charcterization of human hNTH1, a homolog of *Escherichia coli* endonuclease III: Direct identification of Lys-212 as the active nucleophilic residue. *J. Biol. Chem.* **273,** 21585–21593.

Krokan, H. E., Nilsen, H., Skorpen, F., Otterlei, M., and Slupphaug, G. (2000). Base excision repair of DNA in mammalian cells. *FEBS Lett.* **476,** 73–77.

Krokan, H. E., Standal, R., and Slupphaug, G. (1997). DNA glycosylases in the base excision repair of DNA. *Biochem. J.* **325,** 1–16.

Le Page, F., Randrianarison, V., Marot, D., Cabannes, J., Perricaudet, M., Feunteun, J., and Sarasin, A. (2000). Transcription coupled repair of 8-oxoguanine in murine cells: The oggl protein is required for repair in nontranscribed sequences but not in transcribed sequences. *Proc. Natl. Acad. Sci. USA* **97,** 8397–8402.

Lavrovsky, Y., Chatterjee, B., Clark, R. A., and Roy, A. K. (2000). Role of redox-regulated transcription factors in inflammation, aging and age-related diseases. *Exp. Gerontol.* **35,** 521–532.

Lu, R., Nash, H. M., and Verdine, G. L. (1997). A DNA repair enzyme that excises oxidatively damaged guanines from the mammalian genome is frequently lost in lung cancer. *Curr. Biol.* **7,** 397–407.

Meneghini, R. (1997). Iron homeostasis, oxidative stress, and DNA damage. *Free Radical Biol. Med.* **23,** 783–792.

Minowa, O., Arai, T., Hirano, M., Monden, Y., Nakai, S., Fukuda, M., Itoh, M., Takano, H., Hippou, Y., Aburatani, H., Masumura, K., Nohmi, T., Nishimura, S., and Noda, T. (2000). Mmh/Ogg1 gene inactivation results in accumulation of 8-hydroxyguanine in mice. *Proc. Natl. Acad. Sci. USA* **97,** 4156–4161.

Ocampo, M. T., Chaung, W., Marenstein, D. R., Chan, M. K., Altamirano, A., Basu, A. K., Boorstein, R. J., Cunningham, R. P., and Teebor, G. W. (2002). Targeted deletion of mNth1 reveals a novel DNA repair enzyme activity. *Mol. Cell. Biol.* **22,** 6111–6121.

Parkins, C. S., Stratford, M. R., Dennis, M. F., Stubbs, M., and Chaplin, D. J. (1997). The relationship between extracellular lactate and tumour pH in a murine tumour model of ischaemia-reperfusion. *Br. J. Can.* **75,** 319–323.

Ramana, C. V., Boldogh, I., Izumi, T., and Mitra, S. (1998). Activation of apurinic/ apyrimidinic endonuclease in human cells by reactive oxygen species and its corelation with their adaptive response to genotoxicity of free radicals. *Proc. Natl. Acad. Sci. USA* **95,** 5061–5066.

Takao, M., Kanno, S., Kobayashi, K., Zhang, Q. M., Yonei, S., Van Der Horst, G. T., and Yasui, A. (2002a). A back-up glycosylase in Nth1-knockout mice is a functional Nei (endonuclease VIII) homologue. *J. Biol. Chem.* **277,** 42205–42213.

Takao, M., Kanno, S., Shiromoto, T., Hasegawa, R., Ide, H., Ikeda, S., Sarker, A. H., Seki, S., Xing, J. Z., Le, X. C., Weinfeld, M., Kobayashi, K., Miyazaki, J., Muijtjens, M., Hoeijmakers, J. H., van der Horst, G., and Yasui, A. (2002b). Novel nuclear and mitochondrial glycosylases revealed by disruption of the mouse Nth1 gene encoding an endonuclease III homolog for repair of thymine glycols. *EMBO J.* **21**, 3486–3493.

Tomkinson, A. E., and Mackey, Z. B. (1998). Structure and function of mammalian DNA ligases. *Mutat. Res.* **407**, 1–9.

Wallace, S. S. (1998). Enzymatic processing of radiation-induced free radical damage in DNA. *Radiat. Res.* **150**, S60–S79.

Ward, J. F. (1994). The complexity of DNA damage: Relevance to biological consequences. *Int. J. Radiat. Biol.* **66**, 427–432.

Whitehouse, C. J., Taylor, R. M., Thistlethwaite, A., Zhang, H., Karimi-Busheri, F., Lasko, D. D., Weinfeld, M., and Caldecott, K. W. (2001). XRCC1 stimulates human polynucleotide kinase activity at damaged DNA termini and accelerates DNA single-strand break repair. *Cell* **104**, 107–117.

Wiederhold, L., Leppard, J. B., Kedar, P., Karimi-Busheri, F., Rasouli-Nia, A., Weinfeld, M., Tomkinson, A. E., Izumi, T., Prasad, R., Wilson, S. H., Mitra, S., and Hazra, T. K. (2004). AP-endonuclease-independent DNA base excision repair in human cells. *Mol. Cell* **15**, 209–220.

Zharkov, D. O., Rieger, R. A., Iden, C. R., and Grollman, A. P. (1997). NH$_2$-terminal proline acts as a nucleophile in the glycosylase/AP-lyase reaction catalyzed by *Escherichia coli* formamidopyrimidine-DNA glycosylase (Fpg) protein. *J. Biol. Chem.* **272**, 5335–5341.

[4] Analysis of Base Excision DNA Repair of the Oxidative Lesion 2-Deoxyribonolactone and the Formation of DNA–Protein Cross-Links

By JUNG-SUK SUNG *and* BRUCE DEMPLE

Abstract

DNA base lesions arising from oxidation or alkylation are processed primarily by the base excision repair pathway (BER). The damaged bases are excised by DNA *N*-glycosylases, which generate apurinic/apyrimidinic (AP) sites; AP sites produced by hydrolytic decay of DNA or the spontaneous loss of damaged bases are also processed by BER. Free radicals produce various types of abasic lesions as oxidative damage. This chapter focuses on the analysis of DNA repair and other reactions that occur with the lesion 2-deoxyribonolactone (dL), which has received much attention recently. DNA substrates with site-specific dL lesions are generated by photolysis of a synthetic precursor residue; both small oligonucleotide and plasmid-based substrates can be produced. The dL residue is readily incised by AP endonucleases such as the mammalian Ape1 protein, which would bring the lesion into BER. However, the second enzyme of the

METHODS IN ENZYMOLOGY, VOL. 408   0076-6879/06 $35.00
DOI: 10.1016/S0076-6879(06)08004-9

canonical BER pathway, DNA polymerase β, instead of excising Ape1-incised dL, forms a stable DNA–protein cross-link with the lesion. Such cross-links are analyzed by polyacrylamide gel electrophoresis. Incubation of Ape1-incised dL substrates with mammalian cell-free extracts shows that other proteins can also form such cross-links, although DNA polymerase β appears to be the major species. This chapter presents methods for analyzing the extent of DNA repair synthesis (repair patch size) associated with dL in whole cell extracts. These analyses show that dL is processed nearly exclusively by the long patch BER pathway, which results in the repair synthesis of two or more nucleotides.

Introduction

Oxidized abasic (AP) sites are significant components of the free radical damage incurred by DNA. However, detailed studies of oxidative DNA damage have been dominated by a focus on identifying and quantifying the base modifications, their repair pathways, and their cellular consequences. Free radical damage can involve chemical rearrangement of deoxyribose components of DNA, which yields oxidized AP sites, DNA strand breaks, and covalent DNA–protein cross-links (Demple and DeMott, 2002; von Sonntag, 1991). These abasic lesions can be both cytotoxic and mutagenic.

Oxidized AP sites include the earliest X-ray DNA damage identified chemically, 2-deoxyribonolactone (dL). The dL lesion has been reported as a product in DNA treated with numerous genotoxic agents in addition to ionzing radiation, including UV light, organometallic oxidants such as copper phenanthroline, and the antitumor agent neocarzinostatin (Hashimoto et al., 2001; Kappen and Goldberg, 1992; Pratviel et al., 1991; Sigman et al., 1993). The free radical reaction that mediates dL formation involves initial hydrogen abstraction from the nucleotide C1′ carbon, followed by O_2 addition and base loss (Demple and DeMott, 2002). This C1-oxidized AP site resides in an unbroken DNA strand but is subject to alkaline cleavage in a β-elimination reaction. Thus, dL residues constitute "alkali-labile" lesions. For many years, such basic questions as the possible processing of dL by base excision repair (BER) remained unanswered because of an experimental inability to generate dL selectively and by the lack of specific and sensitive methods for quantifying it.

With the development of methods to generate dL residues site specifically, we have begun to define the unique features of dL during DNA repair. The dL lesion is processed efficiently by the major mammalian AP endonuclease, Ape1, which leaves a 5′-terminal oxidized abasic residue (DeMott et al., 2002; Xu et al., 2003). However, Ape1-incised dL is refractory to a subsequent BER step, excision by DNA polymerase β (Polβ).

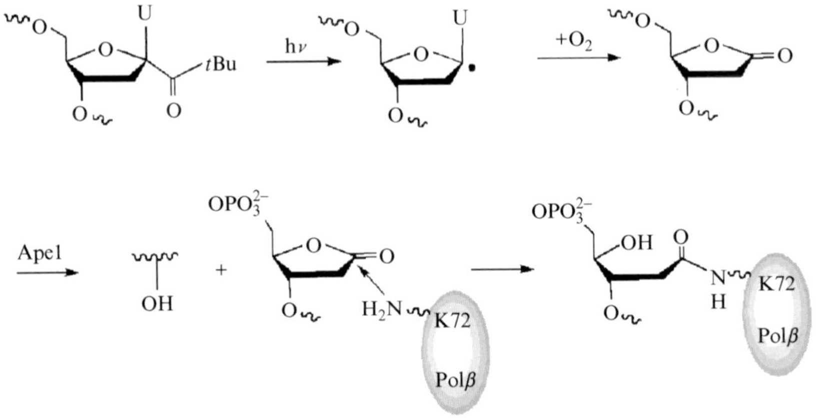

FIG. 1. Selective generation of dL via a photosensitive precursor and dL-mediated DNA–protein cross-linking with Polβ.

Instead, dL forms a stable covalent cross-link with Polβ via the catalytic lysine (residue 72) used for 5′-deoxyribose-5-phosphate (dRP) lyase activity (DeMott et al., 2002; Demple and DeMott, 2002). The cross-linkage is evidently a stable amide bond between the lysine-72 ε-nitrogen and the dL C1 carbonyl (Fig. 1).

In the repair of regular AP sites, removal of the 5′-dRP moiety is a key step in mediating short-patch BER that involves repair synthesis of single nucleotide (Fortini et al., 1998; Srivastava et al., 1998). An alternate BER pathway, long-patch BER, involves strand displacement repair synthesis of at least two nucleotides and excision of the 5′-dRP residue as part of a flap oligonucleotide by the FEN1 nuclease (Huggins et al., 2002; Klungland and Lindahl, 1997). This latter pathway is actually used to process dL residues, thus avoiding formation of the problematic DNA–protein cross-link (Sung et al., 2005). The methods applied to this analysis are presented here.

Preparation of dL-DNA Substrates

Preparation of an Oligonucleotide DNA Substrate Containing a Site-Specific dL Residue

Several chemical methods for the specific generation of dL lesions within DNA oligonucleotides were developed independently by several laboratories (Chatgilialoglu and Gimisis, 1998; Kotera et al., 2000; Lenox *et al.*, 2001; Tronche *et al.*, 1998). All of these methods involve the photolysis of a stable precursor and its conversion to dL at a defined DNA site. For our approach, we have used the 1′-*t*-butylcarbonyl-deoxyuridine precursor developed by Hwang *et al.* (1999), which is inserted during synthesis using standard solid-phase cyanoethyl phosphoramidite chemistry. For much of the work we have used the 30-mer sequence 5′-GTCACGTGCTGCA**X**-ACGACGTGCTGAGCCT-3′, in which X designates the dL precursor residue. We also describe the photolysis of other DNA oligonucleotides of various sizes (17-, 23-, 28-, 50-mers). Selective generation of the radical deoxyuridin-1′-yl via photolysis of the precursor (see details later) is followed by reaction with oxygen, which yields about 90% dL (Fig. 1) and a small amount (about 5%) of DNA strand breaks bearing 3′-phosphate residues (Hwang *et al.*, 1999; Tronche *et al.*, 1998).

Using standard methods (Ausubel *et al.*, 1997), the oligonucleotide containing a site-specific precursor residue is ^{32}P labeled at the 5′ end using T4 polynucleotide kinase and ^{32}P-ATP and is hybridized to a complementary strand to form a 5′-labeled molecule. Alternatively, hybridization to a complementary strand and incorporation by a labeled nucleotide ([α-^{32}P] dCMP in our case) using the exonuclease-free Klenow fragment of DNA polymerase I yield a 3′-labeled molecule. For both procedures, unincorporated ^{32}P nucleotides are removed and duplex DNA containing the dL precursor is eluted using Bio-Spin 20 columns (Bio-Rad), which are equilibrated with neutral H_2O instead of Tris–HCl buffer in order to avoid reaction of the primary amine residues in the Tris base with dL. To generate a site-specific dL lesion, 2–5 pmol of radiolabeled duplex DNA in 50 μl distilled H_2O is placed in a 10×75-mm Pyrex test tube, the top is sealed using Parafilm, and it is subjected to photolysis in a photochemical reactor (RPR-100 from Rayonet Corp., Branford, CT). Photolysis is performed by exposure of DNA to long-wave (350-nm) UV light at 9200 μW/cm^2 under constant cooling to reduce evaporation and prevent degradation of dL.

The conversion of photoprecursor to dL is determined by a DNA cleavage assay that monitors the generation of DNA sites sensitive to scission by an AP endonuclease or hot alkali. Typically, the C1′-*t*-butylcarbonyl precursor on single-stranded DNA oligonucleotides of 17–50 bp can be >90% converted to dL lesions by 30 min of irradiation, whereas the corresponding duplex DNA molecules require at least 120 min irradiation to achieve a similar extent of photoconversion. This decreased reactivity of duplex DNA is attributed to the relative inaccessibility of the C1′ position relative to single-stranded DNA (Hwang and Greenberg, 1999).

Under mild conditions, dL lesions are relatively stable and their spontaneous cleavage occurs with a half-life of ~20 h in single-stranded DNA and 32–54 h in duplex DNA (Zheng and Sheppard, 2004). For our experiments, immediately upon the photoconversion to generate dL residues, the DNA samples are used for experiments.

Verification of a dL Site in DNA

DNA substrates containing a site-specific dL lesion in H_2O are analyzed using a DNA cleavage assay, as used in the analysis of AP sites, by subjecting the substrate to treatment with either AP endonuclease or hot alkali to cleave the abasic site. For the chemical cleavage of dL sites, 10 nM of DNA sample (typically in 5 μl) is removed, treated with 0.3 M NaOH and 30 mM EDTA, and then heated at 70° for 10 min. When single-stranded DNA is being analyzed, only chemical cleavage is employed, as AP endonucleases typically act poorly on single-stranded DNA (Doetsch and Cunningham, 1990).

For enzymatic cleavage of dL or abasic sites generally, DNA samples are diluted to 10 nM to produce an AP endonuclease reaction buffer containing 50 mM HEPEs-KOH (pH 7.5), 20 mM NaCl, 4 mM MgCl$_2$, 0.5 mM dithiothreitol, 5% (v/v) glycerol, and 0.1 mg/ml bovine serum albumin (BSA) and are then incubated with catalytic amounts (1–5 nM) of human Ape1 protein (Masuda *et al.*, 1998; Wilson *et al.*, 1995). When *Escherichia coli* endonuclease IV (Endo IV) is used, MgCl$_2$ is omitted, 2 mM EDTA can be added to the reaction buffer, and 0.4 nM of Endo IV is employed in the reaction. Cleavage reactions are conducted for 30 min at 30°. Because the catalytic activity of Endo IV is resistant to moderate concentrations of EDTA (Levin *et al.*, 1988), this property can be exploited by adding Endo IV to cell-free extracts to cleave dL sites when EDTA-sensitive cellular enzymes such as Ape1 are suppressed, as described in the next section. Such conditions also block many nonspecific nucleases and repair synthesis by DNA polymerases. However, AP lyase and dRP lyase activities are resistant to EDTA (Piersen *et al.*, 2000) and can still react with AP and dL residues; Endo IV eliminated the substrate for the former. Following cleavage by either hot alkali or AP endonuclease, the reaction is terminated by adding NaBH$_4$ to a final concentration of 300 mM, followed by ethanol precipitation and resuspension in 10 μl of denaturing formamide dye buffer (Ausubel *et al.*, 1997). Reaction products are analyzed on 15–20% polyacrylamide/7 M urea DNA sequencing gel electrophoresis.

Analysis of oligonucleotide DNA containing a site-specific dL residue (following photoconversion) results in two different product species depending on the method of cleavage. Enzymatic cleavage of dL in a 3′-end ^{32}P-labeled duplex DNA yields mostly (~85%) an oligonucleotide bearing a dL-5-phosphate residue at its 5′ terminus, which has a slower mobility

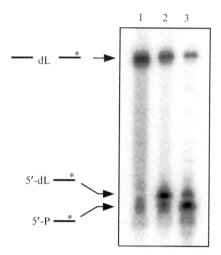

FIG. 2. Cleavage assays for dL residues. A 3' end ^{32}P-labeled duplex oligonucleotide DNA containing a site-specific photo-converted dL lesion (lane 1) was subjected to either enzymatic (lane 2) or chemical (lane 3) cleavage, as described in the text. The resulting DNA products were analyzed by electrophoresis on a 15% polyacrylamide/7 M urea gel.

than the 5'-phosphate product (Fig. 2, lane 2). Generation of small amounts of a 5'-phosphate product with faster mobility reflects instability of both uncleaved dL and AP endonuclease-generated dL-5-phosphate residues during electrophoresis. Consistent with this observation, untreated DNA containing a dL residue also yields small amounts of residual material corresponding to the 5'-phosphate product (Fig. 2, lane 1). Unlike the incision occurring at the 5' side of dL by an AP endonuclease, the chemical cleavage results mostly (~90%) in an incision at the 3' side of dL, yielding an oligonucleotide product bearing a 3'-phosphate residue at its 5' terminus (Fig. 2, lane 3). In both cases, the incision of DNA occurs specifically at the dL residue, which therefore generates oligonucleotide fragments corresponding to the site of the lesion. The unconverted precursor residue is resistant to both enzymatic and chemical treatments.

Preparation of a Plasmid DNA Substrate with a Defined dL Residue

Utilization of closed-circular (form I) plasmid DNA substrates is especially useful for studies comparing short- and long-patch BER pathways. Long-patch BER requires stable binding of proliferating cell nuclear antigen, which requires circular DNA substrates (Biade et al., 1998), or linear molecules with large steric blockages at the ends. Circular molecules may also be useful for analyzing the processing of dL-mediated DNA–protein cross-links (DPCs), which are formed during abortive short-patch BER (see later). For a circular dL substrate, pGEM-3Zf(+) plasmid molecules containing the dL precursor residue are produced by a primer extension reaction, followed by ethidium bromide/cesium chloride gradient centrifugation, and are finally subjected to photoconversion for the dL lesion. A primer (5'-ATCCTCTAGAG**X**CGACCTGCAGG-3'; X=1'-*t*-butylcarbonyl-deoxyuridine) is used with a centrally located precursor residue in order to minimize its possible interference with DNA polymerase or ligase activities. The lesion site [position 39 on the pGEM-3Zf(+) plasmid map] is flanked by various restriction endonuclease recognition sites to facilitate analysis. The primer must be 5'-phosphorylated to allow eventual ligation to complete the closed circular molecule; the 5'-^{32}P label may be introduced at this point. Alternatively, for the analysis of dL-mediated DPC formation, a ^{32}P-labeled nucleotide may be incorporated 11 nucleotides downstream from the dL site by annealing a second, 5'-^{32}P-end-labeled primer (5'-CATGCAAGCTTGAGTATTCTAT-3') immediately downstream from the precursor-containing primer (with an unlabeled 5'-phosphate); the two primers are ligated during DNA synthesis.

Single-stranded circular phagemid pGEM(+) DNA can be produced in large quantities by the aid of a helper phage (R408) infected at an m.o.i. of 100 and purified by the CTAB DNA precipitation method (Sung and Mosbaugh, 2003). A single or both primers are annealed to the pGEM (+) single-stranded DNA to form the primer-template for DNA synthesis. Primer extension reaction mixtures (3030 μl) are prepared containing 200 pmol of primed pGEM DNA, 20 mM HEPES-KOH (pH 7.8), 2 mM dithiothreitol, 10 mM MgCl$_2$, 1 mM ATP, 500 μM each of dATP, dTTP, dCTP, and dGTP, 400 units of T4 DNA polymerase, and 40,000 units of T4 DNA ligase. To avoid diluting the other components while maximizing the yield of circular duplex DNA, the DNA primer-template addition constitutes 20% of the reaction volume, and the T4 DNA polymerase and T4 DNA ligase additions contribute 8.6 and 13.2%, respectively. After incubation for 5 min on ice and 5 min at 25°, the primer extension reaction is carried out for 4 h at 37°. The reaction is terminated with the addition of 535 μl of 0.1 M EDTA.

A sample mixture is prepared by dissolving optical grade CsCl (18.12 g) in 15 ml of TE buffer, combined with the terminated primer extension reaction, adjusted to a final volume of 23.12 ml with TE buffer, and supplemented with 1.88 ml of a 10-mg/ml ethidium bromide solution to achieve a final volume of 25 ml. The total reaction volume can be reduced without affecting resolution to separate form I DNA, but the final concentrations of CsCl and ethidium bromide must be maintained at 0.725 g/ml and 0.752 mg/ml, respectively. The sample is placed into a 5-ml Quick-seal centrifuge tube fitted for a VTi 90 ultracentrifuge rotor (Beckman), and centrifugation is performed at 50,000 rpm for 14 h at 20° without braking in the deceleration step. The form I DNA is collected using a hypodermic needle, extracted five times with 1 ml of 1-butanol-saturated 5 M NaCl, concentrated using a Centricon-30 concentrator (Millipore), and buffer exchanged into H_2O. Typically, ~0.1 mg of DNA is isolated and found to contain >98% form I molecules, as determined by 1% agarose gel electrophoresis (Fig. 3, lane 1).

The purified form I DNA sample (5 pmol) is subjected to photolysis as described in the previous section. Verification of the photoconversion to dL in the plasmid can be done by the enzymatic cleavage assay using AP endonuclease (Ape1 or Endo IV), followed by digestion with *Bam*HI and *Hin*dIII (Fig. 5), which yields a 30-nucleotide DNA fragment containing

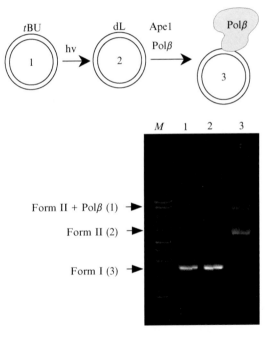

FIG. 3. Plasmid DNA dL substrate and DPC formation. The upper diagram illustrates the progress of the assay. A closed circular pGEM duplex plasmid DNA substrate containing a defined 1′-*t*-butylcarbonyl-deoxyuridine (tBU) residue (lane 1) was subjected to photolysis to generate a site-specific dL residue (lane 2), as described in the text. The dL–plasmid DNA (10 μg/ml) was incubated with 5 nM Ape1 and 300 nM Polβ for 2 h at 30° and analyzed by 1% agarose gel electrophoresis. The numbering of the molecules in the diagram corresponds to the lane numbering on the gel.

intact dL or a shorter DNA fragment following endonuclease incision at the dL site (13 or 17 nucleotides depending on the ^{32}P label position). Rather than monitoring enzyme cleavage that would convert closed circular to nicked plasmid molecules, analysis of the released restriction fragment is necessary because the high light doses used in photoconversion generate unknown photoproducts in the rest of the DNA, which can result in irrelevant strand breaks (J. S. Sung and B. Demple, unpublished data).

Analysis of dL-mediated Spontaneous DNA–Protein Cross-Links

In vitro formation of dL-mediated DPC, mainly by Polβ but also with other unknown proteins, has been identified in mouse and human cell-free extracts (Sung *et al.*, 2005). Whether any of these DPC is repaired is unknown. The methods described in this section focus mainly on the preparation and detection of dL-mediated DPC.

The standard cross-linking reaction mixture contains 50 mM HEPES-KOH (pH 7.5), 20 mM NaCl, 8 mM MgCl$_2$, 0.5 mM DTT, 5% (v/v) glycerol, 0.1 mg/ml BSA, 10 nM 3'-end ^{32}P-labeled DNA substrate containing a site-specific dL residue, and various amounts of purified proteins. Note that both human Ape1 and *E. coli* Endo IV incise rather efficiently at dL residues to leave a 5'-terminal oxidized abasic residue (Xu *et al.*, 2003). In view of the high activity of Ape1 in most mammalian cell types, the incised dL residue may be biologically the most abundant form (Demple and DeMott, 2002). Incision is carried out with catalytic amounts of Ape1 (1–5 nM) or Endo IV (0.4 nM) for 10 min at 30°. This incised material is then incubated at 30° to allow DPC to form, and the reactions are terminated by the addition of SDS–PAGE loading buffer and heating at 100° for 5 min. DPCs and free DNA are resolved on 8–10% SDS–PAGE. Preincised dL residues efficiently form stable DPC with Polβ (Fig. 4A) involving the catalytic lysine-72 of the polymerase N-terminal domain (DeMott *et al.*, 2002). The mobility of the cross-linked species is slower than either the free protein or the free DNA (Fig. 4), with the degree of difference depending on the size of the two species: for an attached 15-mer oligonucleotide, the M_r of the Polβ DPC is $\sim$45,000 compared to M_r $\sim$40,000 for free Polβ.

Cross-linking reactions can be similarly conducted with cell-free extract proteins (Fig. 4B), except that 2 mM EDTA replaces the MgCl$_2$ in order to suppress nonspecific nucleases but allow AP lyase and dRP lyase activities (Piersen *et al.*, 2000). Under these circumstances, it is convenient to use the EDTA-resistant Endo IV enzyme (Levin *et al.*, 1988) so that only a single buffer is necessary. As shown in Fig. 4B, incubation of Endo IV-incised dL-DNA with mouse embryonic fibroblast cell extracts produces several DPC, although the predominant one seems to involve Polβ.

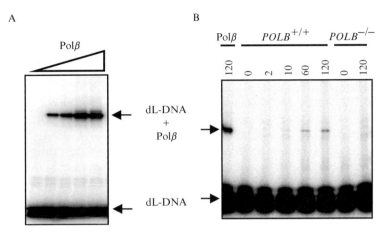

FIG. 4. DPC formed with dL by purified Polβ or cell extract proteins. (A) $3'$-^{32}P-labeled duplex DNA 31-mer (10 nM) containing a dL lesion was incubated in the standard cross-linking reaction mixture with 1 nM Ape1 and increasing amounts (0, 20, 50, 100, 200 nM) of Polβ for 2 h at 30°. (B) A similar reaction, except that 2 mM EDTA was included. The dL–DNA (10 nM) was pretreated with 0.4 nM Endo IV and then incubated with 20 nM Polβ for 120 min (B, lane 1), with 10 μg of $POLB^{+/+}$ mouse embryonic fibroblast extract for 0, 2, 10, 60, and 120 min (lanes 2–6), or with 10 μg of $POLB^{-/-}$ (Polβ-deficient) mouse embryonic fibroblast extract for 0 and 120 min (lanes 7 and 8). The samples were subjected to SDS–PAGE and analyzed using a phosphorimager. The resulting DPC bands are indicated by upper arrows and the free substrate by lower arrows. Adapted from Sung *et al.* (2005).

When the dL residue is on a plasmid DNA molecule, DPCs can be detected using agarose gel electrophoresis. Following incubation of 10 μg/ml of pGME DNA and 300 nM purified Polβ for 60 min at 30° in a standard cross-linking reaction buffer, the products are analyzed on a 1% agarose gel. As shown in Fig. 3, newly formed DPC on the plasmid result in a band with even slower mobility than nicked (form II) DNA.

Analysis of dL-Specific BER

DNA lesions processed by BER seem to go through both short-patch (single nucleotide) and long-patch (multinucleotide) pathways. Through its dRP lyase activity, Polβ is essential for the short-patch pathway in mammalian cells (Fortini *et al.*, 1998; Matsumoto and Kim, 1995; Sobol *et al.*, 1996). The distribution of products resulting from the two BER branches can be determined in at least two different ways, as described here. Both assays rely on the use of a closed circular DNA substrate containing a site-specific dL lesion and repair-proficient whole cell extracts.

Preparation of BER-Proficient Cell-Free Extract

Human HeLa or 293 cells or mouse NIH 3T3 or embryonic fibroblast cells are grown in the appropriate culture medium to ~90% confluency. Cells (~8 × 10^7) are harvested by scraping and resuspended with 2 packed cell volumes of hypotonic lysis buffer: 20 mM HEPES-KOH (pH 8.0), 1 mM EDTA, 1 mM DTT, 1 mM phenylmethylsulfonyl fluoride (PMSF), and 1× protease inhibitor cocktail (Sigma/Aldrich). The cell suspension is mixed, placed on ice for 20 min, and homogenized at 0–4° using a Dounce homogenizer fitted with a Teflon pestle (~20 strokes). After centrifugation of the cell lysate at 20,000g for 15 min at 4°, the supernatant containing soluble whole cell extract proteins is transferred into a Slide-A-Lyzer dialysis cassette (6–8000 MWCO; Pierce) and dialyzed extensively against 20 mM HEPES-KOH (pH 7.6), 100 mM NaCl, 1 mM dithiothreitol, 0.1 mM EDTA, 1 mM PMSF, 10% (v/v) glycerol, and 1× protease inhibitor cocktail (Sigma/Aldrich). The dialysis is necessary to remove deoxynucleotides that can interfere with subsequent analysis.

Determination of dL-Specific BER Patch Size Distribution

The repair patch size assay is based on the resistance of phosphorothioate nucleotides, incorporated during DNA repair synthesis, to degradation by *E. coli* exonuclease III (Exo III) (Sung *et al.*, 2001). The assay scheme is outlined in Fig. 5. For this analysis, the pGEM plasmid contains either a site-specific dL residue or a conventional AP site, and the ^{32}P label is located 11 nucleotides upstream of the lesion, as described in the previous section. BER reactions are performed at 30° for 1 h in the standard BER reaction mixture containing 100 mM HEPES-KOH (pH 7.5), 50 mM KCl, 5 mM MgCl$_2$, 1 mM dithiothreitol, 0.1 mM EDTA, 2 mM ATP, 0.5 mM β-NAD, 20 μM each of α-phosphorothioate dNTPs, 5 mM phosphocreatine, 200 units/ml phosphocreatine kinase, 5–10 μg/ml of the appropriate pGEM DNA substrate, and 0.5 μg of cell-free extract. Following the incubation, the reactions are terminated by adding EDTA to a final concentration of 10 mM, followed by heating at 70° for 3 min. RNase A is then added to 80 μg/ml, and the reaction mixtures are incubated for 10 min at 37°. Each reaction is then adjusted to 0.5% SDS and 190 μg/ml proteinase K and is incubated further for 30 min at 37°. The samples are extracted with phenol/chloroform/iosamyl alcohol (25:24:1), ethanol precipitated, and resuspended in 10–20 μl of TE buffer.

To establish the 3′ boundary of the repair patch, the recovered DNA is first digested with *Hind*III to generate linear DNA with a recessed 3′ terminus to facilitate Exo III digestion. Typically, treatment with 100 units of Exo III for 60 min at 37° is sufficient to fully digest 100 ng of untreated

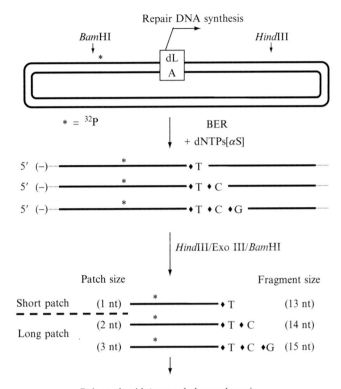

FIG. 5. Scheme for determining BER repair patch size.

control DNA. Degradation of the repaired strand terminates when the exonuclease encounters the first incorporated phosphorothioate nucleotide, which is the last residue incorporated by DNA polymerase downstream of the lesion site. The DNA is next digested with *Bam*HI, which releases ^{32}P-labeled fragments of 12 nucleotides (no repair), 13 nucleotides (repair synthesis incorporating a single nucleotide), 14 nucleotides (two nucleotides incorporated), and so on (Fig. 5). These products are resolved by electrophoresis on 15% polyacrylamide/urea DNA sequencing gels, followed by either autoradiography and densitometry or phosphorimaging. The amount of ^{32}P label in each band is quantified, and the fraction in each patch size is determined by dividing the amount of label in the corresponding band by the total label in all the bands. When such an analysis is applied to dL compared to an AP substrate processed by a mouse embryonic fibroblast cell-free extract (Fig. 6), the oxidized residue yields

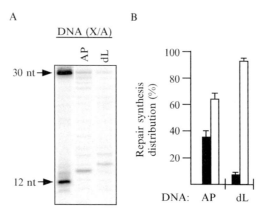

FIG. 6. DNA repair patch size distributions for AP and dL substrates. (A) Standard BER reactions were performed with a pGEM plasmid substrate containing either an AP or a dL site in the presence of α-phosphorothioate dNTPs in place of normal dNTPs and subjected to the repair patch analysis as depicted in Fig. 5 and described in the text. DNA fragments were analyzed on a 15% polyacrylamide/urea DNA sequencing gel. (B) The ^{32}P radioactivity for each band was quantified as described in the text. The proportion of repair products with a single nucleotide patch (black bars) and the sum of those with multinucleotide patches (white bars) are expressed as percentage distributions. Means and SD for three experiments are shown. Adapted from Sung et al. (2005).

very little single nucleotide BER compared to the normal AP site (8% vs 35%, respectively). The small percentage of single nucleotide products seen with dL may be due to a small amount of strand breaks with 3′-phosphates that are also formed during photolysis to generate dL (DeMott et al., 2002; Sung et al., 2005).

Determination of BER Mode in the Complete Repair of dL

Although the BER assay described earlier is useful in determining the repair patch size, it has the disadvantage that it may also score incomplete products. Such products may include those that are complete except for the ligation step or, more problematically, material resulting from strand displacement DNA synthesis without completion of BER. The following assay avoids this pitfall, but involves instead parallel repair reactions that allow only one or the BER mode to occur.

As shown in Fig. 7A, pGEM plasmid DNA substrates have the site-specific DNA lesion located on the immediate 5′ side of an AccI cleavage site or just one nucleotide upstream (5′) of a HincII site. The plasmid DNA substrate is incubated with a cell-free extract in the standard BER reaction as described earlier, except that one or more dNTPs are omitted and a

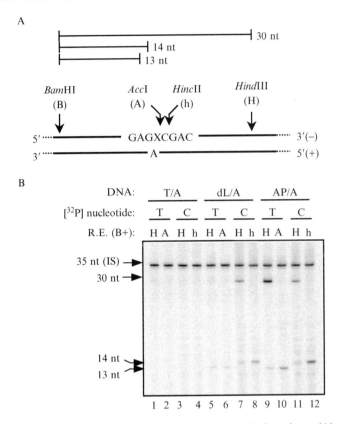

FIG. 7. Determination of BER mode for completely repaired products. (A) A partial nucleotide sequence of the pGEM DNA substrate is shown. X denotes a T, dL, or AP site. The unique restriction digestion sites are indicated by arrows. (B) BER reaction mixtures (100 μl) containing DNA at 5 μg/ml of pGEM (T/A) (lanes 1–4), pGEM (dL/A) (lanes 5–8), or pGEM (AP/A) (lanes 9–12) and 50 μg of wild-type mouse embryonic fibroblast extract were incubated for 30 min at 30° in the presence of 20 μM [α-^{32}P]dTTP (T; lanes 1, 2, 5, 6, 9, and 10) or in a mixture of cold dTTP and [α-^{32}P]dCTP (C; lanes 3, 4, 7, 8, 11, and 12) to allow DNA repair synthesis associated with either short-patch or long-patch BER, respectively. The DNA products were isolated and digested with BamHI (B) together with HindIII (H), AccI (A), or HincII (h), as indicated, followed by electrophoresis on a 15% polyacrylamide/urea DNA sequencing gel. As an internal standard (IS), a 5'-^{32}P-labeled 35-mer oligonucleotide DNA was included in each reaction and used to control for variations associated with DNA isolation and sample loading. Taken from Sung et al. (2005).

selected [α-^{32}P]dNTP is included. After the BER reaction, the recovered DNA is digested with a set of appropriate restriction enzymes and then analyzed by electrophoresis and autoradiography as described earlier. In this case, only the repair products are labeled.

To allow only short-patch BER, $[\alpha\text{-}^{32}P]dTTP$ is supplied as the only nucleotide, required by the template A residue opposite the lesion. Following complete single nucleotide BER, digestion with *Bam*HI and *Hind*III generates a ^{32}P-labeled 30-mer DNA fragment, whereas the unligated product is a ^{32}P-labeled 13-mer DNA fragment (see Fig. 7B, lane 9). Digestion with *Acc*I and *Bam*HI yields a ^{32}P-labeled 13-mer only from the completely repaired DNA (Fig. 7B, lane 10); *Acc*I does not cleave when the lesion lies within its recognition sequence.

For long-patch BER, similar reactions are conducted, but in the presence of $[\alpha\text{-}^{32}P]dCTP$ and unlabeled dTTP. Under these conditions, incorporation of ^{32}P-dCMP requires repair DNA synthesis of at least two nucleotides. Digestion of complete repair products with *Bam*HI/*Hind*III is expected to produce a labeled 30-mer, whereas *Hinc*II/*Bam*HI digestion yields a 14-mer. The patterns for a dL substrate compared to a regular AP site substrate are clearly distinguishable in this assay: the former yields almost no single nucleotide repair product (Fig. 7B).

Acknowledgments

Work in the Demple laboratory has been supported by grants from the NIH (GM40000 and CA71993). We are grateful to our colleagues, especially Dr. M. S. DeMott and Dr. L. Gellon, for helpful discussions.

References

Ausubel, F. M., Brent, R., Moore, D. D., Seidman., J. G., Smith, J. A., and Struhl, K. (1997). "Current Protocols in Molecular Biology." Wiley Interscience, New York.

Biade, S., Sobol, R. W., Wilson, S. H., and Matsumoto, Y. (1998). Impairment of proliferating cell nuclear antigen-dependent apurinic/apyrimidinic site repair on linear DNA. *J. Biol. Chem.* **273**, 898–902.

Chatgilialoglu, C., and Gimisis, T. (1998). Fate of the C-1′ peroxyl radical in the 2′-deoxyuridine system. *Chem. Commun.* 1249–1250.

DeMott, M. S., Beyret, E., Wong, D., Bales, B. C., Hwang, J. T., Greenberg, M. M., and Demple, B. (2002). Covalent trapping of human DNA polymerase beta by the oxidative DNA lesion 2-deoxyribonolactone. *J. Biol. Chem.* **277**, 7637–7640.

Demple, B., and DeMott, M. S. (2002). Dynamics and diversions in base excision DNA repair of oxidized abasic lesions. *Oncogene* **21**, 8926–8934.

Doetsch, P. W., and Cunningham, R. P. (1990). The enzymology of apurinic/apyrimidinic endonucleases. *Mutat. Res.* **236**, 173–201.

Fortini, P., Pascucci, B., Parlanti, E., Sobol, R. W., Wilson, S. H., and Dogliotti, E. (1998). Different DNA polymerases are involved in the short- and long-patch base excision repair in mammalian cells. *Biochemistry* **37**, 3575–3580.

Hashimoto, M., Greenberg, M. M., Kow, Y. W., Hwang, J. T., and Cunningham, R. P. (2001). The 2-deoxyribonolactone lesion produced in DNA by neocarzinostatin and other damaging agents forms cross-links with the base-excision repair enzyme endonuclease III. *J. Am. Chem. Soc.* **123**, 3161–3162.

Huggins, C. F., Chafin, D. R., Aoyagi, S., Henricksen, L. A., Bambara, R. A., and Hayes, J. J. (2002). Flap endonuclease 1 efficiently cleaves base excision repair and DNA replication intermediates assembled into nucleosomes. *Mol. Cell* **10**, 1201–1211.

Hwang, J. T., and Greenberg, M. M. (1999). Kinetics and stereoselectivity of thiol trapping of deoxyuridin-1'-yl in biopolymers and their relationship to the formation of premutagenic alpha-deoxynucleotides. *J. Am. Chem. Soc.* **121**, 4311–4315.

Hwang, J. T., Tallman, K. A., and Greenberg, M. M. (1999). The reactivity of the 2-deoxyribonolactone lesion in single-stranded DNA and its implication in reaction mechanisms of DNA damage and repair. *Nucleic Acids Res.* **27**, 3805–3810.

Kappen, L. S., and Goldberg, I. H. (1992). Neocarzinostatin acts as a sensitive probe of DNA microheterogeneity: Switching of chemistry from C-1' to C-4' by a G/T mismatch 5' to the site of DNA damage. *Proc. Natl. Acad. Sci. USA* **89**, 6706–6710.

Klungland, A., and Lindahl, T. (1997). Second pathway for completion of human DNA base excision-repair: Reconstitution with purified proteins and requirement for DNase IV (FEN1). *EMBO J.* **16**, 3341–3348.

Kotera, M., Roupioz, Y., Defrancq, E., Bourdat, A. G., Garcia, J., Coulombeau, C., and Lhomme, J. (2000). The 7-nitroindole nucleoside as a photochemical precursor of 2'-deoxyribonolactone: Access to DNA fragments containing this oxidative abasic lesion. *Chemistry* **6**, 4163–4169.

Lenox, H. J., McCoy, C. P., and Sheppard, T. L. (2001). Site-specific generation of deoxyribonolactone lesions in DNA oligonucleotides. *Org. Lett.* **3**, 2415–2418.

Levin, J. D., Johnson, A. W., and Demple, B. (1988). Homogeneous *Escherichia coli* endonuclease IV: Characterization of an enzyme that recognizes oxidative damage in DNA. *J. Biol. Chem.* **263**, 8066–8071.

Masuda, Y., Bennett, R. A. O., and Demple, B. (1998). Dynamics of the interaction of human apurinic endonuclease (Ape1) with its substrate and product. *J. Biol. Chem.* **273**, 30352–30359.

Matsumoto, Y., and Kim, K. (1995). Excision of deoxyribose phosphate residues by DNA polymerase beta during DNA repair. *Science* **269**, 699–702.

Piersen, C. E., McCullough, A. K., and Lloyd, R. S. (2000). AP lyases and dRPases: Commonality of mechanism. *Mutat. Res.* **459**, 43–53.

Pratviel, G., Pitie, M., Bernadou, J., and Meunier, B. (1991). Mechanism of DNA cleavage by cationic manganese porphyrins: Hydroxylations at the 1'-carbon and 5'-carbon atoms of deoxyriboses as initial damages. *Nucleic Acids Res.* **19**, 6283–6288.

Sigman, D. S., Mazumder, A., and Perrin, D. M. (1993). Chemical nucleases. *Chem. Rev.* **93**, 2295–2316.

Sobol, R. W., Horton, J. K., Kuhn, R., Gu, H., Singhal, R. K., Prasad, R., Rajewsky, K., and Wilson, S. H. (1996). Requirement of mammalian DNA polymerase-beta in base-excision repair. *Nature* **379**, 183–186.

Srivastava, D. K., Berg, B. J., Prasad, R., Molina, J. T., Beard, W. A., Tomkinson, A. E., and Wilson, S. H. (1998). Mammalian abasic site base excision repair. Identification of the reaction sequence and rate-determining steps. *J. Biol. Chem.* **273**, 21203–21209.

Sung, J. S., Bennett, S. E., and Mosbaugh, D. W. (2001). Fidelity of uracil-initiated base excision DNA repair in *Escherichia coli* cell extracts. *J. Biol. Chem.* **276**, 2276–2285.

Sung, J. S., DeMott, M. S., and Demple, B. (2005). Long-patch base excision DNA repair of 2-deoxyribonolactone prevents the formation of DNA-protein cross-links with DNA polymerase beta. *J. Biol. Chem* **280**, 39095–39103.

Sung, J. S., and Mosbaugh, D. W. (2003). *Escherichia coli* uracil- and ethenocytosine-initiated base excision DNA repair: Rate-limiting step and patch size distribution. *Biochemistry* **42,** 4613–4625.

Tronche, C., Goodman, B. K., and Greenberg, M. M. (1998). DNA damage induced via independent generation of the radical resulting from formal hydrogen atom abstraction from the C1′-position of a nucleotide. *Chem. Biol.* **5,** 263–271.

Von Sonntag, C. (1991). The chemistry of free-radical-mediated DNA damage. *Basic Life Sci.* **58,** 287–317.

Wilson, D. M., Takeshita, M., Grollman, A. P., and Demple, B. (1995). Incision activity of human apurinic endonuclease (Ape) at abasic site analogs in DNA. *J. Biol. Chem.* **270,** 16002–16007.

Xu, Y. J., DeMott, M. S., Hwang, J. T., Greenberg, M. M., and Demple, B. (2003). Action of human apurinic endonuclease (Ape1) on C1′-oxidized deoxyribose damage in DNA. *DNA Repair* **2,** 175–185.

Zheng, Y., and Sheppard, T. L. (2004). Half-life and DNA strand scission products of 2-deoxyribonolactone oxidative DNA damage lesions. *Chem. Res. Toxicol.* **17,** 197–207.

[5] Isolation and Analyses of MutY Homologs (MYH)

By A-LIEN LU-CHANG

Abstract

The base excision repair carried out by the bacterial MutY DNA glycosylase and eukaryotic MutY homolog (MYH) is responsible for removing adenines misincorporated into DNA opposite 7,8-dihydro-8-oxo-guanines (8-oxoG), thereby preventing G:C to T:A mutations. MutY and MYH can also remove adenines from A/G and A/C and can remove guanines from G/8-oxoG mismatches at reduced rates. Biallelic germline mutations in the human *MYH* gene predispose individuals to multiple colorectal adenomas and carcinoma. Four functional assays are usually employed to characterize the MutY and MYH. Gel mobility shift or fluorescence anisotropy assays measures DNA-binding affinity and the apparent dissociation constants. Glycosylase assay determines the catalytic parameters of the enzyme. By using a trapping assay in the presence of sodium borohydride, the protein–DNA covalent intermediate can be identified. The *in vivo* activity of MutY or MYH can be measured by complementation in *Escherichia coli mutY* mutants or fission yeast *Schizosaccharomyces pombe MYH* knockout cells. MutY and MYH interacting proteins can be analyzed by the glutathione *S*-transferase pull-down assay, Far-western, and coimmunoprecipitation. The *in vitro* and *in vivo* activities of MYH can be modulated by several proteins, including mismatch recognition enzymes

METHODS IN ENZYMOLOGY, VOL. 408   0076-6879/06 $35.00

MSH2/MSH6, proliferating cell nuclear antigen, and apurinic/apyrimidinic endonuclease.

Introduction

DNA bases are subjected to oxidative damage from cellular metabolism, as well as from exogenous stimuli, such as ionizing radiation and various chemical oxidants. Such damage, if not repaired, contributes to genome instability and degenerative conditions, including aging and cancer. 7,8-Dihydro-8-oxo-guanine (8-oxoG or GO) is one of the most stable products of oxidative DNA damage and has the most deleterious effects. In *Escherichia coli*, MutT, MutM, MutY, MutS, and Nei (endonuclease VIII) are involved in defending against the mutagenic effects of 8-oxoG lesions (Fig. 1) (reviewed in Hazra *et al.*, 2001; Lu *et al.*, 2001). The MutT protein eliminates 8-oxo-dGTP from the nucleotide pool with its pyrophosphohydrolase activity, whereas the MutM glycosylase (Fpg protein) removes both mutagenic GO adducts and ring-opened purine lesions. When C/GO mismatches are not repaired by MutM, adenines are frequently incorporated opposite GO bases during DNA replication (Michaels and Miller, 1992; Tchou and Grollman, 1993) and can subsequently cause G:C to T:A transversions. MutS and MutY are involved to increase replication fidelity by removing the adenine misincorporated opposite GO or G. The MutS-dependent mismatch repair removes the mismatched A on the daughter DNA strand (reviewed in Modrich and Lahue, 1996). MutY, an adenine and weak guanine DNA glcosylase, is active on A/GO, A/G, A/C, or G/GO mismatches. A/GO mismatches are particularly important biological substrates of MutY glycosylase. Nei can excise GO when GO is opposite a cytosine or adenine during DNA replication and can serve as a backup pathway to repair 8-oxoG in the absence of MutM and MutY (Blaisdell *et al.*, 1999; Hazra *et al.*, 2001).

The N-terminal domain of the *E. coli* MutY protein retains the catalytic activity, and the C-terminal domain of MutY plays an important role in the recognition of GO lesions (Gogos *et al.*, 1996; Li *et al.*, 2000). The X-ray crystal structures of the catalytic domain of *E. coli* MutY(D138N) with bound adenine (Guan *et al.*, 1998) and of *Bacillus stearothermophilus* intact MutY bound to DNA (Fromme *et al.*, 2004) show that MutY distorts the bound DNA substrate and the mismatched A is flipped out of the helix. The action of MutY is characterized by its tight binding to the apurinic/apyrimidinic (AP) site produced after the adenine is removed (Table I) (Chmiel *et al.*, 2001; Li *et al.*, 2000; Porello *et al.*, 1998). The affinity of MutY to AP/GO is particularly strong and is mediated by its C-terminal domain. The AP endonucleases, Endo VI and Exo III, have been shown to

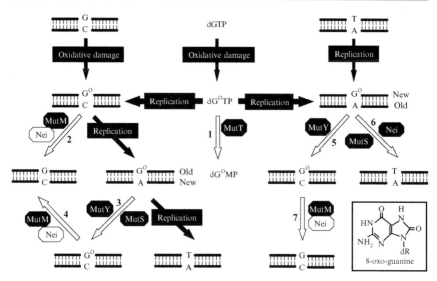

FIG. 1. 8-oxoG repair in *E. coli*. MutT, MutM, MutS, MutY, and Nei (Endo VIII) are involved in defending against the mutagenic effects of 8-oxoG lesions (structure is shown in the inset). The MutT protein hydrolyzes 8-oxo-dGTP (dG°TP) to 8-oxo-dGMP (dG°MP) and pyrophosphate (reaction **1**). GO (G°) in DNA can be derived from oxidation of guanine or misincorporation of dG°TP during replication. The MutM glycosylase removes GO adducts while it is paired with cytosine (reactions **2**, **4**, and **7**). Nei can function as a backup for MutM to remove GO from GO/C. When C/GO is not repaired by MutM, adenines are frequently incorporated opposite GO bases by DNA polymerase III during DNA replication. A/GO mismatches are repaired to C/GO by the MutY- or MutS-dependent pathway (reaction **3**). When dG°TP is incorporated opposite adenine during DNA replication, MutY repair on GO/A can cause more mutation (reaction **5**), whereas GO/A repair by MutS and Nei can reduce mutation (reaction **6**). Adapted from Lu *et al.* (2001).

enhance the turnover of MutY with A/G but not with A/GO substrates (Pope *et al.*, 2002). One central issue concerning MutY function is how MutY searches for the mismatches within a vast excess of normal DNA in the genome. We have proposed a model whereby MutY scans the DNA cooperatively as a dimer or a multimeric complex to locate base-base mismatches (Lee *et al.*, 2004).

The mechanism to defend against the mutagenic effects of 8-oxoG lesions is conserved among organisms (Table II). Human cells process functional homologs of MutT, MutM, MutY, MutS, and Nei. The human homolog of bacterial MutY, hMYH, encodes a DNA glycosylase that excises adenines misincorporated opposite GO or G and removes 2-hydroxyadenines misincorporated with template G (reviewed in Cheadle and Sampson, 2003; Lu *et al.*, 2001). Germline mutations in the *hMYH* gene cause autosomal

TABLE I
APPARENT DISSOCIATION CONSTANTS (K_D) OF *E. COLI* MutY

DNA duplex[a]	K_d (nM)
A/G-20	5.3 ± 0.5^b
A/C-20	15 ± 3^b
A/GO-20	0.07 ± 0.01^b
AP/G-20	2.2 ± 0.3^b
AP/GO-20	0.18 ± 0.11^c
C/GO-20	11.5 ± 3.8^c
T/GO-20	0.11 ± 0.01^c
G/GO-20	0.11 ± 0.01^b
A/G-44	1.8 ± 0.3^b
A/GO-44	0.14 ± 0.01^d
C:G-20	375 ± 80^b
C:G-44	315 ± 49^d

[a] The mismatched duplex DNA substrates are represented by a base-base mismatch followed by the chain length (number of base pairs). C:G represents a homoduplex DNA substrate. Dissociation constants for *E. coli* MutY are derived from Lu *et al.* (1995),[b] Li *et al.* (2000),[c] and Lu and Fawcett (1998).[d]

TABLE II
8-OXOG REPAIR ENZYMES IN DIFFERENT ORGANISMS

E. coli bacteria	Human mammal	*Saccharomyces cerevisiae* baking yeast	*Schizosaccharomyces pombe* fission yeast
MutY	hMYH	No	SpMYH
MutT	hMTH1	ScMTH1	?
MutM	hOGG1	ScOGG1	No
	hOGG2	ScOGG2	
MutS	hMSH2/MSH6	ScMSH2/MSH6	SpMSH2/MSH6
	hMSH2/MSH3	ScMSH2/MSH3	SpMSH2/MSH3
	Others	Others	Others
Nei (Endo VIII)	hNEIL1	No	No
	hNEIL2		
	hNEIL3		

recessive colorectal adenomatous polyposis, which is characterized by multiple adenomas, some of which progress to cancer. Due to the association with colon cancer, great interest has arisen in screening *hMYH* gene mutations and assay of altered hMYH activities.

The MYH activity can be modulated by other proteins. We have shown that MYH is directly associated with proliferating cell nuclear antigen (PCNA), replication protein A (RPA), AP endonuclease (APE1), and hMutSα (MSH2/MSH6) via hMSH6 (Chang and Lu, 2002; Gu et al., 2002; Parker et al., 2001). Association between MYH and PCNA is biologically important for in vivo MYH function (Chang and Lu, 2002). The glycosylase and DNA-binding activities of MYH can be stimulated by APE1 and MutSα (Gu et al., 2002; Yang et al., 2001). We showed that SpMYH is associated with the Schizosaccharomyces pombe checkpoint proteins, Rad9, Rad1, and Hus1 (referred as the 9-1-1 complex) (Chang and Lu, 2005). The 9-1-1 complex has predicted structural homology to the PCNA sliding clamp (Shiomi et al., 2002; Venclovas and Thelen, 2000) and is involved in signaling the DNA damage response of cell cycle arrest or apoptosis (reviewed in Zhou and Elledge, 2000). Moreover, the interaction of MYH-Hus1 is enhanced and DNA damage-induced SpHus1 phosphorylation is dependent of SpMYH expression following oxidative stress (Chang and Lu, 2005). This chapter describes the isolation of MutY and MYH and their four functional assays. Some of these methods have been published elsewhere (Lu, 2000). In addition, three methods employed to analyze MutY and MYH interacting proteins are also described.

Isolation of MutY and MYH Proteins

MutY Protein Expression

MutY and MutY mutant proteins are expressed in E. coli mutY mutant strains. PR70 (Su- lacZ X74 galU galK Sm^r micA68::Tn10kan) and GBE943 [lacIp4000(LacI^q) lacZp4008(Lac L8) srlC-300::Tn10 λ$^-$ IN(rrD-rrnE)1 micA68::Tn10Kan] contain a transposon-inserted mutY gene. GM7724 cell has the chromosomal mutY gene replaced by a transposon and does not contain any MutY activity. The strain CC104/mutY [araΔ (gpt-lac)5 F' (lacI378 LacZ461 proA$^+$B$^+$) mutY::mini-Tn10] contains a lacZ mutation at residue 461 of β-galactosidase and a transposon-inserted mutY gene. DE3 lysogenic strains can be constructed according to the procedures described by Invitrogen when the mutY gene is under the control of the T7 promoter. MutY overexpression plasmid pMYW-1 containing the entire mutY gene in pET11a (Novagen) is suitable for soluble MutY mutant construction (Wright et al., 1999). pGEV-MutY containing the mutY gene fused to the streptococcal protein G (GB1 domain) at its N terminus in pGEV1 (Li and Lu, 2003) should be used to construct less soluble MutY mutants. The E. coli strain harboring the expression plasmid is grown in 6 liters of Luria-Bertani broth (10 g of tryptone, 5 g of yeast extract, and 10 g of NaCl per liter) containing 100 μg/ml ampicillin at 37°. The culture can be grown in three 2-liter media batches in 6-liter flasks using a New Brunswick Scientific C25KC incubator shaker. Protein expression is induced at an OD_{600} of 0.6 by adding isopropyl β-D-thiogalactoside (IPTG) to a final concentration of 0.4 mM to the culture at 20°. Cells are harvested by centrifugation in a SLC-3000 rotor 16 h after induction. At the end, remove as much media as possible and store the cell paste at −80°.

MutY Protein Purification

The MutY and mutant proteins are purified by the method as described (Li and Lu, 2003; Tsai-Wu et al., 1992). Before enzyme purification, all required buffers are prepared (filtered through a 45-μm membrane and autoclaved) and the columns are packed at 4°. All of the column chromatography steps are conducted by a FPLC system at 4°; all buffer solutions are flushed with helium gas; and centrifugation is done at 15,000g for 30 min. Cells (20 g of cell paste) are resuspended in 60 ml of buffer T [50 mM Tris–HCl (pH 7.6), 0.1 mM EDTA, 1 mM dithiothreitol (DTT), and 0.1 mM phenylmethanesulfonyl fluoride (PMSF)] and disrupted with a bead beater (Biospec Products) using 0.1-mm glass beads. Cell debris is removed by centrifugation, and the supernatant is then treated with 5% streptomycin sulfate in buffer T. After stirring for 45 min, the solution is centrifuged, and the supernatant is collected as fraction I. Ammonium sulfate is added to fraction I to a final concentration of 45% (0.26 g/ml). The solution is stirred for 45 min and the protein is precipitated overnight. After centrifugation, the protein pellet is resuspended in 15 ml of buffer T and dialyzed against two changes of 1 liter of the same buffer for 1.5 h each. The dialyzed protein sample is centrifuged, and the supernatant is filtered through a 45-μm membrane and diluted threefold with buffer A [20 mM potassium phosphate (pH 7.4), 0.1 mM EDTA, 10% glycerol, 0.5 mM DTT, and 0.1 mM PMSF] containing 50 mM KCl (fraction II). Fraction II is loaded onto a 50-ml phosphocellulose (Whatman P-11) column, which has been equilibrated with buffer A containing 50 mM KCl. After washing with 100 ml of equilibration buffer, proteins are eluted with a 500-ml linear gradient of KCl (0.05–0.6 M) in buffer A. Fractions eluted at about 0.3 mM KCl are pooled based on the protein profile on 12% polyacrylamide gel in the presence of SDS (SDS–PAGE) (fraction III). These fractions should have a brown color because MutY contains a Fe–S cluster. Fraction III is loaded onto a 20-ml hydroxylapatite column equilibrated with buffer B [10 mM potassium phosphate (pH 7.4), 10 mM KCl, 0.1 mM EDTA, 10% glycerol, 1 mM DTT, and 0.1 mM PMSF]. The flow-through and early elution fractions are pooled and diluted with buffer B to a conductivity equal to 80 mM KCl (fraction IV). Fraction IV is loaded onto a 5-ml heparin high-trap column (Amersham Pharmacia Biotech) equilibrated with buffer A containing 50 mM KCl and 10% glycerol. Upon washing with 10 ml of equilibration buffer, the column is eluted with a 50-ml linear gradient of KCl (0.05–0.6 M) in buffer A containing 10% glycerol. Fractions containing the MutY protein are eluted at about 0.4 mM KCl (fraction V), pooled, and stored in small aliquots at $-80°$. If necessary, MutY can be purified further on a MonoS column (Amersham). After dilution with buffer A to a conductivity equal to 80 mM KCl (fraction IV), fraction V is then applied to a 1-ml MonoS column that has been equilibrated in buffer A containing 50 mM KCl. After washing with 10 ml of equilibration buffer, the protein is eluted with a 20-ml linear gradient of KCl (0.05–0.5 M) in buffer A containing 10% glycerol. Fractions containing the MutY protein eluted at about 0.2 mM KCl are pooled (fraction VI), divided into small aliquots, and stored at $-80°$. The buffer for heparin and MonoS columns can be substituted with TEG buffer [50 mM Tris–HCl (pH 7.6), 0.1 mM EDTA, 10% glycerol, 1 mM DTT, and 0.1 mM PMSF].

The MutY protein concentration can be determined by the Bradford method (Bio-Rad) (Bradford, 1976) or by absorption at 280 nm with an extinction coefficient of 75,000 $M^{-1}cm^{-1}$. A typical yield of wild-type MutY protein is about 5 mg protein per gram of cell paste while some mutant proteins may have lower yields. The iron content of MutY is measured by the method of Kennedy $et\ al.$ (1984). The iron standard is from Sigma-Aldrich Chemical Co. Protein samples (25–50 μg) and iron standards (25–400 ng) are diluted with water to 0.1 ml and mixed with 0.1 ml of reagent A (4.5% SDS in 1.5% saturated sodium acetate) and 0.1 ml of reagent B (4.5% ascorbic acid and 0.15% sodium metabisulfite in 6.7% saturated sodium acetate). The mixtures are incubated at 30° for 15 min and supplemented with 5 μl of reagent C (1.8% Ferene). The absorption at 593 nm is measured.

Purification of Recombinant Eukaryotic MYH Proteins

Recombinant $S.\ pombe$ MYH has been expressed from plasmid pSPMYH11a-4 (Lu and Fawcett, 1998) in GBE943/DE3 cells. SpMYH is purified similarly to MutY, except the fraction from 40 to 70% (0.23–0.44 g/ml) ammonium sulfate precipitation is used. Due the high homology between SpMYH and hMYH, $S.\ pombe$ has been used as a model organism to study the human MYH repair pathway.

Recombinant hMYH expressed from plasmid pET21a-hMYH (Gu and Lu, 2001) in GBE943/DE3 cells has been partially purified. The purification procedure is similar to *E. coli* MutY as described earlier except using 55% (0.33 g/ml) ammonium sulfate precipitation. Due to the lower expression and poor solubility, the yield and purity of hMYH are low. To improve hMYH protein translation, the host cell can be transformed with plasmid pRARE from the Rosetta strain (Invitrogen), which contains many rare tRNA genes. The human MYH protein can be purified as fusion proteins tagged by both streptococcal protein G (GB1 domain) and 6-His in the PR70/DE3/pRARE (*mutY* mutant) host (Bai *et al.*, 2005). The GB1 domain increased the solubility of hMYH and the His tag allowed for affinity purification. His-tagged hMYH fusion proteins are bound to nickel agarose (Qiagen). Cell extracts are incubated with the beads at 4° for 1 h. After washing with buffer N [50 mM potassium phosphate (pH 8.0) and 300 mM NaCl] containing 50 mM imidazole, bound proteins are eluted by buffer N containing 250 mM imidazole. The partially purified proteins are visualized by 10% SDS–PAGE, and fractions containing hMYH are pooled, dialyzed with buffer A containing 50 mM KCl and 10% glycerol, divided into small aliquots, and stored at −80°.

Because it is difficult to purify hMYH to homogeneity, mouse (m) MYH may provide an alternative solution. Yang *et al.* (2001) have cloned the *mMyh* cDNA, lacking the first 28 N-terminal amino acids, into the pQE30 vector (Qiagen). The mouse MYH protein has been purified from *E. coli* CC104/*mutY* containing pQE30Myh and pREP4 (Qiagen) by Pope and David (2005) and Yang *et al.* (2001) using slightly modified procedures. The cell lysate prepared as described for the MutY protein is loaded onto an SP Sepharose column, which is washed with 0.1 M KCl in buffer I [20 mM potassium phosphate (pH 7.6), 10% glycerol, and 0.5 mM PMSF]. MYH is step eluted with 0.5 M NaCl in buffer I. The MYH-containing fractions are pooled, and the NaCl and imidazole are adjusted to a final concentration of 1 M and 20 mM, respectively. The resulting sample is incubated with 2 ml Ni-NTA agarose resin (Qiagen) for 1 h at 4°. The protein-bound Ni-NTA resin is poured into a column and washed with 10 column volumes of buffer I containing 1 M NaCl and 40 mM imidazole. The bound mMYH is eluted with buffer I containing 0.3 M NaCl and 0.2 M imidazole. The eluent from the Ni-NTA column is diluted fivefold with buffer II [20 mM sodium phosphate (pH 7.5), 10 mM EDTA, 5 mM DTT, and 10% glycerol] and applied to a 1-ml heparin column (Amersham). The proteins are eluted with a linear gradient of NaCl (0.1–0.7 M) in buffer II. The intact mMYH is eluted at 0.4 M NaCl. Fractions containing MYH are combined, dialyzed overnight in buffer A containing 50 mM NaCl and 50% glycerol, cleared by centrifugation, and stored at −80° in small aliquots.

Functional Assays of MutY and MYH

Preparation of DNA Substrates

The DNA substrates are as follow:

```
19-mer  5'  CCGAGGAATTXGCCTTCTG    3'
        3'    GCTCCTTAAYCGGAAGACG  5'
40-mer  5'  AATTGGGCTCCTCGAGGAATTXGCCTTCTGCAGGCATGCC          3'
        3'        CCCGAGGAGCTCCTTAAYCGGAAGACGTCCGTACGGGGCC  5'
```

(X= A, C, G, or T and Y=G or GO).

The oligonucleotides are designed to be labeled at one unique 3' or 5' end. Both the 19-mer and 40-mer annealed heteroduplexes can be labeled at the 3' end with $[\alpha\text{-}^{32}P]dCTP$ on the X strand and are converted to 20- and 44-mer, respectively, after the sticky ends are filled in with the Klenow fragment of DNA polymerase I and dATP, dGTP, and dTTP. For 5' end labeling, the top strands are first labeled with $[\gamma\text{-}^{32}P]ATP$ and polynucleotide kinase and then are annealed with the bottom strands. The single-stranded overhangs are filled in with the Klenow fragment and unlabeled deoxynucleotide triphosphates. Detailed methods for the preparation of DNA substrates can be found elsewhere (Lu, 2000). A/GO-containing oligonucleotides are the substrates of MutY and MYH, and the 44-mer duplexes are better substrates for eukaryotic MYH. For fluorescence anisotropy assays, a fluorescein tag (5,6 FAM) is linked to the 5' ends of the top strands of the 19-mer oligonucleotides. The DNA oligonucleotides are synthesized by Integrated DNA Technologies, Inc.

MutY Gel Mobility Shift Assay

MutY has been shown to bind catalytic substrates (A/G, A/GO, A/C, and G/GO), its reaction products, and some noncatalytic substrates, including T/GO and C/GO (Table I) (Li *et al.*, 2000; Lu and Fawcett, 1998; Lu *et al.*, 1995). The MutY protein is diluted to the 0.03–500 nM range in a buffer containing 20 m*M* potassium phosphate (pH 7.4), 1.5 m*M* DTT, 0.1 m*M* EDTA, 50 m*M* KCl, 200 μg/ml bovine serum albumin (BSA), and 50% glycerol. MutY (1 μl) is added to a 20-μl binding reaction mixture containing 20 m*M* Tris–HCl (pH 7.6), 80 m*M* NaCl, 1 m*M* DTT, 1 m*M* EDTA, 2.9% glycerol, 20 ng of poly(dI-dC), and 1.8 fmol of labeled DNA (final concentration is 90 p*M*). After incubation at 37° for 30 min, the mixtures are supplemented with 2 μl of 50% glycerol and analyzed on a 6% polyacrylamide gel in 50 m*M* Tris borate (pH 8.3) and 1 m*M* EDTA

(TBE buffer). To determine K_d values, eight to nine different MutY enzyme concentrations are used to bind DNA substrates in experiments that are repeated at least three times. Bands corresponding to enzyme-bound and free DNA are quantified from PhosphorImager images and K_d values are obtained from analyses by a computer-fitted curve generated by the Enzfitter program (Leatherbarrow, 1987) or from protein concentrations that resulted in 50% of the maximal binding.

Fluorescence Anisotropy Assay

Fluorescence anisotropy is measured using the Beacon 2000 variable temperature fluorescence polarization system (Panvera) equipped with fluorescein excitation (490 nm) and emission (535 nm) filters. Binding reactions are performed as for the gel mobility shift assay except that no glycerol is added, the fluorescein-labeled DNA concentration is 2 nM, and the reaction volume is 100 μl. Each sample is read as blank prior to the addition of Fl-tagged DNA. After adding Fl–DNA, samples are incubated for 1 min before measurement of anisotropy (A_m) and total fluorescence emission (F_{535}). Data obtained by this method provide a better index of solution binding affinity. The fluorescence anisotropy assay is a better choice to measure weak DNA binding than the gel mobility shift assay because such binding is more sensitive to gel sieving (Lee et al., 2004).

MutY Glycosylase and Trapping Assays

The glycosylase assay is carried out in a 10-μl reaction containing 20 mM Tris–HCl (pH 7.6), 80 mM NaCl, 1 mM DTT, 1 mM EDTA, 2.9% glycerol, and 50 μg/ml of BSA. After incubation at 37° for 30 min, the reaction mixtures are supplemented with 1 μl of 1 M NaOH and heated at 90° for 30 min. Five microliters of formamide dye (90% formamide, 10 mM EDTA, 0.1% xylene cyanol, and 0.1% bromphenol blue) is added to the sample, which is then heated at 90° for 2 min and 5 μl of the mixture is loaded onto a 14% polyacrylamide sequencing gel containing 7 M urea. If nonspecific nucleases are a problem, the concentration of EDTA in the reaction can be increased to 5 mM.

Covalent complexes of MutY with DNA substrates are formed by the trapping assay similarly to the glycosylase assay except NaCl and BSA are omitted and 0.1 M NaBH$_4$ is added. A NaBH$_4$ (1 M) stock solution is freshly prepared and is added to the reaction mixture immediately after the enzyme is added. After incubation at 37° for 30 min, 2.5 μl of fivefold concentrated dye buffer containing 25% glycerol, 5% SDS, 155 mM Tris–HCl (pH 6.8), 5% β-mercaptoethanol, and 0.5 mg/ml bromphenol blue is added to the samples, which are heated at 90° for 3 min and separated on a 12% SDS–PAGE.

Assays of SpMYH and hMYH Binding and Glycosylase Activities

The binding and nicking activities of MYH are assayed similar to those of MutY as described earlier except using different buffers and the reactions of SpMYH are incubated at 30°. The SpMYH binding buffer contains 10 mM Tris–HCl (pH 7.6), 40 mM NaCl, 0.5 mM DTT, 5 mM EDTA, 1.45% glycerol, 50 μg/ml BSA, and 5 ng of poly(dI-dC). The hMYH buffer contains 10 mM Tris–HCl, pH 7.4, 20 mM NaCl, 0.5 mM DTT, 2.25 mM EDTA, 5 μM ZnCl$_2$, 1.45% glycerol, 75 μg/ml BSA, and 10 ng of poly(dI-dC). The MYH glycosylase buffer is similar to the MYH binding buffer except no NaCl and poly(dI-dC) are added.

Measurement of E. coli Mutation Frequency

Forward Rifampicin Resistance Assay. Four single colonies are inoculated into LB medium and grown overnight at 37°. The cultures (0.1 ml) are plated onto LB agar plates containing 0.1 mg/ml rifampicin (dissolved in methanol). The plates should be prepared within 1 week and stored in the dark. The cell titer of each culture is determined by plating 0.1 ml of a 10^{-6} dilution onto LB agar plates. Rifampicin-resistant colonies are counted the next day after 37° incubation. The experiments are repeated three times. The ratio of Rif^R cells to total cells is the mutation frequency. The rifampicin resistance mutation assay can be used in any genetic background; however, we have found that CC104/*mutYmutM* with an improper λDE3 lysogen offers a more sensitive assay for *in vivo* complementation because proteins from the T7 promoter are expressed less in this cell than those from the *mutY* own promoter in a wild-type cell (Li and Lu, 2003; Lu and Wright, 2003).

Lac$^+$ Reversion Assay. Six strains of *E. coli* with different mutations at the Glu461 position in the *lacZ* gene have been developed (Cupples and Miller, 1989). The CC104 strain allows detection of G:C-to-T:A transversions by monitoring the Lac$^-$ to Lac$^+$ frequency. Four single colonies are inoculated into LB medium, with or without IPTG, and are grown overnight at 37°. Samples are then plated on lactose minimal medium and titers are determined on glucose minimal medium. Lac$^+$ colonies are counted after 3 days of growth. The experiments are repeated three times.

Measurement of S. pombe *Mutation Frequency*

Five independent yeast colonies are grown to late log phase in Edin-burgh minimal media (EMM) (U.S. Biological) containing 0.1 mg/ml uracil and 0.1 mg/ml of required amino acids. Each culture (0.2 ml) is plated onto EMM agar plates containing 1 mg/ml 5-fluoro-orotic acid (FOA) and 0.1 mg/ml uracil. FOA-resistant colonies are counted after 5 days of growth. The cell titer is determined by plating 0.1 ml of a 10^{-4} dilution onto plates without FOA. The mutation frequency is calculated as the ratio of FOA-resistant cells to the total cells. The measurement should be repeated more than three times.

Protein Interaction Assays

Many methods have been described to investigate and identify protein–protein interactions. This chapter only lists three methods used to analyze MutY and MYH interacting proteins.

GST Pull-Down Assay

The GST pull-down assay is similar to the described procedures (Parker *et al.*, 2001). The hMYH gene is cloned into the pGEX-4T-2 vector (Amer-sham) to express fusion proteins of glutathione *S*-transferase and hMYH (GST-hMYH). The fusion protein is expressed in *E. coli* (BL21Star/DE3) cells (Stratagene). The cell paste, from a 500-ml culture grown as described earlier, is resuspended in 9 ml of buffer G [50 mM Tris–HCl (pH 7.4), 150 mM NaCl, and 2 mM EDTA] containing 0.5 mM DTT and 0.1 mM PMSF and is treated with lysozyme (1 mg/ml) for 30 min at 4°. After sonication, the solution is centrifuged at 10,000g for 20 min. The superna-tant (10 ml) is added to 1 ml of glutathione-Sepharose 4B (Amersham) in buffer G and incubated overnight at 4°. GST fusion proteins bound to the beads are pelleted at 1000g in a tabletop centrifuge for 5 min, washed four times with 2 ml of buffer G, and incubated with 5% BSA in buffer G for 2 h at 4°. The beads are pelleted at 1000g for 5 min and washed four times with 2 ml of buffer G. The beads are suspended in buffer G containing 0.1% sodium azide and a protease inhibitor cocktail (Sigma-Aldrich Chemical Co.) to form 50% slurry and stored at 4°.

GST constructs (300 ng) and GST alone immobilized on glutathione-Sepharose 4B are incubated with 0.1 μg of target proteins or 0.4 mg of nuclear extracts overnight in 200 μl buffer G containing 0.1% NP-40 at 4°. After centrifugation at 1000g, the supernatant is saved and the pellets are washed four times with 1 ml of buffer G at 4°. The pellet is reconstituted in 1× SDS loading buffer and both the supernatant (~10% of total volume) and pellet fractions are resolved on a SDS–PAGE and transferred to a nitrocellulose membrane. The membrane is allowed to react with primary antibody against target proteins. Western blotting is detected by the ECL analysis system from Amersham according to the manufacturer's protocol.

Coimmunoprecipitation of hMYH Interacting Proteins

HeLa whole cell extracts (0.8 mg) or nuclear extracts (0.4 mg) in 1 ml buffer G containing protease inhibitor cocktail (Sigma-Aldrich Chemical Co.) are precleared by adding 30 μl protein G agarose (Invitrogen, Carlsbad, CA) for 1–4 h at 4°. After centrifugation at 1000g, the supernatant is incubated with 1 μg of hMYH antibody overnight at 4°. Protein G agarose (30 μl) is added and incubated for 4–12 h at 4°. After centrifugation at 1000g, the supernatant is saved and the pellet is washed five times with 800 μl of buffer G. Both the supernatant (~10% of total volume) and pellet fractions are resolved on a 10% SDS–PAGE and Western blot analyses are performed. A control is run concurrently without the primary antibody.

Far Western Analysis

This method is useful to determine which component(s) of a protein complex interacts with hMYH. Purified proteins are separated on SDS–PAGE and transferred to a nitrocellulose membrane. The membrane is blocked with 5% low fat milk in phosphate-buffered saline for 1 h and is then incubated with 3 μg/ml partially purified recombinant hMYH at 4° overnight. After extensive washing with blocking solution, the membrane is incubated with hMYH antibody and subjected to Western blotting.

Acknowledgments

This work is supported by Grant GM35132 from the National Institutes of General Medical Science and Grant CA78391 from National Cancer Institute, National Institutes of Health.

References

Bai, H., Jones, S., Guan, X., Wilson, T. M., Sampson, J. R., Cheadle, J. P., and Lu, A. L. (2005). Functional characterization of two human MutY homolog (hMYH) missense mutations (R227W and V232F) that lie within the putative hMSH6 binding domain and are associated with hMYH polyposis. *Nucleic Acids Res.* **33**, 597–604.

Blaisdell, J. O., Hatahet, Z., and Wallace, S. S. (1999). A novel role for *Escherichia coli* endonuclease VIII in prevention of spontaneous G → T transversions. *J. Bacteriol.* **181**, 6396–6402.

Bradford, M. (1976). A rapid and sensitive method for the quantitation of microgram quantities of protein utilizing the principle of protein-dye binding. *Anal. Biochem.* **72**, 248–254.

Chang, D. Y., and Lu, A.-L. (2002). Functional interaction of MutY homolog (MYH) with proliferating cell nuclear antigen (PCNA) in fission yeast, *Schizosaccharomyces pombe*. *J. Biol. Chem.* **277**, 11853–11858.

Chang, D. Y., and Lu, A.-L. (2005). Interaction of checkpoint proteins Hus1/Rad1/Rad9 with DNA base excision repair enzyme MutY homolog in fission yeast, *Schizosaccharomyces pombe*. *J. Biol. Chem.* **280**, 408–417.

Cheadle, J. P., and Sampson, J. R. (2003). Exposing the MYH about base excision repair and human inherited disease. *Hum. Mol. Genet.* **12**, Spec. No. 2, R159–R165.

Chmiel, N. H., Golinelli, M. P., Francis, A. W., and David, S. S. (2001). Efficient recognition of substrates and substrate analogs by the adenine glycosylase MutY requires the C-terminal domain. *Nucleic Acids Res.* **29**, 553–564.

Cupples, C. G., and Miller, J. H. (1989). A set of *lacZ* mutations in *Escherichia coli* that allow rapid detection of each of the six base substitutions. *Proc. Natl. Acad. Sci. USA* **86**, 5345–5349.

Fromme, J. C., Banerjee, A., Huang, S. J., and Verdine, G. L. (2004). Structural basis for removal of adenine mispaired with 8-oxoguanine by MutY adenine DNA glycosylase. *Nature* **427**, 652–656.

Gogos, A., Cillo, J., Clarke, N. D., and Lu, A.-L. (1996). Specific recognition of A/G and A/8-oxoG mismatches by *Escherichia coli* MutY: Removal of the C-terminal domain preferentially affects A/8-oxoG recognition. *Biochemistry* **35**, 16665–16671.

Gu, Y., and Lu, A.-L. (2001). Differential DNA recognition and glycosylase activity of the native human MutY homolog (hMYH) and recombinant hMYH expressed in bacteria. *Nucleic Acids Res.* **29**, 2666–2674.

Gu, Y., Parker, A., Wilson, T. M., Bai, H., Chang, D. Y., and Lu, A.-L. (2002). Human MutY homolog (hMYH), a DNA glycosylase involved in base excision repair, physically and functionally interacts with mismatch repair proteins hMSH2/hMSH6. *J. Biol. Chem.* **277**, 11135–11142.

Guan, Y., Manuel, R. C., Arvai, A. S., Parikh, S. S., Mol, C. D., Miller, J. H., Lloyd, S., and Tainer, J. A. (1998). MutY catalytic core, mutant and bound adenine structures define specificity for DNA repair enzyme superfamily. *Nature Struct. Biol.* **5**, 1058–1064.

Hazra, T. K., Hill, J. W., Izumi, T., and Mitra, S. (2001). Multiple DNA glycosylases for repair of 8-oxoguanine and their potential *in vivo* functions. *Prog. Nucleic Acid Res. Mol. Biol.* **68**, 193–205.

Kennedy, M. C., Kent, T. A., Emptage, M., Merkle, H., Beinert, H., and Munck, E. (1984). Evidence for the formation of a linear [3Fe-4S] cluster in partially unfolded aconitase. *J. Biol. Chem.* **259**, 14463–14471.

Leatherbarrow, R. J. (1987). "Enzfitter: A Non-linear Regression Analysis Program for IBM PC." Elsevier Science, Amsterdam.

Lee, C. Y., Bai, H., Houle, R., Wilson, G. M., and Lu, A.-L. (2004). An *Escherichia coli* MutY mutant without the six-helix barrel domain is a dimer in solution and assembles cooperatively into multisubunit complexes with DNA. *J. Biol. Chem.* **279**, 52653–52663.

Li, L., and Lu, A.-L. (2003). The C-terminal domain of *Escherichia coli* MutY is involved in DNA binding and glycosylase activities. *Nucleic Acids Res.* **31**, 3038–3049.

Li, X., Wright, P. M., and Lu, A.-L. (2000). The C-terminal domain of MutY glycosylase determines the 7,8-dihydro-8-oxo-guanine specificity and is crucial for mutation avoidance. *J. Biol. Chem.* **275**, 8448–8455.

Lu, A.-L. (2000). Repair of A/G and A/8-oxoG mismatches by MutY adenine DNA glycosylase. *In* "DNA Repair Protocols, Prokaryotic Systems" (P. Vaughan, ed.), pp. 3–16. Humana Press, Totowa, NJ.

Lu, A.-L., and Fawcett, W. P. (1998). Characterization of the recombinant MutY homolog, an adenine DNA glycosylase, from *Schizosacchromyces pombe. J. Biol. Chem.* **273,** 25098–25105.

Lu, A-L., Li, X., Gu, Y., Wright, P. M., and Chang, D.-Y. (2001). Repair of oxidative DNA damage. *Cell Biochem. Biophy.* **35,** 141–170.

Lu, A.-L., Tsai-Wu, J.-J., and Cillo, J. (1995). DNA determinants and substrate specificities of *Escherichia coli* MutY. *J. Biol. Chem.* **270,** 23582–23588.

Lu, A.-L., and Wright, P. M. (2003). Characterization of an *Escherichia coli* mutant MutY with a cysteine to alanine mutation at the iron-sulfur cluster domain. *Biochemistry* **42,** 3742–3750.

Michaels, M. L., and Miller, J. H. (1992). The GO system protects organisms from the mutagenic effect of the spontaneous lesion 8-hydroxyguanine (7,8-dihydro-8-oxo-guanine). *J. Bacteriol.* **174,** 6321–6325.

Modrich, P., and Lahue, R. S. (1996). Mismatch repair in replication fidelity, genetic recombination and cancer biology. *Annu. Rev. Biochem.* **65,** 101–133.

Parker, A., Gu, Y., Mahoney, W., Lee, S.-H., Singh, K. K., and Lu, A.-L. (2001). Human homolog of the MutY protein (hMYH) physically interacts with protein involved in long-patch DNA base excision repair. *J. Biol. Chem.* **276,** 5547–5555.

Pope, M. A., and David, S. S. (2005). DNA damage recognition and repair by the murine MutY homologue. *DNA Repair (Amst)* **4,** 91–102.

Pope, M. A., Porello, S. L., and David, S. S. (2002). *Escherichia coli* apurinic-apyrimidinic endonucleases enhance the turnover of the adenine glycosylase MutY with G:A substrates. *J. Biol. Chem.* **277,** 22605–22615.

Porello, S. L., Leyes, A. E., and David, S. S. (1998). Single-turnover and pre-steady-state kinetics of the reaction of the adenine glycosylase MutY with mismatch-containing DNA substrates. *Biochemistry* **37,** 14756–14764.

Shiomi, Y., Shinozaki, A., Nakada, D., Sugimoto, K., Usukura, J., Obuse, C., and Tsurimoto, T. (2002). Clamp and clamp loader structures of the human checkpoint protein complexes, Rad9-1-1 and Rad17-RFC. *Genes to Cells* **7,** 861–868.

Tchou, J., and Grollman, A. P. (1993). Repair of DNA containing the oxidatively-damaged base 8-hydroxyguanine. *Mutat. Res.* **299,** 277–287.

Tsai-Wu, J.-J., Liu, H.-F., and Lu, A.-L. (1992). *Escherichia coli* MutY protein has both N-glycosylase and apurinic/apyrimidinic endonuclease activities on A·C and A·G mispairs. *Proc. Natl. Acad. Sci. USA* **89,** 8779–8783.

Venclovas, C., and Thelen, M. P. (2000). Structure-based predictions of Rad1, Rad9, Hus1 and Rad17 participation in sliding clamp and clamp-loading complexes. *Nucleic Acids Res.* **28,** 2481–2493.

Wright, P. M., Yu, J., Cillo, J., and Lu, A.-L. (1999). The active site of the *Escherichia coli* MutY DNA adenine glycosylase. *J. Biol. Chem.* **274,** 29011–29018.

Yang, H., Clendenin, W. M., Wong, D., Demple, B., Slupska, M. M., Chiang, J. H., and Miller, J. H. (2001). Enhanced activity of adenine-DNA glycosylase (Myh) by apurinic/apyrimidinic endonuclease (Ape1) in mammalian base excision repair of an A/GO mismatch. *Nucleic Acids Res.* **29,** 743–752.

Zhou, B. B., and Elledge, S. J. (2000). The DNA damage response: Putting checkpoints in perspective. *Nature* **408,** 433–439.

[6]   Use of Yeast for Detection of Endogenous Abasic Lesions, Their Source, and Their Repair

By SERGE BOITEUX and MARIE GUILLET

Abstract

Apurinic/apyrimidinic (AP) sites are expected to be one of the most frequent endogenous lesions in DNA. AP sites are potentially lethal and mutagenic. Data shows that the simultaneous inactivation of two AP endonucleases (Apn1 and Apn2) and of the nuclease Rad1-Rad10 causes cell death in *Saccharomyces cerevisiae*. We suggest that the essential function of Apn1, Apn2, and Rad1-Rad10 is to repair endogenous AP sites and related 3'-blocked single strand breaks. This data led us to conclude that the burden of endogenous AP sites is not compatible with life in absence of DNA repair. This chapter describes two genetic assays to investigate origin, repair, and biological consequences of endogenous AP sites in yeast. The first assay relies on genetic crosses and tetrad analysis and uses the *apn1 apn2 rad1* triple mutant. The *apn1 apn2 rad1* triple mutant is unviable; however, it can form microcolonies. By means of genetic crosses, *apn1 apn2 rad1 x* quadruple mutants are generated. The size of the colonies formed by each quadruple mutant is compared to that of the *apn1 apn2 rad1* triple mutant. Three classes of genes (*x*) were identified: (i) genes whose inactivation aggravates the phenotype (reduces microcolony size), such as *RAD9, RAD50, RAD51, RAD52, MUS81*, and *MRE11*; (ii) genes whose inactivation alleviates the phenotype, such as *UNG1, NTG1*, and *NTG2*; and (iii) genes whose inactivation is neutral, such as *MAG1* or *OGG1*. The second assay uses the *apn1 apn2 rad14* triple mutant, which is viable but exhibits a spontaneous mutator phenotype. This mutant was used in a colethal screen. This assay allowed the identification of mutation in DNA repair genes such as *RAD1* or *RAD50*, as well as a mutation in the *DUT1* gene coding for the dUTPase, which has impact on the formation of AP sites in DNA. A model that summarizes our present and puzzling data on the origin and repair of endogenous AP sites is also presented.

AP Sites in DNA and Related 3'-Blocked and 5'-Blocked Single Strand Breaks Form an Abundant and Deleterious Class of Endogenous DNA Damage

Cellular DNA is continuously damaged by endogenous and exogenous reactive species. The outcome of DNA damage is generally adverse,

0076-6879/06 $35.00
DOI: 10.1016/S0076-6879(06)08006-2

contributing to degenerative processes such as aging and cancer (Barnes and Lindahl, 2004; Hoeijmakers, 2001). Apurinic/apyrimidinic (AP) sites are expected to be one of the most frequent endogenous lesions in DNA and present a strong block to DNA replication. It has been calculated that the spontaneous hydrolysis of the N-glycosylic bond results in the formation of about 10,000 AP sites per day per mammalian cell (Lindahl and Nyberg, 1972). AP sites can also result from the removal of damaged or inappropriate bases by DNA N-glycosylases in the course of the base excision repair (BER) process (Barnes and Lindahl, 2004; Boiteux and Guillet, 2004). Damaged DNA bases can arise in several ways, most importantly by methylation, oxidation, and deamination of normal bases, yielding a variety of lesions such as N^7-methylguanine (7meG), 8-oxo-7,8-dihydroguanine (oxoG), or uracil (U). Inappropriate bases such as uracil, hyoxanthine, and xanthine can be incorporated by DNA polymerases during replication and repair. The vast majority of damaged and inappropriate bases in DNA is removed by DNA N-glycosylases yielding AP sites (Barnes and Lindahl, 2004). A study indicates that normal human liver cells present a steady-state level of about 50,000 AP sites per genome (Nakamura and Swenberg, 1999). This value suggests a steady-state level of about 250 AP sites per genome in yeast. Figure 1A summarizes the various events at the origin of endogenous AP sites in DNA. In addition to being abundant, AP sites can cause cell death by blocking DNA replication and transcription (Boiteux and Guillet, 2004; Prakash and Prakash, 2002). AP sites can also cause mutations, yielding mainly single base-pair substitution in *Escherichia coli* and yeast (Boiteux and Guillet, 2004; Otterlei *et al.*, 2000; Prakash and Prakash, 2002). Furthermore, cleavage of AP sites by AP endonucleases or by DNA N-glycosylases/AP lyases results in the formation of single strand breaks (SSBs) with $3'$- or $5'$-blocked ends that cannot be used as substrates by DNA polymerases or DNA ligases, respectively (Fig. 1B). Moreover, $3'$- or $5'$-blocked SSBs can be converted into double strand breaks (DSB) after DNA replication (Caldecott, 2001; Guillet and Boiteux, 2002). This chapter discusses AP sites and $3'$-blocked SSBs as a single class of lesions.

To counteract the deleterious action of AP sites, organisms have developed multiple strategies (Boiteux and Guillet, 2004; Friedberg *et al.*, 1995). They prevent the incorporation of unusual or damaged bases by sanitizing the pool of deoxyribonucleotides-triphosphates (dNTPs) by the enzymatic hydrolysis of DNA precursors such as dUTP, dITP, or 8-oxo-dGTP. If incorporated into DNA, the modified base moieties are removed by DNA N-glycosylases to yield AP sites. It should be noted that dUMP incorporation is likely the major source of the formation of AP sites in yeast (Guillet and Boiteux, 2003). In addition, organisms efficiently remove AP sites using

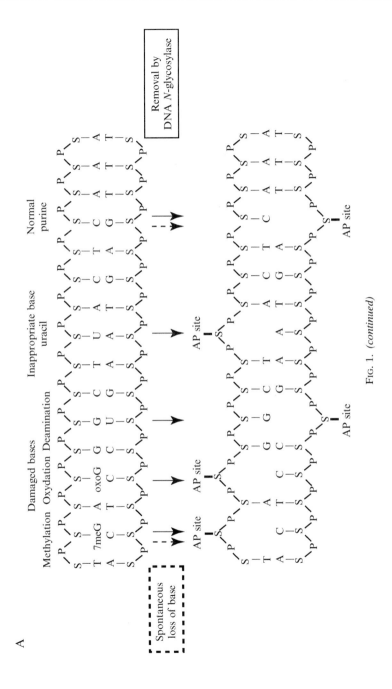

FIG. 1. (continued)

B

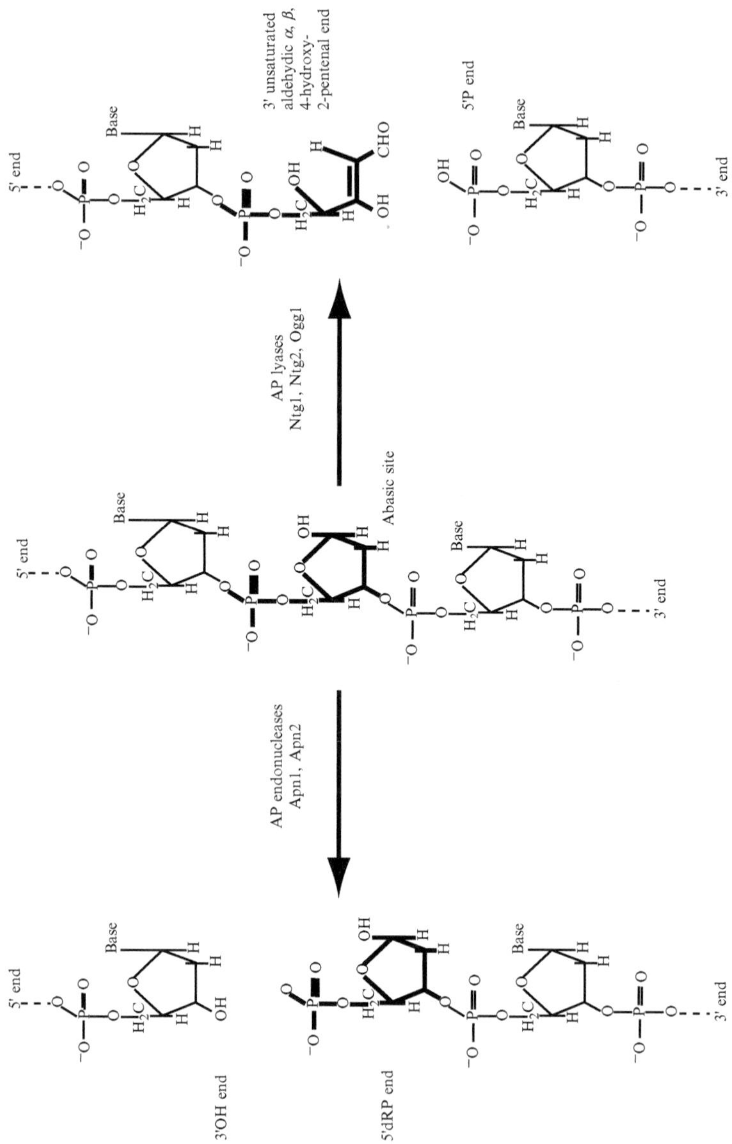
5' end
Base
Abasic site
3' end
AP endonucleases
Apn1, Apn2
AP lyases
Ntg1, Ntg2, Ogg1
3'OH end
5'dRP end
3' unsaturated
aldehydic α, β,
4-hydroxy-
2-pentenal end
5'P end
CHO

a variety of DNA repair pathways, such as base excision repair (BER) and nucleotide excision repair (NER). This chapter describes genetic assays developed in our laboratory whose aim is to investigate the formation and repair of endogenous AP sites in the budding yeast *Saccharomyces cerevisiae*, a simple eukaryote that can be used as a paradigm for most essential DNA transactions (Resnick and Cox, 2000).

Endogenous AP Sites Cause Cell Death in the Absence of Apn1, Apn2, and Rad1-Rad10

The BER pathway, mediated by AP endonucleases (Apn1 and Apn2), is the primary defense against AP sites and 3'-blocked SSBs in *S. cerevisiae* (Boiteux and Guillet, 2004). Apn1 and Apn2 catalyze the hydrolytic cleavage of the phosphodiester backbone at the 5' side of an AP site yielding SSBs with 3'-OH and 5'-dRP ends (Fig. 1B). Apn1 and Apn2 are also endowed with a 3'-phosphodiesterase activity removing 3'-blocking groups such as 3'-phosphate (3'-P), 3'-phosphoglycolate (3'-PGA), or 3'-dRP (Boiteux and Guillet, 2004). Importantly, SSBs with 3'-dRP result from the cleavage of AP sites under alkaline condition or catalyzed by DNA *N*-glycosylases such as Ntg1, Ntg2, and Ogg1 (Fig. 1B). The *apn1 apn2* double mutant is viable and does not exhibit evident growth defects. However, it exhibits a high sensitivity to agents that generate AP sites and a weak spontaneous mutator phenotype (Table I). The relatively mild phenotype of an *apn1 apn2* double mutant is unexpected, considering that AP sites are the most abundant endogenous lesions in DNA. This might be explained by the presence of overlapping DNA repair pathways. Indeed, additional inactivation of the *RAD14* gene, to yield the *apn1 apn2 rad14* triple mutant, abolishes NER and strongly increases the sensitivity to the lethal effect of methyl-methane-sulfonate (MMS) compared to an *apn1 apn2* double mutant (Torres-Ramos *et al.*, 2000). Furthermore, an *apn1*

FIG. 1. Origin and chemical nature of endogenous AP sites and associated 3'- and 5'-blocked SSBs in DNA. (A) AP sites in DNA are generated by spontaneous hydrolysis of the *N*-glycosylic bond of normal purines or damaged bases and by excision of normal, inappropriate and damaged bases by DNA *N*-glycosylases. P, phosphate; S, sugar (deoxyribose). (B) Central: Chemical structure of a regular AP site in DNA. Left: 5'-blocked SSB with a 5'-deoxyribose-phosphate end (5'-dRP). Right: 3'-blocked SSB with a 3'-unsaturated aldehydic (α,β-4-hydroxy-2-pentenal) end (3'-dRP). Because 3'- or 5'-blocked SSBs result primarily from the cleavage of AP sites by AP lyases and AP endonucleases, we suggest that these lesions constitute a single class. Lesions are in double-stranded DNA; because of space constraints, only one DNA strand is represented. (See color insert.)

TABLE I
PROPERTIES OF MUTANTS AFFECTED IN THE REPAIR OF AP SITES IN
SACCHAROMYCES CEREVISIAE

Relevant genotype	Viability	Spontaneous mutagenesis[a] (Can^R per 10^7 cells)
Wild type	Yes	9 (1)
Δ *apn1*	Yes	18 (2)
Δ *apn2*	Yes	8 (1)
Δ *rad14*	Yes	22 (2)
Δ *ung1*	Yes	38 (4)
Δ *apn1* Δ *apn2*	Yes	31 (3)
Δ *apn1* Δ *apn2* Δ *rad14*	Yes	133 (15)
Δ *apn1* Δ *apn2* Δ *rad1*	No	NA[b] NA[b]
Δ *apn1* Δ *apn2* Δ *rad14* Δ *ung1*	Yes	127 (14)
Δ *apn1* Δ *apn2* Δ *rad1* Δ *ung1*	Yes	266 (30)

[a] Determination of spontaneous mutation frequencies as described previously by Gellon *et al.* (2001). All strains used in this study are derivatives of the wild-type strain FF18733 (*MATa, leu2-3-112, trp1-289, his7-2, ura3-52, lys1-1*). Yeast strains were grown to saturation in YPD medium at 30° for 3 days. Three series of 11 cultures were inoculated in YPD at a starting density of about 100 cells/ml and allowed to grow for 2–3 days at 30°. Cell density was measured by plating dilutions on YPD plates and counting the colonies after 3 days. The quantification of canavanine-resistant mutants (Can^R) was determined after plating of aliquots of each culture on YNBD plates containing 60 μg/ml of canavanine sulfate (Research Organics Inc). Plates were incubated 4–5 days before counting. Frequencies are the ratio of Can^R mutants per ml/viable cells per ml. Values are medians of ≥30 independent cultures. Values in parentheses are the fold induction relative to WT.
[b] Not applicable because not viable.

apn2 rad14 triple mutant exhibits a sevenfold higher spontaneous mutation frequency than an *apn1 apn2* double mutant (Table I). Taken together, this data led us to conclude that inactivation of BER and NER is not sufficient to impair viability; however, it reveals the biological impact (spontaneous mutagenesis) of endogenous DNA damage.

However, the inactivation of the *RAD1* or *RAD10* gene to yield *apn1 apn2 rad1* or *apn1 apn2 rad10* triple mutants reveals synthetic lethality (Guillet and Boiteux, 2002). Although unable to form visible colonies, an *apn1 apn2 rad1* triple mutant forms microcolonies of about 300 dead cells, whereas a wild-type colony contains more than 10^7 cells (Table I and Fig. 2). These results show that the simultaneous inactivation of BER, NER, and Rad1-Rad10 nuclease causes cell death in *S. cerevisiae*. The known functions of Apn1, Apn2, and Rad1-Rad10 point to DNA damage at the origin of the lethality. To test this hypothesis, we have expressed bacterial AP endonucleases (Nfo [Endo IV] or Xth [Exo III]) into an *apn1 apn2 rad1* triple mutant. Results show that the bacterial AP endonucleases efficiently suppress the synthetic lethality of mutations in *APN1, APN2,* and *RAD1* (Guillet and Boiteux, 2002). Therefore, we suggest that the essential function of Apn1, Apn2, NER, and Rad1-Rad10 is to repair endogenous AP sites and 3′-blocked SSBs. This data strongly suggests that the burden of spontaneous AP sites is not compatible with life in absence of DNA repair, which is also consistent with the proposition that AP sites are deleterious and one of the most frequent spontaneous lesions in DNA.

Use of the *apn1 apn2 rad1* Triple Mutant to Study the Origin and Repair of Endogenous AP Sites in DNA

The aim of our research is to assess the formation and the biological impact of endogenous DNA damage in eukaryotic cells. The challenge was to measure biological events in cells that are, by definition, not exposed to an exogenous agent. Because of the efficiency of DNA repair, the impact of endogenous DNA damage cannot be investigated in WT cells. Therefore, we used the *apn1 apn2 rad1* triple mutant as a tool to dissect origin, repair, and biological consequences of spontaneous AP sites in yeast. The experimental strategy is based on classical yeast genetic and consists in the generation of quadruple mutants *apn1 apn2 rad1 x*, where (x) is a gene of interest a priori involved in the metabolism of AP sites (Fig. 2). The assay measures the aggravation or alleviation of the phenotype (size of microcolonies) of quadruple mutants compared to that of the *apn1 apn2 rad1* triple mutant. Briefly, haploid mutant strains were crossed to generate diploids heterozygous for the genes of interest: *APN1*/Δapn1 *APN2*/Δ apn2 *RAD1*/Δrad1 X/Δx. After sporulation, tetrad analysis was performed (Fig. 2A). Four days after dissection, the size of colonies (visible) and microcolonies (not macroscopically visible) was analyzed (see legend of Fig. 2). Finally, genotyping of spores was performed to determine the quadruple mutant colonies (Fig. 2A, arrows). From previous data (Guillet and Boiteux, 2002), we knew that the *apn1 apn2 rad1* triple mutant could form microcolonies of about 300 cells, which was used as an internal control (Fig. 2B, left). Figure 2B and C show that the inactivation of *RAD9, MUS81, RAD50, RAD51, RAD52,* or *MRE11* aggravates the phenotype, yielding microcolonies of less than 50 cells compared to about 300 cells in the control. Therefore, we conclude that homologous recombination proteins, as well as DNA damage checkpoint proteins, are involved in the repair of endogenous AP sites (Fig. 3).

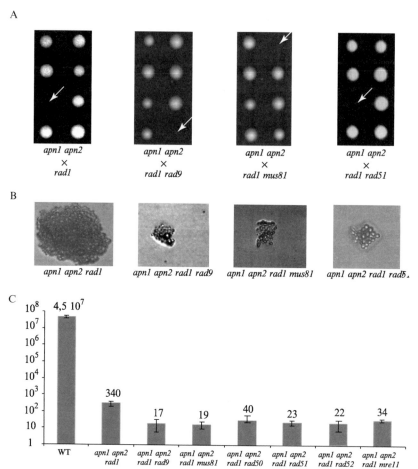

FIG. 2. Genetic assay used to investigate the formation and repair of endogenous DNA damage in *Saccharomyces cerevisiae*. All strains used in this study are derivatives of the wild-type strain FF18733 (*MATa, leu2-3-112, trp1-289, his7-2, ura3-52, lys1-1*). Gene deletions were produced by the polymerase chain reaction (PCR)-mediated one-step replacement technique (Baudin *et al.*, 1993). Yeast strains were grown at 30° in YPD. All media, including agar, were from DIFCO. Presporulation and sporulation procedures were performed as described (Resnick *et al.*, 1983). Micromanipulation and dissection of asci were performed using a Singer MSM System as described (Sherman and Hicks, 1991). The genotype of unviable spores was inferred from the segregation pattern of the three viable spores. When required, PCR analyses on genomic DNA were performed to determine the spore genotype. The number of cells per colony or microcolony was counted by microscopic observation. Alternatively, the number of cells per colony was estimated after suspension in sterile water and counting using a Mallassez cell. For all crosses, cells were counted 4 days after dissection. (A) Tetrad analysis resulting from indicated crosses. Arrows show *apn1 apn2 rad1* triple mutant (left) or *apn1 apn2 rad1 rad9* (middle left), *apn1 apn2 rad1 mus81* (middle right), and *apn1 apn2 rad1 rad51* (right) quadruple mutants. (B) Microscopic views of indicated microcolonies (×40). (C) Average number of cells per microcolony ± SD (at least 10 microcolonies). Note: all alleles used are deletions and are available from the authors.

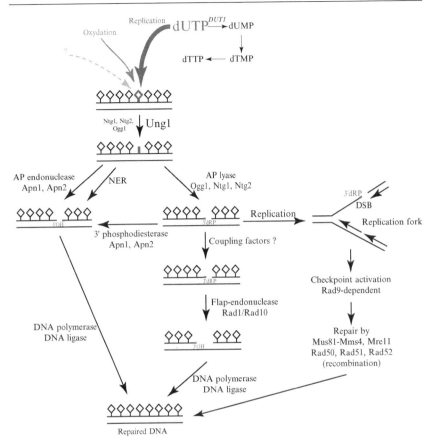

FIG. 3. A model in progress. Genetic assays using *apn1 apn2 rad1* and *apn1 apn2 rad14* triple mutants allowed us to investigate the origin and repair of endogenous DNA damage in yeast. This model deals with our present knowledge that is necessarily incomplete and it applies to rapidly growing cells. Cell death is presumably due to a cascade of events that is initiated by the formation of AP sites in DNA and their subsequent conversion into SSBs and ultimately DSBs. These data demonstrate the impact of normal cellular metabolism on the genetic material in living organisms. (See color insert.)

Using the *apn1 apn2 rad1* microcolonies assay, we have also been able to identify functions that alleviate the phenotype in response to endogenous DNA damage. Because removal of damaged or inappropriate DNA bases by DNA *N*-glycosylases is an important source of AP sites, it was rational to inactivate genes encoding these proteins in the *apn1 apn2 rad1* background. Results show that the inactivation of *MAG1* and *OGG1* does not affect the size of the microcolonies, suggesting that neither methylated bases nor 8-oxoG is a critical source of endogenous AP sites in DNA. In contrast, inactivation of *NTG1* and *NTG2* partially rescues an *apn1 apn2 rad1* triple mutant, yielding unviable microcolonies of about 10^4 cells. Finally, inactivation of *UNG1* in the *apn1 apn2 rad1* background restores the viability (Guillet and Boiteux, 2003). Although viable, the *apn1 apn2 rad1 ung1* quadruple mutant exhibits growth defects and a strong spontaneous mutator phenotype (Table I). In addition, the overexpression of the *DUT1* gene coding for the dUTPase activity suppresses the lethality of the *apn1 apn2 rad1* triple mutant. These results strongly suggest that uracil incorporated in DNA via the dUTP pool by DNA polymerases is a critical source of endogenous AP sites in yeast (Guillet and Boiteux, 2003). However, uracil in DNA is presumably not the only endogenous source of AP sites in DNA. Using the same assay, we observed that growth of an *apn1 apn2 rad1* triple mutant under anaerobic condition allows a significant increase in the size of the microcolonies formed (M. Guillet *et al.*, unpublished result). These last results point to oxidative DNA damage as another source of endogenous AP sites or 3′-blocked SSBs. To conclude, the targeted strategy using the *apn1 apn2 rad1* triple mutant allowed us to identify repair pathways and discuss the origin of AP sites in DNA. It would be of interest to systematically generate quadruple mutants with genes involved in a variety of cellular processes, such as transcription, energetic metabolism, or scavenging of free radicals and other reactive species.

Use of the *apn1 apn2 rad14* Triple Mutant to Study the Origin and Repair of Endogenous AP Sites in DNA

The rationale for this second assay that uses the *apn1 apn2 rad14* triple mutant is very similar to that described in the previous section. It is also based on genetic crossing and tetrad analysis. However, this assay allowed an untargeted approach to identify functions involved in the origin and repair of AP sites in yeast. Although viable, an *apn1 apn2 rad14* strain exhibits a strong spontaneous mutator phenotype that is presumably the hallmark of endogenous AP sites (Table I). The assay screened for mutations that impair the viability of an *apn1 apn2 rad14* triple mutant (colethal screen). The genes identified could be essential for the Rad1-Rad10 repair pathway to be functional or genes that avoid the formation of large amounts of AP sites. To identify such mutations, we used the red/white color colony assay as described previously (Koshland *et al.*, 1985). First, we expressed the *xth* gene encoding the major AP endonuclease of *E.coli* under the control of a *GAL1* promoter on a centromeric yeast plasmid carrying the *ADE3* gene in an *apn1 apn2 rad14 ade2 ade3* mutant (M. Guillet *et al.*, unpublished result). Second, we generated mutations in this mutant by exposing the strain to UVC light. We expected that in the presence of galactose, Xth would suppress the lethality of *apn1 apn2 rad14 ade2 ade3 x* mutants. Finally, we screened for plasmid-dependent strains by selecting red nonsectored colonies on galactose-containing plates. Moreover, because the *xth* gene is under the control of a *GAL1* promoter, the candidates should have a strong growth defect on glucose-containing plates but not on galactose-containing plates. Out of 22,000 clones screened, three mutants were selected. Two of them were identified as *apn1 apn2 rad14 rad1* and *apn1 apn2 rad14 rad50* quadruple mutants. Another candidate was an *apn1 apn2 rad14 dut1-1* quadruple mutant harboring a missense mutation in a highly conserved residue in the Dut1 protein, an enzyme that hydrolyses dUTP into dUMP (M. Guillet *et al.* unpublished result). It should be noted that the *rad50* and *dut1-1* mutations lead to synthetic sickness rather than synthetic lethality. Therefore, this screening allowed us to identify two classes of mutants, one affected in DNA repair (*rad1* and *rad50*) and another affected in a function that can modulate the production of AP sites (*dut1-1*) (Fig. 3).

Conclusions and Model

To the best of our knowledge, the synthetic lethality of mutations in *APN1, APN2,* and *RAD1* is the first experimental evidence that endogenous DNA damage causes cell death in the absence of DNA repair. These results, in turn, strongly suggest that DNA repair mechanisms should have emerged very early in the course of evolution. They also point to AP sites as a critical source of DNA damage. Figure 3 summarizes our views about origin and repair of endogenous DNA damage in yeast. This model is based on results obtained using the two assays described in this chapter. The model proposes that dUMP incorporation is a major source of endogenous AP sites. Oxidative DNA damage is another source of AP sites and related lesions. Obviously other sources of AP sites, such as dITP incorporation, cannot be excluded (Fig. 3, top). In WT cells, AP sites are primarily repaired by the BER pathway that is initiated by Apn1 or Apn2, with NER as a backup activity. If AP endonucleases are absent or saturated, AP sites can also be incised by AP lyases such as Ntg1, Ntg2, and Ogg1, yielding 3′-blocked SSBs, which are more toxic than AP sites themselves (Guillet and Boiteux, 2002). This part of the model explains why inactivation of AP lyases (Ntg1 and Ntg2) in the *apn1 apn2* background is beneficial in terms of survival. In such cells, AP sites are presumably tolerated by translesion DNA synthesis using DNA polymerases such as Rev3, Rev1, or Rad30 (Prakash and Prakash, 2002). In WT cells, 3'-blocked ends at SSBs are processed by the 3'-phosphodiesterase activity of Apn1 and Apn2 or by the 3'-flap endonuclease activity of Rad1-Rad10 (Guzder *et al.*, 2004). Afterward, the DNA repair process is completed by a DNA polymerase and a DNA ligase (Fig. 3, left). When Apn1, Apn2, and/ or Rad1-Rad10 is absent or saturated, AP sites and 3'-blocked SSBs accumulate in DNA and lead to the formation of DSBs that activate the *RAD9*-dependent G2/M checkpoint. During cell cycle arrest, DNA tails with 3'-blocked ends are released by the action of the Mus81-Mms4 endonuclease, allowing replication fork restart via a recombination-dependent mechanism. This model can explain the role of genes such as *RAD9*, *MUS81*, *RAD50*, *RAD51*, *RAD52*, or *MRE11* in the repair of endogenous AP sites in yeast (Fig. 3, right). This model also suggests interfaces among detection, signaling, and repair of DNA damage to prevent genetic instability (Rouse and Jackson, 2002).

Acknowledgments

This work was supported by CNRS and CEA. MG was supported by a fellowship from the Association pour la Recherche contre le Cancer (ARC).

References

Barnes, D. E., and Lindahl, T. (2004). Repair and genetic consequences of endogenous DNA base damage in mammalian cells. *Annu. Rev. Genet.* **38,** 445–476.

Baudin, A., Ozier-Kalogeropoulos, O., Denouel, A., Lacroute, F., and Cullin, C. (1993). A simple and efficient method for direct gene deletion in *Saccharomyces cerevisiae*. *Nucleic Acids Res.* **21,** 3329–3330.

Boiteux, S., and Guillet, M. (2004). Abasic sites in DNA: Repair and biological consequences in *Saccharomyces cerevisiae*. *DNA Repair (Amst.)* **3,** 1–12.

Caldecott, K. W. (2001). Mammalian DNA single-strand break repair: An X-ra(y)ted affair. *Bioessays* **23,** 447–455.

Friedberg, E., Walker, G., and Siede, W. (1995). "DNA Repair and Mutagenesis." ASM Press, Washington, DC.

Gellon, L., Barbey, R., Auffret van der Kemp, P., Thomas, D., and Boiteux, S. (2001). Synergism between base excision repair, mediated by the DNA glycosylases Ntg1 and Ntg2, and nucleotide excision repair in the removal of oxidatively damaged DNA bases in *Saccharomyces cerevisiae*. *Mol. Genet. Genomics* **265,** 1087–1096.

Guillet, M., and Boiteux, S. (2002). Endogenous DNA abasic sites cause cell death in the absence of Apn1, Apn2 and Rad1/Rad10 in *Saccharomyces cerevisiae*. *EMBO J.* **21,** 2833–2841.

Guillet, M., and Boiteux, S. (2003). Origin of endogenous DNA abasic sites in *Saccharomyces cerevisiae*. *Mol. Cell. Biol.* **23,** 8386–8394.

Guzder, S. N., Torres-Ramos, C., Johnson, R. E., Haracska, L., Prakash, L., and Prakash, S. (2004). Requirement of yeast Rad1-Rad10 nuclease for the removal of 3′-blocked termini from DNA strand breaks induced by reactive oxygen species. *Genes Dev.* **18**, 2283–2291.

Hoeijmakers, J. H. (2001). Genome maintenance mechanisms for preventing cancer. *Nature* **411**, 366–374.

Koshland, D., Kent, J. C., and Hartwell, L. H. (1985). Genetic analysis of the mitotic transmission of minichromosomes. *Cell* **40**, 393–403.

Lindahl, T., and Nyberg, B. (1972). Rate of depurination of native deoxyribonucleic acid. *Biochemistry* **11**, 3610–3618.

Nakamura, J., and Swenberg, J. A. (1999). Endogenous apurinic/apyrimidinic sites in genomic DNA of mammalian tissues. *Cancer Res.* **59**, 2522–2526.

Otterlei, M., Kavli, B., Standal, R., Skjelbred, C., Bharati, S., and Krokan, H. E. (2000). Repair of chromosomal abasic sites *in vivo* involves at least three different repair pathways. *EMBO J.* **19**, 5542–5551.

Prakash, S., and Prakash, L. (2002). Translesion DNA synthesis in eukaryotes: A one- or two-polymerase affair. *Genes Dev.* **16**, 1872–1883.

Resnick, M. A., and Cox, B. S. (2000). Yeast as an honorary mammal. *Mutat. Res.* **451**, 1–11.

Resnick, M. A., Game, J. C., and Stasiewicz, S. (1983). Genetic effects of UV irradiation on excision-proficient and -deficient yeast during meiosis. *Genetics* **104**, 603–618.

Rouse, J., and Jackson, S. P. (2002). Interfaces between the detection, signaling, and repair of DNA damage. *Science* **297**, 547–551.

Sherman, F., and Hicks, J. (1991). Micromanipulation and dissection of asci. *Methods Enzymol.* **194**, 21–37.

Torres-Ramos, C. A., Johnson, R. E., Prakash, L., and Prakash, S. (2000). Evidence for the involvement of nucleotide excision repair in the removal of abasic sites in yeast. *Mol. Cell. Biol.* **20**, 3522–3528.

[7] Activities and Mechanism of DNA Polymerase β

By WILLIAM A. BEARD, RAJENDRA PRASAD, and SAMUEL H. WILSON

Abstract

DNA polymerase β plays an essential role in the base excision repair pathway necessary to cleanse the genome of simple base lesions and abasic sites. Abasic sites arise in DNA from spontaneous base loss (depurination) and DNA-damage specific glycosylases that hydrolyze the *N*-glycosidic bond between the deoxyribose and the damaged base. DNA polymerase β contributes two enzymatic activities: DNA synthesis and deoxyribose-phosphate removal through nucleotidyl transferase and lyase mechanisms, respectively. The active site for each of these activities resides on a distinct domain of the protein: 31-kDa polymerase domain and amino-terminal 8-kDa lyase domain. The simple organization of each domain and the ability to assay each activity have hastened our understanding of the

METHODS IN ENZYMOLOGY, VOL. 408
0076-6879/06 $35.00
DOI: 10.1016/S0076-6879(06)08007-4

faithful replication of DNA during repair synthesis and the flux of inter-mediates through single nucleotide base excision repair and its alternate pathways.

Introduction

Cells contain multiple and overlapping DNA repair pathways that are essential for maintaining the integrity of genomic DNA. The DNA repair pathway known as "base excision repair"(BER) protects the genome by removing simple base lesions and abasic sites arising from a variety of exogenous and endogenous DNA-damaging agents. Spontaneous and en-zymatic removal of damaged bases through hydrolytic N-glycosidic bond cleavage leads to a mutagenic BER intermediate referred to as an abasic or apurinic/apyrimidinic (AP) site. The repair of an AP site requires four coordinated enzymatic activities: (i) strand incision by AP endonucle-ase, (ii) removal of the resulting 5'-deoxyribose phosphate (dRP) backbone of the AP site by DNA polymerase β (pol β) associated lyase activity, (iii) single nucleotide gap-filling DNA synthesis by pol β, and (iv) ligation of the resulting nick by DNA ligase. Thus, two of the four activities are contributed by pol β. A "knockout" of the pol β gene in mice results in embryonic lethality, indicating an essential role of pol β during fetal development (Sobol et al., 1996). More importantly, the hypersensitivity of these pol β null mouse embryonic fibroblasts toward monofunctional DNA-alkylating agents provides compelling evidence for the cellular role of pol β in BER.

DNA polymerase β is a member of the X family of DNA polymerases (Delarue et al., 1990). It is found in all vertebrate species as a 39-kDa protein lacking intrinsic 3'- or 5'-exonuclease activities, but containing 5'-dRP lyase and AP lyase activities (Prasad et al., 1998). In light of its size, the enzyme is considered the simplest naturally occurring cellular DNA polymerase and is an ideal model for studies of nucleotidyl transfer-ase and lyase reaction mechanisms. Mammalian pol β can be expressed in Escherichia coli at high levels, hastening biophysical and kinetic character-ization (Beard and Wilson, 1995a). The recombinant proteins from human and rat are fully active in DNA synthesis and possess substrate specificity and catalytic properties similar to those of the natural enzymes (Abbotts et al., 1988; Osheroff et al., 1999a; Patterson et al., 2000).

Controlled proteolytic or chemical cleavage of pol β indicated that it is folded into discrete domains (Beard and Wilson, 1995a). Subsequently, the X-ray crystallographic structure of the ternary substrate complex defined the location of these domains in relation to the global structure of the polymerase and substrates (Sawaya *et al.*, 1997). DNA polymerase β is composed of two domains: an amino-terminal 8-kDa lyase (residues 1–90) and a 31-kDa polymerase domain (residues 91–335). When bound to single nucleotide-gapped DNA, pol β forms a donut-like structure. The 5′-phosphate in the single nucleotide gap is bound in a lysine-rich pocket of the lyase domain that includes the active site nucleophile, Lys72 (Deterding *et al.*, 2000). The 3′-OH of the primer terminus is situated near conserved aspartates (190, 192, and 256) of the polymerase domain that coordinate two divalent metals (Mg^{2+}). The catalytic metal lowers the pK_a of the 3′-OH of the growing primer terminus while the nucleotide-binding metal coordinates the triphosphate moiety, hastening binding of the incoming nucleotide. Additionally, the nucleotide-binding metal assists PP_i dissociation. Both metals are believed to stabilize the proposed penta-coordinated transition state of the nucleotidyl transferase reaction. The template strand is radically bent as it exits the polymerase active site, thereby permitting the polymerase to sandwich the nascent base pair between the primer terminal base pair and α-helix N (Beard *et al.*, 2004).

Kinetic Characteristics of DNA Synthesis

Kinetic Mechanism

DNA polymerases utilize a similar kinetic mechanism (Fig. 1). However, the magnitude of each step is dependent on the specific polymerase as well as DNA sequence. Steady-state kinetic analyses indicate that pol β follows an ordered addition of substrates (Tanabe *et al.*, 1979). After binding a DNA substrate (Fig. 1, step 1), DNA polymerases preferably bind a nucleoside triphosphate (dNTP) that preserves Watson–Crick hydrogen bonding as dictated by the template base (step 2). Upon binding the correct dNTP, the ternary complex undergoes numerous conformational changes (Bose-Basu *et al.*, 2004; Kim *et al.*, 2003; Vande Berg *et al.*, 2001; Yang *et al.*, 2002) (step 3). In some instances, these conformational changes limit the rate of nucleotide insertion so that the chemical step (step 4) is not rate limiting. In other instances, some aspect of the chemical step can be rate limiting (Joyce and Benkovic, 2004). The identity of the rate-limiting conformational change is not known. Following chemistry, the ternary product complex undergoes a conformational change (step 5) that facilitates PP_i release (step 6). At this point, the extended product (DNA_{+1}) is released (single nucleotide insertion) or serves as substrate DNA for another round of insertion (processive DNA synthesis).

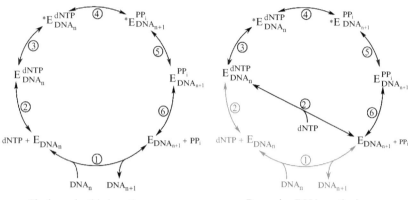

Single-nucleotide insertion   Processive DNA synthesis

FIG. 1. General reaction pathways for DNA synthesis. (Left) The kinetic scheme for the insertion of a single nucleotide. After binding DNA (step 1), a nucleoside triphosphate (dNTP) binds (step 2). The ternary substrate and product complexes undergo several conformational rearrangements and, in some instances, one or more of these nonchemical steps could be rate determining. These conformational changes are shown as single steps (steps 3 and 5) that surround the chemical step (step 4). The postchemical conformational change may facilitate PP_i release (step 6). The kinetic significance of structurally observed conformational transitions for pol β upon ligand binding is not known. (Right) The kinetic path shows the continued synthesis of DNA without the polymerase dissociating from DNA (processive DNA synthesis).

Kinetic Approaches

There are three standard kinetic approaches used to characterize the kinetic steps of DNA polymerases: (i) steady state, (ii) presteady state, and (iii) single turnover. These approaches differ from one another by the assay concentration of enzyme and the ratio of polymerase to substrate DNA. Traditionally, a steady-state approach employs a low concentration of enzyme relative to DNA (E $\ll$ DNA). In contrast, a presteady state approach uses a high concentration of enzyme to follow the formation and decay of intermediates (high E, E $<$ DNA). Finally, a single turnover analysis employs a high concentration of enzyme to saturate substrate DNA to follow catalytic events at the active site without interference from catalytic cycling (high E, E $\gg$ DNA). These latter two approaches often require a rapid mixing and quenching instrument.

Using defined template–primer oligonucleotides it is possible to measure the kinetics of nucleotide insertion by all three techniques. For a well-defined mechanism, the results of all three techniques should be self-consistent (i.e., predictive of the results from other approaches). The k_{cat}

and K_m values ascertained from a steady-state approach must be interpreted in the context of a defined mechanism (Beard and Wilson, 1995b). Only in this way can these values be interpreted reliably. For a polymerase that binds DNA tightly (slow dissociation rate constant, k_{-1}) and inserts a nucleotide rapidly (k_3 or $k_4 \gg k_{-1}$), a burst of product formation occurs during the first turnover (Johnson, 1992). Subsequent turnovers occur at a steady-state rate that is equivalent to k_{-1}. The burst phase is referred to as the presteady state. The amplitude of the burst is equivalent to the active enzyme fraction, and the time course of the burst can be described by a single exponential. The rate constant describing the exponential phase is hyperbolically dependent on dNTP concentration ($K_{d,dNTP}$, step 2) (Beard and Wilson, 1995b; Johnson, 1992). The maximal rate of this phase (i.e., saturating dNTP) is defined as k_{pol} (step 3 or 4).

A presteady-state analysis of pol β-dependent single nucleotide gap filling demonstrates that the presteady-state and steady-state phases of the reaction are not well separated (Vande Berg et al., 2001). This is due to the similar magnitudes of k_{pol} and k_{-1} resulting in partially rate-limiting steps during the course of the reaction. An alternate approach to define k_{pol} and $K_{d,dNTP}$ utilizes the single turnover experiment so that catalytic cycling does not complicate the time course. This is the most common approach used to kinetically characterize wild-type and mutant forms of pol β.

DNA polymerase β prefers to bind gapped DNA substrates that have a 5′-phosphate on the downstream strand in the gap (Prasad et al., 1994). In the absence of this downstream strand, pol β has very low DNA-binding affinity. In contrast, tight DNA binding does not require the upstream strand that would normally be extended. This result suggests that pol β targets short DNA gaps in DNA through the 5′-phosphate on the downstream portion of the gap. This is consistent with the processive gap-filling DNA synthesis observed with short (<5 nucleotide) DNA gaps (Singhal and Wilson, 1993). Surprisingly, pol β can extend template–primers that do not have a downstream strand (i.e., not gapped). This is due to the increased DNA-binding affinity of pol β for these substrates in the presence of the correct nucleotide (unpublished data). Single turnover analysis indicates that pol β prefers a single nucleotide-gapped DNA substrate with kinetic parameters k_{pol}, $K_{d,dCTP}$, and $K_{d,DNA}$ of 10 s^{-1}, 6 μM, and 20 nM (Vande Berg et al., 2001). It should be noted that these parameters are dependent on DNA sequence and solution conditions. Accordingly, kinetic parameters should be scrutinized carefully before attempting to draw conclusions from data published from different laboratories.

To determine whether a conformational change (step 3) or chemistry (step 4) limits k_{pol}, the rate constant for the incorporation of an α-thio-substituted dNTP analog is usually examined and compared to that for the natural nucleotide. Due to the lower electronegativity of sulfur relative to oxygen, a significant decrease in rate upon sulfur substitution would suggest that chemistry is rate limiting. Model studies with phosphate triesters indicate a large elemental effect upon substitution of sulfur at a nonesterified position, whereas studies with phosphate diesters indicate a smaller decrease in rate upon sulfur substitution (Herschlag *et al.*, 1991). The thio-elemental effect observed for pol β is only 2.1, suggesting that a step other than chemistry may be rate limiting (Vande Berg *et al.*, 2001). However, this interpretation must be tempered, as the intrinsic elemental effect is not known and there are steric considerations imposed by the polymerase active site, in addition to the electronegativity of sulfur, that could influence the observed rate (Joyce and Benkovic, 2004).

Another approach commonly used to identify if a nonchemical step limits nucleotide insertion is to compare the apparent concentration of the ternary substrate complex formed when the reaction is quenched with acid (pulse-quench) compared to when it is chased with nucleotide (pulse-chase; follow radioactive dNTP incorporation and chase with unlabeled dNTP). If a dNTP-bound complex exists where the nucleotide is not in rapid equilibrium with free dNTP (e.g., complex between steps 2 and 3) and the chemical equilibrium (step 4) permits accumulation of the isomerized ternary substrate complex, then more product will be formed with the chase protocol than with the acid quench. This central substrate complex will be quenched immediately with acid, but if it is not in equilibrium with free dNTP, then this complex can form product in the presence of cold dNTP. However, because the apparent burst for pol β is partially rate limited by at least two steps and it appears it has a strong commitment to product formation, very little isomerized ternary complex can accumulate, even if a rate-limiting conformational change limits nucleotide insertion. Structural characterizations of pol β in different liganded states clearly identify a number of conformational changes that must occur before the transition state is achieved. Indeed, the apparent binding affinity for gapped DNA is nearly 200-fold tighter in the presence of a correct nucleotide than in its absence, indicating that a conformational change has occurred (Beard *et al.*, 2004).

Steady-state kinetic characterization of mouse pol β failed to demonstrate a reversal of the polymerization reaction (i.e., pyrophosphorolysis) in the presence of PP_i with activated DNA. However, PP_i was inhibitory for DNA synthesis (Tanabe *et al.*, 1979). Pyrophosphorolysis can be measured on nicked DNA; however, its rate is diminished significantly on single nucleotide-gapped DNA (unpublished data). This suggests that the primer terminus of a one nucleotide-gapped substrate does not enter the polymerase active site readily; an event required to add PP$_i$ to the terminal primer residue.

Fidelity

DNA polymerases must select and incorporate a complementary dNTP from a pool of structurally similar molecules to preserve Watson–Crick base pairing.

The fidelity of DNA synthesis is a measure of substrate specificity and can be calculated from the catalytic efficiencies for alternate substrates: Fidelity = $([k_{cat}/K_m]_{correct} + [k_{cat}/K_m]_{incorrect})/(k_{cat}/K_m)_{incorrect}$ where k_{cat}/K_m is catalytic efficiency for the correct or incorrect nucleotide. The catalytic efficiency determined from a steady-state characterization, k_{cat}/K_m, is equivalent to k_{pol}/K_d determined by a transient-state approach (presteady state or single turnover). Fidelity is a measure of the total number of insertions that occur before a misinsertion event. In some instances it is useful to express substrate specificity as relative misinsertion efficiency $(k_{cat}/K_m)_{incorrect}/(k_{cat}/K_m)_{incorrect}$ (Goodman *et al.*, 1993).

Although the fidelities of DNA polymerases vary widely, the catalytic efficiencies by which they insert the incorrect nucleotides are only weakly dependent on the identity of the polymerase. These differences in fidelities are due to the divergent abilities of polymerases to insert the right nucleotide (Beard *et al.*, 2002); low-fidelity enzymes insert the correct nucleotide slowly, whereas high-fidelity DNA polymerases insert the correct nucleotide rapidly (Beard and Wilson, 2003; Beard *et al.*, 2002). Accordingly, structure–function studies of DNA polymerases must interpret site-directed mutagenesis kinetic results in terms of absolute catalytic efficiencies rather than overall fidelity. If the fidelity of a mutant polymerase is altered, then the catalytic efficiencies for the alternate substrates is affected *differentially* by the protein modification, indicating that the side chain of interest provides different interactions for the correct and incorrect incoming nucleotide. The error rate for pol β single nucleotide gap filling is about 1 error/3000 nucleotides synthesized (Osheroff *et al.*, 1999a) and is even greater with larger gaps (Osheroff *et al.*, 1999b). This represents a moderate fidelity in terms of the wide spectrum of fidelities exhibited by natural exonuclease-deficient DNA polymerases (Beard *et al.*, 2002).

General Considerations

The assay conditions should be considered in some detail. Generally, it is best to attempt to mimic physiological conditions: pH 7.4, temperature 37°, and ionic strength 150 mM. We typically employ a minimal reaction mix consisting of 50 mM Tris–HCl, pH 7.4, 100 mM KCl, and 5 mM MgCl$_2$. Although these conditions may not provide optimum enzyme activity, they better reflect conditions that the polymerase would naturally encounter. It is appropriate to modify these conditions only after careful consideration. For example, a lower reaction temperature may be appropriate if a mutant form of pol β is not stable at 37°. If the reaction temperature is altered, the pK_a of Tris is also altered ($-0.031/°$) and needs to be taken into account.

In some instances, 5 mM dithiothreitol (DTT) and 50–100 μg/ml bovine serum albumin (BSA) are included in the reaction mixture to stabilize the enzyme. DTT is used to protect enzyme sulfhydryl groups from oxidation, whereas BSA serves to stabilize dilute solutions of enzymes. In this regard, substrates are also known to stabilize enzyme stocks. Although we do not find these approaches necessary for wild-type enzymes, mutant enzymes could benefit from these supplemental additions.

DNA synthesis is routinely followed by examining the size of the DNA product. This is achieved by radioactively labeling the DNA with [γ-^{32}P] ATP on the 5′ end with polynucleotide kinase or the 3′ end with [α-^{32}P] dNTP. The products of DNA synthesis with 5′-labeled primers and non-labeled dNTPs are separated by gel electrophoresis corresponding to size and typically quantified by phosphoimagery. In contrast, DNA synthesis products that employ radioactive dNTPs can be collected by DNA precipitation or filter binding and quantified by scintillation counting. These protocols are outlined in detail elsewhere (Beard and Wilson, 1995b).

Measurement of dRP Lyase Activity

Removal of the 5′-dRP Backbone of an AP Site

It is generally accepted that the 5′-dRP group is removed from the BER intermediate by the dRP lyase activity carried in the 8-kDa amino-terminal domain of pol β (Deterding *et al.*, 2000; Matsumoto and Kim, 1995; Piersen *et al.*, 1996). This activity controls the flux of DNA damage through monofunctional glycosylase-initiated BER (Srivastava *et al.*, 1998) and is a critical determinant for the cytotoxicity of monofunctional DNA-methylating agents (Sobol *et al.*, 2000). This enzymatic reaction is mediated through a β-elimination mechanism involving nucleophilic attack by an enzyme primary amine on C1′ of the deoxyribose, proton abstraction at C2′, and double bond formation between C2′ and C3′ (Deterding *et al.*, 2000; Feng *et al.*, 1998; Prasad *et al.*, 1998). The ε-amino group of Lys72 was identified as the nucleophile in the attack on C1′ forming a Schiff

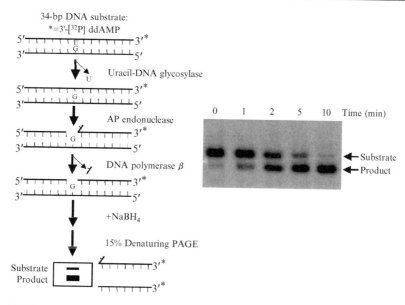

FIG. 2. DNA polymerase β-dependent dRP lyase activity assay. A schematic outline of substrate preparation and the activity assay are illustrated. A uracil-containing oligonucleotide is 3'-radiolabeled with [α-^{32}P]ddATP and TdT. The radiolabeled oligonucleotide is annealed with a complementary strand to generate a central G–U mispair. This duplex is treated with uracil DNA glycosylase to remove uracil (U) and AP endonuclease to incise the abasic site on its 5' side. This generates an upstream strand with a 3'-hydroxyl and a radiolabeled downstream strand with a 5'-deoxyribose phosphate residue. This strand serves as the substrate for the dRP lyase reaction. The substrate and product DNA can be separated by denaturing gel electrophoresis; removal of the dRP residue results in a shorter faster-migrating band that can be visualized and quantified by phosphoimagery. Results of a typical time course (10 min) generated with 10 nM pol β and 50 nM incised DNA are also shown. The position of the substrate and product bands is indicated.

base intermediate (Deterding *et al.*, 2000). Formation of the Schiff base intermediate is thought to trigger a cascade of events ultimately resulting in elimination of the dRP group at C3'. The mechanism for this pol β-catalyzed reaction, beyond the general scheme noted here for β eliminations, is not well understood and is under investigation (Prasad *et al.*, 2005).

DNA Substrate for dRP Lyase Activity Determination

A synthetic 34-bp DNA substrate that contains uracil at position 16 is often used for a dRP lyase assay (Fig. 2). The DNA sequence of the oligonucleotides is as follows:

5′-CTGCAGCTGATGCGC<u>U</u>GTACGGATCCCCGGGTAC-3′,
   U strand
5′-GTACCCGGGGATCCGTACGGCGCATCAGCTGCAG-3′,
   template strand

The U strand is labeled on its 3′ end with terminal deoxynucleotidyl transferase (TdT) using [α–^{32}P]ddATP and annealed to the complementary template strand by heating the solution to 90° for 3 min, followed by slow cooling to 25°. The ^{32}P-labeled duplex DNA is separated from unreacted labeled ddATP using a microspin G-25 column according to the manufacturer's suggested protocol and stored at −30°.

The ^{32}P-labeled uracil-containing DNA (100 nM) is pretreated with 10 nM uracil DNA glycosylase (UDG) in a buffer that contains 50 mM HEPES, pH 7.5, 20 mM KCl, and 2 mM DTT, and 1 mM EDTA. The reaction mixture is incubated at 37° for 20 min, supplemented with 5 mM MgCl$_2$ and 10 nM AP endonuclease (APE), and incubated an additional 20 min. Due to the labile nature of the UDG/APE-treated DNA, the substrate is prepared immediately before use in the dRP lyase activity assay.

Measurement of dRP Lyase Activity

The dRP lyase assay is typically performed using a reaction mixture (10 μl) containing 50 mM HEPES, pH 7.5, 20 mM KCl, 2 mM DTT, 1 mM EDTA, and 20–50 nM preincised ^{32}P-labeled AP site containing DNA substrate. The reaction is initiated by adding pol β to a final concentration of 10 nM and the reaction mixture is incubated at 37°. Aliquots can be removed at time intervals, transferred to an ice bath, and the DNA stabilized by the addition of 2 M NaBH$_4$ to a final concentration of 340 mM. The incubation is continued for 30 min at 0–1°. The stabilized DNA products are recovered by ethanol precipitation in the presence of 0.1 μg/ml tRNA and then resuspended in 10 μl of gel-loading buffer. After incubation at 75° for 2 min, the reaction products are separated by electrophoresis in a 15% polyacrylamide gel containing 8 M urea in TBE. The gels are scanned with a PhosphorImager and data analyzed by ImageQuant software.

Product Analysis

The dRP lyase activity protocol and typical results of polyacrylamide gel electrophoresis of the dRP lyase reaction product are shown in Fig. 2. DNA polymerase β released the dRP group from the cleaved abasic site DNA substrate in a time-dependent manner. The release of the 5′ sugar phosphate product from the 3' ^{32}P-labeled substrate is visualized by the appearance of a radiolabeled band migrating slightly faster than the substrate (Fig. 2). It is important to note that the substrate itself is moderately unstable under the reaction conditions used so it is necessary to analyze control mixtures incubated without enzyme for each incubation period.

Measurement of *In Vitro* BER

General Considerations

Measurements of BER capacity in cellular extracts can be grouped into two categories based on the structure of the DNA substrate: assays employing short "linear" oligonucleotide substrates and those that employ "circular" plasmid substrates. The oligonucleotide approach is used commonly to probe single nucleotide (Singhal *et al.*, 1995) and alternate long-patch BER (Prasad *et al.*, 2000, 2001). These BER pathways are distinguished by the size of the DNA synthesis repair patch. The plasmid-based assay is useful to probe proliferating cell nuclear antigen-dependent long-patch BER even though this type of substrate is typically used at subsaturating concentrations (Biade *et al.*, 1998; Dianov, 2003; Fortini *et al.*, 1998; Horton *et al.*, 2000). In contrast, oligonucleotide substrate concentrations can be easily varied and used near saturating levels (Srivastava *et al.*, 1998). Hence, rates of DNA repair obtained with oligonucleotide and plasmid substrates are difficult to compare directly. Until the recent advent of methods for preparation of plasmid substrates in large quantities (Wang and Hays, 2002), plasmid-based assays were conducted at relatively low substrate concentration (i.e., pM). Recent advances in plasmid preparation techniques promise to eliminate this deterrent. However, virtually all of the plasmid-based BER measurements reported to date have been at subsaturating levels of substrate DNA, making the results difficult to compare with BER assays employing oligonucleotide substrates at much higher concentrations. A second important distinguishing feature of these alternate BER assays is product characterization to differentiate repair by the two subpathway, single nucleotide and long-patch BER. These two subpathways often occur in parallel in the same extract, and the ratio between them depends on a variety of factors, such as enzyme level, including variant forms of these enzymes, and properties of the substrate DNA itself (Dianov *et al.*, 1999; Klungland and Lindahl, 1997; Prasad *et al.*, 2000). In general terms, a modification in the 5'-dRP group that renders it resistant to β elimination by the pol β dRP lyase will promote long-patch BER over the single nucleotide pathway (Horton *et al.*, 2000).

Assay for In Vitro *BER Capacity*

The oligonucleotide substrate approach to measure BER in a cell extract or with purified proteins is outlined in Fig. 3. A 35-bp DNA substrate containing uracil at position 16 in one strand is obtained by annealing a uracil-containing oligonucleotide with its complementary strand containing a guanine residue opposite the uracil. The BER process

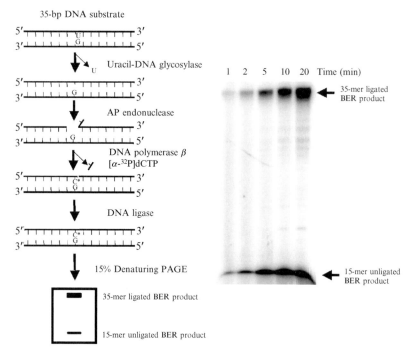

FIG. 3. DNA polymerase β-dependent BER activity assay. An unlabeled uracil-containing oligonucleotide is annealed with a complementary strand to generate a central G–U mispair at position 15. This duplex is incubated with a cellular or nuclear extract and $[\alpha\text{-}^{32}\text{P}]\text{dCTP}$ to provide the necessary enzymatic activities to accomplish BER. Uracil DNA glycosylase will remove uracil (U) and AP endonuclease incises the abasic site on its 5' side. DNA polymerase β will label the repaired DNA strand by adding a radiolabeled dCMP residue, thereby generating a repair intermediate (15-mer). When the 5'-dRP residue is removed, the nicked substrate can be ligated to generate a full-length product (35-mer). In the BER scheme, note that the two distinct activities of pol β (DNA synthesis and dRP removal) are illustrated as a single step. DNA synthesis is much more rapid than dRP removal and therefore precedes generation of a ligatable substrate for DNA ligase. Accordingly, the 15-mer intermediate is observed routinely. BER intermediates and product DNA can be separated by denaturing gel electrophoresis. A typical time course (20 min) for an *in vitro* BER reaction generated with 10 μg/ml of bovine testis nuclear extract, 250 nM DNA, and 2.2 μM $[\alpha\text{-}^{32}\text{P}]\text{dCTP}$ is also shown. The position of the 15-mer BER intermediate and 35-mer product bands is indicated.

is initiated by removal of the damaged or inappropriate base, in this case uracil opposite the G residue, as illustrated in Fig. 3. After the uracil base is removed by UDG, the abasic or AP site is cleaved by APE at the 5' side of the abasic site. This results in a one nucleotide gap with 3'-hydroxyl and 5'-dRP groups at the margins. DNA polymerase β first inserts the missing nucleotide, dCMP, to fill the gap and then excises the 5'-dRP residue. Finally, a DNA ligase seals the nicked BER intermediate completing BER (Fig. 3). Repair of a DNA lesion-containing duplex substrate can be followed easily by measuring incorporation of radiolabeled dNMP into the lesion-containing strand.

Substrate Preparation

A common lesion found in genomic DNA is the consequence of cytosine deamination that results in uracil situated opposite guanine. The uracil residue is removed by BER and can be measured easily. A typical BER substrate can be constructed with synthetic 35-mer oligonucleotides so that a uracil residue (position 15, Fig. 3) is positioned opposite guanine in the complementary strand. The 3' end is blocked with a 3'-deoxynucleotide (X). A typical sequence utilized in our BER reactions is as follows: 5'-GCCCTGCAGGTCGAUTCTAGAGGATCCCCGGGTAX-3', U strand; 5'-GTACCCGGGGATCCTCTAGAGTCGACCTGCAGGGC-3', template strand.

Lyophilized oligonucleotides are resuspended in water, and concentrations are determined from their UV absorption at 260 nm. The U and template strands are annealed by heating a solution of 10 μM of each strand to 90° for 3 min, followed by slow cooling to 25°. The annealed mixture is dispensed in aliquots and stored at −30°.

BER Assay

The proteins for the BER reaction can be provided by a nuclear extract such as bovine testis nuclear extract (Singhal *et al.*, 1995) or cellular extract (Sobol *et al.*, 1996). Alternatively, BER can be reconstituted using purified recombinant human enzymes, namely DNA glycosylase, AP endonuclease, pol β, and DNA ligase I (Srivastava *et al.*, 1998). The repair reaction (20 μl) is assembled on ice and contains 50 mM HEPES, pH 7.5, 20 mM KCl, 2 mM DTT, 1 mM EDTA, 4 mM ATP, 5 mM diTris phosphocreatine, 100 μg/ml creatine phosphokinase, 0.5 mM NAD, 10 mM MgCl$_2$, 250 nM 35-bp duplex DNA, and 2.2 μM [α-^{32}P]dCTP (specific activity, 1 × 10^6 dpm/pmol). The repair reaction is initiated by the addition of cell extracts or purified enzymes and is incubated at 37°. Aliquots (5 μl) are withdrawn at specified time periods and the reaction is terminated by the addition of an equal volume (5 μl) of DNA gel-loading buffer. The quenched reaction mixture is incubated at 75° for 2 min, and the reaction products are separated by electrophoresis in a 15% polyacrylamide gel containing 8 M urea in TBE. The gels are scanned with a PhosphorImager, and data are analyzed by ImageQuant software.

A typical time course illustrating the accumulation of an intermediate and product is shown in Fig. 3. Incorporation of [α-^{32}P]dCMP during the BER reaction was carried out on the 35-bp duplex DNA containing a uracil residue at position 15. The ^{32}P-labeled bands corresponding to 35- and 15-mer DNA molecules represent the complete repair product and the gap filled but unligated repair intermediate, respectively. In this example, the ratio of incorporation into these two bands is roughly one to one after a 20-min incubation.

Reagents

Stock Solutions

> 10× BER buffer: 500 mM HEPES, pH 7.5, 200 mM KCl, 10 mM EDTA, 20 mM DTT
>
> Dilution buffer: 50 mM HEPES, pH 7.5, 100 mM KCl, 0.1 mM EDTA, 1 mM DTT, 100 μg/ml BSA, 20% (v/v) glycerol
>
> 10× TBE: 890 mM Tris-base, 890 mM boric acid, 2 mM EDTA, pH 8.3
>
> Polyacrylamide: 40% (w/v), 19:1, acrylamide:bis-acrylamide
>
> MgCl$_2$: 1 M filter-sterilized solution
>
> NAD: 500 mM, dispensed in aliquots and stored at −80°
>
> dNTP: 100 mM, dispensed in aliquots and stored at −80°
>
> diTris-phosphocreatine: 100 mM, dispensed in aliquots and stored at −80°
>
> Creatine phosphokinase: 2 mg/ml, dispensed in aliquots and stored at −80°
>
> ATP: 100 mM, dispensed in aliquots and stored at −80°
>
> Gel-loading buffer: 20 mM EDTA, 95% formamide, 0.02% bromphenol blue, and 0.02% xylene cyanol
>
> Sodium borohydride: 2 M solution, freshly prepared in water

DNA and Enzymes

Synthetic oligonucleotides purified by high-pressure liquid chromatography can be obtained from a number of commercial sources (Oligos Etc, Inc. and Midland Certified Reagent Company Inc.). [α-^{32}P]ddATP (3000 Ci/

mmol), [α-^{32}P]dCTP (3000 Ci/mmol), dNTP solutions, diTris-phosphocreatine, creatine phosphokinase, ATP, and microspin G-25 columns are available from GE Healthcare. Terminal deoxynucleotidyl transferase is available from Promega (Madison, WI). Recombinant human pol β (Beard and Wilson, 1995a), AP endonuclease (Strauss *et al.*, 1997), uracil DNA glycosylase (Slupphaug *et al.*, 1995), and DNA ligase I (Wang *et al.*, 1994) can be overexpressed and purified as described or, in some instances, are available from commercial sources (e.g., Trevigen and R&D Systems).

Acknowledgment

This research was supported by the Intramural Research Program of the NIH, National Institute of Environmental Health Sciences.

References

Abbotts, J., Sen Gupta, D. N., Zmudzka, B., Widen, S. G., Notario, V., and Wilson, S. H. (1988). Expression of human DNA polymerase β in *Escherichia coli* and characterization of the recombinant enzyme. *Biochemistry* **27**, 901–909.

Beard, W. A., and Wilson, S. H. (1995a). Purification and domain-mapping of mammalian DNA polymerase β. *Methods Enzymol.* **262**, 98–107.

Beard, W. A., and Wilson, S. H. (1995b). Reverse transcriptase. *In* "A Practical Approach: HIV Volume 2" (J. Karn, ed.), pp. 15–36. IRL Press, New York.

Beard, W. A., and Wilson, S. H. (2003). Structural insights into the origins of DNA polymerase fidelity. *Structure* **11**, 489–496.

Beard, W. A., Shock, D. D., and Wilson, S. H. (2004). Influence of DNA structure on DNA polymerase β active site function: Extension of mutagenic DNA intermediates. *J. Biol. Chem.* **279**, 31921–31929.

Beard, W. A., Shock, D. D., Vande Berg, B. J., and Wilson, S. H. (2002). Efficiency of correct nucleotide insertion governs DNA polymerase fidelity. *J. Biol. Chem.* **277**, 47393–47398.

Biade, S., Sobol, R. W., Wilson, S. H., and Matsumoto, Y. (1998). Impairment of proliferating cell nuclear antigen-dependent apurinic/apyrimidinic site repair on linear DNA. *J. Biol. Chem.* **273**, 898–902.

Bose-Basu, B., DeRose, E. F., Kirby, T. W., Mueller, G. A., Beard, W. A., Wilson, S. H., and London, R. E. (2004). Dynamic characterization of a DNA repair enzyme: NMR studies of [*methyl*-^{13}C]methionine-labeled DNA polymerase β. *Biochemistry* **43**, 8911–8922.

Delarue, M., Poch, O., Tordo, N., Moras, D., and Argos, P. (1990). An attempt to unify structures of polymerases. *Protein Eng.* **3**, 461–467.

Deterding, L. J., Prasad, R., Mullen, G. P., Wilson, S. H., and Tomer, K. B. (2000). Mapping of the 5'-2-deoxyribose-5-phosphate lyase active site in DNA polymerase β by mass spectrometry. *J. Biol. Chem.* **275**, 10463–10471.

Dianov, G. L. (2003). Monitoring base excision repair by *in vitro* assays. *Toxicology* **193**, 35–41.

Dianov, G. L., Jensen, B. R., Kenny, M. K., and Bohr, V. A. (1999). Replication protein A stimulates proliferating cell nuclear antigen-dependent repair of abasic sites in DNA by human cell extracts. *Biochemistry* **38**, 11021–11025.

Feng, J.-A., Crasto, C. J., and Matsumoto, Y. (1998). Deoxyribose phosphate excision by the N-terminal domain of the polymerase β: The mechanism revisited. *Biochemistry* **37,** 9605–9611.

Fortini, P., Pascucci, B., Parlanti, E., Sobol, R. W., Wilson, S. H., and Dogliotti, E. (1998). Different DNA polymerases are involved in the short- and long-patch base excision repair in mammalian cells. *Biochemistry* **37,** 3575–3580.

Goodman, M. F., Creighton, S., Bloom, L. B., and Petruska, J. (1993). Biochemical basis of DNA replication fidelity. *Crit. Rev. Biochem. Mol. Biol.* **28,** 83–126.

Herschlag, D., Piccirilli, J. A., and Cech, T. R. (1991). Ribozyme-catalyzed and nonenzymatic reactions of phosphate diesters: Rate effects upon substitution of sulfur for a nonbridging phosphoryl oxygen atom. *Biochemistry* **30,** 4844–4854.

Horton, J. K., Prasad, R., Hou, E., and Wilson, S. H. (2000). Protection against methylation-induced cytotoxicity by DNA polymerase β-dependent long patch base excision repair. *J. Biol. Chem.* **275,** 2211–2218.

Johnson, K. A. (1992). Transient-state kinetic analysis of enzyme reaction pathways. *In* "The Enzymes: Mechanisms of Catalysis" (D. S. Sigman, ed.), Vol. XX, pp. 1–61. Academic Press, New York.

Joyce, C. M., and Benkovic, S. J. (2004). DNA polymerase fidelity: Kinetics, structure, and checkpoints. *Biochemistry* **43,** 14317–14324.

Kim, S.-J., Beard, W. A., Harvey, J., Shock, D. D., Knutson, J. R., and Wilson, S. H. (2003). Rapid segmental and subdomain motions of DNA polymerase β. *J. Biol. Chem.* **278,** 5072–5081.

Klungland, A., and Lindahl, T. (1997). Second pathway for completion of human DNA base excision-repair: Reconstitution with purified proteins and requirement for DNase IV (FEN1). *EMBO J.* **16,** 3341–3348.

Matsumoto, Y., and Kim, K. (1995). Excision of deoxyribose phosphate residues by DNA polymerase β during DNA repair. *Science* **269,** 699–702.

Osheroff, W. P., Beard, W. A., Wilson, S. H., and Kunkel, T. A. (1999a). Base substitution specificity of DNA polymerase β depends on interactions in the DNA minor groove. *J. Biol. Chem.* **274,** 20749–20752.

Osheroff, W. P., Jung, H. K., Beard, W. A., Wilson, S. H., and Kunkel, T. A. (1999b). The fidelity of DNA polymerase β during distributive and processive DNA synthesis. *J. Biol. Chem.* **274,** 3642–3650.

Patterson, T. A., Little, W., Cheng, X., Widen, S. G., Kumar, A., Beard, W. A., and Wilson, S. H. (2000). Molecular cloning and high-level expression of human polymerase β cDNA and comparison of the purified recombinant human and rat enzymes. *Protein Expr. Purif.* **18,** 100–110.

Piersen, C. E., Prasad, R., Wilson, S. H., and Lloyd, R. S. (1996). Evidence for an imino intermediate in the DNA polymerase β deoxyribose phosphate excision reaction. *J. Biol. Chem.* **271,** 17811–17815.

Prasad, R., Batra, V. K., Yang, X.-P., Krahn, J. M., Pedersen, L. C., Beard, W. A., and Wilson, S. H. (2005). Structural insight into the DNA polymertase β deoxyribose phosphate lyase mechanism. *DNA Repair* **4,** 1347–1357.

Prasad, R., Beard, W. A., and Wilson, S. H. (1994). Studies of gapped DNA substrate binding by mammalian DNA polymerase β: Dependence on 5'-phosphate group. *J. Biol. Chem.* **269,** 18096–18101.

Prasad, R., Beard, W. A., Strauss, P. R., and Wilson, S. H. (1998). Human DNA polymerase β deoxyribose phosphate lyase: Substrate specificity and catalytic mechanism. *J. Biol. Chem.* **273,** 15263–15270.

Prasad, R., Dianov, G. L., Bohr, V. A., and Wilson, S. H. (2000). FEN1 stimulation of DNA polymerase β mediates an excision step in mammalian long patch base excision repair. *J. Biol. Chem.* **275**, 4460–4466.

Prasad, R., Lavrik, O. I., Kim, S. J., Kedar, P., Yang, X. P., Vande Berg, B. J., and Wilson, S. H. (2001). DNA polymerase β-mediated long patch base excision repair: Poly(ADP-ribose) polymerase-1 stimulates strand displacement DNA synthesis. *J. Biol. Chem.* **276**, 32411–32414.

Sawaya, M. R., Prasad, P., Wilson, S. H., Kraut, J., and Pelletier, H. (1997). Crystal structures of human DNA polymerase β complexed with gapped and nicked DNA: Evidence for an induced fit mechanism. *Biochemistry* **36**, 11205–11215.

Singhal, R. K., Prasad, R., and Wilson, S. H. (1995). DNA polymerase β conducts the gap-filling step in uracil-initiated base excision repair in a bovine testis nuclear extract. *J. Biol. Chem.* **270**, 949–957.

Singhal, R. K., and Wilson, S. H. (1993). Short gap-filling synthesis by DNA polymerase β is processive. *J. Biol. Chem.* **268**, 15906–15911.

Slupphaug, G., Eftedal, I., Kavli, B., Bharati, S., Helle, N. M., Haug, T., Levine, D. W., and Krokan, H. E. (1995). Properties of a recombinant human uracil-DNA glycosylase from the *UNG* gene and evidence that *UNG* encodes the major uracil-DNA glycosylase. *Biochemistry* **34**, 128–138.

Sobol, R. W., Horton, J. K., Kühn, R., Gu, H., Singhal, R. K., Prasad, R., Rajewsky, K., and Wilson, S. H. (1996). Requirement of mammalian DNA polymerase β in base excision repair. *Nature* **379**, 183–186.

Sobol, R. W., Prasad, R., Evenski, A., Baker, A., Yang, X. P., Horton, J. K., and Wilson, S. H. (2000). The lyase activity of the DNA repair protein β-polymerase protects from DNA-damage-induced cytotoxicity. *Nature* **405**, 807–810.

Srivastava, D. K., Vande Berg, B. J., Prasad, R., Molina, J. T., Beard, W. A., Tomkinson, A. E., and Wilson, S. H. (1998). Mammalian abasic site base excision repair: Identification of the reaction sequence and rate-determining steps. *J. Biol. Chem.* **273**, 21203–21209.

Strauss, P. R., Beard, W. A., Patterson, R. A., and Wilson, S. H. (1997). Substrate binding by human apurinic/apyrimidinic endonuclease indicates a Briggs–Haldane mechanism. *J. Biol. Chem.* **272**, 1302–1307.

Tanabe, K., Bohn, E. W., and Wilson, S. H. (1979). Steady-state kinetics of mouse DNA polymerase β. *Biochemistry* **18**, 3401–3406.

Vande Berg, B. J., Beard, W. A., and Wilson, S. H. (2001). DNA structure and aspartate 276 influence nucleotide binding to human DNA polymerase β: Implication for the identity of the rate-limiting conformational change. *J. Biol. Chem.* **276**, 3408–3416.

Wang, H., and Hays, J. B. (2002). Mismatch repair in human nuclear extracts: Quantitative analyses of excision of nicked circular mismatched DNA substrates, constructed by a new technique employing synthetic oligonucleotides. *J. Biol. Chem.* **277**, 26136–26142.

Wang, Y. C., Burkhart, W. A., Mackey, Z. B., Moyer, M. B., Ramos, W., Husain, I., Chen, J., Besterman, J. M., and Tomkinson, A. E. (1994). Mammalian DNA ligase II is highly homologous with vaccinia DNA ligase: Identification of the DNA ligase II active site for enzyme-adenylate formation. *J. Biol. Chem.* **269**, 31923–31928.

Yang, L., Beard, W. A., Wilson, S. H., Broyde, S., and Schlick, T. (2002). Polymerase β simulations suggest that Arg258 rotation is a slow step rather than large subdomain motions *per se*. *J. Mol. Biol.* **317**, 679–699.

[8] Direct Removal of Alkylation Damage from DNA by AlkB and Related DNA Dioxygenases

By Barbara Sedgwick, Peter Robins, and Tomas Lindahl

Abstract

The cytotoxic alkylation lesions 1-methyladenine (1-alkyladenine) and 3-methylcytosine are removed efficiently from DNA by direct damage reversal, catalyzed by the *Escherichia coli* AlkB protein and its human homologs ABH2 and ABH3. The enzymes act by oxidative demethylation, employing Fe(II) and α-ketoglutarate as cofactors, and release the methyl moiety as formaldehyde. The isolation of these enzymes from overproducing cells is described, as well as the preparation of radioactively labeled substrates and procedures for enzyme assays. Functionality *in vivo* is examined by complementation of the low survival of alkylated single-stranded DNA bacteriophage in an *E. coli alkB* mutant.

The main DNA lesions generated by simple alkylating agents in single-stranded DNA are 7-methylguanine, 1-methyladenine (1-meA) and 3-methylcytosine (3-meC) (Bodell and Singer, 1979). The 7-methylguanine adduct, which is also formed in similar amounts in double-stranded DNA, is tolerated well because it is neither markedly cytotoxic nor miscoding. In contrast, the sites of alkylation to 1-meA and 3-meC are protected in duplex DNA by the complementary strand so these lesions are only induced in transiently occurring single-stranded regions, e.g., in transcription bubbles and at replication forks. The 1-meA and 3-meC lesions are toxic because they prevent hydrogen bonding with a complementary nucleotide. They have been found to be repaired by direct damage reversal that is operational on both single- and double-stranded DNA, employing an iron-catalyzed mechanism of oxidative demethylation.

The most studied enzyme of this type is the *E. coli* AlkB dioxygenase. The *alkB* gene product was initially characterized as a DNA repair function that counteracts toxic alkylation damage (Kataoka *et al.*, 1983). The level of AlkB protein in *E. coli* is very low, but is inducible by alkylating agents in a response regulated by the Ada protein (Sedgwick and Lindahl, 2002). Attempts to develop an *in vitro* assay for AlkB failed for many years, in retrospect because the wrong DNA substrate, as well as incorrect cofactors, was used. Transfection experiments with single- and double-stranded DNA established that AlkB acts on lesions generated only in single-stranded DNA, implicating 1-meA and 3-meC residues as the likely

METHODS IN ENZYMOLOGY, VOL. 408   0076-6879/06 $35.00
DOI: 10.1016/S0076-6879(06)08008-6

substrates (Dinglay *et al.*, 2000). Bioinformatics on folding and sequence analysis indicated that AlkB is a member of the family of Fe(II) and α-ketoglutarate-dependent dioxygenases (Aravind and Koonin, 2001). Using this new information, appropriate reaction mixtures for AlkB could be designated, although the unexpectedly high and sharp pH optimum of pH 8.0 temporarily delayed progress (Falnes *et al.*, 2002; Trewick *et al.*, 2002). Thus, AlkB functions as a 1-meA- (and 3-meC-) DNA dioxygenase, removing the toxic methyl group by oxidative demethylation to restore the native unmethylated form of DNA, with the release of the methyl group as formaldehyde (Fig. 1). AlkB also acts, albeit more slowly, on the minor alkylation lesions 1-methylguanine and 3-methylthymine (Delaney and Essigmann, 2004; Falnes, 2004; Koivisto *et al.*, 2004).

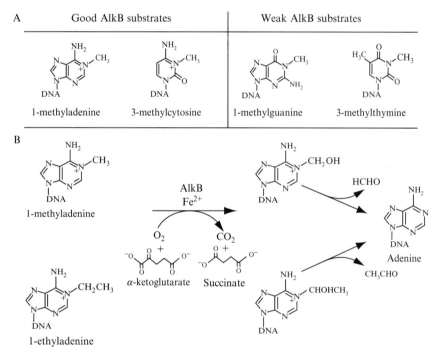

FIG. 1. (A) Methylated bases in DNA that are directly dealkylated by the AlkB protein. The major lesions 1-methyladenine and 3-methylcytosine are good substrates compared with the minor lesions 1-methylguanine and 3-methylthymine. (B) The AlkB reaction mechanism. AlkB requires α-ketoglutarate and dioxygen as cosubstrates. The α-ketoglutarate is oxidized to generate succinate and CO_2. The methyl group of the methylated base is oxidized and released as formaldehyde, which directly regenerates the unmodified base. Ethyl adducts are released as acetaldehyde.

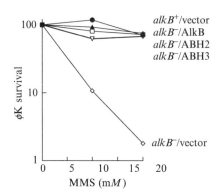

FIG. 2. Complementation of an *E. coli alkB* mutant by human ABH2 and ABH3. Reactivation of MMS-treated φK single-stranded DNA phage was monitored in an *alkB* mutant expressing ABH2 or ABH3. Host strains were *alkB⁺* or *alkB22* derivatives of *E. coli* AB1157 *gyrA*. Plasmid pBAR32 encodes AlkB protein, pGST.ABH2 encodes GST-tagged ABH2, pGST.ABH3 encodes GST-tagged ABH3, and pGEX6P-1 is the vector. ●, *alkB⁺*/ pGEX6P-1; ◇, *alkB22*/pGEX6P-1; ▲, *alkB22*/pBAR32; □, *alkB22*/pGST.ABH2; ▽, *alkB22*/ pGST.ABH3 (from Duncan *et al.*, 2002).

Human cells contain two functional homologs of AlkB, called ABH2 and ABH3 (Aas *et al.*, 2003; Duncan *et al.*, 2002; Sedgwick, 2004) (Fig. 2). These enzymes have similar substrate specificities and are mainly confined to the cell nucleus. It is presently unclear why two similar enzymes are present. ABH2 shows some preference for methylated poly(dA) over methylated poly(dC) and also prefers a double-stranded substrate, while the reverse is valid for ABH3. Somewhat surprisingly, measurements of ABH2 and ABH3 mRNA levels by Northern blots did not establish a correlation with cell proliferation; ABH2 is high in liver, and ABH3 in spleen, but mRNA encoding these enzymes was present at detectable levels in all tissues and cell lines investigated (Duncan *et al.*, 2002). ABH3, but apparently not ABH2, can demethylate 1-meA in RNA (Aas *et al.*, 2003), but the physiological relevance of this activity remains uncertain. *E. coli* AlkB efficiently repairs 1-ethyladenine residues in ethylated DNA and also acts on short oligonucleotides substrates and even on a mononucleotide with a 5'-phosphate, but ABH2 and ABH3 show poor activity with such substrates (Koivisto *et al.*, 2003). Furthermore, *E. coli* AlkB binds to single-stranded DNA, but less efficiently to double-stranded DNA (Dinglay *et al.*, 2000; Mishina *et al.*, 2004b).

In addition to ABH2 and ABH3, the human genome encodes an additional six proteins related to AlkB. They have been called ABH1 and ABH4–8, and the open reading frames are shown schematically in

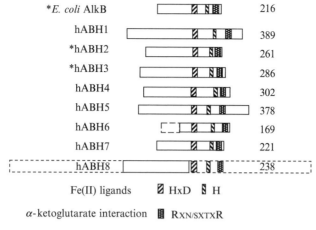

FIG. 3. Human homologs of the AlkB protein. Eight human amino acid sequence homologs of the *E. coli* AlkB protein have been identified (P. Bates and B. Sedgwick, unpublished data) (Kurowski *et al.*, 2003). hABH2 and hABH3 have DNA repair functions closely related to those of AlkB. No such activity has been demonstrated for hABH1 (Aas *et al.*, 2003; Duncan *et al.*, 2002). Further studies of the role of hABH1, and also of the more recently described hABH4 to hABH8, are in progress. Conserved amino acid sequences that ligand Fe(II) and interact with α-ketoglutarate are indicated. The total number of amino acids is indicated to the right. The hABH6 sequence appears incomplete at the amino terminus when compared with *Mus musculus* (NP_932144) and *Pan troglodytes* (XP_512606) homologs, as indicated by dashed lines. Similarly, hABH8 appears to be incomplete when compared with *P. troglodytes* (XP_522172, 1010 amino acids) and *Canis familiaris* (XP_546542, 1078 amino acids) homologs, which have additional amino-terminal sequences and also a SAM-dependent methyltransferase domain C-terminal of the AlkB homolog. The extent of the *P. troglodytes* homolog is indicated by dashed lines. Database accession codes are AlkB, NP_416716; hABH1, AAF01478; hABH2, NP_001001655; hABH3, NP_631917; hABH4, NP_060091; hABH5, BAA91078; hABH6, NP_116267.1; hABH7, NP_115682; hABH8, NP_620130.

Fig. 3. Their functions are presently unknown, and most of them are inactive in standard assays with an alkylated single-stranded DNA substrate. ABH1 does not even bind to DNA (Mishina *et al.*, 2004b) and currently appears an unlikely candidate for a DNA repair enzyme.

Purification of AlkB, ABH2, and ABH3

AlkB, ABH2, and ABH3 proteins with six histidines, GST, or FLAG tags have been overexpressed and purified from *E. coli* (Aas *et al.*, 2003; Dinglay *et al.*, 2000; Duncan *et al.*, 2002; Falnes *et al.*, 2002; Trewick *et al.*, 2002). His- and FLAG-tagged AlkB proteins have similar activities (Trewick *et al.*, 2002). His-tagged-ABH2 and -ABH3 had greater activity when isolated from insect cells by approximately 50- and 5-fold, respectively, although posttranslational modification in eukaryotic cells has not been observed (Koivisto *et al.*, 2003). The methods documented here describe purification of His-tagged apoproteins, with Fe(II) being added back in the reaction mixture.

Purification of His-Tagged AlkB, ABH2, and ABH3 Proteins
  from E. coli

To purify AlkB, inoculate *E. coli* carrying the cloned IPTG-inducible *alkB* gene (Dinglay *et al.*, 2000) in 1 liter Luria broth containing 50 μg/ml carbenicillin and incubate at 37° with aeration. When the culture reaches A_{600} 0.3 to 0.4, add isopropyl-β-D-thiogalactoside (IPTG) to 1 mM to induce *alkB* gene expression and incubate at 37° for 4 h. Collect cells by centrifugation at 10,000g for 15 min, and wash the cell pellet once with cold phosphate-buffered saline. Resuspend the cells in 25 ml 50 mM HEPES-KOH, pH 8, 300 mM NaCl, 2 mM EDTA, 2 mM 2-mercaptoethanol, and 5% glycerol. Disrupt the cells in a French press or by addition of lysozyme to a concentration of 2 mg/ml followed by incubation at 0° for 30 min and sonication. Clarify lysates by centrifugation at 17,000g 4° for 30 min and dialyse against the cell resuspension buffer but lacking EDTA. Recover the dialysate, add imidazole, pH 8, to 5 mM and load onto a 1-ml packed volume of nickel-nitrilotriacetic acid agarose (Qiagen) preequilibrated with 50 mM HEPES-KOH, pH 8, 100 mM NaCl, 2 mM 2-mercaptoethanol, 5% glycerol, and 5 mM imidazole. Wash the column with 20 column volumes of this buffer and repeat with buffer containing 40 mM imidazole. Continue to wash with 10 column volumes of buffer containing 60 mM imidazole and then elute the AlkB protein with buffer containing 250 mM imidazole, collecting fractions from these two latter steps. Analyze the protein content of the fractions by SDS–PAGE and gel staining. Pool fractions containing AlkB protein and dialyze into 30 mM potassium phosphate, pH 7.5, 300 mM NaCl, 2 mM dithiothreitol, and 50% glycerol. The expected yield of AlkB protein from a 1-liter culture is 5 mg. Divide the enzyme solution into small aliquots and keep at −70° for long-term storage. Once thawed, keep the protein at −20° for short-term storage.

His-tagged-ABH2 and -ABH3 proteins can be overexpressed and purified from *E. coli* BL21-CodonPlus (Stratagene) using a similar method to that for AlkB protein, except that during induction the incubation temperature is reduced to 25° to improve the yield of soluble enzyme (Duncan *et al.*, 2002).

Purification of His-Tagged ABH2 and ABH3 Proteins from Baculovirus-Infected Sf9 Cells

Infect 10^8 *Sf9* insect cells with recombinant baculovirus (Koivisto *et al.*, 2003) at a multiplicity of infection of 5. Grow for 3 days when expressing ABH2 and 4 days when expressing ABH3. Collect cells by centrifugation at 1000g for 10 min. Resuspend in 5 pellet volumes of lysis buffer, 25 m*M* Tris–HCl, pH 8, 150 m*M* NaCl, 0.1% Nonidet P-40, and 1 m*M* imidazole and incubate at 0° for 20 min. Clarify the lysates by centrifugation at 14,000g 4° for 15 min and load onto a 50-μl packed volume of nickel-nitrilotriacetic acid agarose. Wash with 500 μl lysis buffer lacking Nonidet P-40 but containing 10 m*M* imidazole, followed by 500 μl buffer containing 40 m*M* imidazole, and elute the bound protein with buffer containing 250 m*M* imidazole. Pool fractions containing ABH2 or ABH3 protein, dialyze, and store as for AlkB. Expected yields from 10^8 baculovirus-infected insect cells are about 25 μg ABH2 and 100 μg ABH3. If deemed necessary, the proteins may be purified further by Mono S cation-exchange gradient chromatography.

Purification of Untagged AlkB

Untagged AlkB has been overexpressed and isolated as the AlkB-Fe(II) complex with the cofactor α-ketoglutarate also bound. The procedure involves batch use of DEAE-cellulose and chromatography on Sepharose- and Mono S cation-exchange columns and should be completed within 5 h. The protein–metal complex is isolated as pink/purple fractions (Mishina *et al.*, 2004a).

^{14}C-Methylated Substrates

Substrates can be prepared by incubation of polynucleotides with ^{3}H-labeled methyl methanesulfonate (MMS) or *N*-methylnitrosourea, but such radioactive alkylating agents are expensive, labile, and not readily available. Instead, we recommend the use of [^{14}C]methyl iodide, which works well for the alkylation of polynucleotides in the absence of proteins and yields the same alkylated bases as MMS.

To prepare substrates containing [1-^{14}C]methyladenine or [3-^{14}C] methylcytosine, dissolve 50 units (1.8 mg) of poly(dA) in 2 ml 10 m*M* sodium cacodylate, pH 7, or poly(dC) in 2 ml 50 m*M* HEPES-KOH, pH 8. Cool 1 mCi [^{14}C]methyl iodide, supplied in a break-seal ampoule (Amersham Biosciences), on dry ice for 1 h. Transfer to 0° and add the poly (dA) or poly(dC) to the top of the glass ampoule. Carefully break the seal and mix the polymer solution with the condensed methyl iodide. Incubate at 30° for 6 h. Recover the radiolabeled polymer from the ampoule. Add sodium acetate (pH 7) to 0.3 M and precipitate the polynucleotide with 3 volumes of cold ethanol at −70° for 1 h. Centrifuge at 15,000g for 15 min and discard the radioactive supernatant. Add 10 ml 95% ethanol to the pellet and repeat the centrifugation. Repeat this wash step twice with 10 ml 80% ethanol and then air dry the pellet for 10 min. Dissolve the pellet in 1.5 ml 10 mM HEPES-KOH, pH 8, divide into small aliquots, and store at −70°.

A ^{14}C-ethylated substrate was similarly prepared by treating poly(dA) with [^{14}C]ethyl iodide (ICN) and incubating at 37° for 18 h.

A substrate containing [3-^{14}C]methylthymine was generated by dissolving 25 units (0.9 mg) poly(dT) in 1 ml 200 mM Na$_2$HPO$_4$ adjusted to pH 10.8 and exposing to 1 mCi [^{14}C]methyl iodide [as described for poly(dA)] at 20° for 6 h.

Double-stranded substrates containing ^{14}C-1-meA, ^{14}C-3-meC, or ^{14}C-3-meT were formed by annealing the radiolabeled methylated polynucleotide to an equimolar amount of nonradioactive, nonmethylated complementary polynucleotide in 10 mM HEPES-KOH, pH 8, at 20° for 1 h. Because poly(dG) has a low solubility, anneal radiolabeled poly(dC) to poly(dI).

Escherichia coli AlkB, but not human ABH2 and ABH3, can demethylate small oligonucleotides, with methylated 5′-dAMP being the minimal substrate. Nevertheless, AlkB is more efficient with a polymer substrate (Koivisto *et al.*, 2003).

Assays

Standard Assay Using ^{14}C-Alkylated Substrates

Stock solutions of 200 mM α-ketoglutarate and ascorbate are aliquoted and stored at −20°. Fe(NH$_4$)$_2$(SO$_4$)$_2$·6H$_2$0 (10 mM) is prepared just before use and is added last to the reaction mixture. Bovine serum albumin (BSA) is EDTA and DNase free. Calf thymus carrier DNA (2 mg/ml) should be fully dissolved in 10 mM Tris–HCl, pH 8, 1 mM EDTA, heat denatured at 95° for 10 min, and then cooled rapidly to 0°. Shear the carrier DNA by forcing several times through a medium syringe needle.

Incubate 1 pmol AlkB protein in a 100-μl reaction mixture containing ^{14}C-methylated poly(dA) or poly(dC) (2 μg, 2000 cpm), 50 mM HEPES-KOH, pH 8, 1 mM α-ketoglutarate, 2 mM ascorbate, 75 μM Fe (NH$_4$)$_2$(SO$_4$)$_2$·6H$_2$O, and 50 μg/ml BSA at 37° for 15 min. Stop the reaction by the addition of EDTA to 10 mM. Precipitate the substrate by adding NaCl to 100 mM and denatured carrier DNA to 0.2 mg/ml; mix well and add 2 volumes cold ethanol. Incubate at −70° for 30 min and then centrifuge at 15,000g for 15 min. Without disturbing the pellet, recover the top 80% of the supernatant and monitor ethanol-soluble radioactive material by scintillation counting.

To assay for ABH2 activity, reduce the buffer pH to 7.5 and the Fe(II) concentration to 25 μM. For both ABH2 and ABH3, increase the amount of protein to 10 pmol and the incubation time to 30 min.

To assay for activity on ^{14}C-methylated poly(dT) change the buffer in the reaction mix to 50 mM MES-KOH, pH 6, for AlkB protein and pH 6.5 for ABH3 and increase the incubation time to 30 min. ABH2 has a very weak activity on this single-stranded substrate but is active on double-stranded methylated poly(dT)/poly(dA) at pH 6.5 (Koivisto et al., 2004).

Analysis of ^{14}C-Methylated Bases Remaining in the Substrate by HPLC

To determine which methylated bases are repaired by AlkB or its homologs, incubate the purified protein with ^{14}C-methylated DNA or poly(dA) in the standard assay conditions and retrieve the ethanol-precipitated product. To analyze methylated purines remaining in the substrate, treat the precipitate with 0.1 M HCl at 95° for 1 h. Add 1 μg unlabeled methylated purines as markers (1-meA, 3-meA, and 7-meA, and also 7-meG and 3-meG if using DNA) before HPLC analysis. Load onto an HPLC Whatman Partisil 10 cation-exchange column in 0.1 M ammonium formate, pH 3.6, and apply a gradient of methanol from 20 to 40%. Record elution positions of markers by A_{260} readings and monitor released ^{14}C-labeled purines by scintillation counting of fractions (Trewick et al., 2002).

3-Methylcytosine is the only methylation product of poly(dC) (Singer and Grunberger, 1983). To determine whether 3-methylcytosine has been removed from ^{14}C-methylated poly(dC), hydrolyze the ethanol precipitate with 90% formic acid in a sealed glass ampoule at 180° for 20 min. Add unlabeled 3-methylcytosine as a marker and analyze the released products by HPLC as described earlier but run isocratically in 2.5% methanol (Trewick et al., 2002).

Direct Reversion of a Methylated DNA Base to the Unmodified Form

Oligonucleotides, a 41-mer TTTTT(ATTTTTT)$_5$ or a single-stranded PCR product containing [^{3}H]cytosines similarly interspersed between thymine residues (Duncan et al., 2002; Trewick et al., 2002), are heavily methylated so that >50% of the adenine or cytosine residues is converted to 1-methyladenine or [3-methyl-^{3}H]cytosine. The oligonucleotides are composed of mainly thymine residues because they are rarely methylated compared with adenine or cytosine at neutral pH (Singer and Grunberger, 1983). Nevertheless, use of a ^{3}H-labeled substrate to monitor reversal of [3-methyl-^{3}H]cytosine to [^{3}H]cytosine generated a clearer HPLC profile after formic acid hydrolysis.

To heavily methylate the oligonucleotides, treat eight times with 50 mM dimethyl sulfate in 75 mM sodium cacodylate, pH 7.4, at 30° for 2 h. Between each treatment, centrifuge through a Sephadex G-25 minicolumn equilibrated in the same buffer. To determine the level of methylation achieved, acid hydrolyze the material and monitor the released 1-methyladenine or [3-methyl-^{3}H]cytosine residues by HPLC as outlined earlier. To demonstrate direct reversion of 1-methyladenine to adenine or 3-methylcytosine to cytosine, incubate the heavily methylated oligonuceotides with AlkB in standard reaction mixtures. Precipitate with ethanol, acid hydrolyze the precipitate, and analyze the released products by HPLC (Duncan *et al.*, 2002). When using the nonradiolabeled substrate, do not add markers directly to the sample but run these separately.

Identification of the [^{14}C]Aldehyde Released

AlkB and its functional homologs oxidize the alkyl groups on 1-alkyladenine and 3-alkylcytosine in DNA and release them as an aldehyde (Fig. 1). To characterize or quantify the aldehyde released, incubate AlkB with a ^{14}C-alkylated poly(dA) substrate in the standard reaction mixture. Precipitate the substrate with ethanol, centrifuge, and recover the supernatant. Derivatize the ethanol-soluble material with dinitrophenylhydrazone (DNPH) (Houlgate *et al.*, 1989). Add 1 volume 0.25% DNPH in acetonitrile and 0.5% perchloric acid. Centrifuge the sample at 1000g for 5 min, dry the supernatant under vacuum, and redissolve in 50% methanol. Analyze the DNPH derivatives by reversed-phase HPLC on a Phenomenex Hypersil column using a linear 50–90% methanol gradient in water. Collect fractions and quantify radioactive material by scintillation counting. Use DNPH derivatives of formaldehyde and acetaldehyde as markers and monitor these at A_{254} (Duncan *et al.*, 2002).

Assaying ABH2/ABH3 in Crude Cell Extracts

The activity of ABH2 and ABH3 has not been detected in crude human cell extracts. Moreover, 1 pmol purified recombinant ABH2 protein is 50% inhibited by a small amount of crude extract (5 μg protein) in a standard 100-μl assay. Because AlkB and its homologs bind to single-stranded DNA (Dinglay *et al.*, 2000; Mishina *et al.*, 2004b), this inhibition appears to be due to fragments of cellular DNA and RNA in the extract. To reveal

ABH2/3 activity, the cell extract was fractionated by Mono S cation-exchange chromatography. Fractions eluting at high NaCl concentrations (>500 mM) contained the ABH2/3 proteins and were active in the repair of ^{14}C-methylated poly(dA).

General Assays for α-Ketoglutarate-Dependent Dioxygenases

These methods may be useful if, for example, the methylated substrate is not available in a radiolabeled form or has not been clearly defined, as is the case for several human homologs (Fig. 3).

1. Measure the consumption of O_2 using an oxygen electrode (Trewick et al., 2002).
2. Assay the release of $^{14}CO_2$ from α-keto[1-^{14}C]glutarate (Welford et al., 2003) or [1-^{14}C]succinate from α-keto[5-^{14}C]glutarate (Falnes et al., 2002).

In Vivo Assay: Survival of Alkylated Single-Stranded DNA Bacteriophage in E. coli AlkB Mutants

The AlkB protein repairs toxic DNA damage generated in single-stranded DNA. A characteristic of AlkB mutants is consequently defective reactivation of methylated single-stranded DNA bacteriophages (Dinglay et al., 2000). This phenotype has been used to determine whether AlkB can repair DNA damage induced by a variety of alkylating agents (Duncan et al., 2002; Koivisto et al., 2003), to identify human homologs of alkB that complement the defective reactivation (Duncan et al., 2002) (Fig. 2), and to demonstrate that AlkB and ABH3 can reactivate a MMS-treated RNA phage (Aas et al., 2003). Two single-stranded DNA phages M13 and ϕK (Kodaira et al., 1996) have been used in reactivation experiments. M13 is easier to manipulate, but ϕK was useful in complementation experiments for the following reasons.

1. M13 requires an F' host whereas ϕK does not. We found that F' derivatives of E. coli DE3 lysogens, which are required to overexpress genes cloned into several different expression vectors, were susceptible to cell lysis when making competent cells or on incubation of cultures with IPTG. An F$^-$ host carrying DE3 was therefore used with ϕK phage or, alternatively, a vector that did not require DE3 for expression was used with M13.

2. Plasmids carrying an f1 origin of DNA replication, e.g., pET28, convey cellular resistance to M13 phage. ϕK should therefore be used with these vectors.

Survival of Alkylated M13 Phage

MMS is used here to exemplify this method.

Requirements: *E. coli* F' *alkB*+ and *alkB*− strains such as BS141 (*E. coli* K12 AB1157/F'*proAB*+ *lacI*^{Q} *lacZ*ΔM15*Tn10*) and BS143 (as BS141 but *alkB117::Tn3*);

M13 bacteriophage stock of known titer; 25 to 30 LB agar plates; aliquots of 3 ml molten 0.8% LB agar at 48°; aliquots of 900 μl 1× M9 salts (Miller, 1992), 10 m*M* MgSO$_4$ (M9/Mg). For all phage manipulations, use micropipette tips containing filters to avoid pipette contamination.

Inoculate strains into LB broth containing antibiotics required to maintain selection of the F' factor. Incubate at 37° until the cultures reach A_{600} 0.8. Begin preparation of the MMS-treated M13 phage approximately 1 h before the cultures reach A_{600} 0.8, i.e., at approximately A_{600} 0.2. Dilute the M13 stock in M9/Mg to 2×10^8 pfu/ml. Dilute the MMS stock (11.4 *M*) to 100 m*M* by adding 10 μl MMS to 1.13 ml M9/Mg at room temperature. Flick the tube until all globules of MMS dissolve. Dilute the MMS further (see Table I).

Add 100 μl M13 phage (2×10^8 pfu/ml) to the MMS dilutions in tubes 1 to 4. The phage concentrations are now 10^8 pfu/ml and MMS treatment doses 0, 5, 10, and 15 m*M*.

Incubate at 30° for 30 min. Add 100 μl treated phage to 900 μl M9/Mg to obtain 10^{-1} phage dilutions. Flick to mix. Serial dilute further to obtain 10^{-2}, 10^{-3}, 10^{-4}, and 10^{-5} dilutions. Plate various dilutions in duplicate with the *E. coli* cultures at $A_{600} \sim 0.8$. Add 200 μl culture and 100 μl of the phage dilutions to 3 ml molten LB agar at 48°. Pour onto LB agar plates at room temperature and carefully swirl to mix and spread evenly. Allow the agar to set and invert the plates. Incubate overnight at 37°. Count the pfu and calculate percentage survival for each MMS dose. Plot log percentage survival against MMS dose (see Table II).

TABLE I
MMS DILUTIONS

Tube	Mix 100 m*M* MMS (μl)	With M9/MgSO$_4$ (μl)	MMS concentration (m*M*)
1	0	100	0
2	10	90	10
3	20	80	20
4	30	70	30

TABLE II
PHAGE DILUTIONS

Tube	Final MMS dose (mM)	Phage dilutions to plate on an AlkB⁺ strain	Phage dilutions to plate on an AlkB⁻ strain
1	0	10^{-5}	10^{-5}
2	5	10^{-5}	10^{-5} 10^{-4}
3	10	10^{-5}	10^{-4} 10^{-3}
4	15	10^{-5}	10^{-3} 10^{-2}

Survival of Methylated ϕK Phage and Complementation by Putative AlkB Homologs

Requirements: HK82 (AB1157 *alkB22 gyrA*)/vector, HK82/constructs of interest, and a positive control HK80(AB1157 *gyrA*)/vector or HK82/construct expressing *alkB* (see Fig. 2); ϕK stock of known titre; LB plates containing 5 mM CaCl₂; 3-ml aliquots of 0.8% LB molten agar containing 5 mM CaCl₂ at 48°; aliquots of 900 μl M9/Mg.

Inoculate cultures of the various strains in LB broth containing the appropriate selective antibiotics and incubate to A_{600} 0.5 to 0.8. Add IPTG to 100 μM and incubate at 25° for 3 h or apply alternative conditions required for inducing recombinant gene expression. Treat the ϕK phage with MMS (as for M13). Dilutions may have to be adjusted depending on the titer of the ϕK stock. Plate various dilutions with the various strains in the molten 0.8% agar. Immediately incubate at 37°. The ϕK phage replicates rapidly and can form countable plaques in 3 to 4 h at 37°. For convenience, the incubator can be switched off after 2 h (use a timer) and the plates left overnight in a cooling incubator.

References

Aas, P. A., Otterlei, M., Falnes, P. O., Vagbo, C. B., Skorpen, F., Akbari, M., Sundheim, O., Bjoras, M., Slupphaug, G., Seeberg, E., and Krokan, H. E. (2003). Human and bacterial oxidative demethylases repair alkylation damage in both RNA and DNA. *Nature* **421,** 859–863.

Aravind, L., and Koonin, E. V. (2001). The DNA-repair protein AlkB, EGL-9, and leprecan define new families of 2-oxoglutarate- and iron-dependent dioxygenases. *Genome Biol.* **2,** Epub., RESEARCH0007.

Bodell, W. J., and Singer, B. (1979). Influence of hydrogen bonding in DNA and polynucleotides on reaction of nitrogens and oxygens toward ethylnitrosourea. *Biochemistry* **18,** 2860–2863.

Delaney, J. C., and Essigmann, J. M. (2004). Mutagenesis, genotoxicity, and repair of 1-methyladenine, 3-alkylcytosines, 1-methylguanine, and 3-methylthymine in alkB *Escherichia coli. Proc. Natl. Acad. Sci. USA* **101,** 14051–14056.

Dinglay, S., Trewick, S. C., Lindahl, T., and Sedgwick, B. (2000). Defective processing of methylated single-stranded DNA by *E. coli alkB* mutants. *Genes Dev.* **14,** 2097–2105.

Duncan, T., Trewick, S. C., Koivisto, P., Bates, P. A., Lindahl, T., and Sedgwick, B. (2002). Reversal of DNA alkylation damage by two human dioxygenases. *Proc. Natl. Acad. Sci. USA* **99,** 16660–16665.

Falnes, P. O. (2004). Repair of 3-methylthymine and 1-methylguanine lesions by bacterial and human AlkB proteins. *Nucleic Acids Res.* **32,** 6260–6267.

Falnes, P. O., Johansen, R. F., and Seeberg, E. (2002). AlkB-mediated oxidative demethylation reverses DNA damage in *Escherichia coli. Nature* **419,** 178–181.

Houlgate, P. R., Dhingra, K. S., Nash, S. J., and Evans, W. H. (1989). Determination of formaldehyde and acetaldehyde in main stream cigarette smoke by high-performance liquid chromatography. *Analyst.* **114,** 355–360.

Kataoka, H., Yamamoto, Y., and Sekiguchi, M. (1983). A new gene (*alkB*) of *Escherichia coli* that controls sensitivity to methyl methane sulfonate. *J. Bacteriol.* **153,** 1301–1307.

Kodaira, K.-I., Oki, M., M., K., Kimoto, H., and Taketo, A. (1996). The virion proteins encoded by bacteriophage fK and its host-range mutant fKhT: Host-range determination and DNA binding properties. *J. Biochem.* **119,** 1062–1069.

Koivisto, P., Duncan, T., Lindahl, T., and Sedgwick, B. (2003). Minimal methylated substrate and extended substrate range of *Escherichia coli* AlkB protein, a 1-methyladenine-DNA dioxygenase. *J. Biol. Chem.* **278,** 44348–44354.

Koivisto, P., Robins, P., Lindahl, T., and Sedgwick, B. (2004). Demethylation of 3-methylthymine in DNA by bacterial and human DNA dioxygenases. *J. Biol. Chem.* **279,** 40470–40474.

Kurowski, M. A., Bhagwat, A. S., Papaj, G., and Bujnicki, J. M. (2003). Phylogenomic identification of five new human homologs of the DNA repair enzyme AlkB. *BMC Genomics* **4,** 48–53.

Miller, J. H. (1992). "A Short Course in Bacterial Genetics." Cold Spring Harbor Laboratory Press, Cold Spring Harbor, NY.

Mishina, Y., Chen, L. X., and He, C. (2004a). Preparation and characterization of the native iron(II)-containing DNA repair AlkB protein directly from *Escherichia coli. J. Am. Chem. Soc.* **126,** 16930–16936.

Mishina, Y., Lee, C. H., and He, C. (2004b). Interaction of human and bacterial AlkB proteins with DNA as probed through chemical cross-linking studies. *Nucleic Acids Res.* **32,** 1548–1554.

Sedgwick, B. (2004). Repairing DNA-methylation damage. *Nature Rev. Mol. Cell. Biol.* **5,** 148–157.

Sedgwick, B., and Lindahl, T. (2002). Recent progress on the Ada response for inducible repair of DNA alkylation damage. *Oncogene* **21,** 8886–8894.

Singer, B., and Grunberger, D. (1983). "Molecular Biology of Mutagens and Carcinogens: Reactions of Directly Acting Agents with Nucleic Acids," pp. 45–96. Plenum Press, New York.

Trewick, S. C., Henshaw, T. F., Hausinger, R. P., Lindahl, T., and Sedgwick, B. (2002). Oxidative demethylation by *Escherichia coli* AlkB directly reverts DNA base damage. *Nature* **419,** 174–178.

Welford, R. W. D., Schlemminger, I., McNeill, L. A., Hewitson, K. S., and Schofield, C. J. (2003). The selectivity and inhibition of AlkB. *J. Biol. Chem.* **278,** 10157–10161.

[9] Purification and Characterization of DNA Photolyases

By Gwendolyn B. Sancar and Aziz Sancar

Abstract

Members of the photolyase/cryptochrome family of blue-light photo-receptors are monomeric proteins of 50–70 kDa that contain two noncovalently bound chromophores/cofactors: either folate or deazaflavin, which act as a photoantenna, and a two electron-reduced FAD, which acts as a catalytic cofactor. DNA photolyases bind their substrates with high affinity and specificity and subsequently use blue light as a cosubstrate for the *in situ* conversion of ultraviolet-induced cyclobutane pyrimidine dimers and (6-4) photoproducts to canonical bases, thereby restoring the integrity of DNA. The determinants for binding, as well as the mechanism of the photolysis reaction, have been studied extensively using highly purified enzyme. In contrast, neither the substrate nor the reaction catalyzed by the closely related cryptochromes has been identified. This chapter describes methods used to purify DNA photolyases from a variety of organisms using an *Escherichia coli* overexpression system, as well as the properties of the purified enzymes and some of the assays commonly used to study DNA binding and repair by these enzymes *in vitro*.

Introduction

Historical Perspective

The field of DNA repair enzymology began in the late 1950s when Rupert and colleagues discovered that ultraviolet (UV)-inactivated transforming DNA could be reactivated by mixing it with an *Escherichia coli* cell-free extract and exposing the mixture to white light (Rupert *et al.*, 1958). This study and subsequent work by Rupert firmly established that an enzyme, called photoreactivating enzyme (DNA photolyase, EC 4.1.99.3), present in *E. coli* (and baker's yeast) extracts, binds to UV-damaged DNA in a light-independent manner and, while complexed to DNA, absorbs a blue-light photon that initiates the repair reaction. Following repair the enzyme dissociates from the DNA:

METHODS IN ENZYMOLOGY, VOL. 408   0076-6879/06 $35.00
Copyright 2006, Elsevier Inc. All rights reserved.   DOI: 10.1016/S0076-6879(06)08009-8

$$E + S \underset{k_2}{\overset{k_1}{\rightleftharpoons}} ES \underset{k_3 = k_p^1}{\overset{h\nu}{\rightarrow}} E + P$$

As apparent from this scheme, the repair reaction follows standard Michaelis–Menten kinetics with one important exception: catalysis is absolutely dependent upon light. An alternative view of the reaction scheme is to consider the reaction a bisubstrate process that proceeds in a sequential and ordered manner. From a practical standpoint, the light dependence of catalysis makes it possible to investigate the catalytic step by flash photolysis over a time span ranging from 100 femtoseconds (10^{-13} s) to milliseconds (10^{-3} s).

The major lesion induced in DNA by UV is the cyclobutane pyrimidine dimer and photolyase repairs DNA by splitting the cyclobutane ring to generate two canonical pyrimidines, a process referred to as "direct repair" to differentiate it from base and nucleotide excision repair mechanisms (Setlow et al., 1965; Wulff and Rupert, 1962). Photolyase is not an abundant enzyme; an E. coli cell contains only 10–20 molecules of enzyme (Harm et al., 1968; Kavaklı and Sancar, 2004). As a consequence, purification and extensive characterization of the enzyme became possible only after the advent of recombinant DNA technology. The gene encoding E. coli photolyase was cloned in 1976 (Sancar and Rupert, 1978a), which eventually led to the large-scale purification and biochemical, spectroscopic, photophysical, and structural characterization of the enzyme (Sancar, 1994, 2003, 2004a; Sancar and Sancar, 1988).

Photolyase/Cryptochrome Family

The classical photolyase first discovered in E. coli and then Saccharomyces cerevisiae and many other organisms repairs cyclobutane pyrimidine dimers (Pyr<>Pyr) only; it cannot repair the second major UV photoproduct, the (6-4) pyrimidine-pyrimidone adduct. Almost four decades later Todo and co-workers (1993) identified a new photolyase that specifically repairs (6-4) photoproducts in Drosophila melanogaster. This enzyme has high sequence identity to classical photolyase (Todo et al., 1996), as does a protein called cryptochrome, encoded by a gene that regulates blue light responses in Arabidopsis thaliana (Ahmad and Cashmore, 1993). Cryptochrome, however, has no DNA repair activity (Lin et al., 1995; Malhotra et al., 1995). In 1995 an expressed sequence tag in the human genome database was reported to be a photolyase homolog (Adams et al., 1995); however, subsequent investigation demonstrated that neither this protein nor a newly discovered second homolog has photolyase activity (Hsu et al., 1996). Hsu and co-workers suggested that these proteins, which they called hCRY1 and hCRY2, might be blue-light photoreceptors for the circadian clock in humans. Although not the subject of this work, substantial evidence suggests that this is indeed the case (Sancar, 2004b). Thus, the photolyase/cryptochrome family of enzymes encompasses three groups of blue-light photoreceptors: photolyase (or CPD photolyase), which repairs pyrimidine dimers; (6-4) photolyase, which repairs (6-4) photoproducts; and cryptochrome, which regulates growth and development in plants and the circadian clock in humans and other animals.

Photolyase/cryptochrome family members have been found in all three kingdoms: archaea, eubacteria, and eukaryia; however, the distribution among species is sporadic. Photolyase is found in *E. coli*, *S. cerevisiae*, *Drosophila melanogaster*, opossum, and some animal viruses (where it is incorporated in the virion), but it is absent from *Bacillus subtilis*, *Schizosaccharomyces pombe*, and placental mammals, including raccoons and humans. The (6-4) photolyase is present in *D. melanogaster*, *Xenopus laevis*, rattlesnake, zebrafish, and *Arabidopsis thaliana*, but not in *Caenorhabditis elegans* or humans and has not been detected so far in eubacteria and archaeae. In contrast, cryptochrome has been found in several eubacteria, including *Vibrio cholerae* (two enzymes), *A. thaliana* (three enzymes), zebrafish (five enzymes), *D. melanogaster* (one enzyme), humans and mice (two enzymes in each), but not in archaeae or in *C. elegans*. Thus, *D. melanogaster* contains all three members of the photolyase/cryptochrome family, *C. elegans* lacks all three, humans have cryptochrome but no photolyase or (6-4) photolyase, and *E. coli* has photolyase but no (6-4) photolyase or cryptochrome. Phylogenetic trees based on the hundreds of photolyase/cryptochrome sequences present in public databases have been constructed (Brudler *et al.*, 2003; Cashmore, 2003; Kleine *et al.*, 2003; Partch *et al.*, 2005). Although there is no consensus about the conclusions emerging from these analyses, one view is that photolyase and cryptochrome have a single progenitor. A primordial organism may have employed a photosensory pigment to detect light and regulate its physiology and location in synchrony with the 24-h periodicity (circadian = about a day) of the geophysical light–dark cycle so as to minimize exposure of the organisms's DNA to damage. The same pigment may have been used to repair DNA damage that would inevitably occur under these conditions. Later mutation and selection, in some cases accompanied by gene duplication, could have produced enzymes that perform one or the other function more efficiently, thus giving rise to the present-day photolyases, which repair DNA, and cryptochromes, which regulate the daily oscillations in the physiology and movement of organisms. Following a brief summary of the structure and function of photolyases, this chapter deals exclusively with methods and procedures for purification and characterization of these enzymes.

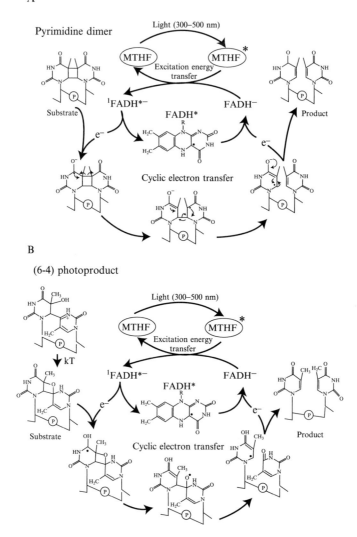

FIG. 1. Reaction mechanism of (A) pyrimidine dimer photolyases and (B) (6-4) photolyases. The pyrimidine dimer or (6-4) photoproduct flips out of the center of the DNA helix and into a cavity in the enzyme, bringing the substrate close to the flavin chromophore. In the case of the MTHF and deazaflavin Pyr<>Pyr photolyases, the chromophores absorb a photon and transfer the excitation energy to FADH⁻, producing ¹(FADH⁻)*. An electron is transferred from ¹(FADH⁻)* to the dimer, initiating a 2 + 2 cycloreversion reaction that restores both the pyrimidines to their original undamaged state. FADH⁻ is regenerated by back electron transfer from the dimer. The mechanism for the (6-4) photolyases is thought to be

Structure of Photolyases

The photolyase/cryptochrome blue-light photoreceptors are monomeric proteins of 50–70 kDa with two noncovalently bound chromophores/cofactors. The catalytic chromophore is FAD in the two electron-reduced and -deionized form, FADH$^-$ (Kim *et al.*, 1993; Payne *et al.*, 1987; Sancar *et al.*, 1987a). The second chromophore functions as a photoantenna and is 5,10-methenyltetrahydrofolate (MTHF) in most photolyases and cryptochromes; however, a few species, including *Anacystis nidulans*, *Streptomyces grisius*, and *Methanobacterium thermoautotrophicum*, utilize 8-hydroxy-5-deazariboflavin (8-HDF) as the second chromophore (Eker *et al.*, 1990; Kiener *et al.*, 1989).

The crystal structures of photolyases from three organisms, *E. coli* (Park *et al.*, 1995), *A. nidulans* (Tamada *et al.*, 1997), and *Thermus thermophilus* (Komori *et al.*, 2001), have been solved, as have the structures of cryptochromes from *Synechocystis* (Brudler *et al.*, 2003) and *A. thaliana* (Cry1) (Brautigam *et al.*, 2004). In addition the crystal structure of the *A. nidulans* photolyase-product complex has been solved at high resolution (Mees *et al.*, 2004). Remarkably, although these enzymes demonstrate only about 20–25% sequence identity and some bind MTHF while others bind 8-HDF, the traces of C_α backbone atoms are superimposable among the family members. The structure of *E. coli* photolyase is shown in Fig. 1 and is representative of most features of the entire family. The enzyme is composed of two domains (Fig. 1A): an N-terminal α/β domain (residues 1–131) and a C-terminal α-helical domain (residues 204–471). The two domains are connected by a long and structured interdomain loop. The photoantenna chromophore (MTHF in Fig. 1) is located in a shallow groove between the two domains, while the FADH$^-$ cofactor is deeply buried within the α-helical domain and is tightly bound by hydrogen bonds and ionic interactions with at least 14 amino acids. The distance between the photoantenna and the catalytic chromophore is 16.8 Å center to center. A surface potential representation of the enzyme (Fig. 1B) reveals the presence of a positively charged groove running nearly two-thirds of the length of the molecule and, lying in the approximate center of the groove, a hole leading to the flavin cofactor. These features immediately suggest that

---

similar except that thermal conversion (kT) of the photoproduct to the oxetane intermediate occurs upon formation of the ES complex (Kim *et al.*, 1994; Zhao *et al.*, 1997). This may be facilitated by protonation of the 3′ residue in the photoproduct upon binding (Hitomi *et al.*, 2001). The oxetane intermediate is the substrate for the photochemical reaction.

the positively charged residues lining the groove interact with negatively charged phosphates in the DNA backbone to orient the dimer at the hole. Remarkably, this hole has the correct dimensions and polarity to accommodate a pyrimidine dimer; however, the only way that a dimer can enter the hole is for it to flip out of the axis of the DNA helix ("dinucleotide flipping"). Studies on site-specific mutants of *S. cerevisiae* and *E. coli* photolyases (Li and Sancar, 1990; Vande Berg and Sancar, 1998), as well as the recent crystal structure of the *A. nidulans* photolyase-product complex (Mees *et al.*, 2004), support this dinucleotide flip model for substrate binding by photolyases. In contrast, in the cryptochrome structure the positively charged groups leading to the hole have been replaced with negatively charged residues (Brautigam *et al.*, 2004), consistent with the low-affinity binding to DNA seen with these proteins. This charge difference and the presence of a C-terminal extension in many cryptochromes are the only major structural differences that have been noted thus far between photolyases and cryptochromes.

Reaction Mechanism

Photolyase locates damage by three-dimensional diffusion (Sancar *et al.*, 1987c) and then flips the dimer into the active site to form a stable ES complex (Li and Sancar, 1990; Mees *et al.*, 2004; Park *et al.*, 1995; Vande Berg and Sancar, 1998). Depending on the source of the enzyme, the equilibrium association constant for ES complex formation ranges from 10^8 to 10^9 M^{-1} (Sancar, 1990). The photochemical reaction (Fig. 2A) is initiated when the photoantenna chromophore (MTHF or 8-HDF) absorbs a blue photon and transfers the excitation energy to the FADH$^-$ cofactor by Förster resonance energy transfer (lifetime = 50–100 ps). The excited singlet state flavin transfers an electron to the Pyr<>Pyr (lifetime = 170 ps) to generate FADH^0 neutral radical and Pyr<>Pyr$^{-\bullet}$ anionic radical. The cyclobutane ring of the dimer radical spontaneously rearranges to yield two canonical pyrimidines and the flavin radical is restored by back electron transfer (lifetime = 560 ps). The repaired dinucleotide flips out of the enzyme and into the DNA duplex and the enzyme dissociates from the DNA ($k_{off} \sim 1s^{-1}$).

The repair of (6-4) photoproduct poses a more challenging chemical problem because formation of the photoproduct involves not only sigma bond formation between C6 of the 5' base and C4 of the 3' base, but also transfer of the C4-OH (or NH) group of the 3' base to C5 of the 5' base of the adduct. Thus, simple cleavage of the C6-C4 sigma bond still leaves two damaged bases, whereas it has been shown that (6-4) photolyases restore the damaged bases to their canonical forms (Kim *et al.*, 1994). A model to

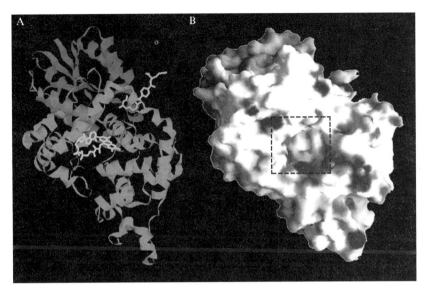

FIG. 2. Structure of *E. coli* photolyase. (A) A ribbon diagram of the backbone and the two chromophores: MTHF is shown in cyan and FAD is shown in yellow. The α/β nucleotide-binding domain is colored red and the helical FAD binding domain is colored green. (B) Electrostatic surface potential surrounding the substrate-binding site of *E. coli* photolyase. Note the central cavity leading from the surface of the enzyme to the interior and the surrounding band of positive electrostatic potential (blue areas). Areas of high negative potential are colored red (Park *et al.*, 1995). (See color insert.)

account for this proposes that certain parts of the repair pathway are directly comparable to repair of Pyr<>Pyr by photolyases (Fig. 2B). According to this model, upon binding to substrate the enzyme positions the photoproduct into the active site cavity by dinucleotide flipping (Zhao *et al.*, 1997) and then a thermal reaction converts the photoproduct to an oxetane (azetidine) four-membered ring intermediate, which is not unlike a cyclobutane ring (Kim *et al.*, 1994). The presence of two His residues in the active site appears to facilitate this conversion (Hitomi *et al.*, 2001). From this point the photochemical reaction proceeds in a manner analogous to cyclobutane photolyases. The quantum yield of cyclobutane photolyases is in the range of 0.7–0.9 whereas that of (6-4) photolyases is in the range of 0.05–0.1, suggesting that back electron transfer from unrepaired (6-4) photoproducts or other side reactions competes efficiently with the repair reaction. Clearly more photochemical and photophysical work on (6-4) photolyases is needed to gain better insight into the reaction mechanism.

Overexpression and Purification of Photolyases

Photolyases are present in low concentrations in all organisms convenient for study and therefore overexpression through genetic manipulation is the first and crucial step in obtaining enzyme of sufficient purity, quantity, and concentration for structural, physical, or mechanistic studies. Systems based on *E. coli* and highly controlled inducible promoters carried on plasmid vectors generally produce moderate to high levels of soluble, active protein. A commonly encountered problem is that high-level expression of photolyase leads to progressive selection for mutations that reduce the expression level. The following procedures describe conditions for growing and storing strains that can circumvent these problems. Readers may wish to refer to Schleicher *et al.* (2005) for a description of *E. coli* photolyase overexpression in *B. subtilis*. Using procedures developed for purification of the enzyme from *E. coli* overexpressing cells, these workers report a twofold higher yield compared to the homologous system.

Folate Class Photolyases

Photolyases from *E. coli* and *S. cerevisiae* are the prototypes for this class of enzyme. We have included purification protocols for both enzymes as they demonstrate the different approaches that may be necessary to purify even relatively closely related enzymes.

E. coli Pyr<>Pyr Photolyase

EXPRESSION. Plasmid pMS969 (Tet^R Amp^R Phr$^+$) (Sancar *et al.*, 1983), a derivative of ptac12 (Amann *et al.*, 1983), is propagated in any RecA$^-$ strain carrying *Flac i^Q*. We routinely use UNC523 (*phr::kan uvrA::Tn10*) (Malhotra *et al.*, 1995) and maintain selection for the plasmid through the addition of 50 μg/ml of ampicillin to all growth media. When preparing DNA from this plasmid, take care to prevent cultures from entering stationary phase for prolonged periods as this invariably leads to selection of plasmids expressing *phr* at lower levels. Optimal levels of expression are obtained using freshly transformed cells. Inoculate a single colony into 5 ml of Luria broth (LB) and incubate at 37° with shaking for 8 h. Use 1 ml of this culture to prepare an overnight culture consisting of 100 ml of LB, also incubated with shaking at 37°. The next morning, add 10 ml of the overnight culture to five 2-liter Erlenmeyer flasks, each containing 1 liter of LB. Incubate the cultures with vigorous shaking at 37° until the $A_{600} = 0.6$–0.8 (3.5–4 h), at which time add 2.0 ml of 0.5 M isopropyl-β-D- thiogalactopyranoside (IPTG) to each flask. Continue incubation for another 4 h, at which point the photolyase typically constitutes 10–15% of total cellular protein. Harvest the cells by centrifugation at 5000g for 20 min, wash once with

10 ml of phosphate-buffered saline per liter of starting culture, combine the cells into 50-ml centrifuge tubes to contain no more than the equivalent of 3 liter of cells per tube, centrifuge again, freeze the cell pellet in a dry ice–ethanol bath, and store at −80°.

CELL LYSIS. All subsequent steps are performed at 4°. The fractionation described assumes beginning with pellets from 5 liters of cells. Although we initially purified the enzyme under yellow safelights (Sylvania Gold fluorescent lamps), we have found that purification under normal room intensity illumination is acceptable. Thaw the cell pellets in ice water and then add 10 ml of ice cold lysis buffer [50 mM Tris (pH 8.0), 100 mM NaCl, 1 mM EDTA, 10% (w/v) sucrose, 10 mM β-mercaptoethanol] containing 1 mg/ml lysozyme per liter of starting culture. Gently resuspend the cells and then incubate them on ice overnight or on a rotator at 4°. The next morning, complete the lysis step by sonication (we use 6 × 15-s sonication cycles generated by a 60 Sonic Dismembrator [Fisher Scientific] using the standard tip set at an output of 15). Centrifuge the sonicate at 35,000 rpm for 90 min in a precooled Ti70 rotor using Oak Ridge tubes previously chilled overnight to −20°. Combine all of the supernatants taking care to avoid disturbing the pellet.

AMMONIUM SULFATE PRECIPITATION. Measure the volume of the combined supernatant and place it in a chilled flask with a sterile stir bar in an ice-filled beaker on top of a stir plate in a cold room. Slowly add, with stirring, 0.43 g of enzyme-grade ammonium sulfate per milliliter of supernatant over the course of 45–60 min. Stir an additional 45–60 min, transfer to cooled centrifuge tubes, and spin at 12,000g for 10 min. Discard the supernatant, and dissolve the pellet in the minimum volume of equilibration buffer [50 mM Tris (pH 7.5), 100 mM KCl, 1 mM EDTA, 10% glycerol, 10 mM β-mercaptoethanol] necessary to dissolve most of the material (usually 25–50 ml). Dialyze at 4° against 2 liter equilibration buffer with a buffer change at 4 h. If the dialysate appears cloudy, centrifuge it at 12,000g for 10 min before the next step. Alternatively, the ammonium sulfate pellet can be dissolved in buffer and the salt removed using a Bio-Rad P-6DG column (exclusion size 6000 Da)(Gindt et al., 1999).

BLUE SEPHAROSE CHROMATOGRAPHY. Dilute the dialysate with an equal volume of equilibration buffer and load it onto a 100-ml Blue Sepharose CL-6B column (20 × 2.5 cm) in the same buffer at a flow rate of 0.15 ml/min. Wash with two column volumes of wash buffer (equilibration buffer containing 10 mM ATP) at a flow rate of 0.4–0.5 ml/min and then elute the enzyme with high salt elution buffer [50 mM Tris (pH 7.5), 2 M KCl, 1 mM EDTA, 10% glycerol, 10 mM β-mercaptoethanol] at a flow rate of 0.4–0.5 ml/min, collecting 5-ml fractions. Fractions containing photolyase are discerned easily by their blue color due to the presence of the neutral blue flavin radical (FADH⁰), an oxidized form of the catalytic chromophore FADH⁻ that develops upon cell lysis (Payne *et al.*, 1987). Concentrate the fractions to 5–6 ml using a Centriprep YM-30 concentrator (Amicon).

BIO GEL P-100 CHROMATOGRAPHY. Load the sample onto a 45 × 1.6-cm Bio Gel P-100 column equilibrated with 67 mM KPi buffer (67 mM potassium phosphate [pH 6.8], 1 mM EDTA, 10% glycerol, 10 mM β-mercaptoethanol) and then develop the column with 500 ml of the same buffer. Collect 5-ml fractions throughout and monitor for blue color. Pool all blue fractions.

HYDROXYLAPATITE CHROMATOGRAPHY. Load the blue fractions from the Bio Gel column at a flow rate of 0.2 ml per minute onto a 20-ml (1.6 × 10 cm) hydroxylapatite column equilibrated with 67 mM KPi buffer. Wash with 3–5 ml of the same buffer and then elute at a flow rate of 0.4 ml per minute with a gradient of 67–330 mM KPi buffer (67 mm to 330 mM potassium phosphate [pH 6.8], 1 mM EDTA, 10% glycerol, 10 mM β-mercaptoethanol). Collect 3-ml fractions, pool the blue fractions, dialyze overnight against 200 volumes of EC storage buffer (50 mM Tris [pH 8.0], 50 mM NaCl, 1 mM EDTA, 50% [v/v] glycerol, 10 mM dithiothreitol [DTT]) and then quick-freeze in small aliquots in an ethanol–dry ice bath. Store at $-80°$.

YIELD, PURITY, AND CHROMOPHORE COMPOSITION. The procedure described earlier typically yields 15–25 mg of >98% pure photolyase from 5 liters of culture (~850 mg total protein [Sancar *et al.*, 1984]). The quality of the enzyme may be gauged initially by the color of the preparation, although it is imperative to characterize further the absorbance spectrum and activity as described later. Preparations that appear deep blue are of good quality. Oxidation of the neutral blue flavin radical chromophore to FAD$_{ox}$ yields preparations displaying a green or yellow color depending on the extent of oxidation.

RAPID PURIFICATION METHODS. Photolyase can be obtained in high yield using more rapid methods than the procedure just described. Two of these methods are described later. The enzyme preparations obtained are often of sufficient quality for use as repair reagents, but may not be of sufficient purity for some photophysical experiments.

Three-step purification (Gindt *et al.*, 1999). Five liters of cells are grown, induced, collected, and lysed as described earlier. Following collection of the ammonium sulfate precipitate, suspend the proteins in 10 ml of equilibration buffer. The precipitate will not go into solution completely at this point. Transfer the mix to a dialysis bag and dialyze for at least 4 h with one buffer change against 1 liter of equilibration buffer. The dialysate, which should now be clear, is loaded onto a 50-ml Blue Sepharose column. Wash the column with 300 ml of equilibration buffer and then with 300 ml of wash buffer. Elute remaining bound proteins, including photolyase, by washing the column with high salt elution buffer. Pool the blue fractions and dialyze against 20–50 volumes of 33 mM KPi buffer (33 mM potassium phosphate [pH 6.8], 1 mM EDTA, 10% glycerol, 10 mM β-mercaptoethanol) with one buffer change. Apply the dialysate to a 10-ml hydroxylapatite column equilibrated with 33 mM KPi buffer minus glycerol, wash the column with 75 ml of 33 mM KPi buffer, and develop the column with a 75-ml gradient of 33–330 mM KPi buffer, followed by a 20-ml wash with 330 mM KPi buffer. Run aliquots of the blue fractions on SDS–polyacrylamide gels to assess the purity of the peak and side fractions. This procedure customarily yields up to 10 mg from peak fractions that are $\sim$ 90% pure. Dialyze the enzyme into enzyme storage buffer as described earlier.

One-day purification (Kay *et al.*, 1999). Following growth and collection of cells from 5 liters of culture as described earlier, suspend the cells in 25 ml (per 17.5 g of cells) of buffer A (50 mM HEPES [pH 7.0], 100 mM NaCl, 10% sucrose [w/v], 10 mM DTT), lyse by sonication, and centrifuge as described earlier; apply the supernatant from the centrifugation to a Blue Sepharose CL-6B column (2.5 $\times$ 9 cm). After washing with 5–10 column volumes, develop the column with a gradient of 0.1–2 M KCl in buffer B (50 mM HEPES [pH 7.0], 10% [v/v] glycerol, 10 mM DTT). Combine blue-colored fractions, precipitate with ammonium sulfate (0.43 g/ml), and harvest the precipitate as described earlier. Dissolve the pellet in buffer B containing 50 mM NaCl, apply the solution to a HiPrep 26/10 desalting column (Pharmacia), and elute with the same buffer. Combine blue-colored fractions and place on a heparin Sepharose CL-6B (2.5 $\times$ 8 cm) column, wash with 5 column volumes, and elute with a gradient of 0.1–1 M KCl in buffer B. Pool the blue fractions, concentrate the protein by ultrafiltration using C–30 membranes from Amicon, add glycerol to a final concentration of 50% (v/v), and store at $-70°$. This entire purification procedure can be carried out in 1 day beginning with frozen cells.

S. cerevisiae Pyr<>Pyr Photolyase

Despite the extensive similarities between *S. cerevisiae* and *E. coli* enzymes, both the behavior of the yeast enzyme during purification and the presence of proteases that attack the yeast enzyme but not the bacterial enzyme necessitate a different strategy for purification. In particular, lysis of cells in the presence of ammonium sulfate followed by fractionation on phenyl Sepharose is necessary to inhibit and subsequently remove a protease that cleaves $\sim$2 kDa from the amino terminus of Phr1.

EXPRESSION. Plasmid pCB1241 carries the *S. cerevisiae* gene under the control of the *tac* promoter on a Tet[R] derivative of pUNC09 (Sancar and Smith, 1988). We routinely propagate the plasmid in *E. coli* strain CSR603 (*recA1 uvrA6 phr1*) (Sancar and Rupert, 1978b) containing $F'laci^Q$; how-ever, any RecA⁻ Phr⁻ strain carrying $F'laci^Q$ is suitable. Colonies from freshly transformed strains are checked for the presence of a ~60-kDa band on SDS–10% polyacrylamide gels (Sancar *et al.*, 1987b), and the best overproducers are grown to $A_{600} = 0.6$–0.9 and stored at $-20°$ in 40% glycerol. Stocks are viable for 6–12 months. For overexpression, inoculate a 40-ml culture of LB containing 20 μg/ml of tetracycline with 0.4 ml of the stock culture and incubate at 37° for 12–16 h with shaking. The next morning, inoculate each of four 1-liter cultures with 10 ml of the overnight and continue incubation and shaking until the culture reaches $A_{600} = 0.8$–1.0. Add IPTG to a final concentration of 1 mM and incubate for a further 12–16 h. To harvest the cells, cool them rapidly in an ice water bath, centrifuge at 4° and 5000g for 20 min, wash once in ice-cold lysis buffer [50 mM Tris (pH 7.5), 1 mM EDTA, 10% (w/v) sucrose, 10 mM β-mercaptoethanol, ammonium sulfate at 20% of saturation], and resus-pend the cells in 100 ml of the same buffer. Freeze the suspension in 20-ml aliquots in an ethanol–dry ice bath and store at $-80°$.

CELL LYSIS. All steps are performed at 4° under yellow safe light. Use prechilled rotors, centrifuge tubes, glassware, and buffers throughout. Thaw the cell suspension on ice overnight and then lyse the cells on ice by sonication (ten 15-s passes with a Branson Sonifier Model 350 microtip with a power setting of 70%). Remove cellular debris and unlysed cells by sequential centrifugations: 32,000g for 30 min followed by 120,000g for 1 h. Measure the volume of the supernatant and transfer to a fresh beaker.

AMMONIUM SULFATE PRECIPITATION. Over the next hour slowly add with stirring 230 mg of enzyme-grade ammonium sulfate per milliliter of super-natant. The final concentration of ammonium sulfate is thus 60% of satu-ration. Continue stirring for another hour and then centrifuge the solution at 32,000g for 30 min. Dissolve the pellet in 25–30 ml of YPL buffer A (lysis buffer lacking sucrose) and dialyze overnight against 2 liters of buffer. Centrifuge the dialysate for 30 min at 32,000g to remove insoluble debris.

PHENYL SEPHAROSE CHROMATOGRAPHY. Apply the supernatant to a 25-ml phenyl Sepharose CL-4B column in buffer A. After washing the column with 125 ml of YPL buffer A, elute bound proteins with a 250-ml linear gradient of YPL buffer A to 50% ethylene glycol in YPL buffer B [50 mM Tris (pH 7.5), 1 mM EDTA, 10 mM β-mercaptoethanol]. Collect 4-ml fractions. Photolyase elutes at approximately 17% ethylene glycol and can be identified on SDS–10% polyacrylamide gels as a band migrating with a M_r of 60,000 Da. (Note that the yeast enzyme is not blue because its flavin chromophore remains reduced.) It is often advisable at this stage to confirm the identity of the enzyme using an activity assay, particularly if the level of overexpression is suboptimal, as there is at least one other major protein of similar mass that elutes just before photolyase in the gradient. We routinely use 1 μl of eluate from the suspect peak fraction in the transformation assay described later in this chapter. Alternatively, one can add a clearly visible quantity of previously purified photolyase to selected fractions to serve as a marker for the authentic band. Pool the photolyase-containing fractions and then dialyze overnight against 2 liters of G20 equilibration buffer.

BLUE SEPHAROSE CHROMATOGRAPHY. Apply the dialysate to a 20-ml Blue Sepharose CL-6B column in G20 equilibration buffer [50 mM Tris (pH 7.5), 100 mM KCl, 1 mM EDTA, 20% glycerol, 10 mM β-mercaptoethanol]. Wash the column with 100 ml of wash buffer (G20 equilibration buffer containing 10 mM ATP) and develop it with a 150-ml linear gradient of G20 equilibration buffer to the same buffer containing 1.5 M KCl. Photolyase elutes at approximately 0.3 M KCl and is easily identified by SDS–polyacrylamide gel electrophoresis. Pool the desired fractions and dialyze overnight with one buffer change against 2 liters of G20 equilibration buffer.

The elution profiles of full-length photolyase and the amino-terminal cleaved form, Phr1*, largely overlap on this and subsequent chromatography steps; however, it is possible to obtain from the leading or lagging edges of the elution profiles preparations that are >80% of one form or the other (Sancar *et al.*, 1987b). We have not observed significant differences in the activity of these two forms of the enzyme and usually combine them for subsequent processing.

DNA CELLULOSE CHROMATOGRAPHY. At this point the enzyme preparation usually contains one main contaminating band of about M_r 20 kDa in addition to photolyase. This contaminant is removed by DNA cellulose chromatography. Apply the dialysate to a 10-ml denatured DNA cellulose column in G20 equilibration buffer, wash with 100 ml of the same buffer, and elute photolyase with a 50-ml linear gradient of G20 equilibration buffer to the same buffer containing 0.5 M KCl. Collect 1-ml fractions, identify photolyase-containing fractions by SDS–polyacrylamide gel electrophoresis, pool the fractions, and dialyze against 1 liter equilibration buffer for 4 h.

DEAE CELLULOSE/BLUE SEPHAROSE TANDEM CHROMATOGRAPHY. After DNA cellulose chromatography the enzyme preparation is heavily contaminated with material absorbing at 260 nm, presumably DNA leaching off of the DNA cellulose column. In addition the enzyme is dilute. To remove the UV-absorbing material and concentrate the enzyme, two columns are used, a 3-ml DEAE cellulose column and a 3-ml Blue Sepharose column, both in G20 equilibration buffer. The columns are arranged in tandem so that the effluent from the DEAE column feeds into the Blue Sepharose column. The dialysate is applied to the DEAE column, which is subsequently washed with 20 ml of the same buffer to wash the photolyase through the column, leaving the UV-absorbing material behind. After detaching the DEAE column and allowing the buffer head on the Blue Sepharose column to drain to the top of the resin, the Blue column is washed with a further 15 ml of G20 equilibration buffer and then the photolyase is eluted by the addition of 15 ml of the same buffer containing 1.5 M KCl. Collect 0.5-ml fractions and identify those containing photolyase by SDS–PAGE. Pool the desired fractions and dialyze overnight against 250 ml of YPL storage buffer with one buffer change. Freeze in small aliquots in an ethanol–dry ice bath and store at $-80°$.

YIELD, PURITY, AND CHROMOPHORE COMPOSITION. The procedure described earlier typically yields 2–4 mg of >98% pure photolyase from 4 liter of culture. The lower yield compared to the *E. coli* enzyme reflects both a lower level of expression (~8% of total cellular protein) and a lower yield during the purification procedure (~10%). These disadvantages are offset for some studies by the fact that the flavin chromophore of the *S. cerevisiae* enzyme is more stable to oxidation and the folate chromophore is more tightly bound than is the case for the *E. coli* enzyme. Thus the enzyme as isolated usually contains a full, stoichiometric complement of both chromophores in their catalytically active oxidation states.

Deazaflavin Class Pyr<>Pyr Photolyase

Photolyase from *A. nidulans* is the most extensively characterized of the deazaflavin class of enzymes. The protocol that we use routinely to overexpress and purify the enzyme is similar to that used to purify the yeast and *E. coli* enzymes (Malhotra *et al.*, 1992). The *phr* gene is expressed under the control of a *tac* promoter on plasmid pUNC1993 in *E. coli* strain CSR603 F'*laci*^{Q} using conditions described earlier for the *E. coli* enzyme, except that induction is with 1 mM IPTG and postinduction incubation at 37° continues for 10 h prior to harvesting the cells. These conditions routinely lead to the production of 5% of total cellular protein as the *A. nidulans* photolyase. The fractionation described assumes beginning with 5 liters of cells. After the cells are pelleted by centrifugation, resuspend them in 100 ml of lysis buffer [50 mM Tris (pH 8.0), 100 mM NaCl, 1 mM EDTA, 10% (w/v) sucrose, 10 mM β-mercaptoethanol, containing ammonium sulfate to 20% saturation], quick freeze, and store the cells at

$-80°$. Following the procedures described for the *S. cerevisiae* photolyase, thaw and lyse the cells by sonication, centrifuge to remove cellular debris, concentrate the proteins by the addition of ammonium sulfate to 60% of saturation, and collect the pellet by centrifugation. Suspend the pellet in 70 ml of equilibration buffer (50 m*M* Tris [pH 7.5], 100 m*M* KCl, 1 m*M* EDTA, 20% glycerol, 10 m*M* β-mercaptoethanol), and dialyze overnight against 1 liter of the same buffer with one buffer change.

Blue Sepharose Chromatography. Load the dialysate at a rate of 0.4 ml/ min onto a 200-ml Blue Sepharose CL-6B column equilibrated in equilibration buffer [50 m*M* Tris (pH 7.5), 100 m*M* KCl, 1 m*M* EDTA, 20% glycerol, 10 m*M* β-mercaptoethanol], wash the column with 600 ml of the same buffer, and elute the bound proteins with 400 ml of high salt elution buffer (equilibration buffer containing 1.5 *M* KCl), collecting 5-ml fractions. Locate the fractions containing photolyase by SDS–PAGE, looking for a pronounced $M_r \sim 47$-kDa band against a background of many other bands. Pool the photolyase-containing fractions and dialyze overnight against 1 liter of G20 equilibration buffer with one buffer change.

DNA Cellulose Chromatography. Load the dialysate onto a 25-ml single-strand DNA cellulose column in G20 equilibration buffer, wash the column with 100 ml of the same buffer, and develop the column with a 100-ml gradient of G20 equilibration buffer to the same buffer containing in 1.0 *M* KCl. Collect 2-ml fractions. At this point, fractions containing photolyase usually exhibit a blue-purple color typical of the flavin neutral radical. The presence of photolyase can be confirmed by SDS–PAGE. Pool and concentrate the fractions in a Centriprep C-30 filter to a final volume of 2 ml.

BioGel P-100 Chromatography. Load the enzyme preparation onto a Bio Gel P-100 column (45×1.6 cm) equilibrated with 67 m*M* KPi buffer [67 m*M* potassium phosphate (pH 6.8), 1 m*M* EDTA, 10% glycerol, 10 m*M* β-mercaptoethanol]. Develop the column with 160 ml of the same buffer, collect 5-ml fractions, and pool the photolyase-containing fractions and concentrate by ultrafiltration. Dialyze the enzyme into storage buffer [50 m*M* Tris (pH 7.5), 50 m*M* NaCl, 1 m*M* EDTA, 50% (v/v) glycerol, 10 m*M* DTT], freeze, and store as described earlier.

Yield, Purity, and Chromophore Composition. The final preparation of enzyme is greater than 90% pure as judged by SDS–PAGE, and the final yield is approximately 3%. Thus, beginning with 5 liters of cell culture, one can obtain ~ 5 mg of enzyme upon completion of purification. While enzymes from the deazaflavin class are active when purified from *E. coli* (due to the presence of the flavin chromphore/cofactor), full activity is achieved only by postpurification supplementation with deazaflavin, as described later. This is because *E. coli* does not produce deazaflavin (Malhotra *et al.*, 1992).

(6-4) Photolyases

Unlike the photolyases described earlier, problems with insolubility are often encountered when purifying (6-4) photolyases from *E. coli* over-expressing strains. For this reason, all published procedures involve expressing the photolyase as a fusion protein, which both increases its solubility and speeds purification (Nakajima *et al.*, 1998; Todo *et al.*, 1997; Zhao *et al.*, 1997). This section describes the system used to purify (6-4) photolyase from the cloned *D. melanogaster* cDNA (Zhao *et al.*, 1997).

Purification of MBP-Phr(6-4) Fusion Protein. Plasmid pXZ1997, a derivative of pMal-c2 (New England Biolabs) containing the *D. melanogaster phr(6-4)* cDNA fused in frame to the *malE* gene encoding maltose-binding protein (MBP), is propagated in *E. coli* strain UNC523 (*phr::kan uvrA::Tn10*) selecting for ampicillin resistance. We routinely culture 2 liter of cells in LB to $A_{600} = 0.6$–0.8, induce synthesis of the fusion protein by adding IPTG to 0.3 mM, and then continue incubation of the culture for 6 h prior to harvesting the cells by centrifugation. Prepare the cell extract and purify the fusion protein through a 20-ml amylose column as described by the manufacturer (New England Biolabs) and identify the fractions containing the fusion protein by SDS–PAGE, looking for a $M_r \sim 110$-kDa band. At this point the protein is >90% pure. Further purification can be obtained by applying the eluted material to a 10-ml heparin-agarose column equilibrated with 100 ml of equilibration buffer [50 mM Tris (pH 7.5), 100 mM KCl, 1 mM EDTA, 10% glycerol, 10 mM β-mercaptoethanol]. After washing the column with 100 ml of the same buffer, elute the enzyme using a 50-ml gradient of equilibration buffer to the same buffer containing 0.5 M KCl. Concentrate the enzyme using a Centricon 30 filter apparatus (Amicon). At this point the enzyme can either be dialyzed into *E. coli* photolyase storage buffer, quick frozen, and stored at $-80°$ or be treated with factor Xa protease to remove the MBP moiety.

Removal of MBP. The reaction mixture consists of 50 μg of purified MBP-Phr and 5 μg of factor Xa in a 100-μl reaction in 20 mM Tris (pH 8.0), 100 mM NaCl, and 2 mM CaCl$_2$. Incubate the mixture for 4 h at $22°$ and then load it onto a 2-ml heparin agarose column washed previously with 50 ml of equilibration buffer. After loading the sample, wash with 50 ml of equilibration buffer and then elute the photolyase from the column using a 10-ml gradient of 0.1–0.5 M KCl in equilibration buffer. Concentrate the enzyme, dialyze, and store at $-80°$.

Spectroscopic Properties and Characterization of Purified Photolyases

Photolyases exhibit distinctive absorbance spectra reflecting the presence of a flavin chromophore and a "light-harvesting" second chromophore, which, *in vivo*, are bound in 1:1 stoichiometry with the apoenzyme. In the active holoenzymes, the near UV absorbance peak of the catalytic-reduced flavin chromophore (FADH$^-$, $\lambda_{max} = 366$ nm) is usually obscured by the absorbance spectrum of the second chromophore, which can be either 5,10-methenyltetrahydrofolate ($\lambda_{max} = 377$–415 nm) or 8-hydroxy-7,8-didemethyl-5-deazariboflavin ($\lambda_{max} = 440$ nm).

Folate-Containing Enzymes

Absorbance Spectra. The near UV-visible absorbance spectrum of highly purified *S. cerevisiae* photolyase (Fig. 3A) is the archetype for a folate class photolyase as it exists *in vivo* and in its fully active state *in vitro*. The single peak at 377 nm ($\varepsilon_{377} = 28.1 \times 10^3$) reflects overlapping absorbance peaks of the folate (5,10-methenyltetrahydrofolate, $\lambda_{max} = 385$ nm) and reduced flavin (FADH$^-$, $\lambda_{max} = 366$ nm) chromophores, which are present in 1:1 stoichiometry. However, the yeast enzyme is unusual in that, compared to other folate class enzymes that have been characterized thus far, its flavin chromophore is resistant to oxidation and its folate chromophore remains tightly bound during purification. The near UV-visible absorbance spectrum of *E. coli* photolyase, shown in Fig. 3C, is more typical of a purified folate class enzyme, displaying a 384-nm peak ($\varepsilon_{384} = 29.5 \times 10^3$) due to both MTHF and flavin blue neutral radical (FADH^0), as well as FADH^0 absorption bands at 480, 580, and 625 nm. The radical appears during cell lysis and enzyme purification (Payne *et al.*, 1987). Further oxidation produces enzyme preparations with a distinct bright yellow color due to the presence of FAD$_{ox}$; preparations containing this form display absorbance peaks at 384 and 450 nm only (Fig. 3D). Both the blue neutral radical and the FAD$_{ox}$ forms of the enzyme are proficient in binding their substrates but are catalytically inert until activated either chemically (Jorns *et al.*, 1987a; Payne *et al.*, 1990; Sancar *et al.*, 1987a) or photochemically (Heelis and Sancar, 1986; Payne *et al.*, 1987).

The precise stoichiometry of the folate chromophore can vary enormously between enzyme preparations, leading to large variations in absorbance in the 377- to 415-nm range. For example, *E. coli* photolyase purification schemes typically yield an enzyme that has lost 30–50% of this cofactor (Hamm-Alvarez *et al.*, 1989; Johnson *et al.*, 1988). An enzyme saturated with the folate chromophore can be obtained by supplementing the preparation with additional MTHF (Hamm-Alvarez *et al.*, 1990). The

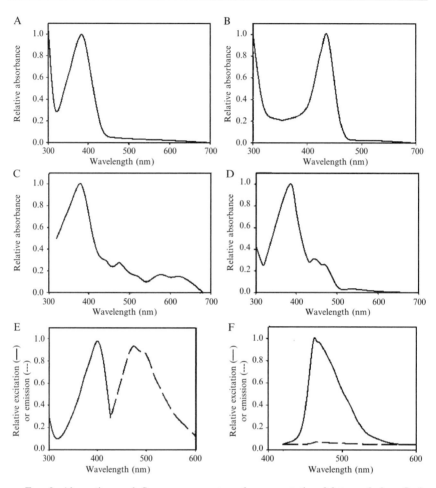

FIG. 3. Absorption and fluoresence spectra of representative folate and deazaflavin photolyases. (A and B) Absorption spectra of the catalytically active forms of *E. coli* (FADH⁻ and MTHF) and *A. nidulans* (FADH⁻ and 8-HDF) photolyases, respectively (Malhotra *et al.*, 1992; Payne and Sancar, 1990). (C) Spectrum of *E. coli* photolyase as it typically appears following purification (Sancar *et al.*, 1987a); the flavin chromophore has been oxidized to the flavin neutral blue radical (FADH⁰) and MTHF is partially depleted. (D) Spectrum of catalytically inert *S. cerevisiae* photolyase following purification and prolonged storage at 4° (Sancar *et al.*, 1987b); the flavin chromophore has been fully oxidized to FAD_{ox} but the enzyme still retains a full complement of the MTHF chromophore. (E) Fluoresence excitation (solid line) and emission (dashed line) spectra of *E. coli* photolyase containing FADH⁻ and MTHF. (F) Fluoresence emission spectrum of *A. nidulans* photolyase (for excitation at 418 nm) containing 8-HDF only (solid line) or 8-HDF and FADH⁻ (dashed line) (Malhotra *et al.*, 1992).

folate chromophore can also be selectively destroyed. Intense light in the 250- to 400-nm range quantitatively converts enzyme-bound MTHF to products that no longer bind to the enzyme or absorb in the near-UV (Heelis *et al.*, 1987; Johnson *et al.*, 1988). When illumination is performed in the presence of reducing agents such as DTT, $FADH^0$ is also converted to the catalytic chromophore $FADH^-$ permitting studies of photolysis by the flavin chromophore only (Payne and Sancar, 1990). Alternatively, folate can be preserved during photoreduction of $FADH^0$, using 630 nm band pass filters (Payne and Sancar, 1990), or dithionite can be used to chemically reduce $FADH^0$ to $FADH^-$ without affecting the folate (Hamm-Alvarez *et al.*, 1989; Jordan and Jorns, 1988; Jorns *et al.*, 1987b). Studies with these forms of the enzyme and their activated counterparts have demonstrated the essential role of reduced flavin in the photolysis reaction (Heelis *et al.*, 1987; Heelis and Sancar, 1986; Jorns *et al.*, 1987b; Sancar *et al.*, 1987a).

Fluoresence Spectra. The primary fluorescent species in most enzyme preparations is enzyme-bound MTHF, which exhibits a single emission peak at 470–480 nm following excitation at 374–384 nm (Fig. 3E) (Sancar, 1992; Sancar *et al.*, 1987b). $FADH^0$ is not fluorescent and, while the catalytic chromophore $FADH^-$ does fluoresce weakly at 505 nm when excited at 366 nm, this is seen only as a shoulder on the MTHF peak in high resolution spectra or when the folate chromophore has been depleted or destroyed. Enzyme containing FAD_{ox} exhibits excitation maxima at 375 and 450 nm and a strong emission peak at 520 nm. All three oxidation states of FAD are strong quenchers of MTHF fluoresence (Heelis *et al.*, 1987; Heelis and Sancar, 1986), providing some of the first evidence for energy transfer from the folate chromophore to the flavin chromophore.

Deazaflavin-Containing Enzymes

Absorbance Spectra. The signature absorption spectrum of this class of enzymes displays a strong peak at 440 nm ($\varepsilon_{440} = 40{,}000$–$50{,}000$) due to the presence of enzyme bound 8-hydroxy-5-deazariboflavin (8-HDF) (Fig. 3B). In high-resolution spectra, the presence of a shoulder at 370 nm as well as peaks at 580 and 625 nm indicates that $FADH^0$ is also present in preparations of purified enzyme. As is the case with folate-containing enzymes, there is variability in the susceptibility of the enzyme-bound flavin chromophore to oxidation (Eker *et al.*, 1990; Kiener *et al.*, 1989). Deazaflavin-containing enzymes purified from natural sources bind 8-HDF with high affinity (Eker *et al.*, 1990; Kiener *et al.*, 1989). However, an enzyme purified from *E. coli* overexpression systems does not contain this chromophore as the bacterium neither produces nor takes up deazaflavin.

A holoenzyme containing both chromophores is obtained by incubating the E-FADH^0 form of the enzyme with 8-HDF (Malhotra *et al.*, 1992), as described later.

Fluoresence Spectra. These enzymes exhibit fluoresence excitation and emission maxima at 420 and 460 nm, respectively, when 8-HDF and either FADH^0 or FADH$^-$ are bound (Fig. 3F) (Kim *et al.*, 1992). However, the intensity of emission is strongly quenched compared to an enzyme containing only 8-HDF (Malhotra *et al.*, 1992), consistent with highly efficient energy transfer from singlet excited state deazaflavin to the flavin chromophore. This supports the model that, like folate, deazaflavin acts as a photoantenna that absorbs light energy and then transfers that energy to the flavin chromophore for catalysis.

Determining Chromophore Stoichiometry

Photolyases bind their chromophores in 1:1 stoichiometry *in vivo*; however, this stoichiometry can be perturbed during enzyme purification, by overexpression in heterologous systems, by physiological or genetic factors, or intentionally by the investigator. The efficiency of photoreactivation is dependent on the stoichiometry of the chromophores, thus for quantitative studies an estimate of the percentage saturation with chromphore is essential. Because absorbance spectra of the enzyme-bound chromophores overlap, it is necessary to denature the enzymes to determine accurately the amount of each chromophore present. We routinely use the following procedure.

Purified enzyme at a concentration of approximately 1 mg/ml is dialyzed from storage buffer into standard buffer (50 mM Tris–HCl [pH 7.5], 100 mM NaCl, 1 mM EDTA, 5% glycerol, 10% sucrose, 1 mM DTT) at 4°. A complete absorbance spectrum is measured over the range from 250 to 700 nm, the enzyme is warmed to 20°, and then SDS is added to a final concentration of 0.8%. At this concentration of SDS the folate, deazaflavin, and flavin chromophores are released from the enzyme. Two hours later the spectrum is recorded again. Because free MTHF breaks down to 10-formyltetrahydrofoloate, which does not absorb at $\lambda > 300$ nm, enzymes from the folate group will display only peaks at 375 and 450 nm, diagnostic of FAD$_{ox}$, in the near-UV/visible range, whereas deazaflavin-containing enzymes will exhibit an additional peak at 418 nm due to liberation of 8-HDF. Dialyze the enzyme solution overnight against standard buffer containing 0.8% SDS and determine the protein concentration from the extinction coefficient of the apoenzyme calculated from the predicted amino acid composition. The molar ratio of the flavin chromphore is determined based on the 450-nm absorbance reading taken prior to dialysis using a value of $\varepsilon_{450}FAD_{ox} = 11.3 \times 10^3$ M^{-1} cm^{-1}. For deazaflavin enzymes, the stoichiometry of 8-HDF is determined using $\varepsilon_{418} = 3.75 \times 10^4$ M^{-1} cm^{-1}. The molar ratio of MTHF is obtained using a value of $\varepsilon_{370-380} = 24,495 \times M^{-1}$ cm^{-1} (Johnson *et al.*, 1988) after subtracting the absorbance contribution of enzyme-bound FADH^0 or FADH$^-$ ($\varepsilon_{380} = 6.0 \times 10^3$ M^{-1} cm^{-1} or $\varepsilon_{384} = 4.6 \times 10^3$ M^{-1} cm^{-1}, respectively [Jorns *et al.*, 1990; Payne and Sancar, 1990]) from the initial readings on native enzyme.

Reconstituting Photolyase from Apoenzyme and Chromophores

The flavin cofactor of the *E. coli* enzyme can be removed by mild denaturation in high salt at low pH (2 *M* KBr, pH 4.0 [Payne *et al.*, 1990]). Conversely, the enzyme can be reconstituted with FAD or its analogs by incubating the enzyme (∼1 mg/ml) on ice for at least 2 h with a twofold molar excess of FAD analog or FAD$_{ox}$ in photolyase storage buffer. Unbound chromophore is removed by passage through a spin column containing Sephadex G-50 fine resin equilibrated with the same buffer (Payne *et al.*, 1990). Subsequently FAD$_{ox}$ must be reduced by chemical or photochemical methods as described earlier to reconstitute active enzyme. This procedure has been extremely useful in studying spectra of the chromophores and reaction requirements for the photolysis reaction. Unfortunately, for reasons that are not presently understood, it is not possible to supplement or reconstitute heterologous photolyases that, when expressed in *E. coli*, contain no detectable flavin (*M. thermoautotrophicum* enzyme) or grossly substoichiometric amounts (1–10%) of the flavin chromophore (*Drosophila melanogaster* enzyme).

Deazaflavin-containing enzymes must also be reconstituted with this chromophore as *E. coli* does not produce deazaflavin. The procedure for reconstitution is straightforward and consists of incubating the enzyme in photolyase storage buffer with a twofold molar excess of the deazaflavin chromophore, followed by removal of free chromophore as described for reconstitution with FAD. Neither denaturation nor reduction after reconstitution is required (Malhotra *et al.*, 1992).

Supplementation with folate requires synthesis of 5,10-methenyltetrahydrofolate. This is accomplished by incubating a 1 *M* solution of folinic acid (5-formyltetrahydrofolate) in 0.01 *N* HCl at 4° for 1–3 days (Hamm-Alvarez *et al.*, 1989; Payne and Sancar, 1990). Conversion to the 5,10-methenyl form is monitored as a quantitative shift in the absorbance peak at 266 to 358 nm. When conversion is complete, MTHF at 2- to 10-fold molar excess is added to photolyase in storage buffer and incubated on ice in the dark for 30 min. Unbound folate (mostly in the form of 10-formylfolate) is removed as described earlier for the flavin chromophore.

Assays for DNA Binding and Repair by Photolyase

Photolyases bind pyrimidine dimers with high specificity and affinity in linear and covalently closed circular DNA, single-stranded DNA, and oligonucleotides as small as oligo(dT$_4$) (Baer and Sancar, 1989; Husain et al., 1987; Jorns et al., 1985; Rupert and To, 1976; Setlow and Bollum, 1968; Sancar et al., 1985). Low-affinity binding can be detected using high concentrations of shorter oligonucleotides, such as T<p>TpT, T<p>Tp, or even the thymine base dimer T<>T (Kim and Sancar, 1991; Liuzzi et al., 1989; Witmer et al., 1989), although binding is not sufficiently stable to be easily detectable for these substrates. In addition, photolyases repair DNA in the presence of up to 10 mM EDTA and function over a wide range of pH and ionic strength, with optimal activity at pH 6–8 and $\mu = 10$–100 mM (Eker et al., 1988, 1990; Sabourin and Ley, 1988; Sancar et al., 1984). These properties have led to the development of assays to detect and quantify photolyase activity under a variety of conditions, including in cell-free extracts. In addition, binding can be studied independently of repair, as the latter step requires light while the former step does not. In all cases, assays are performed under yellow safelights (General Electric or Sylvania gold fluorescent bulbs or equivalent) to prevent uncontrolled photoreactivation. The photolytic step can be carried out under several different conditions that allow the investigator to measure different properties of the enzyme and its reaction. At low fluences the rate of dimer repair is limited by the absorbance cross section of the chromophores and the quantum yield of the photolytic reaction, providing a measure of the efficiency with which an incident photon is used to carry out repair. Flash photolysis, performed by exposing ES complexes to camera flashes of sufficient intensity to repair all substrate in ES complexes, can be used to measure the kinetics of DNA binding and of repair and under steady-state or single turnover conditions; repair is accomplished in 1 ns while a typical camera flash lasts ~1 ms. This section surveys the assays used for measuring DNA binding and repair by photolyases.

DNA-Binding Assays

Nitrocellulose Filter-Binding Assay. This is a modification of the technique originally developed for studying the binding of the Lac repressor to its operator (Riggs and Bourgeois, 1968) and relies on the fact that, under the conditions of the assay, DNA is not retained efficiently on the filters unless it is bound with high affinity by a protein. As modified for photolyase, the substrate is ^{3}H-labeled pBR322 DNA containing an average of 5 pyrimidine dimers per molecule, produced by irradiating the plasmid DNA with 254 nm of light; the average dimer number is quantified by transformation assay. A rule of thumb is 100 J/m^2 of 254 nm radiation produces one photoproduct [0.8 Pyr<>Pyr and 0.2 (6-4) photoproduct] in 1 kbp of DNA. Photolyase (1–70 nM) is incubated with 1.25–2.5 nM substrate in 110 μl of assay buffer [50 mM Tris (pH 7.5), 1 mM EDTA, 10 mM β-mercaptoethanol, 100 μg/ml bovine serum albumin (BSA), 100 mM NaCl] at 20° for 1 h and then filtered in 50-μl aliquots through nitrocellulose filers (Schleicher and Schull BA85, 24 mm) previously equilibrated with assay buffer. The filters are washed, dried, and the counts retained are determined by scintillation counting (Sancar et al., 1985).

The filter-binding assay has been used for qualitative comparison of the substrate-binding activity under different conditions of pH and ionic strength and with substrates of varied molecular structure (Madden and Werbin, 1974; Sancar et al., 1985). With modifications, the assay has also been used to obtain accurate quantitative data on photolyase-binding affinity and repair (Sancar et al., 1987c). Ideally the assay should display a strict correlation between the number of photolyase-Pyr<>Pyr complexes (ES) present in solution and the number of DNA molecules retained on the nitrocellulose filter. However, in practice, molecules with a single photolyase-Pyr<>Pyr complex may not be retained efficiently (Sancar et al., 1985). A second complication is that in irradiated plasmid DNA the number of pyrimidine dimers is not constant on each substrate molecule, but rather varies according to Poisson distribution. Thus it is often necessary to determine the number of photolyase molecules required to retain a single substrate molecule as well as to correct for the effects of the Poisson distribution of Pyr<>Pyr. Although beyond the scope of this work, this issue is dealt with extensively in Sancar et al. (1985, 1987c). An example of a quantitative binding isotherm obtained using this analysis is shown in Fig. 4A.

Electrophoretic Mobility Shift Assays (EMSA). EMSAs have been used extensively in conjunction with oligonucleotides of defined sequence to study the interaction between photolyases and their substrates. This method has proven ideal for defining the roles that dimer composition and sequence context play in substrate recognition independent of the effect of these parameters on the efficiency of photolysis. Although concern has been expressed about the validity of equilibrium association constants determined using EMSA, the binding constants for interaction between yeast and *E. coli* photolyases and T<>T embedded in oligonucleotides of 18 bp or longer are in close agreement with those obtained by

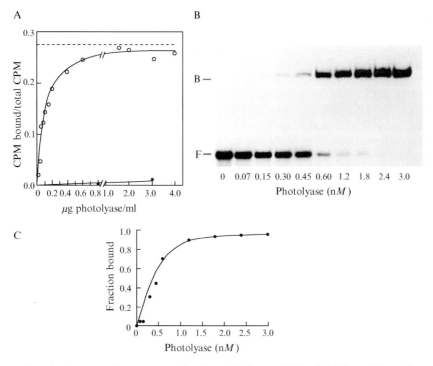

A

B

C

Fig. 4. Assays used to measure photolyase binding to DNA. (A) Nitrocellulose filter-binding assay. ^{3}H-labeled pBR322 DNA irradiated to produce an average of 0.85 dimers/genome was incubated at 3.2 mg/ml with increasing concentrations of photolyase in reaction buffer in the absence of photoreactivating light. ES complexes were captured on nitrocellulose filters and quantified by scintillation counting of labeled DNA. Filled symbols are data from control reactions with nonirradiated DNA. The dashed line indicates the maximum binding expected for a population of plasmid molecules with 0.85 dimers/genome and a ES filter-binding efficiency of 0.34 (Sancar *et al.*, 1985). (B and C) Electrophoretic mobility shift assay (EMSA). 0.1 n*M* ^{32}P-labeled duplex DNA containing T(6-4)C at a defined site was incubated with increasing concentrations of photolyase prior to separation by EMSA. (B) Bound DNA, F, free DNA. (C) The binding isotherm derived from data shown in B (Zhao *et al.*, 1997).

other methods (Baer and Sancar, 1993; Husain and Sancar, 1987; Sancar, 1990), and thus one can with confidence obtain quantitative information about the thermodynamics of photolyase–substrate interaction using this technique.

Pyr<>Pyr and (6-4) photoproduct-containing oligonucleotide substrates for EMSA have been constructed by direct irradiation of short oligodeoxynucleotides (Banerjee *et al.*, 1988; Svoboda *et al.*, 1993), which, following HPLC purification, are incorporated into longer double-stranded oligonucleotides by hybridization and ligation. Another successful strategy has been the synthesis of building blocks for the site-specific incorporation of DNA photoproducts by automated oligonucleotide synthesis (Iwai et al., 1999; Taylor et al., 1987). ES complexes are formed by incubating ^{32}P-labeled substrate and enzyme in assay buffer supplemented with 6% (v/v) glycerol for 20–45 min to allow complex formation and then complexes are separated from nonbound substrate in 5–6% polyacrylamide gels run at 35 mA; it is important to prerun the gels at the same amperage for at least 20 min prior to loading the sample. Total running time will depend on the length of the substrate and the size of the gel. We find that the free double-stranded 43-bp substrate separates well from that bound by yeast photolyase during a 90-min run at 35 mA in 20-cm-long 6% gels (Baer and Sancar, 1993). The precise composition of the running buffer may need to be varied to optimize complex stability; however, in most cases gels prepared and run in 1× TBE (90 mM Tris, 90 mM boric acid, 2.5 mM EDTA) work reasonably well. Free and bound substrate are then quantified using a β scanner or phosphorimager. An EMSA and derived binding isotherm used to obtain the equilibrium-binding constant for the D. melanogaster (6-4) photolyase:(6-4) photoproduct interaction is shown in Fig. 4B and C.

The EMSA as described earlier has been used in both kinetic- and equilibrium-binding studies of photolyase–substrate interactions (Baer and Sancar, 1989; Husain and Sancar, 1987; Li and Sancar, 1991; Zhao et al., 1997). In addition to direct measurement of enzyme–substrate interactions, this assay can also be used to indirectly measure binding through a competition assay. For example, inclusion of increasing amounts of nondamaged DNA with a fixed amount of photolyase and dimer substrate quantitatively decreases the number of radiolabeled complexes seen in the EMSA, allowing determination of the nonspecific binding constant K_{NS} for nonsubstrate DNA. Because photolyases show at least a 10^5-fold greater affinity for their substrates than for undamaged DNA (Sancar, 1990, 2003; Zhao et al., 1997), the EMSA can also be used in determining the quantum yield for the photolytic reaction; in this case repair of substrate leads to release of DNA from the enzyme (Payne and Sancar, 1990).

DNA Footprinting. Interactions between photolyases and DNA have been characterized at nucleotide resolution using DNA footprinting techniques. ^{32}P-end-labeled oligonucleotide substrates containing lesions at defined locations are incubated with concentrations of the appropriate photolyases sufficient to bind >80% of substrate under nonphotoreactivating conditions. Two types of approaches are used, protection and interference. In protection experiments, ES complexes are formed and then exposed to reagents such as DNase I or methidiumpropyl-EDTA-Fe(II), which attack the phosphodiester backbone, or to dimethyl sulfate, which methylates purine bases preferentially. Regions bound by photolyase are protected from attack, and thus when the DNAbackbone is cleaved at the site of modification and the strands are separated on a denaturing polyacrylaminde gel and analyzed by autoradiography, the area protected by the enzyme will appear blank compared to DNA that was treated in the absence of the enzyme. In interference experiments the oligonucleotide substrate is first treated with modifying agents such as dimethyl sulfate or ethylnitrosourea, which ethylates phosphate oxygens, and then the DNA is incubated with photolyase and the free and bound DNA are separated on a nondenaturing polyacrylamide gel. Modification at sites normally contacted by photolyase will inhibit binding and these molecules will migrate as free DNA. Free and bound DNA are excised from the gel and treated with piperidine to break the DNA strand at the site of modification, and the products are resolved on a 12% denaturing polyacrylamide gel and autoradiographed. Sites important for binding appear as bands of increased intensity in the "free" DNA lane and depleted in the "bound" lane. Readers interested in using footprinting techniques with photolyases should consult Husain *et al.* (1987) and Baer and Sancar (1989).

Repair Assays

Transformation Assay. Transformation of *Hemophilus influenza* with UV-irradiated DNA was initially developed to study photoreactivation (Rupert *et al.*, 1958) by an enzyme in crude cell extracts. More recently the sensitivity of this assay has been improved through the use of *E. coli* strain CSR603 (*phr1 recA1 uvrA6*), which is completely deficient in repair Pyr<>Pyr, in conjunction with plasmid pBR322 (Sancar *et al.*, 1984, 1987c). This assay is useful both for kinetic and equilibrium measurements of photolyase binding and repair and also for quantifying the amount of substrate used in other methods described earlier and later in this chapter.

Substrate DNA is prepared by irradiating a stirring solution of plasmid DNA at 20 μg/ml in 10 mM Tris (pH 7.4), 10 mM NaCl, and 1 mM EDTA. A fluence of 120–150 J/m^2 from a germicidal lamp is sufficient to introduce 4–6 Pyr<>Pyr per molecule. The precise number of lethal lesions per molecule is determined by transforming chemically competent CSR603 with 200 ng of the irradiated DNA and comparing the number of tetracycline-resistant transformants to that obtained with unirradiated plasmid DNA at the same concentration using the relationship $-\ln(S/S_0) =$ photoproduct/molecule, where S_0 is the number of colonies obtained following transformation with unirradiated control DNA and S is the number of colonies obtained with irradiated DNA. For repair assays the irradiated DNA, or unirradiated control, is incubated with photolyase in assay buffer [e.g., 50 mM Tris (pH 7.5), 1 mM EDTA, 10 mM β-mercaptoethanol, 100 μg/ml BSA, 100 mM NaCl for *E. coli* photolyase] and then the mixture is exposed to photoreactivating light filtered through glass (a petri dish lid, glass plate, or test tube wall is sufficient). Depending on the experiment, photoreactivation may be carried out under continuous illumination supplied by a black light (GE or Sylvania BLB, $\lambda_{max} = 365_{nm}$) at either high or low fluence rates or by conventional camera flash units for single turnover experiments. For example, continuous illumination at a high fluence rate and with limiting enzyme can be used to measure enzyme turnover number, as shown in Fig. 5A (Sancar *et al.*, 1984, 1987b), while single flash experiments provide a direct measure of the number of photolyase molecules active in both binding and photolysis (Sancar *et al.*, 1987c). Following photoreactivation, 10–20 μl of the mixture (100–200 ng of DNA) is used to transform CSR603 to tetracycline resistance. After 24 h of growth at 37° in the dark, the number of dimers per plasmid molecule is calculated as described earlier except that S is the number of colonies obtained with photoreactivated DNA.

Absorption. Unlike pyrimidine monomers, Pyr<>Pyr do not absorb significantly at 260 nm due to loss of the C5-C6 double bond. This fact has been exploited in the development of spectrophotometric assays for repair of Pyr<>Pyr by photolyases. The most extensively used substrate is oligo (dT$_{18}$). The substrate is prepared by irradiating a 250-μl solution of oligo (dT$_{18}$)[60 μM in 10 mM Tris (pH 7.2), 0.5 mM EDTA] at $\lambda = 254$ nm in a quartz cuvette (Jorns *et al.*, 1985) or by acetone photosensitized irradiation at 320 nm (Kim and Sancar, 1991). Shorter oligonucleotides can also be used; however, the kinetics and final extent of repair decline for oligonucleotides shorter than oligo(dT$_8$). For preparing substrate with 254 nm radiation, a 1-ml cuvette is laid on its side and placed on a rotating platform 5 cm below two Sylvania G15T8 bulbs. Irradiation for approximately 6 min converts 35% of thymine monomers to dimers, calculated using an extinction coefficient $\varepsilon_{260} = 8.3 \times 10^3$ M^{-1} cm^{-1} for thymine monomers in oligonucleotides. This stock solution is used to prepare reactions containing 2.6 μM irradiated oligo(dT$_{18}$) in assay buffer from which β-mercaptoethanol has been omitted. Photolyase is added over a concentration range of 30–200 nM, incubated in the dark for 5 min, and then exposed in a stoppered quartz cuvette to light from two black lights (Sylvania or G.E. BLB) at a distance of 5 cm. Dimer repair displays linear kinetics until approximately 90% of dimers are repaired (usually 10–50 min) and is

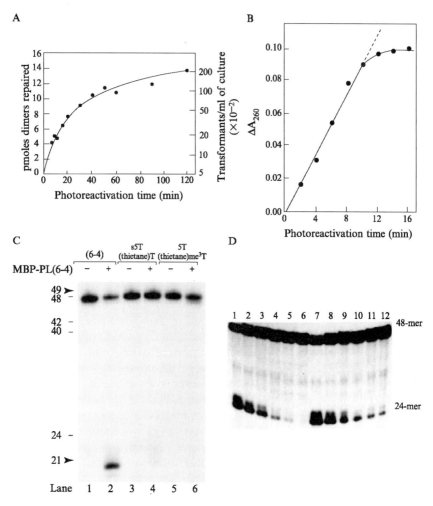

FIG. 5. Methods used to quantitate repair of dimers of (6-4) photoproducts by photolyases. (A) Transformation assay. One picomole of purified yeast photolyase was mixed with 16.8 pmol of pyrimidine dimers in pBR322 DNA (4.5 dimers/molecule) in 600 μl of assay buffer and the mixture was exposed to continuous illumination from a black lamp (10 J/m^2/s at 365 nm). Samples were taken at the indicated times and used for the transformation assay. The initial slope of the curve yields the turnover number of the enzyme (Sancar et al., 1987b). (B) Spectrophotometric (absorbance) assay. Photolyase-mediated repair of pyrimidine dimers in UV-irradiated oligo(dT)$_{18}$ monitored by increased absorbance as a function of photoreactivation time (Jorns et al., 1985). (C) Restriction site restoration assay. Repair of photoproducts monitored by restriction enzyme site restoration. 49-mers with an MseI site containing a (6-4) photoproduct, a s5T(thietane)T, or a 5T(thietane)me^{3}T in place of TT in the recognition sequence were mixed at 0.1 nM concentration with 0.75 nM

calculated based on the increase in absorbance at 260 nm (Fig. 5B). This assay has also been used to explore the influence of base composition and pentose moiety on repair, in this case using poly(dU), poly(rU), UpU, oligo $(dC)_{12-18}$, and CpC nucleotides prepared by acetone photosensitized irradiation (Kim and Sancar, 1991). Thus far it has not proven practical to use this assay to follow repair of (6-4) photoproduct due to difficulty in obtaining adequate amounts of pure substrate (Smith and Taylor, 1993).

Restriction Site Restoration. Digestion of DNA by type II restriction enzymes is often blocked by the presence of pyrimidine dimers or (6-4) photoproducts within the recognition sequence of the enzyme. In addition, despite cleavage of the phosphodiester backbone between two pyrimidines in a dimer, the DNA ends will not separate because the covalent bonds linking the pyrimidine bases will continue to link the ends via the bases. This fact has been exploited in the development of a sensitive assay for the repair of these photoproducts by photolyases. A ^{32}P-labeled oligonucleotide is constructed using the methods described earlier such that it contains a T<>T or T(6-4)T at the TT sequence in a unique *Mse*I recognition site TTAA . *Mse*I normally incises between the two T's but the presence of either lesion blocks incision or end separation (Kim *et al.*, 1994; Smith and Taylor, 1993). This defined substrate is incubated for 30 min at 20° at a concentration of 100 p*M* with 100–750 p*M* photolyase in 50 μl of assay buffer and then exposed to photoreactivating light. (Exposure to 365 nm of radiation from a black light at a fluence rate of 2 mW/cm^2 for 45 min repairs most or all damage, whereas exposure to lower fluence rates and various wavelengths can be used for kinetic and action spectrum studies.) The DNA is then extracted with phenol, ethanol precipitated, dissolved in *Mse*I digestion buffer, and digested with 2 units of *Mse*I at 37° for 90 min. The digested DNA is analyzed on a 12% DNA sequencing gel, and the fraction of dimers repaired is determined by band excision and Cerenkov counting, phosphorimaging, or β scanner (Kim *et al.*, 1994; Zhao *et al.*, 1997). Typical results of this assay are shown in Fig. 5C.

D. melanogaster (6-4) photolyase, incubated in the dark for 20 min, and then exposed to 366 nm light to effect photoreactivation. Following digestion with *Mse*I the DNA was analyzed on a 12% polyacrylamide gel. *Mse*I cleavage produces a single labeled fragment 21 bp in length (Zhao *et al.*, 1997). (D) T4 endonuclease–sensitive site assay. A 48-bp-long oligonucleotide (1.25 n*M*) containing a pyrimidine dimer at a defined position was mixed with 12 μ*M* photolyase and exposed to various fluences of 405 nm light. Dimer repair was quantified by incubation with T4 Endo V Den V. Incision at the dimer produces the 24-mer band. Lanes 1 and 7, no photoreactivation; lanes 2–5 and 8–12, duplicate samples with increasing fluences of photoreactivating light (Payne and Sancar, 1990).

Enzyme-Sensitive Site Assay. T4 endonuclease V (T4 endoV) and *M. luteus* UV pyrimidine dimer glycosylase/endonuclease bind pyrimidine dimers in double-stranded DNA and incise both the glycosidic bond linking the 5' base in the dimer to its sugar and the phosphodiester bond between the two nucleotides in the dimer-containing strand. This generates a nick in the damaged strand. These enzymes have been used in a wide variety of assays to quantify dimer repair, including repair by photolyase. Typically, covalently closed circular UV-irradiated plasmid DNA (e.g., 70 fmol of covalently closed circular UV-irradiated pBR322 DNA containing 1.7 dimers per molecule) is incubated first with photolyase and either kept in the dark or exposed to photoreactivating light, phenol extracted, and treated with T4 endoV in 20 μl of 10 mM Tris (pH 7.4), 10 mM NaCl, and 1 mM EDTA at 37° for 1 h. Incision by the nuclease converts the DNA to the open circular form. These two forms are separated readily by electrophoresis in agarose gels, and thus repair can be measured by comparing the amount of uncleaved UV-irradiated plasmid DNA (the P_0 class of the Poisson distribution) in T4 endoV-digested samples taken pre- and post–treatment with photolyase.

The usefulness of the T4 endoV assay as compared to the restriction site restoration assay is that dimers can be detected in any sequence context in the former whereas only dimers in AATT are detected in the latter. However, the plasmid-based assay is by its nature limited by the Poisson distribution of dimers in the plasmid substrate and the fact that the size of the P_0 class decreases quickly, both of which can limit the sensitivity of the assay. A more sensitive assay has been developed using a ^{32}P-labeled oligonucleotide containing a pyrimidine dimer at a defined location (Payne and Sancar, 1990). The oligonucleotide substrates are constructed using either of the strategies described earlier for the preparation of EMSA substrates (Banerjee *et al.*, 1988; Svoboda *et al.*, 1993; Taylor *et al.*, 1987), exposed to photolyase and photoreactivating light in assay buffer, and then extracted with phenol and ether and ethanol precipitated prior to incubation with T4 endo V. This latter series of steps is essential as any remaining photolyase will bind to any unrepaired dimers and strongly inhibit binding and incision by the endonuclease. The precipitated DNA is dissolved in 12 μl of 10 mM Tris (pH 7.4), 10 mM NaCl, and 1 mM EDTA and then 8 μl of the following mixture is added: 2.5 mM DTT, 250 μg/ml BSA, 125 mM Tris-HCl, pH 8.0, 250 mM NaCl, 2.5 mM EDTA, and 5 units T4 endo V. Following incubation at 37° for 30 min the samples are lyophilized, dissolved in 10 μl of formamide containing 0.05% bromphenol blue and 0.05% xylene cyanol, and analyzed on a 12 % sequencing gel in parallel with a nonphotoreactivated sample and an undamaged control. Incision by T4 endo V will result in the appearance of one or more new bands, depending on the location of the ^{32}P label and the dimer, and photoreactivation will reduce the intensity of the bands generated by T4 endo V cleavage, as seen in Fig. 5D. These differences are quantified easily by band excision and Cerenkov counting, phosphorimaging, or using a β scanner.

Photophysical Methods. In addition to the biochemical assays described here, the binding of photolyase to its substrate and the repair of Pyr<>Pyr have been analyzed using both steady-state and time-resolved fluoresence spectroscopy and ultrafast fluorescence and absorption spectroscopy, which make it possible to capture the reaction intermediates in the subnanosecond time frame (Heelis *et al.*, 1990; Kao *et al.*, 2005; Langenbacher *et al.*, 1997; MacFarlane and Stanley, 2003; Okamura *et al.*, 1991). The basis of ultrafast spectroscopy is to initiate the reaction with a laser pulse of fs–ps duration (the pump) and then, after a short time interval, to deliver a second pulse (the probe) that is used to measure the appearance and disappearance of reaction intermediates by their specific spectroscopic properties. Further details of these assays are outside of the scope of this chapter but may be obtained by consulting the works cited previously.

Acknowledgments

We dedicate this chapter to Dr. Claud S. Rupert for his discovery of and pioneering work on photolyase and for laying the foundation of the field of DNA repair enzymology. This work is supported by National Institutes of Health Grants GM31258 (GS) and GM31082 (AS).

References

Adams, M. D., Kerlavage, A. R., Fleischmann, R. D., Fuldner, R. A., Bult, C. J., Lee, N. H., Kirkness, E. F., Weinstock, K. G., Gocayne, J. D., White, O., *et al.* (1995). Initial assessment of human gene diversity and expression patterns based upon 83 million nucleotides of cDNA sequence. *Nature* **377,** 3–174.

Ahmad, M., and Cashmore, A. R. (1993). HY4 gene of *A. thaliana* encodes a protein with characteristics of a blue-light photoreceptor. *Nature* **366,** 162–166.

Amann, E., Brosius, J., and Ptashne, M. (1983). Vectors bearing a hybrid *trp-lac* promoter useful for regulated expression of cloned genes in *Escherichia coli. Gene* **25,** 167–178.

Baer, M., and Sancar, G. B. (1989). Photolyases from *Saccharomyces cerevisiae* and *Escherichia coli* recognize common binding determinants in DNA containing pyrimidine dimers. *Mol. Cell. Biol.* **9,** 4777–4788.

Baer, M. E., and Sancar, G. B. (1993). The role of conserved amino acids in substrate binding and discrimination by photolyase. *J. Biol. Chem.* **268,** 16717–16724.

Banerjee, S. K., Christensen, R. B., Lawrence, C. W., and LeClerc, J. E. (1988). Frequency and spectrum of mutations produced by a single *cis, syn* thymine-thymine cyclobutane dimer in a single-stranded vector. *Proc. Natl. Acad. Sci. USA* **85,** 8141–8145.

Brautigam, C. A., Smith, B. S., Ma, Z., Palnitkar, M., Tomchick, D. R., Machius, M., and Deisenhofer, J. (2004). Structure of the photolyase-like domain of cryptochrome 1 from *Arabidopsis thaliana*. *Proc. Natl. Acad. Sci. USA* **101**, 12142–12147.

Brudler, R., Hitomi, K., Daiyasu, H., Toh, H., Kucho, K., Ishiura, M., Kanehisa, M., Roberts, V. A., Todo, T., Tainer, J. A., and Getzoff, E. D. (2003). Identification of a new cryptochrome class: Structure, function, and evolution. *Mol. Cell* **11**, 59–67.

Cashmore, A. R. (2003). Cryptochromes: Enabling plants and animals to determine circadian time. *Cell* **114**, 537–543.

Eker, A. P. M., Hessels, J. K. C., and van de Velde, J. (1988). Photoreactivating enzyme from the green alga *Scenedesmus acutus*: Evidence for the presence of two different flavin chromophores. *Biochemistry* **27**, 1758–1765.

Eker, A. P. M., Kooiman, P., Hessels, J. K. C., and Yasui, A. (1990). DNA photoreactivating enzyme from the Cyanobacterium *Anacystis nidulans*. *J. Biol. Chem.* **265**, 8009–8015.

Gindt, Y. M., Vollenbroek, E., Westphal, K., Sackett, H., Sancar, A., and Babcock, G. T. (1999). Origin of the transient electron paramagnetic resonance signals in DNA photolyase. *Biochemistry* **38**, 3857–3866.

Hamm-Alvarez, S., Sancar, A., and Rajagopalan, K. V. (1989). Role of enzyme-bound 5, 10-methenyltetrahydropteroylpolyglutamate in catalysis by *Escherichia coli* DNA photolyase. *J. Biol. Chem.* **264**, 9649–9656.

Hamm-Alvarez, S., Sancar, A., and Rajagopalan, K. V. (1990). The folate cofactor of *Escherichia coli* DNA photolyase acts catalytically. *J. Biol. Chem.* **265**, 18656–18662.

Harm, W., Harm, H., and Rupert, C. S. (1968). Analysis of photoenzymatic repair of UV lesions in DNA by single light flashes. II. *In vivo* studies with *Escherichia coli* and bacteriophage. *Mutat. Res.* **6**, 371–385.

Heelis, P. F., Okamura, T., and Sancar, A. (1990). Excited-state properties of *Escherichia coli* DNA photolyase in the picosecond to millisecond time scale. *Biochemistry* **29**, 5694–5698.

Heelis, P. F., Payne, G., and Sancar, A. (1987). Photochemical properties of *Escherichia coli* DNA photolyase: Selective photodecomposition of the second chromophore. *Biochemistry* **26**, 4634–4640.

Heelis, P. F., and Sancar, A. (1986). Photochemical properties of *Escherichia coli* DNA photolyase: A flash photolysis study. *Biochemistry* **25**, 8163–8166.

Hitomi, K., Nakamura, H., Kim, S. T., Mizukoshi, T., Ishikawa, T., Iwai, S., and Todo, T. (2001). Role of two histidines in the (6-4) photolyase reaction. *J. Biol. Chem.* **276**, 10103–10109.

Hsu, D. S., Zhao, X., Zhao, S., Kazantsev, A., Wang, R. P., Todo, T., and Sancar, A. (1996). Putative human blue-light photoreceptors hCRY1 and hCRY2 are flavoproteins. *Biochemistry* **35**, 13871–13877.

Husain, I., and Sancar, A. (1987). Binding of E. coli DNA photolyase to a defined substrate containing a single T<>T dimer. *Nucleic Acids Res.* **15**, 1109–1120.

Husain, I., Sancar, G. B., Holbrook, S. R., and Sancar, A. (1987). Mechanism of damage recognition by *Escherichia coli* DNA photolyase. *J. Biol. Chem.* **262**, 13188–13197.

Iwai, S., Mizukoshi, T., Hitomi, K., and Todo, T. (1999). Chemical synthesis of oligonucleotides containing the (6-4) photoproduct at the thymine-cytosine site and its repair by (6-4) photolyase. *Nucleosides Nucleotides* **18**, 1325–1327.

Johnson, J. L., Hamm-Alvarez, S., Payne, G., Sancar, G. B., Rajagopalan, K. V., and Sancar, A. (1988). Identification of the second chromophore of *Escherichia coli* and yeast DNA photolyases as 5,10-methenyltetrahydrofolate. *Proc. Natl. Acad. Sci. USA* **85**, 2046–2050.

Jordan, S. P., and Jorns, M. S. (1988). Evidence for a singlet intermediate in catalysis by *Escherichia coli* DNA photolyase and evaluation of substrate binding determinants. *Biochemistry* **27**, 8915–8923.

Jorns, M. S., Sancar, G. B., and Sancar, A. (1985). Identification of oligothymidylates as new simple substrates for *Escherichia coli* DNA photolyase and their use in a rapid spectrophotometric enzyme assay. *Biochemistry* **24,** 1856–1861.

Jorns, M. S., Baldwin, E. T., Sancar, G. B., and Sancar, A. (1987a). Action mechanism of *Escherichia coli* DNA photolyase. II. Role of the chromophores in catalysis. *J. Biol. Chem.* **262,** 486–491.

Jorns, M. S., Wang, B., and Jordan, S. P. (1987b). DNA repair catalyzed by *Escherichia coli* DNA photolyase containing only reduced flavin: Elimination of the enzyme's second chromophore by reduction with sodium borohydride. *Biochemistry* **26,** 6810–6816.

Jorns, M. S., Wang, B., Jordan, S. P., and Chanderkar, L. P. (1990). Chromophore function and interaction in *Escherichia coli* DNA photolyase: Reconstitution of the apoenzyme with pterin and/or flavin derivatives. *Biochemistry* **29,** 552–561.

Kao, Y. T., Saxena, C., Wang, L., Sancar, A., and Zhong, D. (2005). Direct observation of thymine dimer repair by DNA photolyase. *Proc. Natl. Acad. Sci. USA* **102,** 14724–14728.

Kavaklı, I. H., and Sancar, A. (2004). Analysis of the role of intraprotein electron transfer in photoreactivation by DNA photolyase *in vivo*. *Biochemistry* **43,** 15103–15110.

Kay, C. W., Feicht, R., Schulz, K., Sadewater, P., Sancar, A., Bacher, A., Mobius, K., Richter, G., and Weber, S. (1999). EPR, ENDOR, and TRIPLE resonance spectroscopy on the neutral flavin radical in *Escherichia coli* DNA photolyase. *Biochemistry* **38,** 16740–16748.

Kiener, A., Husain, I., Sancar, A., and Walsh, C. T. (1989). Purification and properties of *Methanobacterium thermoautotrophicum* DNA photolyase. *J. Biol. Chem.* **264,** 13880–13887.

Kim, S. T., Heelis, P. F., and Sancar, A. (1992). Energy transfer (deazaflavin–>FADH2) and electron transfer (FADH2–>T <> T) kinetics in *Anacystis nidulans* photolyase. *Biochemistry* **31,** 11244–11248.

Kim, S.-T., Malhotra, K., Smith, C. A., Taylor, J.-S., and Sancar, A. (1994). Characterization of (6-4) photoproduct DNA photolyase. *J. Biol. Chem.* **269,** 8535–8540.

Kim, S.-T, and Sancar, A. (1991). Effect of base, pentose, and phosphodiester backbone structures on binding and repair of pyrimidine dimers by *Escherichia coli* DNA photolyase. *Biochemistry* **30,** 8623–8630.

Kim, S.-T., Sancar, A., Essenmacher, C., and Babcock, G. T. (1993). Time-resolved EPR studies with DNA photolyase: Excited-state FADH^0 abstracts an electron from Trp-306 to generate FADH$^-$, the catalytically active form of the cofactor. *Proc. Natl. Acad. Sci. USA* **90,** 8023–8027.

Kleine, T., Lockhart, P., and Batschauer, A. (2003). An Arabidopsis protein closely related to *Synechocystis* cryptochrome is targeted to organelles. *Plant J.* **35,** 93–103.

Komori, H., Masui, R., Kuramitsu, S., Yokoyama, S., Shibata, T., Inoue, Y., and Miki, K. (2001). Crystal structure of thermostable DNA photolyase: Pyrimidine-dimer recognition mechanism. *Proc. Natl. Acad. Sci. USA* **98,** 13560–13565.

Langenbacher, T., Zhao, X., Bieser, G., Heelis, P. F., Sancar, A., and Michel-Beyerle, M. E. (1997). Substrate and temperature dependence of DNA photolyase repair activity examined with ultrafast spectroscopy. *J. Am. Chem. Soc.* **119,** 10532–10536.

Li, Y. F., and Sancar, A. (1990). Active site of *Escherichia coli* DNA photolyase: Mutations at Trp277 alter the selectivity of enzyme without affecting the quantum yield of photorepair. *Biochemistry* **29,** 5698–5706.

Li, Y. F., and Sancar, A. (1991). Cloning, sequencing, expression and characterization of DNA photolyase from *Salmonella typhimurium*. *Nucleic Acids Res.* **19,** 4885–4890.

Lin, C., Robertson, D. E., Ahmad, M., Raibekas, A. A., Jorns, M. S., Dutton, P. L., and Cashmore, A. R. (1995). Association of flavin adenine dinucleotide with the *Arabidopsis* blue light receptor CRY1. *Science* **269,** 968–970.

Liuzzi, M., Weinfeld, M., and Paterson, M. C. (1989). Enzymatic analysis of isometric trithymidylates containing ultraviolet light-induced cyclobutane pyrimidine dimers. *J. Biol. Chem.* **264**, 6355–6363.

MacFarlane, A. W., and Stanley, R. J. (2003). Cis-syn thymidine dimer repair by DNA photolyase in real time. *Biochemistry* **42**, 8558–8568.

Madden, J. J., and Werbin, H. (1974). Use of membrane binding technique to study the kinetics of yeast deoxyribonucleic acid photolyase reactions: Formation of enzyme-substrate complexes in the dark and their photolysis. *Biochemistry* **13**, 2149–2154.

Malhotra, K., Kim, S.-T., Batschauer, A., Dawut, L., and Sancar, A. (1995). Putative blue-light photoreceptors from *Arabidopsis thaliana* and *Sinapis alba* with a high degree of sequence homology to DNA photolyase contain the two photolyase cofactors but lack DNA repair activity. *Biochemistry* **34**, 6892–6899.

Malhotra, K., Kim, S.-T., Walsh, C. T., and Sancar, A. (1992). Roles of FAD and 8-hydroxy-5-deazaflavin chromophores in photoreactivation by *Anacystis nidulans* DNA photolyase. *J. Biol. Chem.* **267**, 15406–15411.

Mees, A., Klar, T., Gnau, P., Hennecke, U., Eker, A. P., Carell, T., and Essen, L. O. (2004). Crystal structure of a photolyase bound to a CPD-like DNA lesion after *in situ* repair. *Science* **306**, 1789–1793.

Nakajima, S., Sugiyama, M., Iwai, S., Hitomi, K., Otoshi, E., Kim, S.-T., Jiang, C.-Z., Todo, T., Britt, A. B., and Yamamoto, K. (1998). Cloning and characterization of a gene (*UVR3*) required for photorepair of 6-4 photoproducts in *Arabidopsis thaliana*. *Nucleic Acids Res.* **26**, 638–644.

Okamura, T., Sancar, A., Heelis, P. F., Begley, T. P., Hirata, Y., and Mataga, N. (1991). Picosecond laser photolysis studies on the photorepair of pyrimidine dimers by DNA photolyase. 1. Laser photolysis of photolyase–2-deoxyuridine dinucleotide photodimer complex. *J. Am. Chem. Soc.* **113**, 3143–3145.

Park, H.-W., Kim, S.-T., Sancar, A., and Deisenhofer, J. (1995). Crystal structure of DNA photolyase from *Escherichia coli*. *Science* **268**, 1866–1872.

Partch, C. L., Clarkson, M. W., Ozgur, S., Lee, A. L., and Sancar, A. (2005). Role of structural plasticity in signal transduction by the cryptochrome blue-light photoreceptor. *Biochemistry* **44**, 3795–3805.

Payne, G., Heelis, P. F., Rohrs, B. R., and Sancar, A. (1987). The active form of *Escherichia coli* DNA photolyase contains a fully reduced flavin and not a flavin radical, both *in vivo* and *in vitro*. *Biochemistry* **26**, 7121–7127.

Payne, G., and Sancar, A. (1990). Absolute action spectrum of E-FADH$_2$ and E-FADH$_2$-MTHF forms of *Escherichia coli* DNA photolyase. *Biochemistry* **29**, 7715–7727.

Payne, G., Wills, M., Walsh, C. T., and Sancar, A. (1990). Reconstitution of *Escherichia coli* photolyase with flavins and flavin analogues. *Biochemistry* **29**, 5706–5711.

Riggs, A. D., and Bourgeois, S. (1968). On the assay, isolation and characterization of the *lac* repressor. *J. Mol. Biol.* **34**, 361–364.

Rupert, C. S., Goodgal, S. H., and Herriott, R. M. (1958). Photoreactivation *in vitro* of ultraviolet inactivated *Hemophilus influenzae* transforming factor. *J. Gen. Phys.* **41**, 451–471.

Rupert, C. S., and To, K. (1976). Substrate dependence of the action spectrum for photoenzymatic repair of DNA. *Photochem. Photobiol.* **24**, 229–235.

Sabourin, C. L. K., and Ley, R. D. (1988). Isolation and characterization of a marsupial DNA photolyase. *Photochem. Photobiol.* **47**, 719–723.

Sancar, A. (1992). Photolyase: DNA repair by photoinduced electron transfer. *Adv. Electron Transf. Chem.* **2**, 215–272.

Sancar, A. (1994). Structure and function of DNA photolyase. *Biochemistry* **33**, 2–9.

Sancar, A. (2003). Structure and function of DNA photolyases and cryptochrome blue-light photoreceptors. *Chem. Rev.* **103**, 2203–2237.

Sancar, A. (2004a). Photolyase and cryptochrome blue-light photoreceptors. *Adv. Protein Chem.* **69**, 73–100.

Sancar, A. (2004b). Regulation of the mammalian circadian clock by cryptochrome. *J. Biol. Chem.* **279**, 34079–34082.

Sancar, A., and Rupert, C. S. (1978a). Cloning of the *phr* gene and amplification of photolyase in *Escherichia coli*. *Gene* **4**, 295–308.

Sancar, A., and Rupert, C. S. (1978b). Correction of the map location for the *phr* gene in *Escherichia coli* K12. *Mutat. Res.* **51**, 139–143.

Sancar, A., and Sancar, G. B. (1988). DNA repair enzymes. *Annu. Rev. Biochem.* **57**, 29–67.

Sancar, A., Smith, F. W., and Sancar, G. B. (1984). Purification of *Escherichia coli* DNA photolyase. *J. Biol. Chem.* **259**, 6028–6032.

Sancar, G. B. (1990). DNA photolyases: Physical properties, action mechanism, and roles in dark repair. *Mutat. Res.* **236**, 147–160.

Sancar, G. B., Jorns, M. S., Payne, G., Fluke, D. J., Rupert, C. S., and Sancar, A. (1987a). Action mechanism of *Escherichia coli* DNA photolyase. III. Photolysis of the enzyme-substrate complex and the absolute action spectrum. *J. Biol. Chem.* **262**, 492–498.

Sancar, G. B., and Smith, F. W. (1988). Construction of plasmids which lead to overproduction of yeast *PHR1* photolyase in *Saccharomyces cerevisiae* and *Escherichia coli*. *Gene* **64**, 87–96.

Sancar, G. B., Smith, F. W., and Heelis, P. F. (1987b). Purification of the yeast *PHR1* photolyase from an *Escherichia coli* overproducing strain and characterization of the intrinsic chromophores of the enzyme. *J. Biol. Chem.* **262**, 15457–15465.

Sancar, G. B., Smith, F. W., Reid, R., Payne, G., Levy, M., and Sancar, A. (1987c). Action mechanism of *Escherichia coli* DNA photolyase. I. Formation of the enzyme-substrate complex. *J. Biol. Chem.* **262**, 478–485.

Sancar, G. B., Smith, F. W., and Sancar, A. (1983). Identification and amplification of the *E. coli phr* gene product. *Nucleic Acids Res.* **11**, 6667–6678.

Sancar, G. B., Smith, F. W., and Sancar, A. (1985). Binding of *Escherichia coli* DNA photolyase to UV-irradiated DNA. *Biochemistry* **24**, 1849–1855.

Schleicher, E., Hessling, B., Illarionova, V., Bacher, A., Weber, S., Richter, G., and Gerwert, K. (2005). Light-induced reactions of *Escherichia coli* DNA photolyase monitored by Fourier transform infrared spectroscopy. *FEBS J.* **272**, 1855–1866.

Setlow, J. K., and Bollum, F. J. (1968). The minimum size of the substrate for yeast photoreactivating enzyme. *Biochim. Biophys. Acta* **157**, 233–237.

Setlow, R. B., Carrier, W. L., and Bollum, F. J. (1965). Pyrimidine dimers in UV-irradiated poly dI:dC. *Biochemistry* **53**, 1111–1118.

Smith, C. A., and Taylor, J. S. (1993). Preparation and characterization of a set of deoxyoligonucleotide 49-mers containing site-specific *cis*-syn, *trans*-syn-I, (6-4), and Dewar photoproducts of thymidylyl(3′–>5′)-thymidine. *J. Biol. Chem.* **268**, 11143–11151.

Svoboda, D. L., Smith, C. A., Taylor, J.-S., and Sancar, A. (1993). Effect of sequence, adduct type, and opposing lesions on the binding and repair of ultraviolet photodamage by DNA photolyase and (A)BC excinuclease. *J Biol. Chem.* **268**, 10694–10700.

Tamada, T., Kitadokoro, K., Higuchi, Y., Inaka, K., Yasui, A., de Ruiter, P. E., Eker, A. P. M., and Miki, K. (1997). Crystal structure of DNA photolyase from *Anacystis nidulans*. *Nature Struct. Biol.* **4**, 887–891.

Taylor, J.-S., Brockie, I. R., and O'Day, C. L. (1987). A building block for the sequence-specific introduction of *cis*-syn thymine dimers into oligonucleotides: Solid-phase synthesis of TpT{c,s}pTpT. *J. Am. Chem. Soc.* **109**, 6735–6742.

Todo, T., Kim, S.-T., Hitomi, K., Otoshi, E., Inui, T., Morioka, H., Kobayashi, H., Ohtsuka, E., Toh, H., and Ikenaga, M. (1997). Flavin adenine dinucletide as a chromophore of the Xenopus (6-4) photolyase. *Nucleic Acids Res.* **25**, 764–768.

Todo, T., Ryo, H., Yamamoto, K., Toh, H., Inui, T., Ayaki, H., Nomura, T., and Ikenaga, M. (1996). Similarity among the Drosophila (6-4) photolyase, a human photolyase homolog, and the DNA photolyase-blue-light photoreceptor family. *Science* **272**, 109–112.

Todo, T., Takemori, H., Ryo, H., Ihara, M., Matsunaga, T., Nikaido, O., Sato, K., and Nomura, T. (1993). A new photoreactivating enzyme that specifically repairs ultraviolet light-induced (6-4)photoproducts. *Nature* **361**, 371–374.

Vande Berg, B. J., and Sancar, G. B. (1998). Evidence for dinucleotide flipping by DNA photolyase. *J. Biol. Chem.* **273**, 20276–20284.

Witmer, M. R., Altmann, E., Young, H., Begley, T. P., and Sancar, A. (1989). Mechanistic studies on DNA photolyase. 1. Secondary deuterium isotope effects on the cleavage of 2'-deoxyuridine dinucleotide photodimers. *J. Am. Chem. Soc.* **111**, 9264–9265.

Wulff, D. L., and Rupert, C. S. (1962). Disappearance of thymine photodimer in ultraviolet irradiated DNA upon treatment with a photoreactivating enzyme from baker's yeast. *Biochem. Biophys. Res. Commun.* **7**, 237–240.

Zhao, X., Liu, J., Hsu, D. S., Zhao, S., Taylor, J.-S., and Sancar, A. (1997). Reaction mechanism of (6-4) photolyase. *J. Biol. Chem.* **272**, 32580–32590.

[10] Genetic and *In Vitro* Assays of DNA Deamination

By Heather A. Coker, Hugh D. Morgan, and
Svend K. Petersen-Mahrt

Abstract

The DNA deaminase family encompasses enzymes that have been highly conserved throughout vertebrate evolution and which display wide-ranging positive effects upon innate and adaptive immune system and development. Activation-induced cytidine deaminase was identified as a DNA mutator after its necessity in the successful development of high-affinity B cells via somatic hypermutation, class switch recombination, and gene conversion was determined. APOBEC3 exhibits the ability to deaminate retroviral first strand cDNA in a variety of viral infections, including HIV and hepatitis. Recent work has highlighted the potential importance of activation-induced cytidine deaminase (AID) and APOBEC1 in epigenetic reprogramming, and also the role that AID and the APOBECs may

METHODS IN ENZYMOLOGY, VOL. 408   0076-6879/06 $35.00
Copyright 2006, Elsevier Inc. All rights reserved.   DOI: 10.1016/S0076-6879(06)08010-4

have in the development of cancer. In addition to the known activities of these members of the protein family, there are still other deaminases, such as APOBEC2, whose targets and functions are as yet unknown. This chapter provides the details of two assays that have proved to be invaluable in elucidating the exact specificities of deaminases both *in vitro* and in *Escherichia coli*. The application of these assays to future studies of the deaminase family will provide an indispensible tool in determining the potentially diverse functions of the remainder of this family of enzymes.

Introduction

Protein-directed DNA deamination is an exciting new field in molecular biology: whereas we have been aware of the occurrence of spontaneous deamination of cytosine for some time, identification of proteins that deaminate cytosine to form uracil in DNA has only occurred in recent years (Harris *et al.*, 2002; Petersen-Mahrt *et al.*, 2002). Evolutionarily the most ancestral, activation-induced cytidine deaminase (AID) is required to form a functional immune system (Conticello *et al.*, 2005; Muramatsu *et al.*, 2000). Subsequently, other members of the family of DNA deaminases have been shown to be important for the innate immune system, in addition to influencing the spread of retrotransposable elements (Esnault *et al.*, 2005; Sheehy *et al.*, 2002). As deaminases can introduce mutations in DNA, it is also perhaps not surprising that they are now being implicated in the development of cancer. More recently, the expression of AID during early embryogenesis has implicated a role of DNA deaminases in the reprogramming of the epigenetic status of the organism (Morgan *et al.*, 2004).

Due to their emerging importance in the fields of immunology, virology, cancer, and epigenetics, it is vital that one has efficient tools with which to study deaminase activity. In the past, we have developed both an *Escherichia coli* and an *in vitro*-based assay to determine the broad activity, as well as the fine specificity, of these enzymes. This chapter describes in detail how to carry out both such assays.

Escherichia coli Assay of Deamination

Assays of DNA deaminase activity in *E. coli* predate *in vitro* assays requiring purified proteins. The *E. coli* assay exploits the ability of the deaminase to mutate DNA bases of the endogenous *E. coli* genome, scored using particular genes as selectable markers. Verifying the exact nature of the mutation introduced into that gene can provide further additional information on the deaminase. Suppressing DNA repair systems in *E. coli* can also lend support to the action of the deaminase, as may modification of the assay to incorporate exogenous factors.

Mutation Creates a Selectable Marker

There are several genes that, when mutated, generate positive selectable markers (please see Miller, 1972). Due to the possible intrinsic specificities or preferences of a deaminase, it may mutate only a subset of the sites available in a particular gene. Alternatively, a deaminase may not mutate any sites in a particular marker so more than one selectable marker may need to be investigated. It is also possible to modify the system to such an extent as to allow for selection and mutation on an exogenous plasmid (Beale *et al.*, 2004; Pham *et al.*, 2003; Ramiro *et al.*, 2003; Sohail *et al.*, 2003). This chapter focuses on the *E. coli* RNA polymerase gene (*rpoB*) as a marker for deaminase-mediated mutations, using rifampicin (rif) selection.

Deaminase cDNA in an Expression Vector

The deaminase cDNA is transferred to an expression vector containing an inducible promoter. Many such vectors exist; the classic combination of a *tac* promoter and *lacI^q* repressor is combined in the pTrc99A vector (amp^R, ColE1 origin of replication). Its multiple cloning site allows the required deaminase cDNA to be inserted downstream of the promoter and other necessary bacterial transcription and translation elements. Some IPTG inducible systems exhibit low levels of expression in the absence of isopropyl-β-D-thiogalactoside (IPTG), even with lac operon repression. This may be minimized by the addition of 0.2% (w/v) glucose to solid or liquid media.

Bacterial Strain Choice and Relevant Repair Pathways

The product of cytosine deamination, uracil, is predominantly removed via the uracil DNA glycosylase (UDG/*ung*)-dependent base excision repair pathway. The presence of this pathway acting on uracil may therefore have the consequence of affecting the number of C–T transitions observed in the assay. There is, however, the opportunity for assaying DNA deaminase activity in *E. coli* strains lacking *ung* (Harris *et al.*, 2002; Petersen-Mahrt *et al.*, 2002). For example, activity may be assayed in KL16 *E. coli* compared to BW310 (lacking *ung*, but otherwise considered the same genotype as KL16). Evidence that DNA deaminase activity is more severe in the absence of *ung* supports deaminase action on cytidine in DNA. This type of genetic determinism can also be useful when analyzing other potential DNA mutators, where the *E. coli* assay may be adapted for the analysis of mutations derived from noncytosine-based lesions, using strains with deletions in alternative repair pathways (such as *nfi-1* to detect dA deamination (Guo *et al.*, 1997)).

Expression of Additional Proteins

Other activities or factors may be required for deaminase activity that are not normally present in *E. coli*, and the ability to simultaneously express more than one protein in *E. coli* allows one to develop an assay that can incorporate these factors. The second factor is introduced most easily using a second expression vector, with both vectors then being independently selected and their expression verified. In order to assay deaminase activity on methylcytosine, it was necessary to introduce methylcytosine at the potential target sites (Morgan *et al.*, 2004). An inducible expression cassette of the SssI DNA methyltransferase (which methylates CpG residues in a similar fashion to mammalian methylation), generously made available by New England Biolabs, was subcloned into part of pACYC177 containing the p15A ori (compatible with the ColE1 ori of pTrc99A) and kanamycin resistance (Morgan *et al.*, 2004). Mammalian-like CpG methylation is not tolerated in strains of bacteria containing the Mcr/Mrr restriction system so the introduction of SssI requires using a strain lacking both restriction systems (e.g., ER1821 [New England Biolabs]). We established that SssI was functioning by methylation-sensitive restriction enzyme analysis of plasmids in *E. coli*: the plasmid was harvested from the SssI containing *E. coli* and digested with a methylation-sensitive restriction enzyme. Transformation of Mcr$^-$/Mrr$^-$ *E. coli* with this digested plasmid (with undigested plasmid as a control) then indicated the relative extent of plasmid methylation by SssI.

Electrotransformation of Competent E. coli

Many *E. coli* strains suited to protein expression, or possessing other characteristics required by the assay, may not be well suited to chemical transformation, but even two plasmids can be introduced into most strains using electrotransformation (ensure that the DNA is in a low salt solution). It is preferable to transform both plasmids at once to minimize the drift that makes the assay less reproducible following two separate rounds of transformation. At this stage, glucose should be present to prevent induction. Overnight incubation of these plates at less than 37° (e.g., 30–32°) ensures continuous, equivalent growth.

Bacterial Cultures Induced to Express Protein

A general outline of the *E. coli* mutation screen is provided in Fig. 1. Colonies from the plates (as described earlier) are used to set up individual cultures for the induction of protein expression. The frequency of deamination affects the number of cells acquiring resistance, as does the number of cell divisions after the mutation arises. Before the cultures reach saturation, the number of cells in the starting culture should be at a low titer and consistently similar between the different samples. To avoid potential effects of the deaminase (or other factor) on the growth rate of the cells, counting the number of viable cells at the end of the culture (and in some instances during growth) is vital. However, if growth retardation is so severe that the viable count variability among the experimental groups is too extreme, then the use of an 8-h starter culture (without induction, but also without glucose, which would inhibit subsequent induction) followed by an overnight induced culture may be required. Some laboratories have even been able to use a 90-min induction protocol for the mutation assay (Martomo *et al.*, 2005).

Eight to 12 cultures per condition are generally a good experimental setup. For each culture, an entire colony is transferred to a 1.5-ml microcentrifuge tube containing 1 ml of YT or LB media and vortexed. This is then diluted 10^4-fold, and 10 μl is used to inoculate 3 ml of media containing 1 mM IPTG (for induction) and the appropriate plasmid selection antibiotics. If a starter culture is used, then less dilution is warranted: a 1-ml starter culture is inoculated for 8 h, followed by the addition of 2 ml IPTG-containing media. Cultures are incubated at 37° for 16 h with vigorous shaking. For optimal activity, both time and temperature (e.g., 18° for deaminases from cold water fish [Conticello *et al.*, 2005]) can be varied.

Counting Viable Cells and Mutated Cells

Mutation frequency is determined by establishing the number of rif^R colonies vs the number of viable colonies per milliliter. To determine the number of viable cells in the culture, around 50–200 μl of a 10^6 dilution is plated on LB plates (with or without plasmid antibiotic selection). It can be sufficient to count only six to eight plates per condition to obtain an accurate viable count. To determine the number of cells acquiring

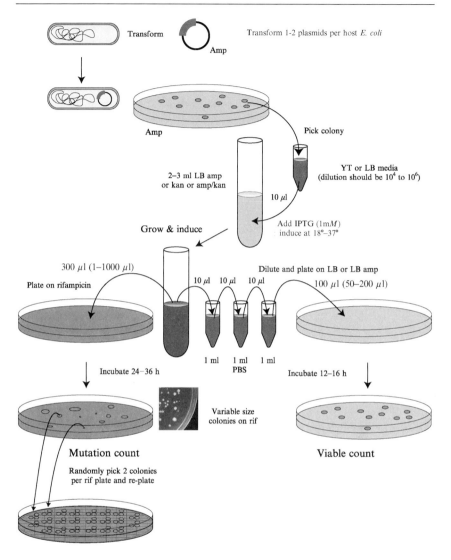

Fig. 1. Schematic of the *E. coli* mutation screen. An outline of the *E. coli* mutation assay using a mutable gene (*rpoB*) as a selection marker for deaminase activity (as described in detail in text). (See color insert.)

mutations that confer resistance, between 1 and 1000 μl of induced culture is plated onto LB low salt plates containing rifampicin (100 μg/ml), ensuring that plates containing rifampicin are protected from light during storage. One can use the same spreader to first spread a diluted viable sample and then the corresponding undiluted original culture without sterilization. For all spreading, the plated volume should be between 50 and 300 μl, potentially requiring dilution or concentration (pulse spin) of the induced culture. Optimal counting accuracy, after an overnight incubation at 37°, is achieved with 100–200 colonies per plate. For rifampicin, a further 8- to 16-h room temperature incubation helps visualize smaller colonies. Colonies should always be counted in an identical way, with respect to time of incubation and size of colony. The mutation frequency under each condition can be determined from the median colony number of rif^R cells per viable cell; this minimizes the effect of an early mutation in a culture resulting in "jack-pot" plate counts. Because the possibility exists that the protein of interest is toxic, one should ensure that the viable count on LB, and LB containing the plasmid selection antibiotic, is equivalent. Scoring on the antibiotic ensures that mutations are detected per cell harboring the plasmid.

Determining the Sequence Context of Mutations

To investigate the qualitative nature of mutations introduced into the selectable gene, colonies are picked from plates containing the selecting agent (e.g., rifampicin). To avoid the "jackpot" effect, only two randomly chosen colonies from each plate are analyzed. The two colonies are respotted onto a master plate also containing rifampicin to ensure that the colony is truly rif^R and also to reduce the number of plates to be stored.

After overnight growth the chosen colonies can be analyzed by conventional sequencing after polymerase chain reaction (PCR) amplification of the implicated gene, e.g., *rpoB* (Beale *et al.*, 2004; Conticello *et al.*, 2005; Harris *et al.*, 2002; Petersen-Mahrt *et al.*, 2002). For amplification of *rpoB*, the oligonucleotides rpopcrf (5'-TTG GCG AAA TGG CGG AAA ACC-3') and rpopcrr (5'-CAC CGA CGG ATA CCA CCT GCT G-3') are used: a 25-μl PCR is performed on a 1/3000 dilution of the colony. The product is purified and resuspended in 125 μl of TE; 5 μl is then used for sequencing with rposeqf (5'-CGT TGG CCT GGT ACG TGT AGA GCG-3') and rposeqr (5'-GGC AAG TTA CCA GGT CTT CTA CG-3').

If only a single site within the target gene is of interest, one can use a simple PCR reaction for the analysis (Morgan *et al.*, 2004). A primer is chosen so that the 3' nucleotide matches only the mutation of interest. In practice, it is necessary to establish stringent PCR conditions using *E. coli* genomic DNA known to contain the mutation. A control PCR should also be carried out to ensure that DNA was present in the original template solution. Both PCRs can be carried out in one reaction tube as a duplex PCR. A crude template solution can be made by transferring a portion of the colony to be analyzed directly into 50 μl of Milli-Q H$_2$O, vortexing, and then using 3 μl of this as template DNA, along with a 95° denaturing and lysing step for 3 min during the initial PCR cycle.

Mutations arising in the selectable marker can in themselves be informative. The spectrum of transitions and transversions should change in the presence of a cytidine deaminase. This increase should be even more extreme with the disruption of DNA repair pathways (Petersen-Mahrt et al., 2002). Sites chosen for mutation, compared to those available for mutation, may also give an indication of the sequence specificity of the deaminase.

In Vitro Assay of Deamination

The in vitro assay requires the use of deaminases that can be produced and purified to a fairly high, pure yield (microgram quantities at least). Even though a number of different methods for producing DNA deaminases have been introduced in the literature (particularly that using baculovirus) (Beale et al., 2004; Bransteitter et al., 2003; Chaudhuri et al., 2003; Muramatsu et al., 1999; Petersen-Mahrt and Neuberger, 2003; Ramiro et al., 2003; Yu et al., 2004), this chapter focuses on the production of tagged DNA deaminases in E. coli, with this having the added advantage of enabling concomitant analysis of a mutator phenotype in the assay described earlier. The following protocols have been used to produce and analyze functional human AID, APOBEC1, APOBEC3G, and also chicken AID, although this chapter concentrates on the production and activity of human AID.

Preparation of Deaminases for Use in the In Vitro Assay

We habitually use a C-terminal his-tagged human AID plasmid under the control of a T7 promotor with kanamycin resistance. We have also successfully modified this procedure to enable protein to be prepared from up to 20 liters of E. coli culture, with protein generated and purified with a N-terminal MBP tag. In the past we also have used untagged protein, purified via ion-exchange chromatography (Petersen-Mahrt and Neuberger, 2003). This allowed us to separate single-stranded DNase activity from the DNA deamination enzymes. To investigate DNA deamination from a crude extract on a single-stranded substrate, we recommend using a random-labeled, heat-denatured DNA. As shown in the past, using random-labeled DNA, the assay is not dependent on the absence of DNase activity (Petersen-Mahrt and Neuberger, 2003), but due to the number of complex interactions and variation in substrate, the presence of DNase activity does not lend the assay to the study of the enzyme in such a controlled fashion as the *in vitro* oligonucleotide assay.

Growth. Transform *E. coli* of the C41(DE3) strain (Miroux and Walker, 1996) with the protein expression plasmid and plate onto an appropriate selective LB plate (all subsequent bacterial growth takes place using nonselective media). Once the transformed bacteria have grown overnight at $37°$, take one colony and resuspend thoroughly in 1 ml $2\times$ TY media. Use this to further inoculate 30 ml $2\times$ TY, which is then grown, shaking at $37°$ until reaching an OD_{600} of 0.4. At this stage, 10 ml of the culture is used to inoculate each of three, 2-liter flasks, each containing 1 liter of $2\times$ TY. These are grown, shaking at $37°$, until reaching an OD_{600} of between 1.0 and 1.5. The flasks should then be immersed in iced H_2O for 15 min prior to induction with a final concentration of 0.4 mM IPTG. In addition, at this stage add $ZnCl2$ to a final concentration of 10 μM. Importantly, holes are made in the foil lids of the flasks to ensure that sufficient air exchange is able to take place during protein production. The cultures should then be incubated, shaking at 250 rpm at $16°$ for 16 h.

Extraction. The 3 liters of bacteria is harvested by centrifugation at 3500g, 10 min at $4°$, the supernatant is removed carefully but thoroughly, and the pellets are resuspended each in approximately 100 ml of ice-cold H_2O before combining. Bacteria are again centrifuged at 3500g, for 10 min at $4°$, before the supernatant is immediately removed and the pellet resuspended in 100 ml of extraction buffer (20 mM MES [pH 6.0], 100 mM NaCl, 50 mM KCl, 5 mM β-mercaptoethanol, 1.6 mM CHAPS, 300 mM L-arginine HCl, filtered and adjusted to pH 6) along with six complete EDTA-free protease inhibitor cocktail tablets (Roche) and 150 mg RNase A (Sigma), which has been heated previously to $80°$ for 10 min in extraction buffer. Keeping the supernatant on ice, sonicate until the cells are completely disrupted (this needs to be optimized depending on the machine used, but we have found that up to 25 min in total, with 15-s pulses at an amplitude of 50, is sufficient). The suspension is then centrifuged at 100,000g, 30 min at $4°$, and the supernatant is filtered using a prechilled 0.22-μm Stericup (Millipore).

In an initial study we determined that greatest protein solubility was obtained using this low pH 6 extraction buffer. Furthermore, adding 300 mM L-arginine HCl, as detailed earlier, appeared to increase the stability of the protein. The addition of RNase A both at this stage and later in the assay itself appears to increase the activity of the protein (Bransteitter *et al.*, 2003).

Purification. Transfer the filtered supernatant into two 50-ml Falcon tubes. Add 1 ml of Ni-NTA beads (Qiagen), prewashed in extraction buffer, to each tube. Incubate the beads with protein at 4°, rotating for 90 min. Centrifuge briefly to pellet beads (1000*g*, 1 min at 4°) before loading the beads onto a Poly-Prep chromatography column (Bio-Rad) fitted with a 21-gauge needle and prewashed with extraction buffer at 4°. Wash the beaded column at 4° with 10 ml of 30 m*M* imidazole and then elute the protein at 4° with 10 ml of 150 m*M* imidazole. Concentrate the eluate using a centrifugal concentrator (Vivaspin) with a M_r 10,000 cutoff (or appropriate for your protein of interest). Follow the manufacturer's instructions to concentrate the eluate to 150 μl or a suitable concentration. We have found that deaminases lose activity when stored at -80° and, over time, even when stored in liquid nitrogen. We would recommend that the concentrated protein be snap-frozen in small aliquots and stored in liquid nitrogen for use within 10 months.

Assay of Deaminase Activity

Oligonucleotides. Using an enzyme assay that is reliant on the presence of free cytosine or cytidine is not only inaccurate, but also does not reflect the true activity of the enzyme. This is predominantly due to the possible contamination of the protein preparation by *E. coli* cytidine deaminases (Beale *et al.*, 2004; Petersen-Mahrt and Neuberger, 2003). The basis of the *in vitro* assay is the deamination of cytosine within labeled oligonucleotides and then the subsequent readout of deamination of those oligonucleotides by cleavage at the sites of deamination. There are no real restrictions on the design of the oligonucleotides; the design simply reflects the particular questions that you wish to answer of the deaminase. The oligonucleotide may contain single or multiple potential target sites. Multiple target sites allow for analysis of the fine specificity of deaminase activity; we found this to be of particular use not only when comparing the preference of AID and APOBEC1 for cytosine compared to methylcytosine but also in determining the exact sequence motif preferences of the deaminases (Morgan *et al.*, 2004). We have found 40-mers to be the most practical length of oligonucleotide if assaying two sites simultaneously, and it is important that the oligonucleotides are HPLC purified to ensure that any prematurely truncated oligonucleotides are removed. Resuspension of the oligonucleotides in TE as usual is perfectly sufficient. An oligonucleotide containing uracil is a useful control for the efficiency/UDG dependence of the reactions in each experiment.

The most convenient *in vitro* assay that we have established depends on 5′ biotinylation of the oligonucleotides to enable subsequent purification by streptavidin-coated magnetic beads. The 3′ end of the oligonucleotide is labeled for visualization; in the past, we have used ^{32}P-labeled oligonucleotides, although we now use 3′-fluorescein-labeled oligonucleotides most frequently. We have also used other fluorescent dyes, such as JOE and TAMRA (PerkinElmer Life Sciences Inc.). While oligonucleotides can be purchased with these labels already attached, end labeling provides a simple means of labeling aliquots of an oligonucleotide with different dyes for the potential of a dual or multiple color readout (although it can be difficult to minimize sufficiently the bleed through of fluorescent emission between the laser channels).

Oligonucleotide Deamination. The exact quantity of a deaminase needed for optimal activity in the *in vitro* assay needs to be determined for each protein preparation by conducting a pilot experiment titrating a range of concentrations. We have found, however, that protein made by our method is generally used at between 0.05 and 0.5 μl. If a protein preparation is particularly good, it may therefore be necessary to dilute the protein further using extraction buffer, e.g., 1/10, and snap freeze again. Refreezing once, in this manner, does not appear to affect the deaminase activity. It is important to use a control protein in each experiment to ensure that spontaneous deamination of the oligonucleotides does not occur during the procedure.

Two microliters of reaction buffer (40 mM Tris pH 8, 40 mM KCl, 50 mM NaCl, 5 mM EDTA, 1 mM dithiothreitol, 10% glycerol), 2.5 mM EDTA, 5 μg RNase A (Sigma) (heated previously to 80° for 10 min in 0.1 M Tris, pH 8), and 2.5 pmol 5′-biotinylated, 3′-FITC-labeled oligonucleotide should be combined to give a final 10-μl reaction once the deaminase is added. The mix is then heated to 90° for 3 min before quenching on ice. (The reaction buffer keeps for up to 6 months if refrigerated, but the dithiothreitol should be replenished monthly.)

The deaminase is thawed on ice and then the appropriate amount is added to the ice-cold reaction mix for each sample in 0.5-ml tubes. Each reaction should be incubated at 37° for 15 min for optimal activity, although we have previously tested other temperatures and incubation times, and it is worth noting that deaminases can still demonstrate (albeit reduced) activity at 18° (Conticello *et al.*, 2005). Each reaction is terminated by adding 100 μl of H$_2$O and then heating at 90° for 4 min before placing on ice.

Purification of Oligonucleotides. Use of biotin-labeled oligonucleotides enables the use of streptavidin-coated magnetic beads for purification of the oligonucleotides. We have modified the purification step to use streptavidin-coated 96-well plates (Roche) and also streptavidin-coated magnetic beads in a 96-well format, but from the point of view of reproducibility we have found that using individual tubes is best.

Eight microliters of resuspended streptavidin-magnetic bead slurry (Dynal) per reaction is washed twice in TEN buffer (10 mM EDTA, 50 mM Tris, pH 7.5, 1 M NaCl) before resuspending in 800 μl TEN per reaction and adding to each protein sample and then transferring to a 1.5-ml microcentrifuge tube. The samples are mixed by rotating at room temperature for 15 min. Use of a Dynal microcentrifuge magnet enables the washing of the bead-bound oligonucleotides: an incubation of 3 min on the magnet is sufficient. The supernatant is then carefully removed by pipette before removing the tube from the magnet and resuspending the beads in 750 μl, prewarmed, 70° TEN. The bead-bound oligonucleotides are then washed by rotating for 5 min at room temperature, before repeating the wash procedure in exactly the same manner with TEN, and then finally with 100 μl room temperature TE, pH 8. The tubes are subject to a pulse spin in a benchtop microcentrifuge before placing on the magnet and removing the supernatant for the final time.

Removal of Uracil from Oligonucleotides. Cleavage of the oligonucleotides at the site of deamination of cytosine to uracil provides the readout of AID activity. The bead-bound oligonucleotides are resuspended in 10 μl of UDG reaction mix (1× UDG buffer, excess UDG [1 μl, 1 U, New England Biolabs]) at 37° for 1 h, mixing the samples gently after 30 min. Contaminant apyriminidic endonuclease within commercial sources of UDG means that a separate cleavage step of the UDG-generated abasic site is not necessary. Twenty microliters of fuscin formamide [0.1% fuscin dye (Sigma), 10 mM Tris, pH 8, in formamide] should then be added to terminate each reaction.

Use of TDG to Resolve a T:G or U:G Mismatch. The flexibility of the *in vitro* assay means that when we investigated deaminase activity on methylcytosine (Morgan *et al.*, 2004), we adopted a different approach to generate a readout of deamination. The deamination of methylcytosine generates a thymine residue and so this event would not be recognized by UDG. However, the resulting T:G base pair is acted on by TDG. Indeed, as TDG recognizes both T:G and U:G mismatches, even an oligonucleotide designed to contain both cytosine and methylcytosine target sites may be processed using TDG and a reverse complementary oligonucleotide.

After purification, the bead-bound oligonucleotides are resuspended in TDG reaction mix, containing 1× buffer (Trevigen) and an excess (5 pmol) of the reverse complement oligonucleotide. Each sample is then heated to 90° for 3 min and the oligonucleotides are allowed to anneal by cooling over 30 min in a hot block to room temperature. An excess of TDG (1 μl,

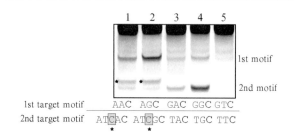

FIG. 2. *In vitro* DNA deamination of FITC-labeled oligonucleotides. Five FITC-labeled oligonucleotides each containing different, multiple deamination target motifs were incubated with AID, annealed with their reverse complement oligonucleotide, and then cleaved at U:G or T:G mismatches with TDG (described in detail in text). Samples were then subject to 20% PAGE-urea and visualized with a fluorescent scanner, enabling the relative efficiency of enzyme activity at each motif to be determined. This allows the fine specificity of deaminase activity to be analyzed; it is evident, for example, from lane 2, that the substrate AGC^{me} (motif 1) is preferred to ATC (motif 2). C* indicates an unmethylated cytosine.

2 U, Trevigen) is then added to each tube (final reaction volume 10 μl), and the samples are incubated at 47° for 1 h. Twenty microliters of fuscin formamide is added to stop each reaction.

Visualization and Quantification of Deamination Assay. In order to visualize the results of the assay, each sample is heated at 95° for 3 min and then quenched on ice before being subject to PAGE–urea electrophoresis on a prerun 15–20% gel at 200 V (for 1–2 h depending on the length of the oligonucleotide and the number of cleavage sites). The fluorescent signal is then detected using fluorescent scanners such as the FLA-5000 (Fuji) (see Fig. 2).

Quantification of the scanned image enables detailed comparisons of the deaminase activity of the different samples, allowing percentage conversion, estimates of protein turnover (when knowledge of microgram of protein per reaction is determined by Coomassie/Western blot studies), and further kinetic studies to be carried out. Percentage conversion may be calculated by the following equation: % conversion = pixal volume of cleaved product (minus background)/(pixal volume of cleaved product (minus background) + pixal volume of uncleaved substrate (minus background)) × 100.

Conclusion

This chapter detailed the two main assays used to determine much of our biochemical knowledge of DNA deaminase activity. The *E. coli* assay has been of particular use in enabling analysis of the activity of deaminases when their purification has not been possible and also in providing an indication of possible *in vivo* deaminase activity. In contrast, the *in vitro* oligonucleotide assay provides the opportunity to compare the activity resulting from different posttranslational modifications made to the deaminases and, although this chapter only detailed the growth of deaminases in *E. coli*, also the activity of deaminases grown in different cell types. Furthermore, using a single oligonucleotide with a defined sequence has been instrumental in enabling us to identify the exact substrate and specificities of different DNA deaminases, its flexibility resulting from the huge range of different oligonucleotides that can be designed to answer specific questions. In addition, the ability to detect different lesions within the oligonucleotides by varying between the use of UDG or TDG expands the potential of this assay even further, providing us with ample tools with which to explore this family of enzymes, whose importance for the survival and evolution of higher organisms is becoming more and more apparent.

Acknowledgments

We acknowledge the contributions of Ian Watts to the development of the oligonucleotide assay and Reuben Harris for introducing the powers of *E. coli* genetics. We apologize to any contributors to the field that we have inadvertently omitted in this review due to space limitations.

References

Beale, R. C., Petersen-Mahrt, S. K., Watt, I. N., Harris, R. S., Rada, C., and Neuberger, M. S. (2004). Comparison of the differential context-dependence of DNA deamination by APOBEC enzymes: Correlation with mutation spectra *in vivo*. *J. Mol. Biol.* **337,** 585–596.

Bransteitter, R., Pham, P., Scharff, M. D., and Goodman, M. F. (2003). Activation-induced cytidine deaminase deaminates deoxycytidine on single-stranded DNA but requires the action of RNase. *Proc. Natl. Acad. Sci. USA* **100,** 4102–4107.

Chaudhuri, J., Tian, M., Khuong, C., Chua, K., Pinaud, E., and Alt, F. W. (2003). Transcription-targeted DNA deamination by the AID antibody diversification enzyme. *Nature* **422,** 726–730.

Conticello, S. G., Thomas, C. J., Petersen-Mahrt, S. K., and Neuberger, M. S. (2005). Evolution of the AID/APOBEC family of polynucleotide (deoxy)cytidine deaminases. *Mol. Biol. Evol.* **22,** 367–377.

Esnault, C., Heidmann, O., Delebecque, F., Dewannieux, M., Ribet, D., Hance, A. J., Heidmann, T., and Schwartz, O. (2005). APOBEC3G cytidine deaminase inhibits retrotransposition of endogenous retroviruses. *Nature* **433,** 430–433.

Guo, G., Ding, Y., and Weiss, B. (1997). nfi, the gene for endonuclease V in *Escherichia coli* K-12. *J. Bacteriol* **179,** 310–316.

Harris, R. S., Petersen-Mahrt, S. K., and Neuberger, M. S. (2002). RNA editing enzyme APOBEC1 and some of its homologs can act as DNA mutators. *Mol. Cell* **10,** 1247–1253.

Martomo, S. A., Fu, D., Yang, W. W., Joshi, N. S., and Gearhart, P. J. (2005). Deoxyuridine is generated preferentially in the nontranscribed strand of DNA from cells expressing activation-induced cytidine deaminase. *J. Immunol.* **174,** 7787–7791.

Miller, J. H. (1972). "Experiments in Molecular Genetics." Cold Spring Harbor Press, Cold Spring Harbor, NY.

Miroux, B., and Walker, J. E. (1996). Over-production of proteins in *Escherichia coli*: Mutant hosts that allow synthesis of some membrane proteins and globular proteins at high levels. *J. Mol. Biol.* **260,** 289–298.

Morgan, H. D., Dean, W., Coker, H. A., Reik, W., and Petersen-Mahrt, S. K. (2004). Aid deaminates 5-methylcytosine in DNA and is expressed in pluripotent tissues: Implications for epigenetic reprogramming. *J. Biol. Chem* **279,** 52353–52360.

Muramatsu, M., Kinoshita, K., Fagarasan, S., Yamada, S., Shinkai, Y., and Honjo, T. (2000). Class switch recombination and hypermutation require activation-induced cytidine deaminase (AID), a potential RNA editing enzyme. *Cell* **102,** 553–563.

Muramatsu, M., Sankaranand, V. S., Anant, S., Sugai, M., Kinoshita, K., Davidson, N. O., and Honjo, T. (1999). Specific expression of activation-inducedcytidine deaminase (AID), a novel member of the RNA-editing deaminase family in germinal center B cells. *J. Biol. Chem.* **274,** 18470–18476.

Petersen-Mahrt, S. K., Harris, R. S., and Neuberger, M. S. (2002). AID mutates *E. coli* suggesting a DNA deamination mechanism for antibody diversification. *Nature* **418,** 99–103.

Petersen-Mahrt, S. K., and Neuberger, M. S. (2003). *In vitro* deamination of cytosine to uracil in single-stranded DNA by apolipoprotein B editing complex catalytic subunit 1 (APOBEC1). *J. Biol. Chem.* **278,** 19583–19586.

Pham, P., Bransteitter, R., Petruska, J., and Goodman, M. F. (2003). Processive AID-catalysed cytosine deamination on single-stranded DNA simulates somatic hypermutation. *Nature* **424,** 103–107.

Ramiro, A. R., Stavropoulos, P., Jankovic, M., and Nussenzweig, M. C. (2003). Transcription enhances AID-mediated cytidine deamination by exposing single-stranded DNA on the nontemplate strand. *Nature Immunol.* **4,** 452–456.

Sheehy, A. M., Gaddis, N. C., Choi, J. D., and Malim, M. H. (2002). Isolation of a human gene that inhibits HIV-1 infection and is suppressed by the viral Vif protein. *Nature* **418,** 646–650.

Sohail, A., Klapacz, J., Samaranayake, M., Ullah, A., and Bhagwat, A. S. (2003). Human activation-induced cytidine deaminase causes transcription-dependent, strand-biased C to U deaminations. *Nucleic Acids Res.* **31,** 2990–2994.

Yu, K., Huang, F. T., and Lieber, M. R. (2004). DNA substrate length and surrounding sequence affect the activation-induced deaminase activity at cytidine. *J. Biol. Chem.* **279,** 6496–6500.

[11]   The Xeroderma Pigmentosum Group C
Protein Complex and Ultraviolet-Damaged
DNA-Binding Protein: Functional Assays for Damage
Recognition Factors Involved in Global Genome Repair

By KAORU SUGASAWA

Abstract

In mammalian nucleotide excision repair (NER) operating throughout
the genome, DNA lesions are recognized by protein factors that specifically bind to the damaged sites. Such damage recognition factors involve the
xeroderma pigmentosum group C (XPC) protein complex and ultraviolet
(UV)-damaged DNA-binding protein (UV-DDB). To assess specific DNA-
binding activities of these factors, it is useful to take advantage of biochemical assays using DNA substrates that contain a defined lesion and/or
artificial structure in a site-specific manner. In addition, it has been shown
that both XPC and UV-DDB are ubiquitylated in response to UV irradiation of cells. This ubiquitylation is mediated by ubiquitin ligase associated
with UV-DDB and is important for the NER process of UV-induced
lesions. Methods for detecting the UV-DDB-dependent ubiquitylation
in vivo and *in vitro* are also described.

Introduction

A protein complex containing the xeroderma pigmentosum group
C (XPC) protein plays an essential role in damage recognition for mam-
malian nucleotide excision repair (NER) operating throughout the genome
(Sugasawa *et al.*, 1998). The XPC complex is a heterotrimer that consists of
XPC, one of the two mammalian homologs of *Saccharomyces cerevisiae*
Rad23p (HR23A or HR23B), and centrin 2 (Araki *et al.*, 2001). The complex
is a structure-specific DNA-binding factor, which binds preferentially to
junctions of double- and single-stranded DNA (Sugasawa *et al.*, 2001, 2002).
Based on such biochemical properties, the complex can recognize and bind to
sites of various helix-distorting base lesions with diverse chemical structures.
Ultraviolet (UV)-damaged DNA-binding protein (UV-DDB) is
another factor involved in damage recognition for global genome NER
(Tang and Chu, 2002). This factor is a heterodimer composed of DDB1 and
DDB2, the latter of which corresponds to the XP group E protein (Chu and
Chang, 1988). UV-DDB binds to UV-induced lesions much more strongly

METHODS IN ENZYMOLOGY, VOL. 408   0076-6879/06 $35.00

than the XPC complex (Batty *et al.*, 2000), whereas its binding to chemical adducts and abnormal DNA structures is less pronounced (Payne and Chu, 1994; Sugasawa *et al.*, 2005). Thus, unlike XPC, UV-DDB appears to specialize in detecting UV lesions.

UV-DDB is associated *in vivo* with the ubiquitin ligase (E3) complex containing cullin 4A, Roc1, and the COP9 signalosome (Groisman *et al.*, 2003). We have shown that the UV-DDB-E3 complex ubiquitylates XPC as well as DDB2; this ubiquitylation is important for the repair process of UV lesions (Sugasawa *et al.*, 2005). This chapter describes experimental procedures currently adopted in our laboratory for the preparation of recombinant protein complexes and for functional assays to assess their damage-specific DNA-binding activities and UV-DDB-dependent ubiquitylation.

Purification of Recombinant Proteins

Although both the XPC complex (Masutani *et al.*, 1994) and UV-DDB (Keeney *et al.*, 1993) can be purified from cultured human cells, relatively large-scale suspension cultures are required as starting material. Therefore, production and purification of recombinant proteins are excellent alternatives. Because some of the subunits (e.g., XPC, DDB1, and DDB2) are very poorly expressed in *Escherichia coli*, the baculovirus expression system is most suitable for this purpose. For construction of recombinant baculoviruses, the Bac-to-Bac system (Invitrogen, Carlsbad, CA) is used routinely in our laboratory, but other systems may be also applicable.

The XPC protein can be expressed either alone or together with other subunits (Araki *et al.*, 2001). HR23 proteins physically stabilize XPC and are essential for its *in vitro* functions. However, when overexpressed in insect cells, the HR23 proteins appear to undergo unidentified posttranslational modifications that are not found in mammalian cells and are consequently detected as multiple bands on SDS–PAGE (Sugasawa *et al.*, 1996). For this reason, we prefer to purify XPC alone and then reconstitute the complex *in vitro* with HR23 and centrin 2, both of which can be overexpressed separately in *E. coli* and purified in large quantities. DDB2 should always be coexpressed with DDB1 because it tends to interact nonspecifically with other proteins and is difficult to purify when expressed alone. We use the pFastBac Dual vector (Invitrogen) for coexpression of the two subunits. When the UV-DDB-E3 complex is required, cullin 4A and Roc1 should be expressed with DDB1 and DDB2 by coinfection with appropriate baculoviruses. Although XPC and UV-DDB can be purified according to the originally described procedures (Keeney *et al.*, 1993; Masutani *et al.*, 1994), the use of epitope tags simplifies purification. We add the FLAG tag to N termini of XPC and DDB1, and we have confirmed that this tag does not significantly interfere with *in vitro* functions of the purified proteins. The following procedures describe the purification of FLAG-tagged proteins.

Materials

EX-CELL 405 medium (JRH Biosciences, Lenexa, KS)
Phosphate-buffered saline (PBS)
NP lysis buffer: 25 mM Tris–HCl (pH 8.0), 1 mM EDTA, 0.3 M NaCl, 10% (v/v) glycerol, 1% (v/v) Nonidet P-40, 1 mM dithiothreitol (DTT), 0.25 mM phenylmethylsulfonyl fluoride (PMSF), protease inhibitor cocktail (Complete, Roche Diagnostics, Basel, Switzerland)
Dialysis buffer: 20 mM sodium phosphate (pH 7.8), 1 mM EDTA, 0.1 M NaCl, 10% (v/v) glycerol, 1 mM DTT, 0.25 mM PMSF
Buffer A: 20 mM sodium phosphate (pH 7.8), 1 mM EDTA, 10% (v/v) glycerol, 0.01% (v/v) Triton X-100, 1 mM DTT, 0.25 mM PMSF
Buffer B: 20 mM sodium phosphate (pH 7.8), 10% (v/v) glycerol, 0.01% (v/v) Triton X-100, 10 mM β-mercaptoethanol, 0.25 mM PMSF
Buffer C: 25 mM Tris–HCl (pH 7.5), 1 mM EDTA, 10% (v/v) glycerol, 0.01% (v/v) Triton X-100, 1 mM DTT, 0.25 mM PMSF.

Cell Culture, Virus Infection, and Extract Preparation

1. Grow HighFive cells at 27° in EX-CELL 405 medium. For each purification, we usually start with monolayer cultures of the cells in twenty 150-mm dishes. At 80–90% confluency, infect the cells with each recombinant virus at a multiplicity of infection (MOI) of 5–10. Incubate further at 27° for ~72 h.

2. Scrape the infected cells into 250-ml conical centrifuge tubes and collect by centrifugation at 100g for 10 min. After washing with PBS, resuspend the cell pellets in an eightfold volume of ice-cold NP lysis buffer by gentle pipetting. Incubate the cell suspension on ice for 30 min with occasional mixing.

3. After centrifugation (20,000g, 20 min), collect the supernatant (clarified extract) and dialyze overnight against dialysis buffer. Significant amounts of precipitate should be visible during dialysis and are subsequently removed by centrifugation (100,000g, 30 min). The resulting clear supernatant (typically ~80 ml) can be frozen in liquid nitrogen and stored at −80° until further purification steps are conducted.

For some recombinant proteins (particularly nuclear proteins), only a small portion of the total expressed protein is solubilized during these extraction procedures. Using harsher extraction conditions may yield larger quantities of the proteins. However, at least in the case of XPC, we have observed that the less soluble fraction of the protein is poorly active. If biochemical assays are undertaken, *do not* try to solubilize too much protein, as the specific activity of the purified protein fraction will be reduced considerably. Coexpression with HR23 and/or centrin 2 does not significantly improve the solubility of XPC.

Purification of the FLAG-XPC Protein

1. All the purification procedures should be carried out at $4°$. We use the ÄKTA Explorer 10S system (Amersham Biosciences, Piscataway, NJ) placed in a cold room, unless otherwise specified. Load the insect cell extract containing FLAG-XPC onto a HiPrep 16/10 heparin FF column (Amersham Biosciences) equilibrated with buffer A containing 0.1 M NaCl. Wash the column successively with buffer A containing 0.1, 0.3 M, and 1 M NaCl (each wash/elution step may require $\sim$100 ml of the buffer). FLAG-XPC should be recovered in the 1 M NaCl fraction.

2. Adjust the 1 M NaCl heparin fraction to 0.3 M NaCl by dilution with buffer A and load onto a column (1 cm diameter $\times$ 9 cm long; bed volume, $\sim$7 ml) packed with anti-FLAG M2 agarose (Sigma, St. Louis, MO) equilibrated with buffer B containing 0.3 M NaCl. The flow rate should be kept low ($\sim$0.1 ml/min) to obtain maximal binding. After loading, wash the column extensively with buffer B containing 1 M NaCl. Reconnect the column to the chromatography system in an inverted position and wash it with buffer B containing 0.3 M NaCl. Elute bound proteins by injecting 10 ml of buffer B containing 0.3 M NaCl and 0.1 mg/ml FLAG peptide (Sigma). Inversion of the column enables the protein to be recovered in a more concentrated form. Identify the fractions containing FLAG-XPC by SDS–PAGE. The eluted protein fraction can be used directly for *in vitro* reconstitution of the complex (see later).

Purification of HR23B and Centrin 2

Human HR23 (hHR23A and B) proteins can be overexpressed in *E. coli* with the conventional pET expression system (Novagen, EMD Biosciences, San Diego, CA) and purified in either a His-tagged or a nontagged form (Masutani *et al.*, 1997). For His-tagged proteins, one-step purification using metal-chelating affinity chromatography is sufficient. Although the two HR23 proteins appear to be indistinguishable in their *in vitro* NER functions (Sugasawa *et al.*, 1997), HR23B may be preferred solely because most XPC *in vivo* is associated with HR23B rather than HR23A (Araki *et al.*, 2001).

Centrin 2 can also be highly expressed in *E. coli* with the pET system. Because centrin 2 is a heat-stable protein, inclusion of a heat treatment step makes it quite easy to purify the protein even in a nontagged form. Detailed procedures for purification of recombinant centrin 2 have been described elsewhere (Araki *et al.*, 2001).

Reconstitution of the XPC Complex

XPC interactions with both HR23B and centrin 2 are very stable; the purified complex has never been found to dissociate unless challenged by denaturing conditions. When XPC is mixed *in vitro* with other recombinant subunits, they form stable complexes rapidly.

To reconstitute the complex, mix FLAG-XPC eluted from the antibody column with a large molar excess of HR23B-His and centrin 2 in buffer B containing 0.3 M NaCl. Incubate on ice for at least 2 h and then load the mixture onto a small heparin column (1-ml HiTrap column; Amersham Biosciences) equilibrated with buffer B containing 0.3 M NaCl. Uncomplexed HR23B-His and centrin 2, as well as the FLAG peptide, should flow through the column, and the reconstituted complex can be eluted with buffer B containing 1 M NaCl. Because complex formation is very efficient, it is usually unnecessary to remove uncomplexed XPC. If desired, however, the complex may be purified further using the His-tag on HR23B.

The purified XPC complex tends to aggregate under conditions of low ionic strength. It is recommended that 0.3 M NaCl (or KCl) always be included in the storage buffer and, if the protein concentration is low, bovine serum albumin (BSA) be included at a concentration of 0.2 mg/ml. The complex should be frozen in liquid nitrogen and stored at $-80°$. When stored in an appropriate buffer, repeated freezing and thawing do not result in any significant loss of activity in our experience.

Purification of UV-DDB

When coexpressed in insect cells, DDB1 tends to be expressed in vast excess over DDB2. If DDB1 is fused to the appropriate tag (we use the FLAG tag), not only the UV-DDB heterodimer but also uncomplexed DDB1 can be purified from the same extract (Sugasawa *et al.*, 2005).

1. Load the insect cell extract onto a heparin column, as described earlier for the purification of XPC. The DDB2-containing complex binds to the column efficiently, while FLAG-DDB1 should flow through. Elute

the bound complex from the heparin column with buffer A containing 0.45 M NaCl instead of 1 M NaCl.

2. Load the flow through, or the 0.45 M NaCl heparin fraction, onto an anti-FLAG M2 agarose column, which should be processed exactly as described for FLAG-XPC. Determine the fractions containing the eluted proteins by SDS–PAGE. Dialyze the pooled fractions against buffer C containing 0.1 M NaCl.

3. To concentrate samples, load the dialyzed fractions onto a Mono Q HR5/5 column (Amersham Biosciences) equilibrated with buffer C containing 0.1 M NaCl. Elute the bound proteins with a linear NaCl gradient (0.1–0.5 M; 20 ml) in buffer C. Both UV-DDB and FLAG-DDB1 should bind to the column and are eluted in an early part of the gradient. Pool the peak fractions, which can be frozen in liquid nitrogen and stored at −80°.

Purification of the UV-DDB-E3 Complex

To obtain the UV-DDB-E3 complex, cullin 4A and Roc1 should be coexpressed with FLAG-DDB1 and DDB2. Initial heparin and anti-FLAG antibody column chromatography can be carried out as for the UV-DDB heterodimer, with the exception of washing the anti-FLAG M2 agarose column with buffer containing 1 M NaCl. E3 subunit association with the complex cannot withstand this salt wash and should be omitted.

Eluates from the anti-FLAG antibody column may contain considerable amounts of heterotrimeric complexes lacking DDB2. To remove such contaminating subcomplexes, the sample should be dialyzed against buffer A containing 0.1 M NaCl and loaded onto a Mono S PC1.6/5 column (Amersham Biosciences, using the SMART system) equilibrated with the same buffer. Only the complex containing DDB2 binds to the Mono S column under these conditions. Elute the bound proteins with 2.4 ml of a 0.1–0.5 M linear NaCl gradient in buffer A. The heterotetrameric UV-DDB-E3 complex elutes at approximately 0.25 M NaCl.

Damage-Specific DNA-Binding Assays

One of the most convenient methods to assess damage-specific DNA-binding activities may be a conventional electrophoretic mobility shift assay (EMSA). For instance, the UV-DDB activity in cell extracts can be detected easily by EMSA using an appropriate DNA fragment that is irradiated heavily with UV (Chu and Chang, 1988). However, the damaged DNA prepared in this manner contains a mixture of various lesions, and the number and structure of the lesions in each DNA fragment are not

known. In addition, it is difficult to control the sites of the lesions generated. For precise determination of binding specificities, highly defined
DNA substrates are required, which ideally should contain only one lesion
per DNA fragment with a fully identified chemical structure at a defined
position.

To prepare such substrates, we have taken advantage of the procedures
originally described by Moggs *et al.* (1996) (Fig. 1). A synthetic oligonucleotide that contains a site-specific lesion is annealed to single-stranded circular
DNA containing a complementary sequence and used as a primer for DNA
synthesis by T4 DNA polymerase. The final nick is sealed by T4 DNA ligase,
and the resulting covalently closed circular DNA is purified either by CsCl-
EtBr density gradient equilibrium or by agarose gel electrophoresis in the

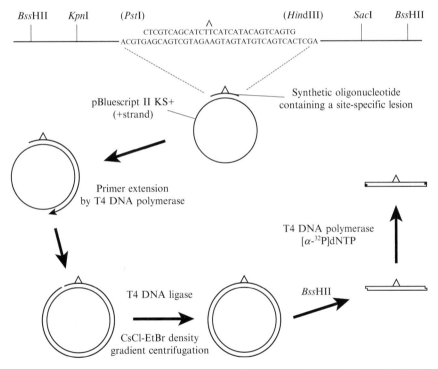

FIG. 1. A scheme for preparation of DNA substrates containing a site-specific lesion
and/or artificial structure. An example of the DNA sequence, which was used for a substrate
containing a UV-induced thymine dimer, i.e., cyclobutane pyrimidine dimer (CPD) or (6-4)
photoproduct (6-4PP)[3], is shown above the plasmid sequence. For EMSA, the double-
stranded circular DNA is digested with *Bss*HII, and recessed 3' termini are filled with T4
DNA polymerase and labeled dNTP.

presence of EtBr. Digestion with appropriate restriction enzymes produces the damage-containing DNA fragment, which can be then labeled and used for EMSA or other DNA-binding assays. By changing the oligonucleotide and the complementary sequence in the single-stranded template, various combinations of lesions with mismatches or artificial structures (bubble, loop, etc.) can be generated (Sugasawa *et al.*, 2001, 2002).

The length of the DNA fragment is an important factor in the binding assays, particularly for assays using the XPC complex. For instance, the apparent affinity of XPC for a double-stranded 30-mer oligonucleotide with a single lesion is much lower than that for a 180-bp fragment containing the same lesion. One may be able to prepare these longer DNA substrates by ligation of multiple synthetic oligonucleotides. However, it would be very difficult to eliminate contamination with DNA that has one or more unligated nicks and the following method is therefore preferred.

Materials

> 10× T4 DNA polymerase buffer (from Takara Shuzo, Otsu, Japan): 330 mM Tris-acetate (pH 7.9), 660 mM potassium acetate, 100 mM magnesium acetate, 5 mM DTT
> TGE buffer: 25 mM Tris-base, 192 mM glycine, 1 mM EDTA
> DNA elution buffer: 0.5 M ammonium acetate, 10 mM magnesium acetate, 0.1% (w/v) SDS
> STE buffer: 10 mM Tris–HCl (pH 8.0), 1 mM EDTA, 0.1 M NaCl
> Protein dilution buffer: buffer C containing 0.3 M NaCl and 0.2 mg/ml BSA
> DNase I stop solution: 20 mM EDTA (pH 8.0), 1% (w/v) SDS, 0.2 M NaCl, 0.1 mg/ml yeast tRNA (Sigma)
> Formamide loading buffer: 90% (v/v) formamide, 10 mM EDTA (pH 8.0), 0.04% (w/v) bromphenol blue, 0.04% (w/v) xylene cyanol FF.

Preparation of Single-Stranded Circular Phagemid DNA

In our laboratory, the pBluescript phagemid vector (Stratagene, La Jolla, CA) is used for preparation of single-stranded template DNA instead of the M13 vector. pBluescript was chosen because it is smaller than M13 (~3 kb vs ~7 kb) and can be converted to a double-stranded form more easily.

1. Using synthetic oligonucleotides, clone the sequence complementary to the lesion-containing oligonucleotide into pBluescript II KS(+). Make sure that the complementary sequence is present in the strand that will be recovered as single-stranded DNA (+ strand of the *lacZ* gene, in

this case). We usually insert the desired sequences into the vector that has been double digested with *Pst*I and *Hin*dIII so that the lesion will be located near the center of the *Bss*HII fragment, which will be routinely used for EMSA (see later).

2. Transform the resulting plasmid into an appropriate *E. coli* strain (e.g., XL-1 Blue) harboring an F' episome. Grow the transformed bacteria in 500 ml of 2× YT medium containing 100 μg/ml ampicillin and then infect with the VCSM13 helper phage (Stratagene) at an MOI of ~10, according to the manufacturer's instructions. After overnight culture in the presence of 70 μg/ml kanamycin, recover secreted phage particles from the medium by standard PEG precipitation.

3. For the following *in vitro* DNA synthesis reaction, the single-stranded DNA templates must be highly purified. We have observed that the standard DNA purification procedures (including proteinase K digestion, phenol/chloroform extraction, and ethanol precipitation) are sometimes insufficient, possibly because of contamination with bacterial genomic DNA fragments. Single-stranded DNA may be purified further by CsCl density gradient centrifugation. Alternatively, phage particles may be purified by banding in a CsCl density gradient before extraction of DNA.

Preparation of Double-Stranded DNA Containing a Site-Specific Lesion

1. To ensure efficient annealing to the single-stranded DNA template, the primer oligonucleotide should be relatively long (we usually use 30-mers), and a lesion, if any, should be located in the middle of the sequence. When an artificial structure containing multiple unpaired bases (e.g., bubble or loop) is expected, the oligonucleotide may need to be longer. The oligonucleotide must be phosphorylated at the 5' end by treatment with T4 polynucleotide kinase in the presence of ATP.

2. Mix the single-stranded circular DNA (30 μg) with a twofold excess of the phosphorylated oligonucleotide (60 pmol) in 100 μl of 2× T4 DNA polymerase buffer. Heat the mixture at 65° for 15 min and then cool slowly to 20°.

3. Add an equal volume of a solution containing 2 m*M* ATP (pH 7.5), 1.2 m*M* each of dNTPs, 0.2 mg/ml BSA, T4 DNA polymerase (50 units, Takara Shuzo), and T4 DNA ligase (20 Weiss units, Takara Shuzo). Incubate the mixture at 25° for at least 6 h (incubations as long as overnight are acceptable).

4. Purify the DNA by phenol/chloroform extraction and ethanol precipitation. Covalently closed circular double-stranded DNA should then be purified by CsCl-EtBr density gradient centrifugation.

Preparation of Radiolabeled DNA Fragments for EMSA

1. Completely digest the double-stranded DNA containing the site-specific lesion (typically 1 μg = $\sim$500 fmol) with *Bss*HII. Purify the DNA by phenol/chloroform extraction and ethanol precipitation. To ensure complete recovery, include 20 μg of glycogen immediately prior to the addition of ethanol.

2. Dissolve the digested DNA in a 25-μl solution containing 1$\times$ T4 DNA polymerase buffer, 100 μM each of dATP, dGTP, and dTTP, $\sim$1 MBq of [α-^{32}P]dCTP ($\sim$110 TBq/mmol), and T4 DNA polymerase (5 units). Incubate the mixture at 20° for 15 min. To completely fill the recessed 3' termini, add 2.5 μl of a 10 mM cold dCTP solution and incubate at 20° for an additional 15 min. After the reaction is stopped by adding 1 μl of 0.5 M EDTA, remove 1 μl of the mixture to measure radioactivity incorporated into the materials that become insoluble in the presence of 5% (w/v) trichloroacetic acid. This will be used for calculating specific radioactivity of the labeled DNA fragments.

3. Purify DNA with phenol/chloroform and pass it through a Microspin G-25 column (Amersham Biosciences) to remove unincorporated nucleotides. Precipitate the eluted DNA with ethanol.

4. Load the recovered DNA onto a 5% native polyacrylamide gel (acrylamide:bis-acrylamide = 19:1, 15 $\times$ 15 $\times$ 0.1 cm) containing TGE buffer. After electrophoresis at 150 V, remove one of the glass plates, but keep the gel attached to the other glass plate. Carefully wrap the gel and glass plate and expose it to X-ray film (a 5- to 10-min exposure should be sufficient). Mark the position of the gel on the film.

5. When the exposed film is developed, a shorter fragment containing the lesion site and a longer fragment derived from the vector should be visible with nearly equal intensities. Align the gel on the film according to the marked position and excise a gel slice containing the shorter DNA band.

6. In a 1.5-ml Eppendorf tube, suspend the gel slice in 0.5 ml DNA elution buffer and carefully crush it with a disposable homogenizer. After adding 0.5 ml of the same buffer, rotate the mixture overnight at room temperature.

7. After brief centrifugation (10,000 rpm, 1 min), save the supernatant containing eluted DNA. Resuspend the pelleted gel pieces with 1 ml of fresh DNA elution buffer, and rotate the mixture again for an additional 3 h. Combine the second eluate with the first.

8. To remove traces of the gel pieces, filter the eluate through a small syringe filter (Millipore Milex-GV; 13 mm diameter, pore size 0.22 μm). DNA should be concentrated by repeated extraction with 2-butanol (to a volume of $\sim$0.4 ml or less) and then precipitated with a 2.5-fold volume of ethanol in the presence of 20 μg glycogen.

9. Dissolve the DNA in 50 μl STE buffer. Remove 1 μl of the solution and measure the radioactivity. A molar concentration of the purified DNA fragments can be calculated by comparing with the previously measured radioactivity.

Electrophoretic Mobility Shift Assay

1. Prepare a master reaction mixture as follows:

Stock solution	Volume/reaction
0.1 M sodium phosphate (pH 7.5), 5 mM EDTA	2 μl
1 M NaCl	1.1 μl
0.1 M MgCl$_2$	0.5 μl
0.1 M DTT	0.1 μl
5 mg/ml BSA (DNase free)	0.2 μl
1% (v/v) Triton X-100	0.1 μl
4 ng/μl covalently closed circular plasmid DNA	1 μl
40% (v/v) glycerol	1 μl
Sterile Milli Q water	2 μl

Multiply the indicated volume of each solution by the number of total reactions required plus two. Deliver 8 μl of the master reaction mixture into Eppendorf tubes.

2. Add 1 μl of the ^{32}P-labeled DNA probe that has been diluted to 3.5 fmol/μl with STE buffer. If all the reactions are to contain the same DNA probe, it may also be included in the master reaction mixture, and the volume to be aliquoted adjusted accordingly.

3. Add 1 μl of purified protein fractions, which are appropriately diluted with protein dilution buffer. The final reaction (10 μl) should contain 0.15 M NaCl. Calculate the master reaction mixture for a protein fraction containing 0.3 M NaCl. If the salt concentration in the protein fraction used is different from 0.3 M, the volume of the 1 M NaCl solution included in the master reaction mixture must be recalculated to provide the 0.15 M final concentration. Incubate the reaction mixtures at 30° for 30 min.

4. After the incubation, load the reaction mixtures in a cold room directly onto a 4% native polyacrylamide gel (acrylamide:bis-acrylamide = 37.5:1, 15 × 15 × 0.1 cm) containing TGE buffer and 5% (v/v) glycerol. Because the samples contain glycerol at a final concentration of 5% (v/v), they can be easily loaded by gravity. Do not include tracking dyes such as bromphenol blue, as the dye itself may affect the mobility of the DNA. Appropriately diluted tracking dyes may be loaded in an empty lane on one side of the gel to monitor electrophoresis.

5. After electrophoresis at 150 V for 3 h in a cold room, dry the gel with a vacuum gel drier. We place a sheet of DE81 chromatography paper (Whatman) between the gel and the 3 MM filter paper. Because the positively charged DE81 paper efficiently traps DNA at neutral pH, both loss of DNA and contamination of the gel drier can be minimized without fixation of the gel. Labeled DNA in the gel can be visualized by autoradiography using conventional X-ray films or imaging analyzers (e.g., BAS systems from Fuji Photofilm) (Fig. 2).

DNase I Footprinting Assay

Once a DNA substrate containing a site-specific lesion is obtained, footprinting assays can be performed to demarcate the region that is protected by binding of the damage recognition factors. The procedures for preparation of labeled DNA probes must be modified for these assays because only one end of either strand needs to be radiolabeled. We first cut the covalently closed circular DNA at a unique site near the lesion with a restriction enzyme that produces 5′ overhangs. The 5′-phosphate groups are then removed with bacterial alkaline phosphatase and labeled with T4 polynucleotide kinase and $[\gamma\text{-}^{32}P]$ATP. The labeled DNA is digested further with a second restriction enzyme on the opposite side of the lesion to produce a DNA fragment of appropriate length (200–300 bp). After filling

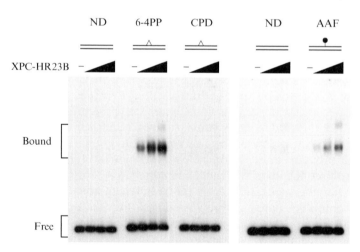

Fig. 2. EMSA showing damage-specific DNA binding of the XPC-HR23B complex. The labeled DNA probes (~180-bp *Bss*HII fragments) contained either a UV-induced photolesion (6-4PP or CPD; left) or an *N*-acetoxy-2-acetylaminofluorene (AAF) adduct (right).

the 3′-recessed termini with T4 DNA polymerase, the labeled DNA fragments are purified by native PAGE, as described for EMSA.

1. Set up binding reactions, as described for EMSA, with the exception that the reaction volume is scaled up to 25 μl. Reactions should be performed with several concentrations of protein complexes, as well as including a negative control without the protein. After incubating at 30° for 30 min, chill the reaction mixtures on ice.

2. Dilute each reaction mixture with 25 μl of a cold 5 mM CaCl$_2$ solution. Add 2 μl DNase I solution, which should be appropriately diluted with sterile Milli Q water immediately before use. The concentration of DNase I must be determined empirically so that desirable DNA ladders can be obtained reproducibly on denaturing PAGE.

3. After addition of DNase I, mix the reaction mixture quickly and incubate at 30° for exactly 2 min. Immediately stop the reaction by adding 50 μl DNase I stop solution. To achieve the same digestion time, multiple samples should not be processed simultaneously, but staggered by appropriate time intervals ($\sim$20–30 s).

4. Purify DNA by phenol/chloroform extraction followed by ethanol precipitation. Dissolve the DNA pellets in 2.5 μl formamide loading buffer. Heat denature the DNA at 95° and then chill on ice and immediately separate by denaturing PAGE (we routinely use a gel that is 20 cm wide, 40 cm long, and 0.25 mm thick). The concentration of polyacrylamide in a gel may vary depending on the length of the DNA probe as well as the distance between the lesion and the labeled end. As a position marker, in one lane of the gel run a sample of Maxam–Gilbert cleavage reaction products (a G ladder alone is usually sufficient) prepared separately using the labeled DNA probe.

5. After electrophoresis, fix the gel for 10 min in 10% (v/v) each of methanol and acetic acid. Dry the gel, and process for autoradiography (Fig. 3).

In Vitro NER Assays

Several *in vitro* assays have been developed to assess NER activities of crude cell extracts and/or purified NER protein factors. All the reaction systems described thus far appear to mimic global genome-type NER in that the presence of the XPC complex is absolutely required for the reactions. Therefore, NER activity of the purified XPC complex can be evaluated by *in vitro* complementation. In contrast, UV-DDB is not essential for such *in vitro* NER reactions, although it reportedly has a stimulatory effect on *in vitro* NER under certain conditions. General methods for setting up *in vitro* NER reactions have been described extensively (Biggerstaff and Wood, 1999; Shivji *et al.*, 1999).

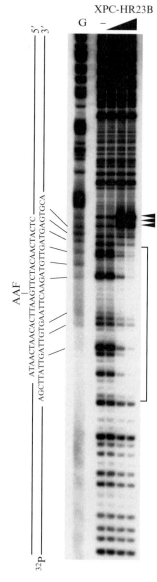

FIG. 3. DNase I footprinting of XPC-HR23B on the DNA substrate containing a site-specific AAF adduct. The DNA probe used was the *Acc65I–BglI* fragment (~280 bp) in which the 5' end of the nondamaged strand was labeled with ^{32}P. Nucleotide sequences around the lesion are shown on the left. The bracket indicates the region protected by XPC-HR23B, whereas sites that became hypersensitive to DNase I are shown by arrowheads. The lane labeled G shows the Maxam–Gilbert sequence ladder.

UV-Induced Ubiquitylation Mediated by the UV-DDB-E3 Complex

We have reported that the XPC protein *in vivo* is reversibly ubiquitylated in response to UV irradiation of cells (Sugasawa *et al.*, 2005). This ubiquitylation absolutely depends on the presence of functional UV-DDB. With the purified UV-DDB-E3 complex, the ubiquitylation of XPC can be reconstituted *in vitro*, where DDB2 and cullin 4A are also ubiquitylated. We have proposed that the UV-induced DNA lesions are efficiently detected first by UV-DDB, and the XPC complex is probably recruited to the damaged sites via a physical interaction with UV-DDB. The UV-DDB-associated E3 is then activated and ubiquitylates not only XPC but also UV-DDB itself. Upon ubiquitylation, the DNA-binding properties of the two factors are altered, which appears to be crucial for transfer of the lesions from UV-DDB to XPC and subsequent initiation of NER. This section describes methods for the detection of UV-DDB-dependent ubiquitylation.

Materials

> Cell lysis buffer: same as NP lysis buffer, except that 10 mM N-ethylmaleimide (NEM) is substituted for DTT
> Polyvinylidene difluoride (PVDF) membrane: Immobilon-P (Millipore, Billerica, MA)
> $10\times$ Ub reaction buffer: 0.5 M Tris–HCl (pH 7.6), 50 mM MgCl$_2$, 2 mM CaCl$_2$, 10 mM DTT
> Ubiquitin-activating enzyme (E1, Boston Biochem, Cambridge, MA)
> Ubiquitin-conjugating enzyme (E2, UbcH5a, Boston Biochem)
> Ubiquitin (Boston Biochem).

UV-Induced Ubiquitylation of XPC In Vivo

1. Culture human fibroblast cells at 37° to near confluence in 60-mm dishes. To detect UV-induced ubiquitylation, the cells need to express not only XPC but also functional UV-DDB. For instance, because cells derived from XP group E patients or Chinese hamsters are deficient in the expression of functional DDB2, XPC would not be ubiquitylated in these cells (Sugasawa *et al.*, 2005).

2. Remove the culture medium by aspiration, and wash the cell monolayers with prewarmed PBS. Irradiate the cells with UVC at 10 J/m^2 under a germicidal lamp (with a 254-nm peak). The UV fluence should be checked immediately before irradiation using a UVX radiometer (UVP Inc., Upland, CA). Add fresh culture medium, and incubate the cells at 37° for appropriate times.

3. After the incubation, add sodium azide to the medium to a final concentration of 0.02% (w/v) and chill the cells on ice. Discard the medium and wash the cells twice with ice-cold PBS. Remove as much PBS as possible.

4. Add 0.5 ml of ice-cold cell lysis buffer into each dish, and lyse the cells on ice for at least 30 min. Occasionally rock the dish to spread the buffer over the entire surface. We include an SH-blocking agent, NEM, in the buffer to inhibit deubiquitylating enzymes during the extraction.

5. Scrape the cell lysate into an Eppendorf tube. Wash the dish with 0.5 ml of the same buffer and combine with the lysate. Centrifuge the mixture at 20,000g for 10 min and recover the clarified supernatant (soluble cell extract), which should contain >90% of the total XPC, regardless of the presence or absence of UV treatment.

6. Determine the protein concentration of the extract according to the method of Schaffner and Weissman (1973) (i.e., the amido black protein assay). The standard Bradford method may not be applicable because of the presence of interfering agents, particularly when the protein concentration is low.

7. Load equal amounts of protein from each extract onto an SDS polyacrylamide gel (use a 7.5 or 8% gel to detect XPC). After electrophoresis, transfer the separated proteins onto a PVDF membrane. Because XPC is quite basic (pI ~9) and a relatively large (~106 kDa) protein, its efficient transfer (particularly of ubiquitylated forms) may need higher voltage and/or longer transfer times than usual. The blotted membrane can be processed according to standard procedures for immunoblotting with appropriate anti-XPC antibodies. However, the percentage of modified XPC protein is usually quite low so that highly sensitive antibodies, as well as detection systems such as chemiluminescence, may be required.

In Vitro *Ubiquitylation Assay*

1. Prepare a master reaction mixture as follows:

Stock solution	Volume/reaction
10× Ub reaction buffer	1.5 μl
1 M NaCl	x μl
0.1 M ATP (pH 7.5)	0.3 μl
5 mg/ml BSA (DNase free)	0.3 μl
1% (v/v) Triton X-100	0.15 μl
0.4 mg/ml E1	0.25 μl
2 mg/ml E2 (UbcH5a)	0.25 μl
10 mg/ml ubiquitin	0.5 μl
Sterile Milli Q water	9.75–x μl

The salt concentration in the final reaction mixture should be 0.1 M NaCl, and the appropriate volume (x) of 1 M NaCl can be calculated once the salt concentration in the protein sample is known. Multiply the indicated volume of each solution by the number of total reactions required plus two. Deliver 13 μl of the master reaction mixture into each Eppendorf tube.

2. Add 1 μl each of the purified XPC and UV-DDB-E3 complexes, which may be appropriately diluted if necessary. If any protein component is to be omitted, the corresponding volume should be replaced by buffer of the same composition. Incubate the reactions at 30° for 30 min.

3. Stop the reactions by adding SDS sample buffer. Analyze the samples by SDS–PAGE and immunoblotting as described earlier.

We have observed that purified XPC tends to aggregate when boiled in the presence of SDS, resulting in a significant loss of signal in immunoblots. This does not occur in crude cell extracts. To detect XPC for *in vitro* ubiquitylation assays, therefore, we add a small amount of a crude cell extract from XPC-deficient cells to each sample. This may not be feasible when ubiquitylation of other proteins in the same reaction is to be examined. In such cases, multiple blots may have to be prepared with aliquots of the samples.

References

Araki, M., Masutani, C., Takemura, M., Uchida, A., Sugasawa, K., Kondoh, J., Ohkuma, Y., and Hanaoka, F. (2001). Centrosome protein centrin 2/caltractin 1 is part of the xeroderma pigmentosum group C complex that initiates global genome nucleotide excision repair. *J. Biol. Chem.* **276**, 18665–18672.

Batty, D., Rapic'-Otrin, V., Levine, A. S., and Wood, R. D. (2000). Stable binding of human XPC complex to irradiated DNA confers strong discrimination for damaged sites. *J. Mol. Biol.* **300**, 275–290.

Biggerstaff, M., and Wood, R. D. (1999). Assay for nucleotide excision repair protein activity using fractionated cell extracts and UV-damaged plasmid DNA. *Methods Mol. Biol.* **113**, 357–372.

Chu, G., and Chang, E. (1988). Xeroderma pigmentosum group E cells lack a nuclear factor that binds to damaged DNA. *Science* **242**, 564–567.

Groisman, R., Polanowska, J., Kuraoka, I., Sawada, J., Saijo, M., Drapkin, R., Kisselev, A. F., Tanaka, K., and Nakatani, Y. (2003). The ubiquitin ligase activity in the DDB2 and CSA complexes is differentially regulated by the COP9 signalosome in response to DNA damage. *Cell* **113**, 357–367.

Keeney, S., Chang, G. J., and Linn, S. (1993). Characterization of a human DNA damage binding protein implicated in xeroderma pigmentosum E. *J. Biol. Chem.* **268**, 21293–21300.

Masutani, C., Araki, M., Sugasawa, K., van der Spek, P. J., Yamada, A., Uchida, A., Maekawa, T., Bootsma, D., Hoeijmakers, J. H. J., and Hanaoka, F. (1997). Identification and characterization of XPC-binding domain of hHR23B. *Mol. Cell. Biol.* **17**, 6915–6923.

Masutani, C., Sugasawa, K., Yanagisawa, J., Sonoyama, T., Ui, M., Enomoto, T., Takio, K., Tanaka, K., van der Spek, P. J., Bootsma, D., Hoeijmakers, J. H. J., and Hanaoka, F. (1994). Purification and cloning of a nucleotide excision repair complex involving the xeroderma pigmentosum group C protein and a human homolog of yeast RAD23. *EMBO J.* **13,** 1831–1843.

Moggs, J. G., Yarema, K. J., Essigmann, J. M., and Wood, R. D. (1996). Analysis of incision sites produced by human cell extracts and purified proteins during nucleotide excision repair of a 1,3-intrastrand d(GpTpG)-cisplatin adduct. *J. Biol. Chem.* **271,** 7177–7186.

Payne, A., and Chu, G. (1994). Xeroderma pigmentosum group E binding factor recognizes a broad spectrum of DNA damage. *Mutat. Res.* **310,** 89–102.

Schaffner, W., and Weissmann, C. (1973). A rapid, sensitive, and specific method for the determination of protein in dilute solution. *Anal. Biochem.* **56,** 502–514.

Shivji, M. K. K., Moggs, J. G., Kuraoka, I., and Wood, R. D. (1999). Dual-incision assays for nucleotide excision repair using DNA with a lesion at a specific site. *Methods Mol. Biol.* **113,** 373–392.

Sugasawa, K., Shimizu, Y., Iwai, S., and Hanaoka, F. (2002). A molecular mechanism for DNA damage recognition by the xeroderma pigmentosum group C protein complex. *DNA Repair* **1,** 95–107.

Sugasawa, K., Okamoto, T., Shimizu, Y., Masutani, C., Iwai, S., and Hanaoka, F. (2001). A multistep damage recognition mechanism for global genomic nucleotide excision repair. *Genes Dev.* **15,** 507–521.

Sugasawa, K., Masutani, C., Uchida, A., Maekawa, T., van der Spek, P. J., Bootsma, D., Hoeijmakers, J. H. J., and Hanaoka, F. (1996). HHR23B, a human Rad23 homolog, stimulates XPC protein in nucleotide excision repair *in vitro*. *Mol. Cell. Biol.* **16,** 4852–4861.

Sugasawa, K., Ng, J. M. Y., Masutani, C., Iwai, S., van der Spek, P. J., Eker, A. P. M., Hanaoka, F., Bootsma, D., and Hoeijmakers, J. H. J. (1998). Xeroderma pigmentosum group C protein complex is the initiator of global genome nucleotide excision repair. *Mol. Cell* **2,** 223–232.

Sugasawa, K., Okuda, Y., Saijo, M., Nishi, R., Matsuda, N., Chu, G., Mori, T., Iwai, S., Tanaka, K., Tanaka, K., and Hanaoka, F. (2005). UV-induced ubiquitylation of XPC protein mediated by UV-DDB-ubiquitin ligase complex. *Cell* **121,** 387–400.

Sugasawa, K., Ng, J. M. Y., Masutani, C., Maekawa, T., Uchida, A., van der Spek, P. J., Eker, A. P. M., Rademakers, S., Visser, C., Aboussekhra, A., Wood, R. D., Hanaoka, F., Bootsma, D., and Hoeijmakers, J. H. J. (1997). Two human homologs of Rad23 are functionally interchangeable in complex formation and stimulation of XPC repair activity. *Mol. Cell. Biol.* **17,** 6924–6931.

Tang, J., and Chu, G. (2002). Xeroderma pigmentosum complementation group E and UV-damaged DNA-binding protein. *DNA Repair* **1,** 601–616.

[12] Purification and Characterization of *Escherichia coli* and Human Nucleotide Excision Repair Enzyme Systems

By Joyce T. Reardon and Aziz Sancar

Abstract

Nucleotide excision repair is a multicomponent, multistep enzymatic system that removes a wide spectrum of DNA damage by dual incisions in the damaged strand on both sides of the lesion. The basic steps are damage recognition, dual incisions, resynthesis to replace the excised DNA, and ligation. Each step has been studied *in vitro* using cell extracts or highly purified repair factors and radiolabeled DNA of known sequence with DNA damage at a defined site. This chapter describes procedures for preparation of DNA substrates designed for analysis of damage recognition, either the 5′ or the 3′ incision event, excision (resulting from concerted dual incisions), and repair synthesis. Excision in *Escherichia coli* is accomplished by the three-subunit Uvr(A)BC excision nuclease and in humans by six repair factors: XPA, RPA, XPC·hR23B, TFIIH, XPF·ERCC1, and XPG. This chapter outlines methods for expression and purification of these essential repair factors and provides protocols for performing each of the *in vitro* repair assays with either the *E. coli* or the human excision nuclease.

Introduction

Nucleotide excision repair (hereafter, excision repair) is the primary pathway for the removal of bulky DNA lesions generated by chemical and physical agents and is conserved from *Escherichia coli* to humans even though, surprisingly, there are no evolutionary relationships between prokaryotic and eukaryotic excision repair factors. The most common environmental source of DNA damage is ultraviolet (UV) irradiation, and the cyclobutane pyrimidine dimer (CPD) is the most frequent UV-generated lesion. Many prokaryotic and eukaryotic organisms have a photolyase that directly repairs CPDs and the less abundant (6-4) photoproduct by monomerizing the dimer in a light-dependent reaction (see Sancar and Sancar, 2006). In humans, however, excision repair is the only known mechanism for the removal of CPDs and (6-4) photoproducts. The basic mechanism of excision repair includes three steps: (1) damage recognition and assembly

METHODS IN ENZYMOLOGY, VOL. 408   0076-6879/06 $35.00
                                                 DOI: 10.1016/S0076-6879(06)08012-8

of repair factors at the damage site, (2) dual incisions in the damaged DNA strand both 5′ and 3′ to the lesion and release of a damage-containing oligomer (excision), and (3) resynthesis to replace the excised oligomer, using the intact complementary strand as a template, and ligation of the repair patch.

The dual incision activity in *E. coli* is carried out by the UvrA, UvrB, and UvrC proteins, and the activity is referred to as the (A)BC excision nuclease or (A)BC excinuclease. UvrA and UvrB form an A_2B_1 hetero-trimer that is the damage recognition complex. The complex locates damage sites with low selectivity and, at sites of damage, UvrB unwinds and kinks the DNA, and UvrB undergoes significant conformational changes leading to formation of a stable UvrB–DNA complex and dissociation of $(UvrA)_2$ from the DNA. UvrC binds with high affinity and specificity to the UvrB–DNA complex and dual incisions are made at the fourth to sixth phosphodiester bond 3′ to the lesion and at the eighth bond on the 5′ side, excising the damage in the form of a 12 to 13 nucleotide-long oligomer (Fig. 1). Following dual incisions, the UvrD helicase releases UvrC and the excised oligomer, and UvrB is displaced by DNA polymerase I concomi-

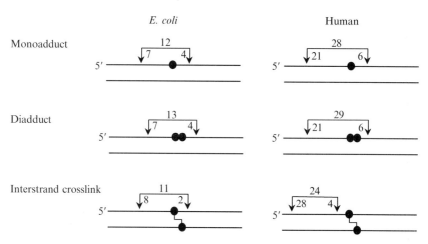

FIG. 1. Incision sites for three types of base adducts processed by *E. coli* and human excision nucleases. Closed circles indicate damaged bases, bracketed arrows show the sizes of primary excision products, and numbers below brackets indicate the number of nucleotides from the incision site to the damaged base. Acetylaminofluorene-guanine and psoralen-thymine are monoadducts, and UV-induced cyclobutane pyrimidine dimers and (6-4) photo-products and cisplatin-adducted guanines are diadducts. Psoralen, mitomycin C, and cisplatin generate interstrand cross-links in addition to intrastrand adducts. The human excision nuclease excises longer oligomers and, in the case of the interstrand cross-link, makes both incisions 5′ to the damaged base and releases a damage-free oligomer.

tant with gap filling to replace the excised oligomer. In the absence of Pol I the gap can be filled by Pol II or Pol III. Finally, the newly synthesized DNA is ligated to complete the repair reaction (Reardon and Sancar, 2005; Sancar and Reardon, 2004). Excision in at least one archaeal species, *Methanobacterium thermoautotrophicum*, is by a dual incision mechanism very similar or identical to that of *E. coli* (Ögrünç *et al.*, 1998).

In humans, excision repair is carried out by six repair factors: RPA, XPA, XPC·hR23B (XPC), TFIIH, XPG, and XPF·ERCC1. RPA, XPA, and XPC (usually in the form of XPC·TFIIH) assemble at the damage site in a random order but in a cooperative manner to form an unstable "closed complex." The helicase subunits of TFIIH (XPB and XPD) unwind the duplex at the damage site; this unwinding is accompanied by conformational changes in RPA, XPA, and XPC·TFIIH that lead to a new set of interactions producing a stable "open complex" called preincision complex 1 (PIC1). This complex is then recognized by XPG that, in an ATP-dependent reaction, enters the complex concomitant with displacement of XPC to form PIC2. XPF·ERCC1 binds PIC2 to form PIC3 in which the dual incisions are made. The so-called damage recognition subunits RPA, XPA, and XPC have low discriminatory power between damaged and undamaged DNA. The ATP hydrolysis-dependent multistep recognition provides a kinetic proofreading mechanism whereby the energy of ATP hydrolysis is converted into information on the status of DNA structure (Reardon and Sancar, 2003, 2004). This is the main mechanism of conferring specificity in human excision repair and, quite likely, in *E. coli* excision repair as well. XPG makes the first incision at the 6th ±3 phosphodiester bond 3′ to the damage followed by the XPF·ERCC1-mediated 5′ incision at the 20th ±5 phosphodiester bond, excising the damage in a 24 to 32 nucleotide-long oligomer (Fig. 1). The excised oligomer and most repair factors dissociate after the dual incision event, leaving RPA at the gap. Polymerase δ/ε and replication proteins RPA/RFC/PCNA fill in the gap, and the newly synthesized DNA is sealed by ligase (Reardon and Sancar, 2005; Sancar and Reardon, 2004).

Excision repair is modulated by higher orders of DNA structure, most notably compaction into chromatin in eukaryotes. In both prokaryotes and eukaryotes, repair is also affected by the binding of regulatory proteins and by replication, recombination, and transcription. The mechanism of transcription-coupled repair is reasonably well understood in *E. coli* and methods for studying this process have been reviewed by Selby and Sancar (2003). In this chapter we confine our discussion to methods for studying basal excision activity in the absence of chromatin compaction and other DNA transactions. There are three basic assays for *in vitro* analysis of excision repair: incision, excision, and resynthesis. The assays were first developed for and used to characterize *E. coli* repair and have been adapted for use with mammalian systems.

Substrates for *In Vitro* Repair Assays

Three general classes of substrates are used to investigate incision of DNA, excision of the damage, and replacement of the excised nucleotides. We first describe the common materials used in making substrates for these assays and then detail the preparation of substrates for the specific assays.

Materials

$[\gamma\text{-}^{32}P]$ATP and $[\alpha\text{-}^{32}P]$dCTP are from MP Biomedicals, GE Healthcare, or Perkin-Elmer Life and Analytical Sciences. T4 DNA polymerase, Klenow fragment of DNA polymerase I, T4 polynucleotide kinase, and T4 DNA ligase are from Invitrogen, New England Biolabs, or Roche. Adenosine triphosphate (ATP) and deoxynucleoside triphosphates (dNTP) are from GE Healthcare, and deoxynucleoside thiotriphosphates (dNTPαS) are from United States Biochemical. Psoralen and cisplatin are from Sigma, and *N*-acetoxy-2-acetylaminofluorene (AAF) is from the NCI Chemical Carcinogen Reference Standard Repository (Midwest Research Institute, Kansas City, MO). Frequently referred to buffers include TBE ($1\times$ is 50 m*M* Tris–borate, pH 8.3, 1 m*M* EDTA) and TEN7.4 ($1\times$ is 10 m*M* Tris–HCl, pH 7.4, 1 m*M* EDTA, 10 m*M* NaCl).

Internally Labeled DNA for the Excision Assay

The principle of the excision assay is that a radiolabel is introduced near the DNA lesion during substrate preparation so that dual incisions (excision) release a labeled, damage-containing oligomer that is detected following electrophoresis and autoradiography or PhosphorImager analysis. We have used circular substrates for this assay, but we find that linear duplexes are suitable for most experiments with either bacterial or mammalian systems.

Circular Substrates

We use pIBI25, or derivatives that were created by inserting nucleotide sequences complementary to specific primers, as templates for second-strand synthesis of double-stranded circular molecules containing one to four uniquely located thymine dimers (T<>T), a psoralen-adducted thymine, or acetylaminofluorene-guanine (AAF-G) with radiolabel 5′ to the damage (Bessho *et al.*, 1997a; Hara and Sancar, 2002; Huang *et al.*, 1992; Selby *et al.*, 1997; Svoboda *et al.*, 1993). Single-stranded DNA is prepared by standard procedures using M13K07 or R408 as the helper phage

(New England Biolabs, Promega). When starting with a new primer–template pair, labeling and second strand synthesis reactions are performed on an analytical scale (e.g., 2–5 pmol template); once experimental conditions are optimized, they can be scaled up. We have prepared 25–100 pmol substrate DNA using the following procedures (Fig. 2A).

Damage-containing oligomers (100 pmol primers) are 5′ end labeled with T4 polynucleotide kinase (50 units) and a 1.2–1.5 molar excess of [γ-^{32}P]ATP (7000 Ci/mmol). After 2–3 h at 37° in the manufacturer's buffer, labeling reactions are terminated by heating at 65°, precipitated with 95% ethanol in the presence of oyster glycogen, and washed with 70% ethanol. Phosphorylated primers are resuspended in 0.1× TEN7.4, and 200 μl annealing mixtures (20 mM Tris–HCl, pH 7.4, 2 mM MgCl$_2$, 50 mM NaCl) are set up, each containing 5–10 pmol single-stranded DNA template and a two- to threefold molar excess of primers. After 1 min at 72–75° the heating block is turned off and mixtures are slowly cooled to <35°. Samples are then placed in an iced water bath, and second strand synthesis reactions (250 μl) are prepared by sequentially adding 20 μl 10× synthesis buffer [100 mM Tris–HCl, pH 7.4, 50 mM MgCl$_2$, 20 mM dithiothreitol (DTT), 10 mM ATP, 5–10 mM each dNTP], T4 DNA ligase (30 units), and T4 DNA polymerase (15 units); reaction volumes are brought to 250 μl by the addition of 50% glycerol. Reactions are first incubated for 5 min in iced water followed by 5 min at room temperature and then 90 min at 37°. Conversion to double-stranded closed circular molecules is monitored by autoradiography after analytical scale agarose gel electrophoresis. Reaction mixtures are combined and extracted once with phenol and once with phenol:chloroform:isoamyl alcohol (25:24:1). Aqueous layers are combined, and a Microcon-30 or Centricon-30 device (Amicon) is used to remove excess oligomers and to concentrate the sample. The concentrated DNA is washed with ~10 volumes 1× TEN7.4. The retentates are collected and closed circles are separated from open circles by centrifugation in a cesium chloride–ethidium bromide gradient (Beckman Ti65 rotor, 40,000 rpm, 48 h). Fractions (150–200 μl) are collected during gradient fractionation and 5-μl aliquots are resolved in agarose gels; fractions containing closed circles are pooled. Ethidium bromide is removed by butanol extraction and cesium chloride is removed by extensive washing with TEN7.4 using a Centricon-30 device. Alternatively, closed circles are recovered by electroelution following separation from open circles in 0.7% agarose gels.

Linear DNA Substrates

Our basic design for linear substrate preparation is ligation of a damaged oligomer with three to seven other partially complementary oligomers to obtain 50- to 200-bp substrates. The 50-bp duplexes are used for electrophoretic mobility shift, incision, and excision assays, the 136- to

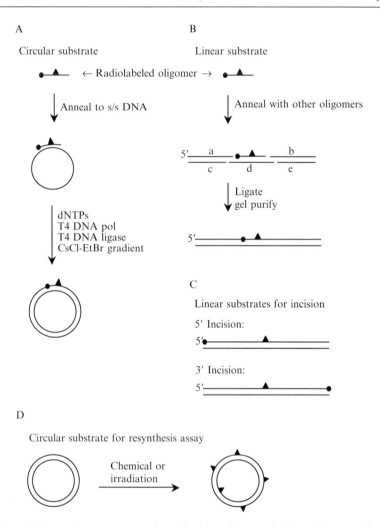

FIG. 2. Substrates for *in vitro* assays for nucleotide excision repair. The positions of radiolabel and DNA lesions are depicted with closed circles and triangles, respectively. (A) Circular substrate with damage at a defined site for the excision assay. The damage-containing oligomer is radiolabeled and annealed to a single-stranded DNA template. After second-strand synthesis and ligation, the synthesis mixture is centrifuged in a CsCl-EtBr gradient to separate covalently closed circles from nicked DNA. (B) Linear substrate for the excision assay. The central, damage-containing oligomer is 5' end labeled and mixed with partially overlapping and complementary (arm) oligomers. After annealing, the oligomers are ligated and substrate DNA is purified by standard elution methods following electrophoretic separation of full-length DNA from incomplete ligation products. (C) Linear substrate for incision assays. These DNA duplexes are prepared as described for excision assay substrates, except for the position of the radiolabel. (D) Circular substrate for the repair synthesis assay. Covalently closed DNA is prepared by standard methods, and DNA free of nicks is purified by CsCl-EtBr gradient centrifugation. The plasmid is then treated with UV or chemicals to generate randomly damaged DNA molecules.

140-bp duplexes for incision and excision assays, and the 200-bp duplexes for excision with nucleosome substrates. We prepare 136- to 140-bp substrates by annealing and ligating six oligomers, one of which contains the damage (Fig. 2B). Damage-containing oligomers, prepared "in house" or obtained from commercial sources (Phoenix BioTechnologies), have been used to make duplexes containing single T<>T, (6–4) photoproduct, psoralen, cisplatin, or AAF adducts, oxidized abasic site, thymine glycol, and 8-oxoguanine (Hara and Sancar, 2003; Huang *et al.*, 1994; Reardon and Sancar, 2003; Reardon *et al.*, 1997a, 1999; van Houten *et al.*, 1986). We introduced cholesterol as a synthetic bulky adduct that can be incorporated into oligonucleotides by automated DNA synthesis, and we made cholesterol-containing substrates using oligomers obtained from Operon Biotechnologies or Midland Certified Reagents (Matsunaga *et al.*, 1995; Mu *et al.*, 1996; Reardon *et al.*, 1997b). Unmodified scaffold ("arm") oligomers, 16–64 nucleotides in length (a to e in Fig. 2B), are obtained from Operon Biotechnologies. This method for preparing linear DNA substrates is versatile: virtually any damage-containing substrate can be prepared by changing the damage-containing oligomer and its complementary sequence, oligomer e in Fig. 2B (Huang *et al.*, 1994). When designing new substrates it is best to have 4–6 nucleotide overhangs for the complementary oligomers to achieve optimal ligation.

We typically prepare this substrate by end labeling 100 pmol of the damaged oligomer with T4 polynucleotide kinase (50 units) and $\sim$150 pmol (1 mCi) [γ-^{32}P]ATP (7000 Ci/mmol) in a 50-μl reaction. In a separate 50-μl reaction mixture, 200 pmol of the arm oligomers is combined and phosphorylated with T4 polynucleotide kinase in buffer containing 10 mM ATP so that 5′ termini are ligatable in subsequent steps. After a 2- to 3-h labeling reaction in the manufacturer's suggested buffer, the two phosphorylation reactions are heat inactivated for 5 min at >65°, combined, and the oligomers are precipitated with 95% ethanol in the presence of 50 μg oyster glycogen. The resulting DNA pellets are washed with 70% ethanol, lyophilized, and resuspended in 15 μl annealing buffer (50 mM Tris–HCl, pH 7.9, 100 mM NaCl, 10 mM MgCl$_2$, 1 mM DTT). After heating 5 min at 70–75°, the heating block is turned off and mixtures are slowly cooled to <30° before placing in iced water for 10 min. We found that the best ligation reactions are achieved in 50-μl reactions with the enzyme and buffer system from Invitrogen: to the 15-μl annealing mixture, we add 8 μl 5 × buffer, ATP to 0.2 mM, T4 DNA ligase (50 units), and water. Ligation mixtures are assembled on ice and are incubated 30–60 min at room temperature and then overnight (12–20 h) at 16°.

The efficiency of ligation is analyzed by autoradiography following resolution of 0.1 μl of the mixture in a 6% polyacrylamide gel

(20:1 acrylamide:bis-acrylamide, 2× TBE) containing 7.7 M urea (sequencing gel). The ligation mixture is lyophilized to less than 10 μl, 50 μl formamide/dye mixture (93% formamide, 10 mM EDTA, 0.013% xylene cyanole, 0.013% bromphenol blue) is added, and the DNA is resolved in a 6% sequencing gel to separate full-length DNA from incomplete ligation products. Substrate DNA is located following autoradiography, excised from the gel, and purified either by electroelution or by the crush-and-soak method using a 0.3 M sodium acetate, pH 5.2, buffer containing 10 mM EDTA and 0.5% SDS. DNA is precipitated with 95% ethanol in the presence of 50 μg oyster glycogen and washed with 70% ethanol. The pellet is dried and resuspended in 25–50 μl annealing buffer such that the radioactivity is >20,000 cpm/μl. Recovered DNA is reannealed by heating 5 min at 70–75° and slowly cooling as described earlier. The DNA concentration and yield are determined by Cerenkov counting or by absorbance at A_{260}. Often a second gel purification is required to obtain highly pure substrate; this is done as described earlier following electrophoresis in a 5% nondenaturing polyacrylamide gel (29:1 acrylamide:bis-acrylamide, 1× TBE). We have used these substrates for as long as detectable radiolabel remains—up to 3 months following substrate preparation.

End-Labeled DNA for Incision and High-Resolution Resynthesis Assays

Linear DNA is prepared as described earlier for the excision assay substrate except different oligomers are radiolabeled (Fig. 2C). For the 5′ incision assay we prepare substrate by labeling the 5′ end of the damaged strand (oligomer a in Fig. 2B). This substrate is also used for the high-resolution resynthesis assay. For 3′ labeling we perform a fill-in reaction with the Klenow fragment: our linear substrates have 1 nucleotide G overhangs at 5′ termini, and the 140-bp duplex is incubated with the Klenow fragment of DNA polymerase I and [α-^{32}P]dCTP to generate a blunt-ended 141-bp duplex with radiolabel at 3′ termini for use in the 3′ incision assay.

Circular Substrates for Low-Resolution Resynthesis Assay

Plasmid DNA is prepared from 2- to 4 liters of starting material using standard alkaline lysis of cell extracts (Maniatis *et al.*, 1982). After centrifugation in a cesium chloride–ethidium bromide gradient, covalently closed DNA is isolated. Ethidium bromide is removed by butanol extraction, and extensive dialysis against 1× TEN7.4 is used to remove cesium chloride. For damaged DNA we use pBR322 (4.3 kb), and for undamaged controls we use M13mp19 (7.2 kb) or pUC19 (2.7 kb). Plasmids are chosen so that each has the same unique restriction endonuclease site for linearization of

DNA after the repair reaction and so that the sizes are sufficiently different that they will separate during electrophoresis (Sibghat-Ullah et al., 1989). Plasmid DNA is treated with UV or chemical agents such as psoralen or cisplatin to introduce random damage (Fig. 2D). For UV damage, 5- to 10-μl droplets of pBR322 (20 μg/ml) are irradiated with 250 J/m^2 using a germicidal lamp emitting predominantly at 254 nm. This procedure introduces $\sim$20 UV photoproducts in each 4.3-kb DNA molecule, 90% of which are T<>T. After irradiation the DNA is analyzed for nicking and repurified if necessary; only plasmid DNA containing >95% closed circles is used in the resynthesis assay. Psoralen-, AAF-, and cisplatin-damaged substrates are prepared by incubating plasmids with these agents and, following appropriate treatments to remove unreacted drugs, DNA is recovered by ethanol precipitation.

Enzymes for Repair Assays

Materials and General Procedures

Chromatographic resins are from Bio-Rad Laboratories, GE Healthcare, Sigma, or IBF. Amylose, glutathione-Sepharose 4B, Ni-NTA, or anti-FLAG M2 resins for affinity purification are from New England Biolabs, GE Healthcare, Qiagen, and Sigma, respectively. All procedures are performed at 4°, and peak fractions are identified by Coomassie blue or silver staining or Western blot analysis, as indicated, after sodium dodecyl sulfate–polyacrylamide gel electrophoresis (SDS–PAGE). Purified E. coli repair factors are dialyzed against buffer A [50 mM Tris–HCl, pH 7.4, 100 mM KCl, 1 mM EDTA, 1 mM DTT, 50% (v/v) glycerol] and stored in small aliquots at −80°. Unless indicated otherwise, human repair factors are similarly stored in buffer B [25 mM HEPES, pH 7.9, 100 mM KCl, 12 mM MgCl$_2$, 0.5 mM EDTA, 2 mM DTT, 12.5–17% (v/v) glycerol].

Escherichia coli Repair Proteins

The UvrA, UvrB, and UvrC coding sequences were subcloned individually into pUNC09 to generate plasmids for protein expression under the control of the *tac* promoter, and these plasmids are used to transform CH296 cells for overproduction of each subunit (Thomas et al., 1985). Procedures were developed for the purification of milligram quantities of protein from 3- to 4 liters of starting material (Fig. 3); these protocols have been described in detail elsewhere (Sancar et al., 1988; Thomas et al., 1985) and are summarized later. More recently, simplified procedures for purifying tagged proteins from thermophilic bacteria have been developed (Jiang et al.,

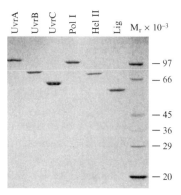

FIG. 3. *Escherichia coli* excision repair proteins. Approximately 0.5 μg of each protein was analyzed by SDS–PAGE followed by Coomassie blue staining. Note that UvrA, UvrB, and UvrC are necessary and sufficient for dual incisions. Helicase II (Hel II), DNA polymerase I (Pol I), and ligase (Lig) are required to complete repair. Numbers to the right of the gel indicate size markers. Reproduced from Husain *et al.* (1985).

2003). This section describes the original purification procedure for the *E. coli* proteins, noting some of the subsequent modifications as necessary.

UvrA

pUNC45-transformed cells are grown in LB medium to A_{600} ~0.8–1.0, induced with 1 mM isopropyl-β-D-thiogalactopyranoside for 8 h, collected by centrifugation, washed with 1× TEN7.4 or phosphate-buffered saline, and frozen at −80°. Cell pellets are thawed overnight on ice, sonicated 10 × 15 s, and cleared of cell debris by centrifugation at 12,000g for 10 min and then at 100,000g for 60 min. The cell lysate is sequentially precipitated with Polymin P and ammonium sulfate prior to chromatography on DEAE-agarose, Blue Sepharose, hydroxylapatite (Bio-gel HT), and a second DEAE-agarose column.

UvrB

pUNC211 is used for expression of UvrB as described earlier for UvrA, but there are two problems with UvrB: following sonication it is specifically cleaved to produce a 78-kDa proteolytic product and a significant amount of UvrB is precipitated with cell debris. To overcome these problems, frozen cells are thawed quickly in a 23° water bath, sonicated, and cleared of cellular debris by centrifugation at 27,000g for 20 min. Although ~90% of UvrB is lost to cleavage/precipitation even under these conditions, sufficient amounts of intact, soluble protein remain for purification. Subsequent research revealed that UvrB is cleaved by the periplasmic

OmpT protease following cell lysis. Thus, transforming pUNC211 into an *ompT⁻* cell line (UNC6200) has completely eliminated the proteolytic cleavage problem, making it possible to purify large quantities of UvrB from 1–2L cells (Lin and Sancar, 1992; Sedgwick, 1989). The clarified cell extract is adjusted to 20% saturated ammonium sulfate by the addition of solid ammonium sulfate and loaded onto a phenyl-Sepharose column that is developed with a gradient of 100 mM KCl buffer containing 20% ammonium sulfate to 50% ethylene glycol. Peak UvrB fractions, eluted in the range of 40–50% ethylene glycol, are located by SDS–PAGE, pooled, and applied to a DEAE-agarose column to obtain purified UvrB.

UvrC

pDR3274 is used for expression of UvrC as described for UvrA, and the clarified extract is applied to an AcA44 gel filtration column. Peak fractions are pooled and chromatographed sequentially on phosphocellulose (P11) and single-stranded DNA cellulose columns. UvrC binds tightly to the latter column and nearly all of the purified protein elutes in a single high-salt fraction. UvrC has low solubility in low ionic strength and, as a rule, we purify the protein in buffers containing at least 0.3 M KCl.

Mammalian Excision Repair Systems

Mammalian Cell Extracts. Most mammalian cell lines that we use are available from ATCC (Manassas, VA) or the NIGMS Human Genetic Cell Repository (Camden, NJ). We have used the method of Manley (1980) to prepare cell-free (whole cell) extracts from HeLa cells, Epstein–Barr-transformed lymphoblastoid cells, fibroblasts, mouse, and Chinese hamster ovary (CHO) cells. We have also used nuclear extracts prepared by the method of Dignam (1983); these extracts are quite active in the repair synthesis assay and are comparable in activity in the excision assay, but we have found that whole cell extract preparations are more consistent and, thus, we routinely use whole cell extracts in our studies. More recently we have prepared both nuclear and whole cell extracts from small numbers of CHO cells using the method of Li (1991); this method is particularly useful when a specific cell strain is not available in large quantities. For routine use we prefer Manley-type extracts prepared from 1- to 2 liters of exponentially growing CHO cells because these cells are easy to culture and typically yield active cell extracts. For purification of proteins we use HeLa cells because they can be cultured more economically on a large scale (50–500 liters) using medium containing 5% fetal bovine serum + 5% horse serum.

Cells are collected by centrifugation at 1000g and washed twice with phosphate-buffered saline, and the packed cell volume (PCV) is determined.

Cells are resuspended in 4 PCV of an ice-cold buffer containing 10 mM Tris–HCl, pH 8, 1 mM EDTA, and 5 mM DTT, and left on ice for 20 min. Cells are lysed with a Dounce homogenizer, using 8–10 strokes with a loose-fitting plunger. The lysate is transferred to an Erlenmeyer flask, and 4 PCV of an ice-cold buffer containing 50 mM Tris–HCl, pH 8, 10 mM MgCl$_2$, 2 mM DTT, 25% sucrose (w/v), and 50% glycerol (v/v) is added while stirring. Saturated ammonium sulfate (1 PCV) is added slowly while stirring and stirring is continued for 30 min. Lysates are centrifuged for 3 h at 100,000g (Beckman Ti 70 rotor). The supernatant is collected carefully, avoiding the cell pellet and viscous material (chromosomal DNA and cell membranes); to avoid these contaminants, ~10% of the lysate is left at the bottom of the tube. The clarified lysate is transferred to an Erlenmeyer flask, ammonium sulfate is added (0.33 g/ml supernatant), and the solution is stirred for 30 min. Precipitated proteins are recovered by centrifugation at 14,500g for 20 min, and the pellet is resuspended in a minimal volume of buffer B containing 100 mM KCl and dialyzed overnight (10–14 h) against the same buffer. Dialysates are centrifuged for 15 min at 14,500g to remove insoluble material, and the supernatant (cell extract) is frozen at −80° in small aliquots. We prefer buffer B containing 12.5% glycerol to avoid glycerol inhibition of repair reactions (see later). The extracts retain activity for at least 10 years and are stable to several freeze/thaw cycles.

Purified Human Repair Factors

PURIFICATION FROM NATIVE SOURCE. We have used different protocols with equal success for the purification of repair factors (Fig. 4) from HeLa

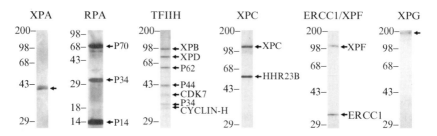

FIG. 4. Human excision repair proteins. The six repair factors necessary and sufficient for dual incision/excision are shown. Purified proteins were analyzed by SDS–PAGE followed by silver staining. Amounts of factors analyzed: XPA, 100 ng; RPA, 800 ng; TFIIH, 50 ng; XPC, 40 ng; XPF·ERCC1, 30 ng; and XPG, 40 ng. Positions of the polypeptides in each repair factor are indicated by arrows. Numbers to the left of each lane indicate size markers. Reprinted with permission from Mu *et al.* (1996). Copyright (1996) The American Society for Biochemistry and Molecular Biology.

cell extracts (Mu *et al.*, 1995, 1996). Here we outline briefly how the repair factors separated during the purification scheme that we used to purify TFIIH, XPC·hR23B, XPF·ERCC1, and XPG from large quantities of HeLa cells (Cellex Biosciences, Minneapolis, MN). XPA and RPA are overexpressed in and purified from *E. coli* and, subsequently, we developed methods to purify microgram quantities of XPC·hR23B, XPF·ERCC1, and XPG from insect cell cultures (see later).

Cell extracts (~8 g, 25 mg/ml) are prepared by the method of Manley *et al.* (1980) from 250 liters of HeLa cells and are applied to a DE52 column. Repair proteins are located by Western blotting and by complementation assay when necessary (Drapkin *et al.*, 1994; Mu *et al.*, 1995). About 20% of cellular proteins bind to this column in 100 mM KCl. The flow through, enriched for repair factors, is collected and applied to an Affi-Gel Blue column that is developed with a 0.1–0.8 M KCl gradient followed by elution with 1.5 M NaSCN. XPC-TFIIH and about half of the XPF·ERCC1 elute in the KCl gradient; XPG, RPA, and the remaining XPF·ERCC1 are eluted with NaSCN. The XPF·ERCC1 that elutes at 0.2–0.4 M KCl is purified further by affinity chromatography using a GST-XPA column (Park *et al.*, 1995). The 1.5 M NaSCN eluates are applied to an SP-Sepharose column; XPG elutes at ~0.4 M KCl and is purified further in Sephacryl S-300 and MonoS (HR 5/5) columns. The XPC-TFIIH containing fractions from the KCl elution of the Affi-Gel Blue column are applied to an SP-Sepharose column that is developed with a KCl gradient; TFIIH elutes at ~0.16 M KCl and XPC at ~0.3 M KCl. This TFIIH is free of other repair factors, except trace amounts of XPC, and is suitable for most repair assays. TFIIH is purified further by sequential chromatography on phenyl-Superose (HR10/10), Sephacryl S-300 (HiLoad 26/60), Mono S (HR 5/5), phenyl-Superose (HR 5/5), and Mono S (HR 5/5) columns. Pure XPC is obtained by chromatography on single-stranded DNA cellulose.

PURIFICATION OF RECOMBINANT PROTEINS

XPA. XPA was subcloned into *E. coli* expression vectors for purification of protein from 2 to 4 liters of cells: pMAL-c2 (New England Biolabs) for maltose-binding protein (MBP) fusion proteins, pGEX18 (GE Healthcare) for glutathione-*S*-transferase (GST) fusion proteins, and pRSET (Invitrogen) for His$_{6X}$ (HIS) fusion proteins (Park and Sancar, 1993; Park *et al.*, 1995). DR153 or BL21DE(3) cells are used for protein expression. We follow the manufacturers' protocols for affinity purification using amylose, glutathione-Sepharose 4B, or Ni-NTA resin. Typically these procedures result in partially purified XPA so we include heparin–agarose as a second chromatographic step to obtain highly purified XPA.

RPA. The p11d-tRPA plasmid that expresses all three subunits is from M.S. Wold (University of Iowa), and RPA is purified as described (Henricksen et al., 1994). We have found that highly purified RPA can be obtained with just an Affi-Gel Blue column provided that the column is washed extensively with 0.8 M KCl and 0.5 M NaSCN prior to elution with 1.5 M NaSCN. Because purified RPA often is insoluble in low salt, this protein is stored in buffer B + 300 mM KCl and diluted into buffer B + 100 mM KCl prior to use.

XPC·hR23B. XPC and hR23B DNA are from F. Hanaoka (Osaka University; Masutani et al., 1994) and were subcloned into the pMAL-c2 expression vector. We expressed intact hR23B in E. coli, but XPC is specifically cleaved during growth in bacterial cells. To circumvent this problem, we subclone DNA into the p2Bac vector (Invitrogen) and generate baculoviruses for expression of the heterodimer, or each subunit alone, in suspension cultures of SF21 cells. While XPC alone is sufficient to reconstitute excision, we typically use the heterodimer construct p2Bac. XPC-HHR23B (Reardon et al., 1996). Highly purified XPC·hR23B is obtained by sequential chromatography on phosphocellulose (P11) and single-stranded DNA cellulose columns. XPC·hR23B (or XPC alone) binds very tightly to single-stranded DNA cellulose and is essentially purified at this step. A DEAE-agarose column is used as a final purification step to remove single-stranded DNA that may have eluted with XPC, which could interfere with repair assays. XPC·hR23B is stored in 20 mM potassium phosphate, pH 7.5, 0.01% Triton X-100, 1 mM EDTA, 1 mM DTT, 300 mM KCl, 10% (v/v) glycerol, and diluted into buffer B + 100 mM KCl prior to use.

TFIIH. Recombinant TFIIH has been purified from insect cells after coinfection with three to nine baculoviruses containing TFIIH DNA. In one study the approach was a double tag strategy (Flag and His tags) for purification of TFIIH and its subcomplexes after coinfection of SF9 cells with the appropriate baculoviruses, each expressing a single subunit, and sequential chromatography using Heparin Ultrogel, cobalt affinity, and anti-FLAG M2 affinity resins (Jawhari et al., 2002; Tirode et al., 1999). A separate study developed a similar double tag strategy and a three-virus baculovirus expression system to infect HighFive cells for purification of nine-subunit TFIIH using Ni-NTA, Sephacryl S-300, and anti-FLAG M2 affinity columns (Fukuda et al., 2001).

XPF·ERCC1. XPF DNA is from Larry H. Thompson (Brookman et al., 1996; Lawrence Livermore National Laboratory) and was subcloned with ERCC1 into p2Bac (Invitrogen) for expression of the heterodimer in insect cells. Recombinant XPF·ERCC1 is purified using GST-XPA affinity chromatography (Bessho et al., 1997b). We have since obtained viruses for the expression of His-XPF and His-ERCC1 (M. Wakasugi and T. Matsunaga,

Kanazawa University). These baculoviruses are used to coinfect suspension cultures of SF21 cells, and pure XPF•ERCC1 is obtained following SP-Sepharose and Ni-NTA chromatography.

XPG. XPG DNA is from S.G. Clarkson (Scherly et al., 1993; University Medical Center, Geneva) and was subcloned into pVL1392 (Invitrogen) for expression in insect cells (Matsunaga et al., 1996). We found that optimal expression is achieved by infecting adherent SF21 cells, and highly purified XPG is obtained following chromatography on phosphocellulose (P11) and Affi-Gel Blue columns. XPG is dialyzed against and stored in buffer B + 150 mM KCl.

DDB. UV-DDB is a heterodimer (p127 + p48, DDB1 + DDB2) with very high affinity for damaged DNA (Keeney et al., 1993), and the small subunit is mutated in XP-E patients (Nichols et al., 1996). XPE cells have essentially normal excision repair activity and this protein appears to have multiple cellular functions not directly related to repair, including apoptosis and E3 ligase activity (Itoh et al., 2003; Shiyanov et al., 1999). UV-DDB has no effect on in vitro excision at low concentration and inhibits at high concentration (Reardon and Sancar, 2003). We obtained pBacPAK8-DDB1 and -DDB2 constructs from S. Linn (University of California, Berkeley) and subcloned tagged DDB2 into the p2Bac vector (Invitrogen) to generate p2Bac-Flag.DDB2.His. Both constructs are used to generate baculoviruses for each subunit alone or the heterodimer following coinfection. DDB1 is purified as described (Nichols et al., 2000), and DDB2 or the UV-DDB heterodimer are purified using anti-Flag immunoaffinity chromatography (Kulaksiz et al., 2005).

In Vitro Repair Assays

Excision Assay

The excision assay (Figs. 5A, 6, and 7) was used to determine that species from the three kingdoms of life, humans (eukarya), E. coli (pro-karya), and M. thermoautotrophicum (archaea), remove damage by making concerted dual incisions bracketing the lesion and releasing the damage in short oligomers. These oligomers are 12 to 13 nucleotides in E. coli (Lin and Sancar, 1991), 24–32 nucleotides in humans, Xenopus, and S. pombe (Huang et al., 1992; Sancar, 1995; Svoboda et al., 1993), and 11 to 12 nucleotides in M. thermoautotrophicum (Ögrünç et al., 1998) (Fig. 6). The principle of the excision assay is that a radiolabel introduced near the DNA lesion during substrate preparation is detected following electrophoresis to separate products from substrate. The excision assay can be carried out with either cell-free extracts or reconstituted excision repair systems. For

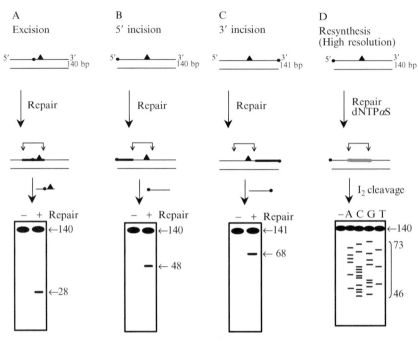

FIG. 5. Schematic illustration of repair assays. This schematic is for the human excision nuclease. It is, however, equally applicable to excision repair in any organism, except sites of incisions, sizes of excision products, and the size of the repair patch might be different in different species. The positions of radiolabel and DNA lesions, monoadduct at nucleotide 70, are depicted with closed circles and triangles, respectively; repair patches are indicated by a gray bar in D. (A–C) Primary excision and incision products are shown. (A) Excision assay. The substrate DNA has an internal radiolabel 5' to the damage. Dual incisions release a radiolabeled, damage-containing oligomer (heavy line in the schematic) of a nominal length of 28 nucleotides (range 24–32 nucleotides) that is detected by autoradiography. (B) 5' incision assay. The substrate DNA is 5' end labeled on the damaged strand. Dual incisions are made during the repair reaction but only DNA fragments resulting from the 5' incision are labeled (heavy line in the schematic) and detectable by autoradiography (nominal position at the 22nd phosphodiester bond 5' to the damage but in the range of 15–25 nucleotides). (C) 3' incision assay. The substrate DNA is 3' end labeled on both strands. Dual incisions are made during the repair reaction but only DNA fragments resulting from the 3' incision in the damaged strand are labeled (heavy line in the schematic) and detectable by autoradiography (nominal position at the 4th phosphodiester bond 3' to the damage but in the range of 3–9 nucleotides). (D) High-resolution resynthesis assay. The substrate DNA is 5' end labeled on the damaged strand and four separate reactions are required, each containing a different dNTPαS. After repair the individual mixtures are treated with iodine to cleave phosphorothioate bonds and, following electrophoresis and autoradiography, the sequence of the repair patch is revealed.

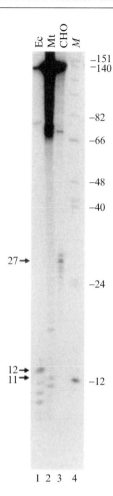

FIG. 6. Dual incision patterns in the three kingdoms. The substrate was a 136-bp duplex containing a centrally located (6-4) photoproduct with a ^{32}P label at the fourth phosphodiester bond 5′ to the photoproduct. Prokaryotes are represented by *E. coli* (Ec), archaebacteria by *M. thermoautotrophicum* (Mt), and eukaryotes by Chinese hamster ovary cells (CHO). *E. coli* releases the photoproduct in 11- to 12-mers, *M. thermoautotrophicum* in the form of 10- to 11-mers, and CHO cells in the form of a 27-mer as the major excision product. Lane 4 (M) contains molecular size markers with length in nucleotides indicated to the right. Reprinted with permission from Ögrünç *et al.* (1998). Copyright (1988) The American Society for Microbiology.

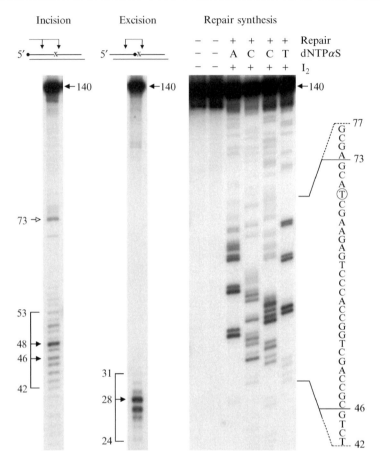

FIG. 7. Probing mammalian excision repair by "incision," "excision," and "repair synthesis" assays *in vitro*. Incision, excision, and high-resolution repair synthesis assays were performed with CHO cell extracts and a 140-bp duplex with a cholesterol substituent at position 70, circled T. In the incision assay, repair factors generate radiolabeled fragments extending from the 5' end of the substrate to the incision site(s), primarily at position 48 with less frequent incision sites covering nearly one turn of the helix; the 73-mer generated by nonspecific nucleases is indicated with an open arrow. In the excision assay with internally labeled DNA, the major excision product is a 28-mer with less frequent species in the 24 to 31 nucleotide range. In the high-resolution resynthesis assay, 5'-terminally labeled DNA is used as a substrate and the repair patch resulting from incorporation of dNTPαS is visualized after partial hydrolysis with iodine. The sequence of the repair patch is shown to the right where the circled T indicates the position of the lesion, at nucleotide 70, that was removed; the prominent repair patch (nucleotides 46–73) is bracketed and dotted lines indicate the less pronounced regions of resynthesis that extend 4 nucleotides in either direction; these are due to the minor 5' and 3' incisions that extend in either direction from the main incision sites. Adapted with permission from Reardon *et al.* (1997b). Copyright (1997) Oxford University Press.

convenience, in *E. coli* we routinely use the reconstituted system and for mammalian excision repair we use both cell-free extract and reconstituted systems; for several other species we have successfully used cell-free extracts. The assays for *E. coli* and humans are described next.

E. coli (A)BC Excinuclease

UvrA, UvrB, and UvrC proteins are diluted in reaction buffer immediately prior to use and are incubated with substrate (Fig. 2B) for 30–60 min at 37° in a 50-μl reaction mixture containing 50 mM Tris–HCl, pH 7.5, 100 mM KCl, 10 mM MgCl$_2$, 1 mM DTT, 2 mM ATP, and 100 μg/ml bovine serum albumin (BSA). Generally, 5–10 nM UvrA, 50–200 nM UvrB, and 20–50 nM UvrC are sufficient for *in vitro* reconstitution of the reaction; each preparation is titrated to determine optimal protein concentrations. Repair reactions are stopped by adding an equal volume of phenol and extracting DNA once with phenol and once with phenol:chloroform: isoamyl alcohol (25:24:1). DNA is precipitated with 95% ethanol in the presence of oyster glycogen and recovered by a 40-min centrifugation at 14,000 rpm, washed with 70% ethanol, dried, and resuspended in 6–10 μl formamide/dye mixture. The DNA is heated 3 min at 95° and resolved in 8–10% sequencing gels. Excision products are visualized by autoradiography and repair is quantitated by PhosphorImager analysis. The extent of repair is determined as a percentage of radioactivity in the excision product size range relative to the signal for full-length substrate.

Human Excision Nuclease

EXCISION WITH CELL-FREE EXTRACTS. Cell extract, 5–20 fmol internally labeled DNA (Fig. 2A and B), and 30–40 ng pBR322 are incubated at 30° in a 25-μl reaction mixture containing 25 mM HEPES, pH 7.9, 45 mM KCl, 4.4 mM MgCl$_2$, 0.4 mM DTT, 0.1 mM EDTA, 4 mM ATP, 200 μg/ml BSA, and 2.5% glycerol (including contributions from 5 μl extract stored in buffer B containing 100 mM KCl + 12.5% glycerol). Excision products are detectable within the first 10 min and increase linearly for 60–120 min; longer incubation times do not increase the level of excision and may result in exonucleolytic degradation of the primary excision products. Including pBR322 (or another covalently closed plasmid) increases the signal-to-noise ratio, presumably by providing a nonradiolabeled substrate for nucleases and nonspecific DNA-binding proteins in the extract. We generally use 50–80 μg extract in a 25-μl reaction ($\sim$2–3 mg/ml); for complementation assays we premix equal amounts (25–40 μg) of extracts from different complementation groups and then incubate with substrate as described. One limiting factor in the amount of cell extract used is the glycerol concentration, as >5–6% glycerol inhibits excision. With efficiently repaired substrates, such as the (6-4) photoproduct or cholesterol, 5 fmol DNA is sufficient. More DNA is used for less efficiently repaired substrates (T<>T or psoralen monoadduct) and for time course experiments where multiple small aliquots are removed from the 25-μl reaction mixture and to compensate for radioactive decay as the substrate ages. After incubation at 30°, reactions are stopped by treating the mixtures with 10 μg proteinase K in the presence of 0.3% SDS for 15 min at 37° and then DNA is extracted, precipitated, and processed as described for Uvr(A)BC.

EXCISION WITH PURIFIED REPAIR FACTORS. Repair factors are preincubated on ice for 10 min and then incubated with substrate (Fig. 2A and B) in a 25-μl reaction mixture containing 35 mM HEPES, pH 7.9, 65 mM KCl, 6.3 mM MgCl$_2$, 0.4 mM DTT, 0.1 mM EDTA, 4 mM ATP, 200 μg/ml BSA, and 5.4% glycerol (including contributions from 8 μl repair factors in buffer B containing 100 mM KCl + 12.5% glycerol). The amounts of repair factors used vary with individual protein preparations and must be optimized by titration. We typically use 25–50 nM XPA, 30–100 nM RPA, 2.5–5 nM XPC·hR23B, 10–20 nM TFIIH, 2–5 nM XPF·ERCC1, and 2–5 nM XPG. The reaction mixtures are incubated for 1–3 h at 30° and then processed as described earlier.

Incision Assays

Incision assays (Figs. 5B, C, and 7) are used in conjunction with or in lieu of excision assays to study repair and were instrumental in determining that (A)BC excinuclease makes dual incisions bracketing the lesion to release damage in short damage-containing oligomers (Sancar and Rupp, 1983) and in demonstrating that XPG and XPF·ERCC1 make the 3′ and 5′ incisions, respectively, in the human excision repair system (Matsunaga *et al.*, 1995). For both *E. coli* and human excision nucleases, incision reactions are conducted as described for the excision assays, except either 5′ or 3′ end-labeled substrates are used (Fig. 2C).

Repair Synthesis Assays

Low-Resolution Repair Synthesis

This assay is often used because substrate preparation is relatively straightforward and the assay does not require high amounts of radio-isotopes. However, repair synthesis is not a direct measure of repair, is prone to high background, and has limited use in probing the individual steps of excision repair. The basis of this low-resolution assay is that damage is introduced into DNA by treating closed circular DNA with chemical agents or UV (Fig. 2D); this randomly damaged DNA is then used as substrate in repair assays with cell extracts that contain all of the necessary factors for damage recognition, excision, and resynthesis (Sibghat-Ullah *et al.*, 1989; Wood *et al.*, 1988). The reaction mixture includes dNTPs, one of which is radiolabeled: during resynthesis the repair patches are labeled and quantitation of incorporated radioactivity serves as a measure of DNA repair. As an internal control for nonspecific incorporation of dNTPs, a second plasmid of different size is included in the reaction mixture.

Cell extract (10 μl, 100–200 μg) is incubated with 200 ng damaged pBR322 and 200 ng undamaged M13mp19 (or pUC19) DNA in a 50-μl reaction mixture containing 35 mM HEPES, pH 7.9, 60 mM KCl, 5.6 mM MgCl$_2$, 0.6 mM DTT, 0.2 mM EDTA, 2 mM ATP, 20 μM dATP, 20 μM dGTP, 20 μM TTP, 8 μM dCTP (+ 2 μCi [α-^{32}P]dCTP), 200 μg/ml BSA, and 2.5–3.4% glycerol. Some investigators include phosphoenolpyruvate/pyruvate kinase in the reaction mixture, but we find that this ATP regeneration system is unnecessary. After a 2- to 3-h incubation at 30°, reactions are stopped by treating with proteinase K/SDS, and DNA is recovered as described earlier. Precipitated DNA is resuspended in 25 μl *Eco*RI buffer, digested for 1 h, and separated in a 1% agarose gel. DNA is visualized by ethidium bromide staining followed by autoradiography, and repair is quantified by Phosphor-Imager analysis. The extent of repair is determined from levels of incorporated radioactivity, normalized for the recovered DNA as determined by scanning the ethidium bromide-stained gel.

High-Resolution Repair Synthesis

This assay (Fig. 5D) is based on the phosphorothioate sequencing method developed by Gish and Eckstein (1988) and was used for precise measurement of repair patches at single nucleotide resolution in both *E. coli* (Sibghat-Ullah *et al.*, 1990) and mammalian systems (Huang *et al.*, 1992; Reardon *et al.*, 1997b). 5' end-labeled DNA (Fig. 2C) is incubated with mammalian cell extracts as described earlier except dNTPs are at 80 μM. Four separate reactions are carried out, each with a different dNTPαS (also at 80 μM). Following excision and resynthesis, DNA is deproteinized, recovered by ethanol precipitation, and resolved in an 8% sequencing gel. Full-length DNA is located by autoradiography and recovered by electroelution and ethanol precipitation. Precipitated DNA is resuspended in 9 μl 20 mM HEPES, pH 7.5, heated for 3 min at 70°, and cooled to room temperature. Iodine is added to 0.5 mM and the reaction mixture is incubated 1 min at 25°. DNA is precipitated with ethanol, resuspended in formamide/dye mixture, and resolved in a 12% sequencing gel without heat prior to loading. The repair patch, generated by partial hydrolysis at phosphorothioate bonds, is visualized after autoradiography (Fig. 7).

Concluding Comments

There is a growing interest in excision repair among scientists with limited or no background in DNA repair. A previous review presented the theoretical principles of various *in vivo* and *in vitro* repair assays for nucleotide excision repair and discussed the advantages and disadvantages of each (Mu and Sancar, 1997). This chapter described the technical details of *in vitro* repair assays developed in our laboratory and it is hoped that they will be useful to the specialists and nonspecialists alike. These assays were instrumental in the discovery of excision by concerted dual incisions in both prokaryotes and eukaryotes, and more recently in Archaea, in defining the minimal set of repair factors required in *E. coli* and human cells, in determining the spectrum of DNA adducts that are substrates for excision repair, and in probing the mechanistic details of excision repair in both systems. These assays are now widely used to further analyze the mechanistic details of excision repair in *E. coli*, yeast, and humans. The assays should be equally useful for studying repair in other organisms because the mechanistic aspects of nucleotide excision repair, if not the individual components, appear to be conserved across species.

Acknowledgment

This work was supported by National Institutes of Health Grant GM32833.

References

Bessho, T., Mu, D., and Sancar, A. (1997a). Initiation of DNA interstrand cross-link repair in humans: The nucleotide excision repair system makes dual incisions 5' to the cross-linked base and removes a 22- to 28-nucleotide-long damage-free strand. *Mol. Cell. Biol.* **17,** 6822–6830.

Bessho, T., Sancar, A., Thompson, L. H., and Thelen, M. P. (1997b). Reconstitution of human excision nuclease with recombinant XPF-ERCC1 complex. *J. Biol. Chem.* **272,** 3833–3837.

Brookman, K. W., Lamerdin, J. E., Thelen, M. P., Hwang, M., Reardon, J. T., Sancar, A., Zhou, Z.-Q., Walter, C. A., Parris, C. N., and Thompson, L. H. (1996). *ERCC4* (*XPF*) encodes a human nucleotide excision repair protein with eukaryotic recombination homologs. *Mol. Cell. Biol.* **16,** 6553–6562.

Dignam, J. D., Lebovitz, R. M., and Roeder, R. G. (1983). Accurate transcription initiation by RNA polymerase II in a soluble extract from isolated mammalian nuclei. *Nucleic Acids Res.* **11,** 1475–1489.

Drapkin, R., Reardon, J. T., Ansari, A., Huang, J.-C., Zawel, L., Ahn, K., Sancar, A., and Reinberg, D. (1994). Dual role of TFIIH in DNA excision repair and in transcription by RNA polymerase II. *Nature* **368,** 769–772.

Fukuda, A., Yamauchi, J., Wu, S.-Y., Chiang, C.-M., Muramatsu, M., and Hisatake, K. (2001). Reconstitution of recombinant TFIIH that can mediate activator-dependent transcription. *Genes Cells* **6,** 707–719.

Gish, G., and Eckstein, F. (1988). DNA and RNA sequence determination based on phosphorothioate chemistry. *Science* **240**, 1520–1522.

Hara, R., and Sancar, A. (2002). The SWI/SNF chromatin-remodeling factor stimulates repair by human excision nuclease in the mononucleosome core particle. *Mol. Cell. Biol.* **22**, 6779–6787.

Hara, R., and Sancar, A. (2003). Effect of damage type on stimulation of human excision nuclease by SWI/SNF chromatin remodeling factor. *Mol. Cell. Biol.* **23**, 4121–4125.

Henricksen, L. A., Umbricht, C. B., and Wold, M. S. (1994). Recombinant replication protein A: Expression, complex formation, and functional characterization. *J. Biol. Chem.* **269**, 11121–11132.

Huang, J.-C., Svoboda, D. L., Reardon, J. T., and Sancar, A. (1992). Human nucleotide excision nuclease removes thymine dimers from DNA by incising the 22nd phosphodiester bond 5′ and the 6th phosphodiester bond 3′ to the photodimer. *Proc. Natl. Acad. Sci. USA* **89**, 3664–3668.

Huang, J.-C., Hsu, D. S., Kazantsev, A., and Sancar, A. (1994). Substrate spectrum of human excinuclease: Repair of abasic sites, methylated bases, mismatches, and bulky adducts. *Proc. Natl. Acad. Sci. USA* **91**, 12213–12217.

Husain, I., Van Houten, B., Thomas, D. C., Abdel-Moncar, M., and Sancar, A. (1985). Effect of DNA polymerase I and DNA helicase II on the turnover rate of UvrABC excision nuclease. *Proc. Natl. Acad. Sci. USA* **82**, 6774–6778.

Itoh, T., O'Shea, C., and Linn, S. (2003). Impaired regulation of tumor suppressor p53 caused by mutations in the Xeroderma pigmentosum *DDB2* gene: Mutual regulatory interactions between p48^{DDB2} and p53. *Mol. Cell. Biol.* **23**, 7540–7553.

Jawhari, A., Uhring, M., Crucifix, C., Fribourg, S., Schultz, P., Poterszman, A., Egly, J. M., and Moras, D. (2002). Expression of FLAG fusion proteins in insect cells: Application to the multi-subunit transcription/DNA repair factor TFIIH. *Protein Expression Purif.* **24**, 513–523.

Jiang, G. H., Skorvaga, M., Van Houten, B., and States, J. C. (2003). Reduced sulfhydryls maintain specific incision of BPDE-DNA adducts by recombinant thermoresistant *Bacillus caldotenax* UvrABC endonuclease. *Protein Expression Purif.* **31**, 88–98.

Keeney, S., Chang, G. J., and Linn, S. (1993). Characterization of a human DNA damage binding protein implicated in xeroderma pigmentosum E. *J. Biol. Chem.* **268**, 21293–21300.

Kulaksiz, G., Reardon, J. T., and Sancar, A. (2005). Xeroderma pigmentosom complementation group E protein (XPE/DDB2): Purification of various complexes of XPE and analyses of their damaged DNA binding and putative DNA repair properties. *Mol. Cell. Biol.* **25**, 9784–9792.

Li, Y., Ross, J., Scheppler, J. A., and Franza, B. R., Jr. (1991). An *in vitro* transcription analysis of early responses of the human immunodeficiency virus type 1 long terminal repeat to different transcriptional activators. *Mol. Cell. Biol.* **11**, 1883–1893.

Lin, J.-J., and Sancar, A. (1991). The C-terminal half of UvrC protein is sufficient to reconstitute (A)BC excinuclease. *Proc. Natl. Acad. Sci. USA* **88**, 6824–6828.

Lin, J. J., and Sancar, A. (1992). (A)BC excinuclease: The *Escherichia coli* nucleotide excision repair enzyme. *Mol. Microbiol.* **6**, 2219–2224.

Maniatis, T., Fritsch, E. F., and Sambrook, J. (1982). "Molecular Cloning: A Laboratory Manual." Cold Spring Harbor Laboratory, Cold Spring Harbor, NY.

Manley, J. L., Fire, A., Cano, A., Sharp, P. A., and Gefter, M. L. (1980). DNA-dependent transcription of adenovirus genes in a soluble whole-cell extract. *Proc. Natl. Acad. Sci. USA* **77**, 3855–3859.

Masutani, C., Sugasawa, K., Yanagisawa, J., Sonoyama, T., Ui, M., Enomoto, T., Takio, K., Tanaka, K., van der Spek, P. J., Bootsma, D., Hoeijmakers, J. H. J., and Hanaoka, F. (1994). Purification and cloning of a nucleotide excision repair complex involving the xeroderma pigmentosum group C protein and a human homologue of yeast RAD23. *EMBO J.* **13**, 1831–1843.

Matsunaga, T., Mu, D., Park, C.-H., Reardon, J. T., and Sancar, A. (1995). Human DNA repair excision nuclease: Analysis of the roles of the subunits involved in dual incisions by using anti-XPG and anti-ERCC1 antibodies. *J. Biol. Chem.* **270**, 20862–20869.

Matsunaga, T., Park, C.-H., Bessho, T., Mu, D., and Sancar, A. (1996). Replication protein A confers structure-specific endonuclease activities to the XPF-ERCC1 and XPG subunits of human DNA repair excision nuclease. *J. Biol. Chem.* **271**, 11047–11050.

Mu, D., Hsu, D. S., and Sancar, A. (1996). Reaction mechanism of human DNA repair excision nuclease. *J. Biol. Chem.* **271**, 8285–8294.

Mu, D., Park, C.-H., Matsunaga, T., Hsu, D. S., Reardon, J. T., and Sancar, A. (1995). Reconstitution of human DNA repair excision nuclease in a highly defined system. *J. Biol. Chem.* **270**, 2415–2418.

Mu, D., and Sancar, A. (1997). DNA excision repair assays. *Prog. Nucleic Acid Res. Mol. Biol.* **56**, 63–81.

Nichols, A. F., Itoh, T., Graham, J. A., Liu, W., Yamaizumi, M., and Linn, S. (2000). Human damage-specific DNA-binding protein p48. *J. Biol. Chem.* **275**, 21422–21428.

Nichols, A. F., Ong, P., and Linn, S. (1996). Mutations specific to the xeroderma pigmentosum group E Ddb⁻ phenotype. *J. Biol. Chem.* **271**, 24317–24320.

Ögrünç, M., Becker, D. F., Ragsdale, S. W., and Sancar, A. (1998). Nucleotide excision repair in the third kingdom. *J. Bacteriol.* **180**, 5796–5798.

Park, C.-H., Mu, D., Reardon, J. T., and Sancar, A. (1995). The general transcription-repair factor TFIIH is recruited to the excision repair complex by the XPA protein independent of the TFIIE transcription factor. *J. Biol. Chem.* **270**, 4896–4902.

Park, C.-H., and Sancar, A. (1993). Reconstitution of mammalian excision repair activity with mutant cell-free extracts and XPAC and ERCC1 proteins expressed in *Escherichia coli.* *Nucleic Acids Res.* **21**, 5110–5116.

Reardon, J. T., Bessho, T., Kung, H. C., Bolton, P. H., and Sancar, A. (1997a). *In vitro* repair of oxidative DNA damage by human nucleotide excision repair system: Possible explanation for neurodegeneration in Xeroderma pigmentosum patients. *Proc. Natl. Acad. Sci. USA* **94**, 9463–9468.

Reardon, J. T., Mu, D., and Sancar, A. (1996). Overproduction, purification, and characterization of the XPC subunit of the human DNA repair excision nuclease. *J. Biol. Chem.* **271**, 19451–19456.

Reardon, J. T., and Sancar, A. (2003). Recognition and repair of the cyclobutane thymine dimer, a major cause of skin cancers, by the human excision nuclease. *Genes Dev.* **17**, 2539–2551.

Reardon, J. T., and Sancar, A. (2004). Thermodynamic cooperativity and kinetic proofreading in DNA damage recognition and repair. *Cell Cycle* **3**, 141–144.

Reardon, J. T., and Sancar, A. (2005). Nucleotide excision repair. *Prog. Nucleic Acid Res. Mol. Biol.* **79**, 183–235.

Reardon, J. T., Thompson, L. H., and Sancar, A. (1997b). Rodent UV-sensitive mutant cell lines in complementation groups 6–10 have normal general excision repair activity. *Nucleic Acids Res.* **25**, 1015–1021.

Reardon, J. T., Vaisman, A., Chaney, S. G., and Sancar, A. (1999). Efficient nucleotide excision repair of cisplatin, oxaliplatin, and bis-aceto-ammine-dichloro-cyclohexyl-amine-platinum(IV) (JM216) platinum intrastrand DNA diadducts. *Cancer Res.* **59**, 3968–3971.

Sancar, A. (1995). Excision repair in mammalian cells. *J. Biol. Chem.* **270**, 15915–15918.

Sancar, A., and Rupp, W. D. (1983). A novel repair enzyme: UVRABC excision nuclease of *Escherichia coli* cuts a DNA strand on both sides of the damaged region. *Cell* **33,** 249–260.

Sancar, A., and Reardon, J. T. (2004). Nucleotide excision repair in *E. coli* and man. *In* "Advances in Protein Chemistry" (W. Yang, ed.), Vol. 69, pp. 43–71. Elsevier Academic Press, San Diego.

Sancar, G. B., and Sancar, A. (2006). Purification and characterization of DNA photolyases. *Methods Enzymol* **408**[9] this volume.

Sancar, A., Thomas, D. C., Van Houten, B., Husain, I., and Levy, M. (1988). Purification and properties of UvrABC excision nuclease and its utilization to detect bulky DNA adducts. *In* "DNA Repair: A Laboratory Manual of Research Procedures" (E. C. Friedberg and P. C. Hanawalt, eds.), Vol. 3, pp. 479–508. Dekker, New York.

Scherly, D., Nouspikel, T., Corlet, J., Ucla, C., Bairoch, A., and Clarkson, S. G. (1993). Complementation of the DNA repair defect in xeroderma pigmentosum group G cells by a human cDNA related to yeast *RAD2*. *Nature* **363,** 182–185.

Sedgwick, B. (1989). *In vitro* proteolytic cleavage of the *Escherichia coli* Ada protein by the ompT gene product. *J. Bacteriol.* **171,** 2249–2251.

Selby, C. P., Drapkin, R., Reinberg, D., and Sancar, A. (1997). RNA polymerase II stalled at a thymine dimer: Footprint and effect on excision repair. *Nucleic Acids Res.* **25,** 787–793.

Selby, C. P., and Sancar, A. (2003). Characterization of transcription-repair coupling factors in *E. coli* and humans. *Methods Enzymol.* **371,** 300–324.

Shiyanov, P., Nag, A., and Raychaudhuri, P. (1999). Cullin 4A associates with the UV-damaged DNA-binding protein DDB. *J. Biol. Chem.* **274,** 35309–35312.

Sibghat-Ullah, Husain, I., Carlton, W., and Sancar, A. (1989). Human nucleotide excision repair *in vitro*: Repair of pyrimidine dimers, psoralen and cisplatin adducts by HeLa cell-free extract. *Nucleic Acids Res.* **17,** 4471–4484.

Sibghat-Ullah, Sancar, A., and Hearst, J. E. (1990). The repair patch of *E. coli* (A)BC excinuclease. *Nucleic Acids Res.* **18,** 5051–5053.

Svoboda, D. L., Taylor, J.-S, Hearst, J. E., and Sancar, A. (1993). DNA repair by eukaryotic nucleotide excision nuclease: Removal of thymine dimer and psoralen monoadduct by HeLa cell-free extract and of thymine dimer by *Xenopus laevis* oocytes. *J. Biol. Chem.* **268,** 1931–1936.

Thomas, D. C., Levy, M., and Sancar, A. (1985). Amplification and purification of UvrA, UvrB, and UvrC proteins of *Escherichia coli*. *J. Biol. Chem.* **260,** 9875–9883.

Tirode, F., Busso, D., Coin, F., and Egly, J.-M. (1999). Reconstitution of the transcription factor TFIIH: Assignment of functions for the three enzymatic subunits, XPB, XPD, and cdk 7. *Mol. Cell* **3,** 87–95.

Van Houten, B., Gamper, H., Hearst, J. E., and Sancar, A. (1986). Construction of DNA substrates modified with psoralen at a unique site and study of the action mechanism of ABC excinuclease on these uniformly modified substrates. *J. Biol. Chem.* **261,** 14135–14141.

Wood, R. D., Robins, P., and Lindahl, T. (1988). Complementation of the Xeroderma pigmentosum DNA repair defect in cell-free extracts. *Cell* **53,** 97–106.

[13] Assays for Transcription Elongation by RNA Polymerase II Using Oligo(dC)-Tailed Template with Single DNA Damage

By Isao Kuraoka and Kiyoji Tanaka

Abstract

A DNA molecule is vulnerable to many types of DNA-damaging agents of endogenous and environmental origins. Although damage to DNA can interfere not only with replication but also transcription, the majority of DNA repair and mutagenesis studies are based on the actions of DNA polymerases in DNA replication. To investigate the actions of RNA polymerase II (RNA-PII) encountering a single DNA lesion on transcription elongation, we employ a transcription elongation assay using purified RNAPII and oligo (dC)-tailed templates containing a DNA lesion at a specific site. This chapter describes an analysis of whether elongating RNAPII stalls at a DNA lesion or whether RNAPII generates mutations in RNA transcripts.

Introduction

When elongating RNA polymerase II (RNAPII) encounters DNA lesions on the transcribed strand in living cells, it basically has to choose between two actions (Doetsch, 2002; Tornaletti, 2005). One action is to stall at the DNA lesion. It is thought that this stalling recruits other DNA repair factors to remove the lesion. If DNA repair (maybe transcription-coupled DNA repair; TCR) does not occur to remove the lesion, which blocks transcription elongation, the stalling of RNAPII continues. Consequently, the cell cannot produce the mRNA transcripts and, if the transcripts are essential, the cell will die. The second action is to bypass the lesion and continue to synthesize the mRNA transcript. Even if RNAPII can bypass a lesion to produce the transcript, the process results in ribonucleotide misincorpration and will generate mutant transcripts. When there is one mutation in the termination codon of an mRNA transcript, the transcript will be subjected to nonsense-mediated mRNA decay (Conti and Izaurralde, 2005; Lejeune and Maquat, 2005). However, when there is no mutation in the termination codon, the mutant transcript generated by damaged transcribed strands may be an important source of mutant proteins, particularly in nondividing cells. Therefore, DNA lesions on transcribed strand are a major challenge to RNAPII transcription.

METHODS IN ENZYMOLOGY, VOL. 408
0076-6879/06 $35.00
DOI: 10.1016/S0076-6879(06)08013-X

To analyze the direct actions of RNAPII on a DNA lesion on the transcribed strand, we employed an *in vitro* transcription elongation assay using purified RNAPII and an oligo(dC)-tailed template containing a single DNA lesion on the transcribed strand at a specific site (Kuraoka *et al.*, 2003; Mei Kwei *et al.*, 2004). This assay can be performed in the absence of any other transcription factors (e.g., *in vitro* transcription assays require TBP, TFIIB, THIIE, THIIF, and TFIIH) and does not need a specific promoter (Yamaguchi *et al.*, 1999). Thus, there is little protein contamination and also few additional effects of other transcription factors on RNAPII in this assay. Only the enzymatic actions of RNAPII on damaged templates are observed.

This chapter describes the generation of an oligo(dC)-tailed template containing a DNA lesion on the transcribed strand at a specific site, an assay for RNAPII transcription elongation, and a technique for the analysis of RNA transcripts generated from damaged templates.

Construction of Closed-Circular Duplex DNA Substrates Containing a DNA Lesion at a Specific Site

To generate an oligo(dC)-tailed template, closed-circular duplex DNA (cccDNA) substrates containing a DNA lesion have to be prepared (Moggs *et al.*, 1996; Shivji *et al.*, 1999). An oligonucleotide containing a DNA lesion of interest is annealed to single-stranded pBluescript DNA modified to obtain a sequence complementary to the oligonucleotide. The cccDNA substrates are synthesized by T4 DNA polymerases and T4 DNA ligases and purified by EtBr-CsCl density gradient centrifugation. The DNA lesion is located on the transcribed strand at a specific site.

Solutions and Materials

> Gel-purified 24-mer oligonucleotides containing a DNA lesion (several DNA lesions phophoramidite are available at Glen-Research)
> Single-stranded pBluescript DNAs are prepared as described (Sambrook and Russell, 2001)

1. For the annealing reaction, a reaction mixture is set up in 200-μl aliquots containing 50 pmol of single-stranded pBluescript DNA, 50 pmol of 5′-phosphorylated oligonucleotide containing a DNA lesion, and 20 μl of 10× T4 DNA polymerase buffer (TaKaRa). Incubate the mixture at 70° for 5 min, 37° for 30 min, 25° for 20 min, and 4° for 20 min.

2. Add 30 μl of 10× T4 DNA polymerase buffer, 50 μl of 0.1% bovine serum albumin (BSA) (TaKaRa), 5 μl of 100 m*M* ATP (Amersham

Biosciences), 100 μl of 2.5 mM dNTP (TaKaRa), 25 μl of T4 DNA polymerase (4 U/μl, TaKaRa), 10 μl of T4 DNA ligase (350 U/μl, TaKaRa), and 80 μl of H$_2$O. Incubate at 37° for 4 h.

3. Purify the synthesized closed-circular duplex DNA substrates by EtBr-CsCl density gradients with a final CsCl concentration of 1.55 g/ml. Dissolve 15 g of CsCl in 9.5 ml of H$_2$O at 37° for 1 h. Mix 500 μl of DNA sample, 1.352 ml of CsCl solution, and 148 μl of a 10-mg/mL EtBr solution (Invitrogen) in a 2-ml Quick-Seal tube. Perform EtBr/CsCl density gradient centrifugation at 90,000 rpm in a TLA 120.2 (Beckman) ultracentrifuge rotor at 18° for ~16 h.

4. Use a handheld UV lamp (312 nm) to visualize the DNA. Collect the lower band of closed-circular duplex DNA with a 21-gauge hypodermic needle and 1-ml syringe in a 1.5-ml tube. Add an equal volume of isoamyl alcohol, vortex for 5 s, and centrifuge at ~15,000 rpm for 1 min at room temperature. Remove and discard the upper phase (pink color). Repeat the extractions until no trace of pink remains. Dilute the DNA solution to 4.0 ml with 10 mM Tris–HCl (pH 8.5).

5. Add the sample to an Amicon Ultra-4 Centrifugal 30K filter device unit, and centrifuge at 4000g for 20 min at 4° to concentrate the DNA to ~50 μl. Add more 10 mM Tris–HCl (pH 8.5) to 4 ml and repeat this step four more times. Quantify the final DNA solution using a spectrophotometer and confirm DNA purity. When there are restriction sites within a DNA lesion site, the presence of the lesion can be confirmed by digestion with a restriction endonuclease. Store the purified DNA in aliquots at −80°.

Preparation of an Oligo(dc)-Tailed Template from Closed-Circular Duplex DNA Substrates

For transcription elongation by RNAPII, an oligo(dC)-tailed template containing a lesion at a specific site on the transcribed strand is generated using EtBr/CsCl density gradient-purified cccDNA substrates. To generate an oligo(dC) tail for RNAPII, we constructed a vector: pBluescript KS– GTG (Kuraoka et al., 2003). In our system, this vector is designed to generate ~130 nucleotide RNA transcripts from a start point of transcription to the damaged region. Purified cccDNA substrates containing a lesion are digested with PstI, and the oligo(dC) tail is added to the 3′ end by terminal deoxynucleotide transferase. After digestion with SmaI, the oligo (dC)-tailed templates containing a lesion are purified.

1. A reaction mixture is set up in 50-μl aliquots. Add 5 μl of 10× restriction endonuclease buffer (TOYOBO) and 5 μl of PstI restriction endonuclease (TOYOBO) to 40 μl of a DNA solution containing the cccDNA (5 pmol) with a lesion at a specific site.

2. Incubate at 37° for 60 min. Add 50 μl of phenol–chloroform–isoamyl alcohol, vortex for 1 min, and centrifuge at 15,000 rpm for 5 min at room temperature. Transfer 45 μl of the upper phase to a new tube. Add 2 μl of pellet paint Co-Precipitant (Novagen) and 5 μl of 3 M sodium acetate and 125 μl of 100% ethanol. Incubate for 2 min at room temperature and centrifuge at 15,000 rpm for 5 min at 4°. Remove the ethanol solution carefully to retain the DNA pellets and rinse 100 μl of 70% ethanol. Dry the DNA pellets.

3. Resuspend the DNA solution with 35 μl of H_2O and add 5 μl of 10× buffer (Amersham Bioscience), 1 mM dCTP (Amersham Bioscience), and 5 μl of terminal deoxynucleotide transferase (Amersham Bioscience). Repeat step 2.

4. Resuspend the DNA solution with 40 μl of H_2O and add 5 μl of 10× restriction endonuclease buffer (TOYOBO) and 5 μl of SmaI restriction endonuclease (TOYOBO). Incubate at 37° for 60 min, add 10 μl of 6× gel loading buffer (TOYOBO), and then purify 3-kbp products using 1% agarose gel electrophoresis. Dilute the DNA to a concentration of 10 ng/μl and store the aliquots at 4° in the dark.

RNAPII Elongation Reaction Using Oligo(dC)-Tailed Template

Figure 1 shows an outline of the assay for RNAPII transcription elongation using an oligo(dC)-tailed template containing a DNA lesion at a specific site.

Hot Labeling Reaction (Fig. 1A)

To investigate whether RNAPII stalls at a lesion of interest, we employ a hot labeling reaction using ^{32}P-UTP and a denatured 6% polyacrylamide gel. When RNAPII stalls, ~130 nucleotide RNA transcripts are observed, generated from the transcription start site to the damaged site in our system (Fig. 1, top). When RNAPII bypasses the lesion, elongated transcripts are observed giving products of around 3 kbp on the denatured gel.

Solutions and Materials

The RNAPII fraction is prepared from HeLa cells as described elsewhere (Hasegawa et al., 2003; Usuda et al., 1991) or purified RNAPII can be purchased from ProteinOne.

NE(−)buffer: 20 mM Tris–HCl (pH 8.0), 20% glycerol, 100 mM KCl, 1 mM EDTA, 1 mM dithiothreitol

NE(+) buffer: NE(−) buffer containing 12.5 mM MgCl$_2$

X50 ACGUTP mixture: 2.5 mM ATP, 2.5 mM CTP, 2.5 mM GTP, 0.5 mM UTP (Amersham Bioscience)

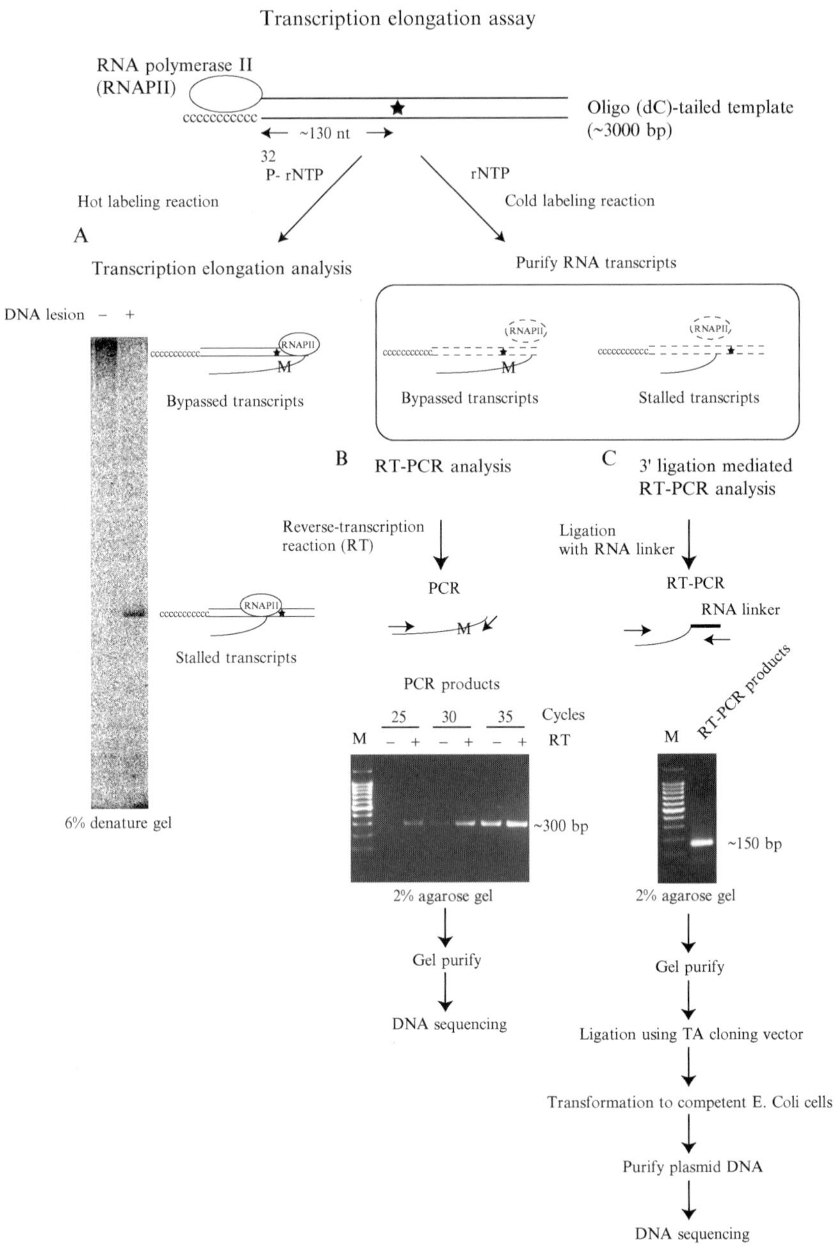

Transcription elongation assay

FIG. 1. Flow diagram of the analysis of RNAPII actions in transcription elongation using an oligo(dC)-tailed template containing a DNA lesion at a specific site.

Stop buffer: 7 M urea, 0.35 M NaCl, 10 mM Tris–HCl (pH 7.4), 10 mM EDTA, 1% SDS

1. Reaction mixtures are set up in 25-μl aliquots. Prepare DNA solutions, Pol II solutions, and NTP solutions as follows. DNA solutions (10 μl): 1 μl of 10 ng/μl of oligo(dC)-tailed template containing a lesion, 0.12 μl of 1 M MgCl$_2$, 2 μl of 40% PEG, and 6.88 μl of H$_2$O. Pol II solutions (4 μL): 1 μl of RNAPII fraction (100 ng/μl), 0.1 μl of 50 mg/ml BSA (Nacalai), 0.2 μl of ribonuclease inhibitor (RNaseIn: Ambion: 20 U/μl), and 2.7 μl of NE(−) buffer. NTP solutions (5 μl): 0.5 μl of ^{32}P-UTP (Amersham Biosciences), 0.5 μl of ×50 ACGUTP mixture, 4 μl of NE(+) buffer, and 5 μl of H$_2$O.

2. Add 20 μl of DNA solution, 8 μl of Pol II solution, and 12 μl of NE(−) buffer to a fresh RNase-free tube.

3. Preincubate the reaction mixtures at 30° for 30 min. Add 5 μl of NTP solution to the reactions. (The elongation reaction will be started by this step.) Incubate for a further 2–60 min.

4. Add 100 μl of stop buffer to the reaction tube, and mix with 100 μl of phenol–chloroform–isoamyl alcohol. Vortex for 5 min and centrifuge for 5 min at maximum speed.

5. Carefully collect 90 μl of the aqueous phase and transfer it to a fresh tube containing 2 μl of yeast tRNA (50 μg/ml), and 10 μl of 3 M sodium acetate. Mix in 250 μl of ice-cold 100% ethanol and then place on dry ice for 20 min.

6. Centrifuge at maximum speed for 15 min at 4° and carefully remove the alcohol. Add 100 μl of 70% ethanol and centrifuge at maximum speed for 5 min at 4°.

7. Carefully remove the alcohol and dry the RNA pellet in a centrifuge under vacuum for 5 min. Resuspend the RNA pellet in 5 μl of formamide loading dye (USB).

8. Heat the samples at 95° for 2 min and load onto a denatured 6% polyacrylamide gel.

9. Transfer the gel to 3MM paper and dry the gel. Visualize by exposing the gel to X-ray film or with a FUJIFILM BAS 2500 bioimage analyzer. An example of the type of image is shown in Fig. 1A.

Cold Labeling Reaction

Cold labeling reactions are employed to analyze RNA transcripts produced from oligo(dC)-tailed templates containing a DNA lesion by RNA-PII. When RNAPII bypasses a lesion, it may misincorporate nucleotides opposite a lesion. For this, we sequence DNA fragments produced by RT-PCR using bypassed transcription products (Fig. 1B). However, when

RNAPII stalls at a DNA lesion, RNAPII-stalled RNA transcripts are generated. To investigate whether the stalled RNAPII has the ability to incorporate nucleotides, we use 3′ ligation-mediated RT-PCR to isolate and sequence the stalled transcripts (Fig. 1C).

Preparation of RNA Transcripts from RNAPII Transcription Elongation Assay Using Oligo(dC)-Tailed Template

Solutions and Materials

Basically, we use the same reagents as in the hot labeling reaction.

1. Reaction mixtures are set up in 50-μl aliquots. Prepare DNA solutions, Pol II solutions, and NTP solutions as follows. DNA solutions (20 μl): 2 μl of 10 ng/μl of oligo(dC) template containing a lesion, 0.24 μl of 1 M MgCl$_2$, 4 μl of 40% PEG, and 13.76 μl of H$_2$O. Pol II solutions (8 μl): 2 μl of RNAPII fraction (100 ng/μl), 0.2 μl of 50 mg/ml BSA, 0.4 μl of RNaseIn, and 5.4 μl of NE(−) buffer. NTP solutions (10 μl): 1 μl of ×50 ACGUTP mixture, 0.5 μl of 2 mM UTP (Amersham Bioscience), 4 μl of NE(+) buffer, and 5 μl of H$_2$O.

2. Add 20 μl of DNA solution, 8 μl of Pol II solution, and 12 μl of NE(−) buffer to a fresh RNase-free tube.

3. Preincubate the reaction mixtures at 30° for 30 min. Add 10 μl of NTP solution to the reaction. Incubate for a further 20–60 min to accumulate RNA transcripts.

4. Add 100 μl of stop buffer to the reaction tubes.

5. Purify transcripts of RNA using an RNeasy minikit (Qiagen). At present, there is no way to purify RNA completely free of DNA, even if the DNA is not visible on an agarose gel. Therefore, we treat the purified RNA transcripts with RNase-free DNase I.

6. Reaction mixtures are set up in 20-μl aliquots. Add 2 μl of 10× DNase buffer (TaKaRa), 0.25 μl of RNaseIn (40 units/μl), 0.1 μl of RNase-free DNase I (5 Kunitz units/μl, TaKaRa), and 17.65 μl of the purified RNA sample (must be less than 1 μg).

7. Incubate at 37° for 30 min. Add 2 μl of 50 mM EDTA. Incubate at 65° for 5 min.

8. Store the DNase I-treated RNAs in aliquots at −20 or −80°.

Analysis of Lesion-Bypassed RNAPII Elongation Transcripts (Fig. 1B)

To examine the nucleotide preference for incorporation opposite a DNA lesion, DNA fragments produced by RT-PCR from RNA transcripts are sequenced. It is important to have good experimental conditions because there is a possibility that the PCR itself will generate a mutation in the RNA and that the PCR may also produce excess DNA fragments from contaminated template DNAs.

Reverse-Transcription Reaction for First-Strand CDNA Synthesis and PCR

1. Reaction mixtures are set up in 12-μl aliquots. Add 1 μl of proper primer (10 pmol/μl), 2 μl of the DNase I-treated RNA solution, and 9 μl of RNase-free H_2O.
2. Incubate at 65° for 5 min and place on ice.
3. Add 4 μl of 5× RT buffer (TOYOBO), 2 μl of 10 mM dNTP solution, 1 μl of ribonuclease inhibitor (10 units/μl, TOYOBO), and 1 μl of reverse transcriptase (4 units/μl, TOYOBO).
4. Incubate at 42° for 60 min and at 85° for 5 min.
5. Store the first-stranded cDNA in aliquots at −20 or −80°.
6. PCR mixtures are set up in 100-μl aliquots. Add 4 μl of the RT solution or the mock solution, 10 μl of 10× PCR buffer (TaKaRa), 8 μl of 2.5 mM dNTP mixture, 6 μl of 25 mM $MgCl_2$, 0.5 μl of primer A (100 pmol/μl), 0.5 μl of primer B (100 pmol/μl), and 0.5 μl of Taq polymerase (5 U/μl, TaKaRa). Also perform a mock reaction using DNase I-treated RNA solutions without a reverse transcription reaction.
7. Perform 25, 30, and 35 cycles: 94° for 30 s, 55° for 30 s, and 72° for 30 s. Check PCR products on the 2% agarose gel. Here, it is important to decide on good PCR conditions (maybe PCR cycles). The PCR products should be detected in the RT solutions, but not the mock solution, because any PCR products of the mock solution must be from contaminated DNA templates. Usually, we use less than 30 cycles (see Fig. 1B, 2% agarose gel).
8. Purify the PCR products from the agarose gel. Sequence the products. Mutations should be detected at sites of DNA lesions.

Analysis of 3′ Ligation-Mediated RT-PCR (Fig. 1C)

To analyze the 3′ end of RNAPII-stalled transcripts on damaged templates, the transcripts must first be ligated with RNA oligonucleotides. The RNA oligonucleotide has 5′-phosphorylated residues for the ligation reaction and 3′ NH_3 residues for preventing self-ligation and ligation with the 5′ of transcripts. The ligated transcripts are amplified by RT-PCR, subcloned into the plasmid vector, and then introduced into competent *Escherichia coli* cells. Sequence data of the plasmid DNA indicate that RNAPII incorporated nucleotides opposite the lesion (Mei Kwei *et al.*, 2004).

Materials

RNA linker: the 5′ end of RNA oligonucleotides is phosphorylated and the 3′ end is NH$_2$ modified.

1. Reaction mixtures are set up in 50-μl aliquots. Add 5 μl of the DNase I-treated RNA solution, 1 μl of RNA linker (20 pmol/μl), 3 μl of 0.1% BSA (TaKaRa), 25 μl of 50% PEG 8000, 5 μl of 10× T4 RNA ligase buffer (TaKaRa), 1 μl of T4 RNA ligase (40 U/μl, TaKaRa), and 10 μl of H$_2$O.
2. Incubate at 16° for 14–16 h.
3. After ligation, incubate at 95° for 3 min.
4. Use 1.4 μl of the ligated transcript solution and a Qiagen OneStep RT-PCR kit (Qiagen). Follow the manufacturer's instructions using primers specific for the 3′-ligated RNA transcripts. Perform a RT reaction: 50° for 30 min, 94° for 30 s, and PCR for 25 cycles: 94° for 30 s, 55° for 30 s, and 72° for 30 s.
5. Add 10 μl of 6× gel-loading buffer to reaction mixtures. Run the 2% agarose gel in 1× TAE buffer and purify the PCR products using the QIAquick gel extraction kit (Qiagen).
6. Ligate the purified PCR products into a TOPO TA cloning vector (Invitrogen) and transform competent *E. coli* cells.
7. Sequence the isolated plasmids. Sequencing data before the anti-RNA linker indicates the position of the 3′ end of RNAPII-stalled transcripts.

Acknowledgments

We thank the past and present members of our laboratory, including Joan Seah Mei Kwei, Katsuyoshi Horibata, Mika Hayashida, and Kyoko Suzuki, for their contributions to the methods described here.

References

Conti, E., and Izaurralde, E. (2005). Nonsense-mediated mRNA decay: Molecular insights and mechanistic variations across species. *Curr. Opin. Cell Biol.* **17,** 316–325.

Doetsch, P. W. (2002). Translesion synthesis by RNA polymerases: Occurrence and biological implications for transcriptional mutagenesis. *Mutat. Res.* **510,** 131–140.

Hasegawa, J., Endou, M., Narita, T., Yamada, T., Yamaguchi, Y., Wada, T., and Handa, H. (2003). A rapid purification method for human RNA polymerase II by two-step affinity chromatography. *J. Biochem. (Tokyo)* **133,** 133–138.

Kuraoka, I., Endou, M., Yamaguchi, Y., Wada, T., Handa, H., and Tanaka, K. (2003). Effects of endogenous DNA base lesions on transcription elongation by mammalian RNA polymerase II. Implications for transcription-coupled DNA repair and transcriptional mutagenesis. *J. Biol. Chem.* **278,** 7294–7299.

Lejeune, F., and Maquat, L. E. (2005). Mechanistic links between nonsense-mediated mRNA decay and pre-mRNA splicing in mammalian cells. *Curr. Opin. Cell Biol.* **17**, 309–315.

Mei Kwei, J. S., Kuraoka, I., Horibata, K., Ubukata, M., Kobatake, E., Iwai, S., Handa, H., and Tanaka, K. (2004). Blockage of RNA polymerase II at a cyclobutane pyrimidine dimer and 6–4 photoproduct. *Biochem. Biophys. Res. Commun.* **320**, 1133–1138.

Moggs, J. G., Yarema, K. J., Essigmann, J. M., and Wood, R. D. (1996). Analysis of incision sites produced by human cell extracts and purified proteins during nucleotide excision repair of a 1,3-intrastrand d(GpTpG)-cisplatin adduct. *J. Biol. Chem.* **271**, 7177–7186.

Sambrook, J., and Russell, D. W. (2001). "Molecular Cloning: A Laboratory Manual," 3rd Ed., pp. 3.30–3.48. Cold Spring Harbor Laboratory Press, Cold Spring Harbor, NY.

Shivji, M. K., Moggs, J. G., Kuraoka, I., and Wood, R. D. (1999). Dual-incision assays for nucleotide excision repair using DNA with a lesion at a specific site. *Methods Mol. Biol.* **113**, 373–392.

Tornaletti, S. (2005). Transcription arrest at DNA damage sites. *Mutat. Res.*

Usuda, Y., Kubota, A., Berk, A. J., and Handa, H. (1991). Affinity purification of transcription factor IIA from HeLa cell nuclear extracts. *EMBO J.* **10**, 2305–2310.

Yamaguchi, Y., Takagi, T., Wada, T., Yano, K., Furuya, A., Sugimoto, S., Hasegawa, J., and Handa, H. (1999). NELF, a multisubunit complex containing RD, cooperates with DSIF to repress RNA polymerase II elongation. *Cell* **97**, 41–51.

[14] *In Vivo* Assays for Transcription-Coupled Repair

By Graciela Spivak, Gerd P. Pfeifer, and Philip Hanawalt

Abstract

This chapter describes the technologies used in our respective laboratories to study the incidence and repair of lesions induced in specific DNA sequences by ultraviolet light, chemical carcinogens, and products of cellular metabolism. The Southern blot method is suitable for analysis of damage and repair in the individual DNA strands of specific restriction fragments up to 25,000 nucleotides in length, whereas the ligation-mediated polymerase chain reaction approach permits analysis of shorter sequences at the nucleotide level. Both methods have unique advantages and limitations for particular applications.

Introduction

In addition to the general excision repair pathways, there are dedicated mechanisms for the removal of some types of lesions from actively transcribed DNA. Transcription-coupled repair (TCR), a specialized pathway for repair in the transcribed strands of active genes, has been demonstrated unequivocally for bulky DNA lesions such as cyclobutane pyrimidine

METHODS IN ENZYMOLOGY, VOL. 408
0076-6879/06 $35.00
DOI: 10.1016/S0076-6879(06)08014-1

dimers (CPD) in bacteria, yeast, and mammalian cells (Mellon, 2005). *In vitro* studies have demonstrated that CPD induce complete arrest of transcription by RNA polymerases from phage, bacteria, and mammals (Scicchitano *et al.*, 2004; Tornaletti and Hanawalt, 1999). TCR requires active transcription (Christians and Hanawalt, 1992); thus the working hypothesis is that arrest of RNA polymerase in the elongation mode serves as a signal to recruit repair complexes. It has been established that lesions that arrest transcription also provide strong signals for accumulation of p53 and apoptosis (Ljungman and Zhang, 1996; Yamaizumi and Sugano, 1994); thus the absence of sunlight-induced tumors in TCR-deficient individuals might be explained by high levels of apoptotic cell death in exposed skin. Novel anticancer therapeutic approaches based on these ideas utilize compounds such as illudin S and Irofulven, which induce DNA damage that is processed exclusively by TCR (Jaspers *et al.*, 2002).

Methods for measuring the incidence and repair of CPD were enhanced greatly when the CPD-specific activity of T4 endonuclease V was characterized (Friedberg *et al.*, 1980). The Southern blot method for the analysis of TCR of UV-induced CPD was developed in the early 1980s (Bohr *et al.*, 1985; Mellon *et al.*, 1987) and used to first define TCR as a pathway targeting repair to damage in transcribed strands of expressed genes. The principle of the method is based on the conversion of DNA lesions into single strand breaks by means of lesion-specific endonucleases (such as T4 endonuclease V), chemical or thermal hydrolysis, exposure to laser irradiation, and so on. Then the approach requires electrophoretic separation of restriction fragments in denaturing agarose gels and hybridization to strand-specific ^{32}P-labeled probes for the particular DNA fragments of interest. DNA strands containing lesions that have been converted into strand breaks are shorter and will migrate farther into the gel than undamaged, full-length strands; as repair takes place, a gradual recovery of signal from full-length strands is observed and quantified. The number of lesions in each strand is calculated using the Poisson expression,

$$\text{single strand breaks} = -\ln(\text{signal from the nicked sample/signal from the control sample}) \tag{1}$$

where the nicked and the control samples are aliquots from each preparation (time point, dose, etc). An example of a Southern blot image from a TCR experiment is shown in Fig. 1.

The ligation-mediated polymerase chain reaction (LMPCR) was originally developed as a method for genomic DNA footprinting and sequencing (Mueller and Wold, 1989; Pfeifer *et al.*, 1989). It was then adapted for the detection of DNA adducts (Pfeifer *et al.*, 1991, 1992). This method also relies on the specific conversion of DNA adducts into strand breaks;

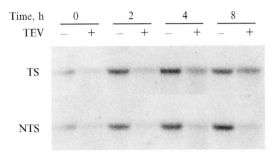

FIG. 1. Removal of UV-induced CPD from the transcribed strand of the *DHFR* gene in Chinese hamster CHO9 cells. Cells were irradiated with 10 J/m² of UVC (254 nm) and lysed immediately or incubated for the times indicated above the lanes. DNA was isolated and restricted with *Kpn*I to yield a 14-kb fragment of *DHFR*. ³²P-labeled riboprobes for the transcribed and the nontranscribed strands of *DHFR* were synthesized using pZH-4 as the template (Mellon *et al.*, 1987). The membrane was probed for one strand, then stripped and probed for the opposite strand. Quantitation of the signal and calculation of the number of CPD indicate that 15, 51, and 83% of the CPD had been removed from the transcribed strand (TS) at 2, 4, and 8 h, respectively, whereas no significant repair of the nontranscribed strand (NTS) was detected (0–10% removal). TEV, T4 phage endonuclease V.

however, an additional requirement for LMPCR is that the breaks must contain an intact nucleotide with a 5′-phosphate group. In a single-sided PCR procedure, gene-specific DNA fragments with a fixed end on one side (defined by a gene-specific primer) and a variable end on the other side (defined by the position of the nick at a lesion) are amplified and detected by hybridization to strand-specific, ³²P-labeled probes.

CPD can be mapped at the DNA sequence level by treatment with T4 endonuclease V, which cleaves the glycosidic bond of the 5′ base in a pyrimidine dimer and also cleaves the sugar-phosphate backbone between the two dimerized pyrimidines; thus, the digestion products still contain dimerized pyrimidine bases at the cleavage sites. The dimer must then be photoreversed with *Escherichia coli* photolyase to allow ligation of the primer (Pfeifer *et al.*, 1992). LMPCR is sensitive enough to measure DNA repair rates for CPD at individual nucleotide positions (Tommasi *et al.*, 2000; Tornaletti and Pfeifer, 1994; Tu *et al.*, 1996). Repair of other DNA lesions has also been measured using LMPCR (Denissenko *et al.*, 1998; Feng *et al.*, 2002; Grishko *et al.*, 1999; Valadez and Guengerich, 2004; Wei *et al.*, 1995; Ye *et al.*, 1998). DNA glycosylases produce abasic sites that can be treated with alkali to result in a strand break. Other base excision repair enzymes that have β-δ lyase activity, such as the *E. coli* Fpg or Nth proteins, produce appropriate strand breaks directly. Bulky DNA adducts can be cleaved with the *E. coli* UvrABC complex, which makes incisions, in most cases four nucleotides 3' to the adduct. These 3' incisions can be detected by LMPCR and this approach has been used for mapping of adducts produced by polycyclic aromatic hydrocarbons (Denissenko *et al.*, 1998). This chapter provides a detailed protocol for analysis of CPD photoproducts and their repair rates by LMPCR. The protocol can be adapted easily for mapping of other types of DNA damage and determining repair.

After the strand cleavage reaction, a gene-specific oligonucleotide primer is annealed to the denatured DNA and is extended to create a blunt end. Then, a ligation reaction provides a common sequence at all 5' ends, allowing exponential PCR to be used for signal amplification. After 18 to 20 PCR amplification cycles, the DNA fragments are separated on sequencing gels, electroblotted onto nylon membranes, and hybridized with a gene-specific probe to visualize the sequence ladders. The general LMPCR procedure is outlined in Fig. 2.

The choice of methodology for the study of incidence and repair of DNA lesions on the individual strands in specific DNA sequences will depend on the end point. The Southern blot is useful for examination of repair levels within particular genes, as the analyzed fragment "averages" over a large region. This method is also relatively insensitive to other lesions that are not converted to strand breaks by the lesion-specific nicking agent. Any other lesions that may interfere with the DNA polymerase reaction can affect the LMPCR approach. However, LMPCR is the method of choice if particular nucleotide context "hot" or "cold" spots are to be analyzed and if correlation with other end points is desired, such as mutation, nucleosome structure, and natural transcription pause sites. The LMPCR method requires 1–2 μg DNA per sample, whereas 10–20 μg per lane is needed to obtain detectable signals in the Southern blot method. The Southern blot method has a "built-in" control, as data for the transcribed and the nontranscribed strands of a particular restriction fragment can be obtained from the same membrane by subsequent hybridizations with the respective riboprobes; for LMPCR analysis of lesions in the opposite strands, separate reactions must be carried out with appropriately designed primers.

Important General Considerations

Detailed protocols describing these methods have been published previously: the Southern blot method (Anson and Bohr, 1999; Bohr and Okumoto, 1988; Spivak and Hanawalt, 1995) and LMPCR (Drouin *et al.*, 1996a; Tornaletti and Pfeifer, 1996). This chapter details the techniques

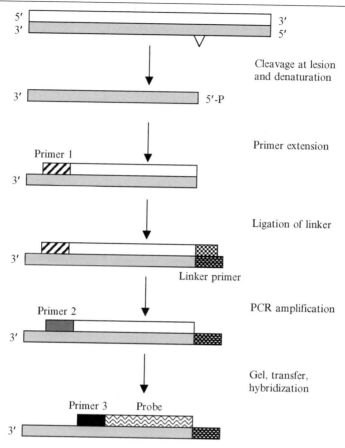

Fɪɢ. 2. Schematic outline of the LMPCR method. Positions of the primers and the hybridization probe are indicated.

that have been used successfully in our respective laboratories for many years.

• Selection of the type and dose of the damaging agent requires careful determinations of cell toxicity and the frequency of lesions induced; a delicate balance must be achieved between the need to induce an average of one lesion per DNA strand in the restriction fragment under examination (Southern blot) or in the sequence being examined (LMPCR) and the need to harvest live cells for analysis of late time points.

• Certain chemicals with long half-lives and/or trapping in cellular compartments continue inducing damage for several hours after removal of the chemical, thus complicating the determination of the initial, "0 time" lesion frequency. In these cases it is important to determine the number of lesions at closely spaced intervals during the early period after treatment.

• Both methods require enzymatic or physical means for nicking the DNA at or near the lesions; the nicking activity should be specific for the lesions to be analyzed and have low background activity on undamaged DNA.

• If the chosen treatment induces alkali-sensitive sites, the denaturing gel electrophoresis step in the Southern blot method may need to be modified [e.g., by using glyoxal denaturation and neutral agarose gels (Drouin et al., 1996b)] to avoid loss of full-length fragments.

• The transcriptional status of each DNA strand within the sequence under examination must be determined experimentally. The analysis of TCR can be complicated in situations where a common promoter is used for divergent transcription in opposite directions (Mellon et al., 1987), when both strands are allegedly transcribed (Venema et al., 1991), or when repair of nontranscribed sequences flanking active genes is affected by transcription (Tolbert and Kantor, 1996).

• Finally, it should be noted that the operation of TCR does not necessarily result in more rapid repair of lesions on the transcribed strand. The efficient global genomic repair of some types of lesions may mask the dedicated TCR pathway. TCR may then be revealed in mutant cells defective in global repair, such as human cells deficient in the XPC gene product, which is not required for TCR of "bulky" adducts.

Southern Blot Method

Cells

We will detail the method as applied to mammalian cells growing in monolayers; this can be adapted for cells that grow in suspension and for cells from other organisms.

It is important to isolate DNA that did not replicate during the repair period to avoid dilution of the signal with daughter, lesion-free DNA strands. This can be achieved by inhibition of replication by the treatment *per se*; by adding inhibitors of replication such as hydroxyurea; or by using a density label such as bromodeoxyuridine in the culture medium after treatment and then separating parental light-density DNA from daughter hybrid-density DNA by means of isopycnic sedimentation in CsCl gradients. Lesions remaining in the parental strand after a round of replication can be analyzed by resolving the respective strands from the hybrid density DNA in denaturing CsCl gradients (Spivak and Hanawalt, 1992).

Radioactive Labeling (Optional)

To prelabel cells, plate them at ~10% of the confluent density and let cells attach for a few hours. Add 0.1–0.5 μCi/ml [^{3}H]thymidine 1 mCi/ml, 80 Ci/mmol and enough nonradioactive thymidine for a final concentration of 10 μM. Alternatively, 0.01 μCi/ml [^{14}C]thymidine 0.1 mCi/ml, 60 Ci/mmol can be used. The appropriate amount of radioactive label must be determined experimentally for each cell type, as cell lines vary in their sensitivity to the radiolytic/toxic effects of these compounds. Optimum labeling conditions should yield ~2000 cpm/μg DNA and cause no significant decrease in cell survival.

Grow cells for two or three generations in radioactive medium. Remove cells from the culture dishes and combine them before replating in noradioactive medium. This step is designed to equally distribute cells that may have different levels of DNA labeling. The final number of culture dishes should be sufficient to yield at least 50 μg DNA per sample to allow duplicate determinations for analysis of repair in single-copy genes. Depending on the dose used, some cell lines will continue to actively replicate their DNA after damage; thus, the relative amount of unreplicated DNA will decrease with time. More dishes should then be prepared for the later repair time points in order to obtain comparable amounts of unreplicated DNA for analysis.

Damage and Repair

Rinse cells twice with warm phosphate-buffered saline (PBS, 137 mM NaCl, 2.75 mM KCl, 15.25 mM Na$_2$HPO$_4$, 1.45 mM KH$_2$PO$_4$).

For UV irradiation, we use a Westinghouse IL782–30 germicidal lamp, which emits primarily 254-nm light. The incident dose must be determined with a photometer; the distance between the lamp and the sample can be adjusted to administer the desired dose in at least 20 sec to allow accurate timing, but for no more than 2 min to avoid dehydration. It is important to ensure uniform illumination along and across the surface of the culture dishes. The lamp should be turned on for ~15 min before use to allow the spectral output to stabilize and left on until all the dishes have been irradiated; a simple shutter such as a sheet of cardboard can be used to shield the cells before and after irradiation. Remove the lids from the dishes while still covered by the shutter; remove the shutter and time the exposure, close the shutter, and replace the lids. Replace culture medium or add lysis solution immediately.

To establish the initial lesion frequency (time = 0), the cells are rinsed twice with *ice-cold* PBS, irradiated, and lysed immediately. Under these conditions, 1 J/m^2 induces an average of 0.007 CPD/kb in each DNA strand.

For treatments with chemical agents, rinse the cells with PBS or other isotonic-buffered solution, add the reagent in buffer, medium with or without serum, as appropriate, and incubate at the temperature and for the time predetermined to induce the desired number of lesions; remove the solution, rinse twice with isotonic buffer to remove all traces of the chemical, and lyse cells immediately for "0 time" samples. Add medium and incubate UV-irradiated or chemically treated cells for the desired repair times. If using density labeling during posttreatment incubation, the medium should contain 10 μM bromodeoxyuridine, 1 μM fluorodeoxyuridine, and 10 μM deoxycytidine.

Extraction and Purification of DNA

Samples are collected by trypsinization and resuspension in cold medium or buffer containing serum to inactivate the trypsin. Collected samples are stored refrigerated or frozen until all the samples have been harvested. Alternatively, samples may be treated at different times and collected all at once. To extract the DNA we have used either organic extraction or salt precipitation methods.

Genomic DNA Precipitation with NaI, SDS, and Isopropanol

This method was adapted from procedures described previously (Greenberg and Ziff, 1984; Hamilton et al., 2001; Nakae et al., 1995; Wang et al., 1994). Cells are pelleted by centrifugation at 2000–3000 rpm for 5 min and resuspended in 1 ml PBS.

Prepare nuclei by adding 5 ml NP-40 lysis solution [10 mM Tris–HCl, pH 7.5, 10 mM NaCl, 3 mM MgCl$_2$, 0.5% (v/v) Nonidet P-40], mixing, and centrifuging as described earlier. To lyse nuclei, add 0.4 ml 1% (w/v) SDS, 5 mM EDTA-Na$_2$, 10 mM Tris–HCl, pH 8.0, and mix. Add proteinase K to 200 μg/ml and incubate at 37° for 1 h, mixing often. To precipitate DNA, add 0.6 ml sodium iodide solution (7.5 M NaI, 20 mM EDTA-Na$_2$, 40 mM Tris–HCl, pH 8.0), mix gently, add 1 ml isopropanol, and mix by inversion until a white precipitate forms. If DNA fibers are visible, spool DNA with a glass rod, immerse in 40% isopropanol and then in 70% ethanol, drain off liquid, and release DNA into a tube with 0.5–3 ml TE buffer (10 mM Tris–HCl, pH 8.0, 1 mM EDTA). Alternatively, spin DNA for 30 min at 5000 rpm in a centrifuge with swinging buckets, decant supernatant, rinse pellet with 5 ml 40% isopropanol, spin 20 min at 5000 rpm, decant, rinse with 5 ml 70% ethanol, spin, decant, air dry tubes upside down, and resuspend DNA with TE buffer. This procedure works well for $\sim 10^7$ fibroblasts per sample; amounts of reagents can be adjusted for different numbers of cells.

Organic Extraction

Samples are collected by trypsinization and resuspension in cold medium; alternatively, lysis solution can be added to the plates. The cells are lysed at 56° for 1 h in 10 mM Tris–HCl, pH 7.5, 1 mM EDTA, and 0.5% SDS to which proteinase K is freshly added to 0.1 mg/ml. Transfer samples to phenol-resistant centrifuge tubes. Wearing protective gloves, glasses, and a lab coat and working in a fume hood, add 1 volume buffer-saturated phenol, cap tubes securely, and incubate at room temperature in a rocking apparatus for 30 min. Centrifuge samples in swinging buckets at 5000 rpm for 30 min at room temperature and remove aqueous phases (and white membrane-like interphases if present) to fresh tubes. Repeat extraction with 1 volume 50% phenol, 48% chloroform, and 2% isoamyl alcohol and then with 1 volume 96% chloroform and 4% isoamyl alcohol. To maximize the yield of DNA, the organic phases can be reextracted with one-third volume TE buffer; the aqueous phases are then pooled before the final extraction.

To precipitate DNA, add 0.25 volume 10 M ammonium acetate, mix, add 2 volume chilled 100% ethanol, and mix by inversion until white precipitate forms. The DNA can be collected by spooling or centrifugation as described earlier, rinsed with 70% ethanol, and air dried.

Resuspend DNA in TE buffer. Determine DNA concentration of appropriate dilutions by A_{260} optical density, taking into account that close to 50% of the purified nucleic acid will be RNA. The high molecular weight DNA may be very viscous and difficult to pipette; cutting off the end of pipette tips to widen the bore may help.

To obtain the desired restriction fragments, add restriction buffer and 2.5 units restriction enzyme/μg of DNA and incubate for 1 h at 37°. Repeat addition of restriction enzyme, and add DNase-free RNase A to 50 μg/ml and incubate for 1 h at 37°. To ensure that restriction is complete, electrophorese small samples (~ 10 μl) on 0.7% agarose gels. To terminate the reaction, add EDTA to 10 mM; store samples at 4°.

Centrifugation in Equilibrium Density Gradients

Place DNA and TE buffer to 4.6 g in 16 × 76-mm polyallomer centrifuge tubes (Beckman). Weigh 6 g CsCl and add to DNA; cover with

Parafilm and incubate at 37° until CsCl is dissolved. The refractive index should be 1.4015 for a density of 1.72 g/ml. Cap tubes, match pairs by weight, and centrifuge at 37,000 rpm, 20° for at least 36 h in a fixed angle rotor. Collect gradients in 30 fractions, 0.2 ml/fraction. To identify fractions containing light- and hybrid-density DNA, small samples ($\sim$10 μl) of each fraction from radioactively labeled cells can be spotted on strips of Whatman #17 filter paper; the strips are washed in cold 5% trichloracetic acid to remove unincorporated nucleotides, rinsed in two changes of 95% ethanol, and dried; and the radioactivity is measured by scintillation counting, or the relative amounts of DNA in the samples can be visualized by staining with ethidium bromide. Pool appropriate fractions and dialyze overnight at 4° against TE buffer containing 0.1 M NaCl. Change buffer to TE and dialyze overnight at 4°. Concentrate the DNA by ethanol precipitation, or change buffer to 1/10 TE, dialyze for no more than 2 h, collect DNA, and concentrate $\sim$10-fold by evaporation. Determine DNA concentration by A_{260}; adjust if necessary.

Alkaline Agarose Gels

Prepare and melt 0.5 % agarose in 1 mM EDTA, keep at 55°. Low electroendosmosis agarose (e.g., SeaKem, Cambrex) gives better mobility and resolution of large DNA fragments. The volume of gel should be adjusted to achieve a $\sim$0.75-cm thickness, i.e., 250 ml for a 25 $\times$ 13.5-cm gel. Level an electrophoresis device and connect ports to a pump (preferably peristaltic) to recirculate the buffer. Pumps that produce vibrations and/or pulses should be avoided, as the gels are easily dislodged from their support. Use combs with teeth $\sim$1.5 $\times$ 9 mm so wells have a 50-μl capacity. When agarose has equilibrated to 55°, add NaOH to 30 mM final concentration, mix (avoiding bubbles), and pour. Remove bubbles with a Pasteur pipette. Let solidify $\sim$2 h in a cold room, $\sim$4 h at ambient temperature.

Pour alkaline running buffer (1 mM EDTA, 30 mM NaOH) over the gel. Remove comb very gently. Fill electrophoresis device with buffer to $\sim$1 mm above the gel; check that pump runs smoothly and that buffer remains level in both the anode and the cathode receptacles.

Sample Preparation and Electrophoresis

Carefully pipette out the amount of DNA to be used in each reaction (mammalian DNA: 1–5 μg DNA per lane if there are multiple copies of the gene of interest per cell, up to 10–20 μg per lane if single copy). It is essential to aliquot exactly the same amount of DNA to two separate tubes from each sample. Add enzyme reaction buffer to each sample and TE buffer if necessary, add the appropriate amount of nicking agent (glycosylase, exinuclease, etc.) to the "+" sample and enzyme storage buffer to the "−" sample from each pair, and incubate at required temperature.

Add one-fifth volume 5× dye: dilute 10× dye (25% Ficoll, 10 mM EDTA, 0.25% bromocresol green) with 1 volume 0.05 N NaOH. Load samples in wells, slowly and steadily to avoid sample loss, reserving one lane for a size marker (e.g., λ phage restricted with *Hind*III). Turn on power supply at 60 V for 15 min, turn pump on, set power supply at 20–35V, ≤100 μAmp, and electrophorese 10–20 h. The optimum voltage and duration should be determined for each restriction fragment size range to be examined.

Gel Washes

Remove gel with wide scoop (Bio-Rad), holding it with a gloved hand (gels are very fragile and slippery), and place on a Pyrex tray; leave the scoop underneath the gel in the tray. All washes are done on orbital shakers at low speed. Use a vacuum aspirator fitted with a plastic pipette to change washes; place pipette tip against the raised lip of the scoop to avoid sucking up the gel.

1. Rinse with H_2O.
2. Wash 40 min with neutralizing solution (0.5 M Tris base, 1.5 M NaCl, pH 7.8) to which ethidium bromide is added to 2 μg/ml.
3. Destain 15 min in H_2O.
4. Place gel on UV transilluminator, wear protective gear against UV radiation (long sleeves, face shield, and gloves); turn on UV light and punch holes with Pasteur pipette at the positions of the size marker bands. Photograph gel. Trim the gel with a scalpel, leaving a ∼1-cm border around the lanes, and measure length and width of the gel.

Steps 5–9 are not necessary if the fragment of interest is <10 kb.

5. Acid washes: 15 min in 0.25 M HCl and repeat. The dye should turn yellow.
6. Rinse with water.
7. Alkaline wash: 40 min in denaturing solution (0.5 M NaOH, 0.39 M NaCl); if performing alkaline transfer, stop here.
8. Rinse with water.
9. Neutralizing wash: 40 min in neutralizing solution.

Transfer of DNA from Gel to Membrane

Several methods can be used for blotting DNA to membranes; we describe a downward capillary system that is economical and easy to set up. Cut five pieces of blotting paper (Schleicher & Schuell GB002) to gel size; use one of them as a template to cut membrane (Hybond N+, Amersham) with a scalpel. Wear gloves and handle membrane with clean forceps.

Wet membrane with H_2O in Pyrex tray; place in a tray with $\sim$100 ml transfer solution: 20× SSPE (22 mM EDTA, 0.2 M NaH_2PO_4, 3.6 M NaCl, adjust pH to 7.7–8.0 with NaOH) or alkaline transfer solution (0.4 N NaOH, 1 M NaCl), let sit 10–15 min. Cut a piece of Whatman 3MM CHR paper the same width and approximately three times as long as the blotting paper; this will be the wick.

Blotting kits can be purchased; an inexpensive kit can be assembled as follows (Fig. 3). Place a stack $\sim$1 inch high of paper towels into a box sitting inside a tray (i.e., Nunc 23 cm^2).

Wet three blotting papers in transfer solution and place on towels. Place membrane atop blotting paper; roll out bubbles with a piece of glass tubing or a short pipette. Carefully slide gel on top of the membrane, wet remaining blotting papers with transfer solution, and place on top of gel; roll out bubbles pressing gently and evenly. Place four waterproof strips (e.g., X-ray film) all around and against the gel to avoid wicking by the S&S filters. Cover with the wick, letting the long ends hang on each side of the gel. Add transfer solution (20× SSPE or alkaline solution) to the larger tray to $\sim$1 cm deep and submerge the ends of the wick into the transfer solution. Transfer should be complete in 2–3 h.

Disassemble blotting stack: discard wick and blotting papers, transfer marker band positions to the membrane with a soft lead pencil, carefully lift gel, label the membrane if desired, and place membrane in a Pyrex tray. The gel can be restained with ethidium bromide to check the extent and homogeneity of the transfer.

If using 20× SSPE for the transfer, an alkaline fixation step is advised. Wet three Whatman #1 filters in $\sim$100 ml 0.4 N NaOH and place on flat enamel or glass tray. Place membrane on wet filters, DNA side up. Let sit 20–60 min.

Rinse membrane 1 min in 5× SSPE; store in a sealed plastic bag or hybridization bottle at 4°.

Prehybridization

Prepare 50× Denhardt's solution: 10 mg/ml each Ficoll type 400, poly-vinylpyrrolidone, and bovine serum albumin (BSA) pentax fraction. Filter through EH celotate 0.45 μm, aliquot, and store at $-20°$.

Prepare prehybridization solution (50% deionized formamide, 5× SSPE, 10× Denhardt's, 1% SDS). Filter through a Millipore-type HA

FIG. 3. Assembly of blotting sandwich. (A) The agarose gel can be seen sitting on top of the membrane and blotting papers over a thick stack of paper towels fitted inside a plastic box. Strips of X-ray film were placed along the sides of the gel. (B) The wick covers the assembled blotting sandwich, drawing transfer solution through the gel and the membrane and into the paper towels. (See color insert.)

0.45-μm membrane and store at 4° for up to a month; warm to 50–60° before use.

Roll membrane in the tray with 5× SSPE, blot off excess liquid, place in hybridization bottle, and unroll the membrane, letting it stick to the inside of the bottle. Up to four membranes can be placed in each bottle; assemble all membranes in the tray with 5× SSPE and roll together before placing them in the bottle. Add 4.75 ml prehybridization solution and

0.25 ml 10 mg/ml salmon sperm DNA; incubate at 42° in a rotisserie-style hybridization oven for at least 2 h.

Preparation of Template for Synthesis of RNA Probe

Plasmid vectors for synthesis of RNA probes are available from several commercial vendors. The genomic restriction fragment of interest is cloned into a vector containing promoters for two promoter-specific RNA polymerases (i.e., T3, T7, and SP6) oriented toward and flanking a multiple cloning site. The construct should be linearized with restriction enzymes to obtain templates for transcription of the opposite strands as shown in the example in Fig. 4 and purified by phenol extraction or by gel filtration with a commercial DNA purification kit.

Determine DNA concentration by A_{260}.

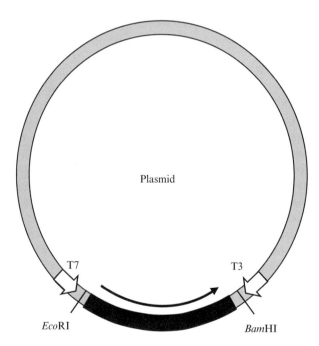

FIG. 4. Schematic representation of a plasmid used to generate templates for riboprobes. A 0.3- to 1.0-kbp fragment of genomic or cDNA is inserted into the multiple cloning site. The black arrow indicates the direction of transcription in the insert. To synthesize probe for the transcribed strand, the plasmid is digested with *Bam*HI, purified, and transcribed with T7 RNA polymerase; for the nontranscribed strand, the *Eco*RI-digested template is transcribed with T3 polymerase. White arrows: promoters for phage DNA polymerases; black arc, insert.

Synthesis of RNA Probe

Use any *in vitro* transcription kit, available from suppliers such as Promega, etc.

To produce ^{32}P-labeled probes, use the ATP, GTP, and UTP provided in the kit and 200 μCi 40 mCi/ml [^{32}P] CTP; add unlabeled CTP if the final concentration is less than 12 μM. Equivalent procedures can be used to synthesize fluorescent or chemiluminescent probes.

Note: if the restriction produces a 3′ protruding end in the template, convert it to a blunt end using the 3′→5′ exonuclease activity of Klenow DNA polymerase: set up a standard transcription reaction minus the ribonucleotides and RNA polymerase, add 5 units DNA Klenow (RNase free), incubate at 22° for 15 min, add nucleotides and RNA polymerase, and proceed as described earlier.

To remove unincorporated ribonucleotides, add nuclease-free H_2O to 100 μl and use Quick Spin G-50 Sephadex columns (Boehringer Mannheim).

To test ^{32}P incorporation, add 1 μl probe to 100 μl salmon sperm DNA 0.1 mg/ml in a 4- to 5-ml tube; add ice-cold 5% TCA to fill the tube, incubate on ice 10 min, filter through a Millipore type HA 0.45-μm membrane, dry the filter, and determine radioactivity by scintillation counting. If the probe is too "hot" and the counter is saturated, make a 1:10 dilution and proceed as described earlier.

Hybridization

Prepare hybridization solution (60% formamide, 6× SSPE, 1.2× Dernhardt's), filter through a Millipore type HA 0.45-μm membrane, and store at 4° for up to a month.

Mix 3.28 ml hybridization solution, 0.096 ml 10 mg/ml salmon sperm DNA, 0.04 ml 25% SDS, 0.02 ml 10 mg/ml yeast tRNA, 40 × 10^6 cpm ^{32}P-labeled RNA probe, and H_2O to 4 ml total volume. Remove prehybridization solution from bottle and add hybridization mixture. Return bottle to hybridization oven and incubate at 50° (RNA probes) for 24–48 h.

Membrane Washes

Transfer membrane from bottle to a plastic box. The hybridization mixture can be stored at −20° and reused. Wash membrane in two changes of 0.1× SSPE, 0.1% SDS for 15 min at room temperature and then in two changes of 0.5× SSPE, 0.1% SDS for 30 min at 60°; these washes are best incubated in an orbital shaker water bath; alternatively, perform washes in the hybridization bottle.

To increase the stringency of the wash (to remove the probe annealed to sequences that have an increasing degree of homology), lower the salt concentration and/or increase the temperature of the washes up to the melting point of the hybrid of interest.

After the last wash, place membrane on paper towels until visible moisture disappears (do not let it dry out). The positions of the size markers can be highlighted with a ^{14}C-labeled pen.

Place membrane in a vinyl sheet protector and seal with a Seal-o-Matic-type impulse sealer or use shrink wrap, avoiding wrinkles. Expose membrane to X-ray film (if using intensifying screens, expose at $-80°$); the time of exposure needed to obtain sufficient signal must be determined for each experimental setup. Alternatively, expose to a PhosphorImager screen (Bio-Rad, Amersham/Molecular Dynamics) for up to 2 days, scan, acquire image, and analyze with appropriate software or use NIH Image freeware (http://rsb.info.nih.gov/nih-image/Default.html). If necessary, subject membrane to additional washes.

Deprobing

Unwrap membrane; place in 0.4 N NaOH at 42° for 15 min. Then incubate in two changes of 0.2× SSPE, 1.0 % SDS, 0.2 M Tris, pH 7.5, at 42° for 15 min each. Expose membrane to make sure the probe was stripped off. The membrane can be prehybridized or wrapped airtight and stored refrigerated or frozen.

Equipment Needed

In addition to the equipment necessary to grow the cells or organisms to be examined, the laboratory should be equipped with electrophoresis apparatuses, power supplies, peristaltic pumps, transilluminator and camera, plastic or glass boxes and trays for gels washes and blotting, a dedicated area with shields and containers for storage, manipulation and disposal of ^{32}P-labeled materials, hybridization oven and bottles, ovens or water baths for membrane washes, dark room, cassettes and X-ray developer, or PhosphorImager system with exposure docks, screens, scanner, and software. Depending on the experimental design, other equipment may be needed, such as a microcentrifuge, gradient fraction collector, laminar flow hood, scintillation counter, spectrophotometer, evaporator, orbital shaker, and rocking device for phenol extractions.

Ligation-Mediated PCR Method

Irradiation of Cells

Cells that grow as monolayers in petri dishes, such as fibroblasts or keratinocytes, are irradiated with a UVC or UVB lamp after removal of the medium and washing with PBS. Typical UV doses for DNA repair assays of CPD are 10 to 20 J/m^2 of 254-nm light.

DNA Isolation

A standard DNA isolation procedure based on proteinase K treatment and phenol–chloroform extraction is used to prepare high molecular weight DNA. The incubation time in proteinase K is adjusted to 2 h at 37° to minimize depurination, which may lead to unwanted background signals in LMPCR. Dissolve the DNA in TE buffer to a concentration of approximately 0.2 μg/μl. The DNA should be dissolved well before T4 endonuclease V is added.

Cleavage of DNA at CPD

DNA is first incubated with T4 endonuclease V and then with *E. coli* photolyase to create fragments with 5′-phosphate groups and ligatable ends. The UV-irradiated DNA (about 10 μg) is mixed with 10 μl of 10× T4 endonuclease V buffer [500 mM Tris–HCl, pH 7.6, 500 mM NaCl, 10 mM EDTA, 10 mM dithiothreitol (DTT), 1 mg/ml BSA] and a saturating amount of T4 endonuclease V in a final volume of 100 μl. Saturating amounts of T4 endonuclease V can be determined by incubating UV-irradiated (20 J/m^2) genomic DNA with various enzyme dilutions and separating the cleavage products on alkaline agarose gels (see later). Incubate with the enzyme at 37° for 1 h and then add dithiothreitol to a final concentration of 10 mM. Add 5 μg of *E. coli* photolyase under yellow light. Irradiate the samples in 1.5-ml tubes from two 360-nm UVA black lights (Sylvania 15W F15T8) filtered through 0.5-cm-thick window glass for 1 h at room temperature at a distance of 3 cm. Extract once with phenol–chloroform. Precipitate the DNA by adding one-tenth volume of 3 M sodium acetate (pH 5.2) and 2.5 volume of ethanol. Leave on dry ice for 20 min. Centrifuge samples for 10 min at 14,000g at 4°. Wash pellets with 1 ml of 75% ethanol and air dry. Dissolve DNA pellets in TE buffer to a concentration of about 0.5 to 1 μg/μl. Determine the frequency of CPD by running 1 μg of the samples on a 1.5% alkaline agarose gel, prepared by suspending agarose in 50 mM NaCl, 4 mM EDTA and microwaving. Pour the gel. After the gel solidifies, soak it in running buffer (30 mM NaOH, 2 mM EDTA) for at least 2 h. Dilute the DNA sample with 1 volume of loading dye (50% glycerol, 1 M NaOH, 0.05% bromocresol green). Incubate for 15 min at room temperature and load the samples, including a size marker such as *Hind*III-digested λ phage DNA. Run the gel at 40 V for 3 to 4 h. Neutralize the gel by soaking for 60 min in 0.1 M Tris–Cl, pH 7.5. Stain the gel with ethidium bromide (1 μg/ml) for 30 min and then destain in water for 30 min. The average mass distribution on an ethidium-stained gel is two times the average lesion frequency. For example, if the highest stain intensity is around 1 kb, then there is an average of one lesion every 500 bp (Drouin *et al.*, 1996b).

LMPCR Reaction

For LMPCR primer design, determine of the T_m of the oligonucleotide primers using calculators available at a number of Web sites, such as http://www.basic.northwestern.edu/biotools/oligocalc.html. The T_m of primer 1 should be about 10° lower than that of primers 2 and 3; e.g., if the T_m of primer 1 is 45°, the T_m of primers 2 and 3 should be 55 and 60°, respectively. If a specific target area is to be analyzed (e.g., a defined sequence position in a gene), primer 1 should be located approximately 100 nucleotides upstream of this target. Primer 2 can overlap for up to 7 or 8 bases with primer 1. Primer 3 is used to make the single-stranded hybridization probe; it should be on the same strand just 3' to primer 2. Primer 3 should not overlap more than 8–10 bases with primer 2. See Fig. 2 for a schematic illustration of the LMPCR method.

Accurate measurements of DNA concentration before LMPCR are critical for quantitative DNA repair assays. To avoid variations in DNA concentration, approximately the same number of cells should be used as starting material for UV irradiation at each time point, and care should be taken that no material is lost during the DNA isolation procedure. DNA concentrations are measured by A_{260} optical density reading. It is important that the DNA be dissolved completely before these measurements are made.

In a siliconized 1.5-ml tube, mix 1 to 2 μg of cleaved DNA, 0.6 pmol of gene-specific primer 1, and 3 μl of 5× Sequenase buffer (250 mM NaCl, 200 mM Tris–Cl, pH 7.7) in a final volume of 15 μl. Incubate at 95° for 3 min and then at 45° for 30 min. Cool on ice and add 7.5 μl cold, freshly prepared Mg-DTT-dNTP mix (20 mM MgCl$_2$, 20 mM DTT, 0.25 mM of each dNTP). Add 1.5 μl of Sequenase (10 units). Incubate at 48° for 15 min and then cool on ice. It is also possible to use other DNA polymerases in this reaction. In particular, good data have been obtained with thermostable Pfu DNA polymerase (Angers *et al.*, 2001). When using other DNA polymerases, the buffer composition, incubation temperature, and the T_m of primer 1 need to be adjusted accordingly. After incubation with Sequenase, add 6 μl of 300 mM Tris–Cl (pH 7.7) and incubate at 67° for 15 min to inactivate the polymerase. Cool on ice, spin 5 sec, and add 45 μl of freshly prepared ligation mix [13.33 mM MgCl$_2$, 30 mM DTT, 1.66 mM ATP, 83 μg/ml BSA, 3 units of T4 DNA ligase (Promega)] per reaction and 100 pmol of linker per reaction (= 5 μl linker). The double-stranded linker is prepared in 250 mM Tris–Cl (pH 7.7) by annealing a 25-mer (5'-GCGGTGACCCGG-GAGATCTGAATTC, 20 pmol/μl) to an 11-mer (5'-GAATTCAGATC, 20 pmol/μl) by heating to 95° for 3 min and gradually cooling to 4° over a time period of 1–2 h. Linkers can be stored at -20° for several months and thawed on ice before use. Incubate the ligation reaction overnight at 18°. After ligation is complete, incubate for 10 min at 70° to inactivate the ligase. Add 8.4 μl of 3 M sodium acetate (pH 5.2), 10 μg *E. coli* tRNA, and 220 μl ethanol. Put the samples on dry ice for 20 min and centrifuge for 15 min at 4° in an Eppendorf centrifuge. Wash the pellets with 75% ethanol. Remove ethanol residues in a Speed-Vac or by air drying. Dissolve the pellets in 50 μl water and transfer to 0.5-ml PCR tubes. Add 50 μl of freshly prepared 2$\times$*Taq* polymerase mix (20 mM Tris–HCl, pH 8.9, 80 mM NaCl, 0.02 % gelatin, 4 mM MgCl$_2$, 10 pmol of the 25-mer linker oligonucleotide, 10 pmol of gene-specific primer 2, 3 units of *Taq* polymerase, and dNTPs at 0.4 mM each) and mix by pipetting. Cover the samples with 50 μl mineral oil and spin briefly. Cycle 18 to 20 times at 95° for 1 min, 60–66°C (the calculated T_m of primer 2) for 2 min, and 76° for 3 min. To completely extend all DNA fragments and to uniformly add an extra nucleotide using terminal transferase activity of the *Taq* polymerase, an additional *Taq* polymerase step is performed (if this step is omitted, double bands may occur on the sequencing gel). One unit of fresh *Taq* polymerase per sample is added together with 10 μl reaction buffer. Incubate for 10 min at 74°. Stop the reaction by adding sodium acetate to 300 mM, EDTA to 10 mM, and add 10 μg tRNA. Extract with 70 μl of phenol and 120 μl chloroform (premixed). Add 2.5 volume of ethanol and put on dry ice for 20 min. Centrifuge samples for 15 min in an Eppendorf centrifuge at 4°. Wash pellets in 1 ml of 75% ethanol and dry the pellets in a vacuum concentrator.

Sequencing Gel Analysis of LMPCR Products

Dissolve pellets in 1.5 μl water and add 3 μl formamide sequencing gel-loading buffer (94% formamide, 2 mM EDTA, pH 7.7, 0.05% xylene cyanol, 0.05% bromphenol blue). Heat the samples to 95° for 2 min prior to loading. Load only one-half of the samples or less. The gel is 0.4 mm thick and 60 cm long, consisting of 8% polyacrylamide (ratio acrylamide to bis-acrylamide = 29 to 1) and 7 M urea in TBE (0.089 M Tris–borate, 0.089 M boric acid, pH 8.3). To allow identification of the sequence position of the UV-specific bands, include Maxam–Gilbert sequencing standards prepared from genomic DNA as described previously (Pfeifer and Riggs, 1993). Run the gel until the xylene cyanol marker reaches the bottom.

Fragments below the xylene cyanol dye do not hybridize significantly. After the run, transfer the lower portion of the gel (i.e., the bottom 40 cm of it) to Whatman 3 MM paper and cover with Saran wrap. Electroblotting of the gel piece can be performed with a transfer box available from Owl Scientific. Pile three layers of Whatman 17 paper, 43 × 19 cm, presoaked in 90 mM TBE buffer, onto the lower electrode. Squeeze the paper with a roller to remove air bubbles between the paper layers. Place the gel piece covered with Saran wrap onto the paper and remove the air bubbles between the gel and the paper by carefully wiping over the Saran wrap with a soft tissue. Remove the Saran wrap and cover the gel with a nylon membrane cut somewhat larger than the gel piece and presoaked in TBE. Put three layers of presoaked Whatman 17 paper onto the nylon membrane, carefully removing trapped air with a roller. Place the upper electrode onto the paper. Perform the electroblotting procedure at 1.6 A and 12 V. After 30 min, remove the nylon membrane and mark the DNA side. A high ampere power supply is required for this transfer.

As an alternative to electroblotting, the PCR products can be labeled *before* gel electrophoresis by incorporating ^{32}P during a primer extension step using a 5' end-labeled primer 3 produced with T4 polynucleotide kinase and ^{32}P-ATP. This primer is then annealed to the denatured PCR products and extended once or several times with *Taq* polymerase. This method works well in most cases, but produces a somewhat higher background and lower signal compared with the electroblotting approach and exposes the worker to more radioactivity.

After electroblotting, dry the membrane briefly at room temperature. Then cross-link the DNA by UV irradiation. UV irradiation can be performed in a commercially available UV cross-linker. Perform hybridization in rotating 250-ml plastic or glass cylinders in a hybridization oven. Soak the nylon membranes briefly in TBE. Roll them into the cylinders by unspooling them from a thick glass rod or glass pipette so that the membranes stick completely to the inside walls of the cylinders without air pockets. Prehybridization is done with 15 ml of hybridization buffer (0.25 M sodium phosphate, pH 7.2, 1 mM EDTA, 7% SDS, 1% BSA) for 10 min. For hybridization, dilute the labeled probe into 7 ml hybridization buffer. Perform prehybridizations and hybridizations at 62°.

To prepare labeled single-stranded probes, 200 to 300 nucleotides in length, use repeated primer extension using *Taq* polymerase with a single primer (primer 3) and a double-stranded template DNA (Törmänen and Pfeifer, 1992). The template can be either a plasmid containing the sequence of interest that is restriction cut approximately 200 to 300 nucleotides 3' to the binding site of primer 3 or a PCR product containing the target sequence. To prepare the single-stranded probe, mix 50 ng of the respective restriction cut plasmid DNA (or 10 ng of the gel-purified PCR product) with primer 3 (20 pmol), 100 μCi of [^{32}P]dCTP, 10 μM of the other three dNTPs, 10 mM Tris–Cl, pH 8.9, 40 mM NaCl, 0.01% gelatin, 2 mM MgCl$_2$, and 3 units of *Taq* polymerase in a volume of 100 μl. Perform 35 cycles at 95° (1 min), 60–66° (1 min), and 75° (2 min). Recover the probe by phenol/chloroform extraction, addition of ammonium acetate to a concentration of 0.7 M, ethanol precipitation at room temperature, and centrifugation. Add this probe to the hybridization mix.

After hybridization, wash the nylon membrane with 2 liter of washing buffer (20 mM sodium phosphate, pH 7.2, 1 mM EDTA, 1% SDS) at 60°. Perform several washing steps in a dish at room temperature with buffer prewarmed to 60°. After washing, dry the membranes briefly at room

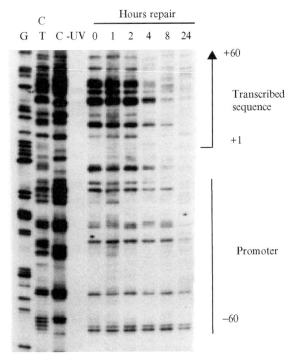

FIG. 5. Removal of UV-induced CPD from the promoter and transcribed strand of the *CDC2* gene in human fibroblasts. Fibroblasts were irradiated with 10 J/m² of UVC (254 nm) and were allowed to repair for various periods of time. The transcribed strand of the human *CDC2* gene, including upstream promoter sequences, was analyzed by LMPCR. Note the faster repair of CPD at sequences downstream of the transcription start site and slow repair in the promoter.

temperature, wrap them in Saran wrap, and expose to a PhosphorImager screen.

Data Analysis

Data analysis is routinely performed by PhosphorImager analysis using a Molecular Dynamics scanner or equivalent equipment. An example of an LMPCR sequencing gel analyzing DNA repair rates in the human *CDC2* gene in human fibroblasts is shown in Fig. 5. Faster repair of CPD is observed in sequences downstream of the transcription start site than in promoter sequences (Tommasi *et al.*, 2000) due to TCR. For quantitation of repair rates, nylon membranes are exposed to the PhosphorImager and radioactivity is determined in all CPD-specific bands of the sequencing gel that show a consistent and measurable signal above background. Background values (from the control lanes without UV irradiation) are subtracted. A repair curve can be established for each CPD position that gives a sufficient signal above background. The time at which 50% of the initial damage is removed can then be determined from this curve.

Acknowledgments

We acknowledge our colleagues who were directly involved in establishing the protocols we have detailed. These include Vilhelm Bohr, Isabel Mellon, Diane Okumoto, Allen Smith, and Silvia Tornaletti.

References

Angers, M., Cloutier, J. F., Castonguay, A., and Drouin, R. (2001). Optimal conditions to use Pfu exo(1/N) DNA polymerase for highly efficient ligation-mediated polymerase chain reaction protocols. *Nucleic Acids Res.* **29,** E83.

Anson, R. M., and Bohr, V. A. (1999). Gene-specific and mitochondrial repair of oxidative DNA damage. *In* "DNA Repair Protocols: Eukaryotic Systems" (D. S. Henderson, ed.), Vol. 113, pp. 257–279. Humana Press, Totowa, NJ.

Bohr, V. A., and Okumoto, D. S. (1988). Analysis of pyrimidine dimers in defined genes. *In* "DNA Repair, a Laboratory Manual of Research Procedures" (E. C. Friedberg and P. C. Hanawalt, eds.), Vol. III, pp. 347–366. Dekker, New York.

Bohr, V. A., Smith, C. A., Okumoto, D. S., and Hanawalt, P. C. (1985). DNA repair in an active gene: Removal of pyrimidine dimers from the DHFR gene of CHO cells is much more efficient that in the genome overall. *Cell* **40,** 359–369.

Christians, F. C., and Hanawalt, P. C. (1992). Inhibition of transcription and strand-specific DNA repair by α-amanitin in Chinese hamster ovary cells. *Mutat. Res.* **274,** 93–101.

Denissenko, M. F., Pao, A., Pfeifer, G. P., and Tang, M. (1998). Slow repair of bulky DNA adducts along the nontranscribed strand of the human p53 gene may explain the strand bias of transversion mutations in cancers. *Oncogene* **16,** 1241–1247.

Drouin, R., Gao, S., and Holmquist, G. (1996b). Agarose gel electrophoresis for DNA damage analysis. *In* "Technologies for Detection of DNA Damage and Mutations" (G. Pfeifer, ed.), pp. 37–43. Plenum Press, New York.

Drouin, R., Rodriguez, H., Holmquist, G., and Akman, S. (1996a). Ligation-mediated PCR for analysis of oxidative DNA damage. In "Technologies for Detection of DNA Damage and Mutations" (G. P. Pfeifer, ed.), pp. 211–225. Plenum Press, New York.

Feng, Z., Hu, W., Chen, J. X., Pao, A., Li, H., Rom, W., Hung, M. C., and Tang, M. S. (2002). Preferential DNA damage and poor repair determine ras gene mutational hotspot in human cancer. *J. Natl. Cancer Inst.* **94,** 1527–1536.

Friedberg, E. C., Ganesan, A. K., and Seawell, P. C. (1980). Purification and properties of a pyrimidine dimer-specific endonuclease from *E. coli* infected with bacteriophage T4. *Methods Enzymol.* **65,** 191–201.

Greenberg, M. E., and Ziff, E. B. (1984). Stimulation of 3T3 cells induces transcription of the c-fos proto-oncogene. *Nature* **311,** 433–438.

Grishko, V. I., Druzhyna, N., LeDoux, S. P., and Wilson, G. L. (1999). Nitric oxide-induced damage to mtDNA and its subsequent repair. *Nucleic Acids Res.* **27,** 4510–4516.

Hamilton, M. L., Guo, Z., Fuller, C. D., Van Remmen, H., Ward, W. F., Austad, S. N., Troyer, D. A., Thompson, I., and Richardson, A. (2001). A reliable assessment of 8-oxo-2-deoxyguanosine levels in nuclear and mitochondrial DNA using the sodium iodide method to isolate DNA. *Nucleic Acids Res.* **29,** 2117–2126.

Jaspers, N. G., Raams, A., Kelner, M. J., Ng, J. M., Yamashita, Y. M., Takeda, S., McMorris, T. C., and Hoeijmakers, J. H. (2002). Anti-tumour compounds illudin S and Irofulven induce DNA lesions ignored by global repair and exclusively processed by transcription- and replication-coupled repair pathways. *DNA Repair* **1,** 1027–1038.

Ljungman, M., and Zhang, F. (1996). Blockage of RNA polymerase as a possible trigger for u.v. light-induced apoptosis. *Oncogene* **13,** 823–831.

Mellon, I. (2005). Transcription-coupled repair: A complex affair. *Mutat. Res.* **577,** 155–161.

Mellon, I., Spivak, G., and Hanawalt, P. C. (1987). Selective removal of transcription-blocking DNA damage from the transcribed strand of the mammalian DHFR gene. *Cell* **51,** 241–249.

Mueller, P. R., and Wold, B. (1989). *In vivo* footprinting of a muscle specific enhancer by ligation mediated PCR. *Science* **246,** 780–786.

Nakae, D., Mizumoto, Y., Kobayashi, E., Noguchi, O., and Konishi, Y. (1995). Improved genomic/nuclear DNA extraction for 8-hydroxydeoxyguanosine analysis of small amounts of rat liver tissue. *Cancer Lett.* **97,** 233–239.

Pfeifer, G. P., Drouin, R., Riggs, A. D., and Holmquist, G. P. (1991). *In vivo* mapping of a DNA adduct at nucleotide resolution: Detection of pyrimidine (6-4) pyrimidone photoproducts by ligation-mediated polymerase chain reaction. *Proc. Natl. Acad. Sci. USA* **88,** 1374–1378.

Pfeifer, G. P., Drouin, R., Riggs, A. D., and Holmquist, G. P. (1992). Binding of transcription factors creates hot spots for UV photoproducts *in vivo. Mol. Cell. Biol.* **12,** 1798–1804.

Pfeifer, G. P., and Riggs, A. D. (1993). Genomic sequencing. *Methods Mol. Biol.* **23,** 169–181.

Pfeifer, G. P., Steigerwald, S. D., Mueller, P. R., Wold, B., and Riggs, A. D. (1989). Genomic sequencing and methylation analysis by ligation mediated PCR. *Science* **246,** 810–813.

Scicchitano, D. A., Olesnicky, E. C., and Dimitri, A. (2004). Transcription and DNA adducts: What happens when the message gets cut off? *DNA Repair* **3,** 1537–1548.

Spivak, G., and Hanawalt, P. (1995). Determination of damage and repair in specific DNA sequences. *Methods: A Companion to Methods in Enzymology* **7,** 147–161.

Spivak, G., and Hanawalt, P. C. (1992). Translesion DNA synthesis in the dihydrofolate reductase domain of UV-irradiated CHO cells. *Biochemistry* **31,** 6794–6800.

Tolbert, D. M., and Kantor, G. J. (1996). Definition of a DNA repair domain in the genomic region containing the human p53 gene. *Cancer Res.* **56,** 3324–3330.

Tommasi, S., Oxyzoglou, A. B., and Pfeifer, G. P. (2000). Cell cycle-independent removal of UV-induced pyrimidine dimers from the promoter and the transcription initiation domain of the human CDC2 gene. *Nucleic Acids Res.* **28,** 3991–3998.

Tornaletti, S., and Hanawalt, P. C. (1999). Effect of DNA lesions on transcription elongation. *Biochimie* **81,** 139–146.

Tornaletti, S., and Pfeifer, G. P. (1994). Slow repair of pyrimidine dimers at p53 mutation hotspots in skin cancer. *Science* **263,** 1436–1438.

Tornaletti, S., and Pfeifer, G. P. (1996). Ligation-mediated PCR for analysis of UV damage. *In* "Technologies for Detection of DNA Damage and Mutations" (G. P. Pfeifer, ed.), pp. 199–209. Plenum Press, New York.

Tu, Y., Tornaletti, S., and Pfeifer, G. P. (1996). DNA repair domains within a human gene: selective repair of sequences near the transcription initiation site. *EMBO J.* **15,** 675–683.

Valadez, J. G., and Guengerich, F. P. (2004). S-(2-chloroethyl)glutathione-generated p53 mutation spectra are influenced by differential repair rates more than sites of initial dna damage. *J. Biol. Chem.* **279,** 13435–13446.

Venema, J., van Hoffen, A., Karcagi, V., Natarajan, A. T., van Zeeland, A. A., and Mullenders, L. H. (1991). Xeroderma pigmentosum complementation group C cells remove pyrimidine dimers selectively from the transcribed strand of active genes. *Mol. Cell. Biol.* **11,** 4128–4134.

Wang, L., Hirayasu, K., Ishizawa, M., and Kobayashi, Y. (1994). Purification of genomic DNA from human whole blood by isopropanol-fractionation with concentrated Nal and SDS. *Nucleic Acids Res.* **22,** 1774–1775.

Wei, D., Maher, V. M., and McCormick, J. J. (1995). Site-specific rates of excision repair of benzo[a]pyrene diol epoxide adducts in the hypoxanthine phosphoribosyltransferase gene of human fibroblasts: Correlation with mutation spectra. *Proc. Natl. Acad. Sci. USA* **92,** 2204–2208.

Yamaizumi, M., and Sugano, T. (1994). UV-induced nuclear accumulation of p53 is evoked through DNA damage of actively transcribed genes independent of the cell cycle. *Oncogene* **9,** 2775–2784.

Ye, N., Holmquist, G. P., and O'Connor, T. R. (1998). Heterogeneous repair of N-methylpurines at the nucleotide level in normal human cells. *J. Mol. Biol.* **284,** 269–285.

[15] TFIIH Enzymatic Activities in Transcription and Nucleotide Excision Repair

By Jean-Philippe Lainé, Vincent Mocquet, and Jean-Marc Egly

Abstract

Transcription and nucleotide excision repair (NER) are two major mechanisms in which the transcription factor TFIIH plays a crucial role. In order to investigate its function, we first described a fast and efficient purification protocol of TFIIH from either HeLa cells or patient cell lines, as well as various *in vitro* enzymatic assays set up in our laboratory. All these enzymatic assays have been adapted to work on immobilized DNA, a powerful tool allowing for sequential protein incubations in various buffer

METHODS IN ENZYMOLOGY, VOL. 408   0076-6879/06 $35.00
DOI: 10.1016/S0076-6879(06)08015-3

conditions, without destabilizing protein complexes bound to the DNA. Runoff transcription assays performed with either whole cell extract or highly purified factors underline the role of TFIIH helicases (XPB and XPD) in the RNA synthesis. Moreover, the requirement of XPB and XPD in NER can also be investigated with various assays corresponding to the different steps of this process. The DNA opening assay (permanganate footprint) highlights DNA unwinding of the double-stranded DNA fragment within the repair complex, whereas the dual incision assay allows for detection of the double cut on both sides of the lesion. The gap-filling reaction following the cuts can be monitored as well with a DNA resynthesis assay. Futhermore, the use of immobilized DNA is of great interest to study the detailed mechanism in which TFIIH plays a central role. This chapter describes the ATP-independent recruitment of TFIIH on the damaged DNA previously recognized by XPC-hHR23B and the sequential arrival and departure of the repair proteins within the NER complex.

I. Introduction

TFIIH is a versatile, multifunctional protein complex engaged in various cellular processes. TFIIH was first discovered to be part of the basal transcriptional machinery where it plays an important role in the initiation of RNA synthesis by the RNA pol II (Coin *et al.*, 1998). It was later linked to nucleotide excision repair (NER), one of the DNA repair pathways (Schaeffer *et al.*, 1993). TFIIH is composed of 10 subunits, of which two contain ATPase and helicase activity (XPD and XPB) and one contains cyclin-dependent kinase activity (cdk7). Three human genetic disorders are connected with mutations in XPB, XPD, and the nonenzymatic subunit p8: xeroderma pigmentosum (XP), Cockayne syndrome (CS), and trichothiodystrophy (TTD) (Giglia-Mari *et al.*, 2004; Lehmann, 2003). Based on the number of phenotypes associated with these different diseases, we set up both transcription and repair *in vitro* assays to investigate the role of each TFIIH subunit, in either mechanism.

Transcription can be divided into several distinct steps, leading to mRNA synthesis. It has been shown that TFIIH is responsible for the DNA opening around the promoter (Holstege *et al.*, 1996) and participates in promoter escape (Bradsher *et al.*, 2000; Moreland *et al.*, 1999). TFIIH also phosphorylates the CTD tail of the largest subunit of RNA pol II (Lu *et al.*, 1992; Roy *et al.*, 1994), allowing for the transition between initiation and elongation, and the recruitment of mRNA processing enzymes. NER eliminates different types of damage distorting the DNA. This "cut-and-paste"-like mechanism requires at least 30 proteins, which participate in the following: (1) damage recognition, (2) open complex formation, (3) dual incision of a 24–32 nucleo-tide damaged fragment, and (4) gap-filling DNA resynthesis and ligation of the last nucleotide incorporated to the extant DNA (Dip et al., 2004). TFIIH enzymatic activities have been shown to be involved in the damaged DNA opening step (Evans et al., 1997). Furthermore, its presence is necessary for recruiting and assembling the various NER proteins at the site of damage (Riedl et al., 2003).

This chapter describes the methodology used to purify and to investi-gate the role of TFIIH in transcription and repair through various in vitro assays.

II. Purification of Human TFIIH from Wild-Type and Patient Cell Lines

The availability of wild-type and mutant TFIIH is of prime importance for investigating the role of TFIIH. Wild-type and mutated TFIIH com-plexes can be obtained either from HeLa cells or from patients' cell lines harboring XP, TTD, or CS phenotypes. The original protocol for TFIIH purification was a time-consuming and expensive technology (Gerard et al., 1991) but was very efficient for obtaining pure TFIIH from large amounts of HeLa cells. For patient-derived cell lines that are difficult to grow, we developed a new procedure optimized for limited amounts of cells.

A. Whole Cell Extract (WCE)

1. All steps should be carried out on ice or at $4°$. Centrifuge the cells (45×10^8) at 5000 rpm for 5 min and discard the supernatant. In order to estimate the volume of the pellet (V1), add 100 ml of cold phosphate-buffered saline (PBS1X) to the centrifuge tube and carefully transfer the pellet to a graduated test tube. Once the volume is measured, wash the cells with 900 ml of cold PBS1X and pellet them again at 5000 rpm for 10 min.

2. For the first ammonium sulfate precipitation, resuspend the cells with 4×V1 of buffer A (Table I) supplemented with 5 mM dithiothreitol (DTT) and a protease inhibitor cocktail (antipain, chymostatin, pepstatin, aprotinin, and leupeptin). Leave on ice for 20 min. Lyse mechanically the cells with 20 strokes in a 25-ml homogenizer/Dounce (Kontes Glass Co.) and dilute the lysate with 4×V1 buffer B supplemented with 2 mM DTT under gentle stirring. Add 1×V1 of saturated ammonium sulfate (4 M) one drop at a time under gentle stirring for 30 min. Centrifuge at 50,000 rpm for 2 h and 30 min to separate the precipitated compounds from the supernatant (Table I).

TABLE I

BUFFER COMPOSITION

Buffer	Composition
A	Tris/HCl, pH 7.9, 10 mM, EDTA 1 mM
B	Tris/HCl, pH 7.9, 50 mM, MgCl$_2$ 10 mM, glycerol 50% sucrose 25%
C	Tris/HCl, pH 7.9, 50 mM, MgCl$_2$ 5 mM, (NH$_3$) 2 SO$_4$ 40 mM, EDTA 0.2 mM, glycerol 15%, DTT 1 mM
D	Tris/HCl, pH 7.9, 10 mM, MgCl$_2$ 5 mM, glycerol 15%, DTT 0.5 mM
E	Tris/HCl, pH 7.9, 50 mM, MgCl$_2$ 5 mM, glycerol 15%, DTT 1 mM, EDTA 0.5 mM, KCl 50 mM
F	Tris/HCl, pH 7.9, 25 mM, MgCl$_2$ 5 mM, glycerol 15%, DTT 0.2 mM, EDTA 0.5 mM NP-40 0.1%
G	Tris/HCl, pH 7.9, 50 mM, EDTA 0.1 mM, glycerol 10%, DTT 0.5 mM, KCl 50 mM
H	Sodium acetate, pH 5.2, 50 mM, SDS 1%
I	Formamide 80%, Tris-borate electrophoresis buffer (TBE) 0.5×, brom-phenol blue 0.02%, xylene cyanol blue 0.02%
J	HEPES-K$^+$, pH 7.6, 10 mM, MgCl$_2$ 1.5 mM, EGTA 0.5 mM, glycerol 10%, DTT 1 mM, and β-glycerol phosphate 10 mM, KCl 50 mM
K	HEPES-K$^+$, pH 7.6, 12.5 mM, MgCl$_2$ 6.25 mM, EDTA 50 mM, glycerol 5%, KCl 50 mM
BW	Tris/HCl, pH 7, 9, 10 mM, EDTA 1 mM, NaCl 2 M
L	25% Ficoll, 1 mM EDTA, 0.4% brom-phenol blue, 0.4% xylene cyanol

3. For the second ammonium sulfate precipitation, clarify the supernatant through gauze and measure its volume (V2). Precipitate the proteins by slowly adding 0.33g×V2 of ammonium sulfate supplemented with 10 μl of 1 M NaOH. Gently stir the mixture for 30 min and centrifuge at 30,000 rpm for 20 min. Resuspend the pellet with 1/10× V2 of buffer C and dialyze it overnight with 5 liter of buffer C. Centrifuge the WCE at 11,000 rpm for 30 min to remove aggregates, snap freeze the supernatant, and store it at $-80°$ or use it immediately for TFIIH purification. At this step we can estimate the total protein concentration of the WCE to be around 10 mg proteins/ml. For reference, the final volume of WCE obtained from 45×10^8 should be around 200 ml.

B. Chromatography

1. Up to 10 mg of proteins can be loaded on 1 ml of a heparin Ultrogel (A4R Sepracor) column. Pour heparin Ultrogel into a 10-ml Poly Prep chromatography gravitational flow column (Bio-Rad) and equilibrate it with buffer D. Centrifuge the WCE to remove aggregates that could obstruct the column and load it onto the heparin column. Wash the column with 2 resin volumes (2 × 1ml) of buffer D. Elute proteins bound to the

heparin column by three sequential steps with 2 resin volumes of buffer D containing 0.2, 0.4, and 1 M KCl. TFIIH is contained in the 0.4 M KCl fraction, which is dialysed with 2 × 2l of buffer E for 2 h.

2. Immunoprecipitation is carried out in batches. Incubate overnight the 2-ml heparin 0.4 M fraction with 400 μg of purified anti-p44 monoclonal antibody (Humbert et al., 1994) cross-linked with 200 μl of protein A-Sepharose Cl-4B beads (Amersham Pharmacia Biotech AB). Spin the mix down at 1000 rpm for 2 min and discard the supernatant. Wash the protein A-Sepharose beads three times with 10× the bead volume of buffer F containing 0.4 M KCl and twice with buffer F containing 0.05 M KCl. TFIIH is finally eluted from the antibody by incubating the beads with 400 μl of buffer F containing 0.05 M KCl, 0.01% Nonidet P-40 (NP-40), and 2 mg/ml of a competitor peptide (corresponding to the N-terminal amino acids of p44 (MDEEPERTKRWEGGYER)) for 6 h. We obtain 10 μg (10 ng/μl) of highly purified TFIIH, which is aliquoted and stored at −80°.

III. DNA Substrate Preparation Procedure

A. Preparation of Biotinylated Ad2MLP Template

A 722 nucleotide long DNA fragment from the pBluescript II-KS(-) phagemid (Stratagene) containing Ad2MLP is generated (Frit et al., 1999).

1. Digest 10 μg of the pBluescript-Ad2MLP with 12.5 μl of the FokI restriction enzyme (4 U/μl) in a mix containing 25 μl of NEB 3 buffer (New England Biolabs) and 2.5 μl of 10 mg/ml bovine serum albumin (BSA) and bring it up to 250 μl with sterilized water. Incubate at 37° for 2 h.

2. Incubate at 75° for 20 min to heat-inactivate FokI and then spin it down.

3. Add 2 μl of the 5 U/μl Klenow enzyme (New England Biolabs) to the reaction along with 3.8 μl of a mix containing 1 mM each of biotin-dUTP (USB), dCTP, and dGTP. Further incubate at 25° for 35 min and then at 75° for 20 min to heat inactivate the Klenow enzyme.

4. Add 10 μl of the 5 U/μl AseI restriction enzyme (New England Biolabs) to the reaction and further incubate at 37° for 2 h. Stop the reaction by adding 50 μl of buffer L.

5. Load the reaction in a 1.3% agarose gel and purify the FokI/AseI DNA fragment with the Qiagen gel purification kit. Elute the DNA fragment with 40 μl of sterilized water.

B. Immobilization of Biotinylated DNA Fragment on Magnetic Beads

1. Use 1 μl of DNA fragment (50 ng/μl stock) for 3 μl of Dynabeads M-280 streptavidin slurry (beads in their storage buffer). For each experiment,

prepare a stock solution of immobilized DNA fragment, which will be further aliquoted in each tube. A 10 sample experiment is described as an example.

2. Wash 30 μl of the slurry three times with buffer BW. Wash the beads by repeatedly collecting them on the inside wall of the tube using a Dynal magnetic particle collector (Dynal MPC), removing the previous buffer by aspiration (while the tube is still placed in the Dynal MPC) and resuspending them in the buffer for the next step.

3. Resuspend the beads in 50 μl with buffer BW. Add 10 μl of the DNA and 40 μl of 10 mM Tris/HCl, pH 7.3. The final NaCl concentration should be 1 M. Incubate at 25° for 20 min using gentle rotation. For proper mixing, 0.02% NP-40 can be added.

4. Wash the beads twice with buffer BW and three times with buffer G. Resuspend the beads with100 μl of buffer G and distribute them (10 μl) in each tube. Note: because the beads tend to fall and gather at the bottom of the tube, a gentle mixing of each tube during each assay is required every 5 to 10 min to resuspend the beads.

IV. Enzymatic Assays

To understand the specific role of the different proteins in either transcription or repair, biochemical tools have been developed. The immobilization of DNA on magnetic beads brought up a wide range of new possibilities and allows one to have an inside look at *in vitro* experiments. Enzymatic assays on beads can be carried out using either a cell-free extract (WCE) or recombinant factors. The optimal amount of each factor added in the *in vitro* assays should be determined for each new batch of proteins. The presence of exonucleases in the WCE such as Ku80/70 can inhibit the reaction due to degradation of the substrate. Therefore, we recommend adding 1 mM wortmannin for each experiment using WCE.

A. Runoff Transcription Assay

TFIIH transcription activity can be followed with an *in vitro* assay measuring the synthesis of a 328 nucleotide transcript using recombinant purified general trancription factors (TFIIA, TFIIB, TFIID/TBP, TFIIE, TFIIF) and purified HeLa RNA pol II. The same assay was performed with mutated TFIIH in order to further investigate the role of its different subunits.

1. Principle. Ad2MLP is currently used as the DNA template to evaluate the transcriptional activity of any transcription factors, such as recombinant TFIIH or TFIIH-containing WCE preparations. In both cases, transcription reactions are performed in two steps: (1) preincubation of the DNA template with either the WCE or the general transcription factors TFIIH and RNA pol II allowing for the formation of the transcription initiation complex and (2) a RNA synthesis step starting upon addition of ribonucleotides allowing for the synthesis of a 328 nucleotide transcript.

Immobilization of the DNA template on magnetic beads does not affect the efficiency of the transcription reaction and allows us to vary the buffer conditions or the protein compositions during these two steps.

2. Transcription Reaction with WCE

a. Place each tube containing the immobilized DNA on the Dynal MPC and remove buffer G just before the experiment. In a 20-μl final reaction volume, incubate 50–75 μg of WCE at 25° for 15 min with buffer G. The final MgCl$_2$ concentration should be 5 mM.

b. Add 5 μl of a mix containing 1 mM each of ATP, GTP, UTP, 20μCi [α-^{32}]PCTP (400 Ci/mmol, Amersham), and 12.5 mM MgCl$_2$ to the 20-μl preincubation reaction and incubate at 25° for 45 min.

c. Add 200 μl of buffer H to the reaction. Extraction of nucleic acids is performed by adding 100 μl of phenol and 100 μl of chloroform and vortexing for 30 s. Spin the samples down for 5 min in a bench centrifuge. Collect the aqueous phase, and precipitate RNAs with 1/20× volume of 5 M ammonium acetate, 2.5× volume of ethanol 100%, and 1.5 μl of 25 mg/ml glycogen. Incubate the samples at −80° for 20 min and centrifuge them at 13,000 rpm at 4° for 25 min.

d. Dry the pellet and resuspend it with buffer I. Boil for 5 min and load onto a 5% denaturing polyacrylamide gel [acrylamide/bisacrylamide (19:1), 7 M urea, 1× Tris–borate, pH 8, electrophoresis buffer]. After electrophoresis, autoradiograph the gel overnight at −80° with an intensifying screen. (Fig. 1).

3. Reconstituted Transcription System

a. Purification of the general transcription factors has already been described (Marinoni *et al.*, 1997). We usually incubate 30 ng of TBP, 15 ng of TFIIB, 100 ng of TFIIEα, 60 ng of TFIIEβ, 40 ng of partially purified TFIIA, 80 ng of TFIIF, 10 ng of partially purified HeLa RNA pol II, and different amounts of TFIIH with bufffer G in a final reaction volume of 20 μl. The MgCl$_2$ concentration should also be 5 mM in this step and brought up to 6.5 mM during incubation with the nucleotides (see Section IV.A.2.b).

b. Refer to Sections IV.A.2.c and IV.A.2.d.

4. Analysis of Results. As a general transcription factor, TFIIH is absolutely required to generate a transcript by the RNA pol II (Coin *et al.*, 1999). Titration of TFIIH on the *Fok*I/*Ase*I substrate leads to an increase of the expected 328 nucleotide length RNA (Fig. 1, lanes 1 to 3). The absence of TFIIH completely prevents transcription (Fig. 1, lane 4).

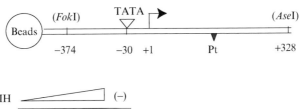

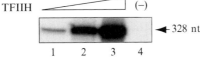

FIG. 1. *In vitro* Run-off Transcription is performed in the presence of all the basal transcription factors (TFIIA, TFIIB, TBP, TFIIE, TFIIF and RNA pol II) and increasing amounts of TFIIH. The length of the corresponding RNA transcripts (328 nucleotides) is indicated. (−) reaction performed in the absence of TFIIH.

B. Nucleotide Excision Repair Assays

Analysis of the mechanism of NER using cell-free extract systems and purified proteins requires suitable DNA substrates containing characterized DNA lesions. As a model lesion, we use the 1,3-intrastrand d(GpTpG)-cisplatin cross-link (Shivji *et al.*, 1999). However, other adducts, such as CPD, 6-4 PP, BPDE, 8 oxoguanine, or thymidine glycol, have been also tested.

Of the different steps of the NER reaction, this chapter describes only (1) the DNA opening step where TFIIH plays a crucial role via its helicase activities, (2) the double cut of the DNA around the lesion, and (3) the DNA resynthesis of the gap (Hansson and Wood, 1989).

1. In vitro *DNA Opening Assay (Permanganate Footprinting Assay).* a. PRINCIPLE. TFIIH contains two ATP-dependent helicase activities located in its XPB and XPD subunits, which will unwind the DNA around the lesion. In a classical helicase assay, we seek to detect the displacement of a single-stranded oligonucleotide from a double-stranded DNA fragment. This method could not be used in our context because of the size of the DNA substrate needed for the assembly of a multiprotein complex. The unwound region is in that case too small to displace a 722 nucleotide long single-stranded fragment. To study the requirement of TFIIH in the DNA opening, we set up the permanganate assay based on the single-stranded region markings: potassium permanganate ($KMnO_4$) is a chemical probe that reacts preferentially with single-stranded thymines of DNA. This oxidation modifies the 5,6 bond of pyrimidines to 5,6 *cis*-diols, which are hydrolyzed by piperidine. The DNA fragment used for this assay should be radiolabeled by adding 3 μl of [α-^{32}P]dCTP (3.3 pmol/μl) instead of 3.8 μl of cold dCTP during the DNA substrate preparation procedure (see Section III.A.3).

    b. *IN VITRO* DNA OPENING/KMnO$_4$ ASSAY.

    1. Place each tube on the Dynal MPC and remove the buffer just before the experiment. Incubate 50 ng of XPC-hHR23B, 100 ng of XPA, 200 ng of RPA, and 100 ng of XPG for 5 min with or without TFIIH (Reconstitued Opening System) with 8.5 μl buffer J and 0.15 μl of 500 mM MgCl$_2$. Bring up the final reaction volume to 20 μl with buffer G. The final MgCl$_2$ concentration should be 4 mM.

    2. Add 5 μl of a mix containing 25 mM ATP, 2.4 μl of 500 mM HEPES-K+, pH 7.6, and 2.1 μl of buffer K and further incubate at 30° for 15 min.

    3. Add 3 μl of a 120 mM KMnO$_4$ solution (Fluka) and incubate the reaction at room temperature for 3 min. Stop the reaction by adding 6 μl of 14.6 M 2-mercaptoethanol and place the tubes on ice for 5 min. Note: The 120 mM KMnO4 solution is prepared from a 1 M stock solution kept at 4° not longer than 2 weeks.

    4. Refer to Section IV.A.2.c.

    5. Wash the pellet with ethanol 80%. Dry and resuspend it with a 100-μl final volume reaction containing 1 mM EDTA, 1 mM EGTA, and 1 M piperidine (Biosolve Ltd) and incubate at 90° for 25 min.

    6. Refer to Sections IV.A.2.c and IV.A.2.d. Load the resuspended pellet onto a 8% denaturating polyacrylamide sequencing gel (Fig. 2).

    c. ANALYSIS OF RESULTS. In the presence of TFIIH, the piperidine cuts specifically at positions −5, −8, and −10, corresponding to the modified thymines that determine DNA opening around the cisplatin (Fig. 2, compare lane 6 to lane 5). In the absence of TFIIH, the profile disappears (Fig. 2, lane 7), highlighting the crucial role of the two helicases in the opening.

    2. In Vitro Dual Incision Assay

    a. PRINCIPLE. This method allows us to visualize the excised damaged DNA fragments once they have been incised by XPG and XPF. After incubation of the DNA template with all the recombinant repair proteins (XPC-hHR23B, XPA, TFIIH, RPA, XPG, and XPF: reconstituted incision system), the damaged single-stranded oligonucleotide is released. A complementary oligonucleotide is then added to create a 5′-overhanged duplex, which serves as a template for a 3′ end-labeling reaction. DNA fragments of different sizes, due to multiple cut positions of XPF, will be detected on the autoradiography.

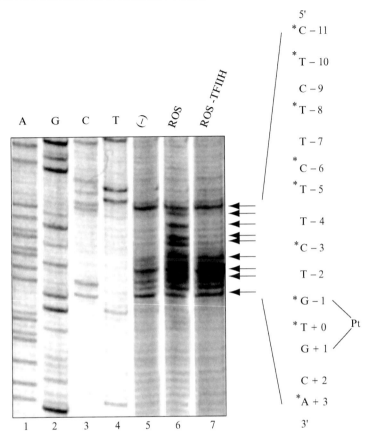

FIG. 2. *In vitro* Permanganate Footprint is performed in the absence (lane 5) or in the presence of XPC-hHR23B, TFIIH, XPA, RPA and XPG (R0S, lane 6). Characteristic KMnO₄ modifications are marked by the arrows. Specifically modified bases are indicated by the asterisk and match with the DNA sequencing (lanes 1 to 4). (−) reaction performed in the absence of DNA repair factors (lane 5).

b. DUAL INCISION REACTION WITH WCE.

1. Place each tube on the Dynal MPC and remove the buffer just before the experiment. Mix 50–75 μg of WCE with 8.5 μl buffer J and bring up the final reaction volume to 20 μl with buffer G. Incubate at 30° for 5 min.

2. Add 5 μl of a mix containing 10 mM ATP, 2.4 μl of 500 mM HEPES-K+, pH 7.6, and 2.1 μl of buffer K and incubate at 30° for not longer than

40 min. Note: the incubation time should not last longer than 40 min to prevent as much as possible the degradation of the released oligonucleotide by some nucleases present in WCE.

3. Add 1.5 μl of a 32 nucleotide oligonucleotide complementary to the excised patch from a 6-ng/μl stock. The 5' overhang is used as a template by the sequenase (DNA polymerase) to incorporate radiolabeled dCMP on the 3' end of the excised fragment.

4. Heat the tubes at 95° for 2 min and spin the tubes down. Let the tubes cool down at room temperature for at least 15 min.

5. Make up a sequenase enzyme (USB)/[α-^{32}P]dCTP mixture such that each reaction contains 0.15 U of sequenase enzyme and 2 μCi of [α-^{32}P] dCTP. Dilute the components with the "sequenase dilution buffer" provided by the manufacturer. Incubate at 37° for 4 min.

6. Add 1.5 μl of dNTP mixture containing 10 μM each of dATP, dGTP, and dTTP and 2 μM of dCTP and incubate at 37° for 12 min.

7. Refer to Sections IV.A.2.c and IV.A.2.d. Load DNA onto a 14% denaturating polyacrylamide sequencing gel (Fig. 3).

c. RECONSTITUTED DUAL INCISION SYSTEM.

1. Purification of the repair factors has already been described (Aboussekhra *et al.*, 1995). Incubate 10 ng of XPC-hHR23B, 20 ng of XPA, 200 ng of RPA, 50 ng of XPG, and 10 ng of XPF-ERCC1 with different amounts of TFIIH for 5 min with 8.5 μl buffer J and 0.15 μl of 500 mM MgCl$_2$. Bring up the final reaction volume to 20 μl with buffer G. The final MgCl$_2$ concentration should be 4 mM.

2. Refer to Section IV.B.2.b.2 but incubate for 90 min.

3. Refer to Sections IV.A.2.c and IV.A.2.d. Load DNA onto a 14% denaturating polyacrylamide sequencing gel.

d. ANALYSIS OF RESULTS. A ladder of bands is characteristic of the NER signal due to the different cut positions by the endonuclease XPF. Increasing amounts of TFIIH are associated with a stronger signal as it participates in the DNA opening and in the recruitement of other NER factors (Fig. 3A, lanes 1 to 3). The importance of this step is underlined by the complete inhibition of the NER signal in the absence of TFIIH.

3. *Analysis of NER Subcomplexes*

a. PRINCIPLE. The binding of DNA to magnetic beads, described in Section III,B, is particularly useful for analyzing intermediate complexes during NER. Using the magnetic rack, the immobilized DNA can be

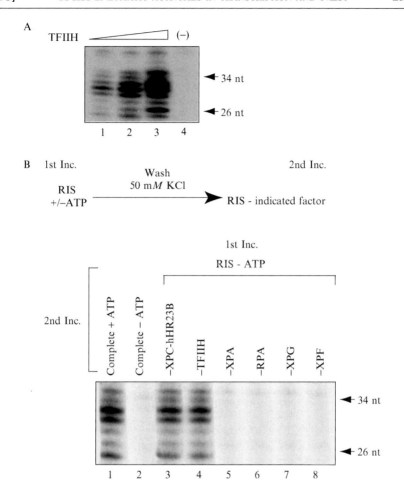

FIG. 3. (A) *In vitro* Reconstituted Dual Incision is performed with XPC-hHR23B, XPA, RPA, XPG, XPF (RIS) and increasing amounts of TFIIH. (−) reaction performed in the absence of TFIIH. (B) ATP-independent Recruitment of TFIIH. All the repair factors and immobilized cis-platinated DNA were incubated without ATP (lanes 3 to 8). Washed DNA beads were then analyzed by a single NER factor omission in a reconstituted dual incision assay (RIS) as indicated. Complete NER was incubated in the presence (lane 1) or the absence (lane 2) of ATP.

separated from the buffer easily. Thus, the same immobilized DNA molecule can be sequentially incubated and washed in various conditions without destabilizing the protein complexes bound to the DNA. The components of the different subcomplexes can be detected either by Western blotting or by a complementation assay. For example, we highlight the formation of an ATP-independent XPC-hHR23B/TFIIH subcomplex during NER by a complementation assay: The reaction occurs in two steps. First the DNA is immobilized onto the magnetic beads and incubated with selected dual incision factors, either in the presence or in the absence of ATP. The supernatant is then removed and the beads are washed. Second, identification of the remaining factors in a particular subcomplex can later be identified by their requirement for an enzymatic assay. In the case of repair proteins, a dual incision assay (see Section IV. B.2) is performed in which the protein of interest is omitted. If the protein was present on the DNA from the first incubation, it could then complement a dual incision assay.

b. COMPLEMENTATION ASSAY.

1. Place each tube on the Dynal MPC and remove the buffer just before incubation. The first incubation mix (except ATP) is the same as described in earlier.

2. Incubate at 30° for 15 min and place the tubes on the Dynal MPC. Remove the supernatant and wash the beads twice with buffer G.

3. Prepare a second mix with ATP as described in Sections IV.B.2.c.1 and IV.B.2.c.2, except that in each tube omit one of the dual incision factors (XPC/hHR23B, TFIIH, XPA, RPA, XPG, or XPF). Incubate the reaction at 30° for 90 min.

4. Refer to Sections IV.B.2.b.3 to IV.B.2.b.6.

5. Refer to Sections IV.A.2.c and IV.A.2.d. Load the DNA onto a 14% denaturating polyacrylamide sequencing gel.

c. ANALYSIS OF RESULTS. Dual incision is an ATP-dependent mechanism, as omission of ATP completely inhibited the reaction (Fig. 3B, lanes 1 and 2). After the first incubation of all the repair proteins in the absence of ATP, complementation assays are carried out. A repair signal is only detected when XPC-hHR23B or TFIIH is omitted from the second incubation (Fig. 3B, lanes 3 and 4, respectively), suggesting that those two proteins are already associated with the damaged DNA from the first incubation in the absence of ATP. Because TFIIH helicase activities are absolutely required for opening the DNA around the lesion to further recruit the repair factors, no repair signal could be detected when XPA, RPA, XPG, or XPF is omitted from the second incubation (Fig. 3B, lanes 5 to 8, respectively).

4. In Vitro DNA Resynthesis Assay

a. PRINCIPLE. The complete NER mechanism results in resynthesis of the gap left behind after the dual incision step. The method described here allows us to follow this gap-filling process using radiolabeled $[\alpha\text{-}^{32}P]dCTP$. The repaired fragment is then cut with *Eco*RI and *Nde*I, which are flanking the lesion and lead to a 94 nucleotide. As a control for the specificity of resynthesis, aphidicolin (a DNA polymerase inhibitor) can be added to the reaction. Moreover, the damaged DNA fragment has been specifically designed to contain the cisplatin (GTG) lesion, which destroys the *Apa*LI restriction site. The repair and the following DNA resynthesis reaction will then recover the *Apa*LI site, allowing for the resynthesized patch to be cut in two smaller fragments of 53 and 41 nucleotides long (Frit *et al.*, 1999).

b. DNA RESYNTHESIS WITH WCE.

1. Place each tube on the Dynal MPC and remove the buffer just before the experiment. Incubate 50–75 μg of WCE at 30° for 5 min as described in Section IV.B.2.a, supplemented with 25 μm each of dATP, dGTP, and dTTP and 4 μCi $[\alpha\text{-}^{32}P]dCTP$. Note: 15 m$M$ aphidicolin can be added to the mixture for the control sample.

2. Refer to Section IV.B.2.c.2.

3. Place the tubes on the Dynal MPC and remove the mixture. Wash the beads twice with buffer BW to remove all proteins from the DNA and then wash the beads again twice with buffer G.

4. Resuspend the beads with a 40-μl final volume mixture containing (4 μl of NEB 4 buffer, 1.5 μl of *Eco*RI [10 U/μl], 0.8 μl of *Nde*I [20 U/μl], 0.4 μl of BSA 1× and 37.3 μl of H$_2$O). Incubate at 37° for 1 h. Note: 1.5 μl of *Apa*LI (10 U/μl) restriction enzyme can be added at this point. Incubate up to 2 h to ensure efficient cut of the DNA fragment.

5. Refer to Sections IV.A.2.c and IV.A.2.d. Load the DNA onto a 8% denaturing polyacrylamide sequencing gel (Fig. 4).

c. ANALYSIS OF RESULTS. After the dual incision, the resynthesis step ensures gap filling and ligation. By cutting with two restriction enzymes (*Eco*RI and *Nde*I), located on each side of the lesion, we can isolate a radiolabeled fragment corresponding to the resynthesized patch (Fig. 4, lane 1). Moreover, with DNA resynthesis restoring the *Apa*LI site, the *Apa*LI restriction enzyme is able to cut the fragment in two (Fig. 4, lane 2). The addition of aphidicolin inhibits DNA synthesis and so confirms the specificity of this experiment (Fig. 4, lane 3). Because the resynthesis step is subsequent to the dual incision step, a lack of TFIIH in the first step of the reaction prevents the DNA synthesis (Fig. 4, lane 4). Any two restriction

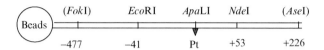

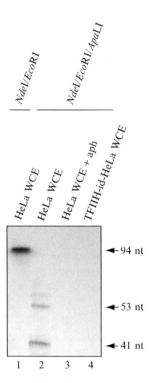

FIG. 4. *In vitro* DNA Resynthesis is performed with HeLa WCE. The newly repaired DNA (top panel) is then cut either by NdeI/EcoRI (lane 1, 94nt) or by NdeI/EcoRI/ApaLI (lanes 2 to 4, 53 and 41 nt). As controls, the cis-platinated DNA is incubated either with an HeLa WCE in the presence of aphidicolin (lane 3) or with a TFIIH-immunodepleted HeLa WCE (lane 4).

enzymes, not encompassing the damage site, can be used to verify that the radiolabeled fragment is a result of a gap-filling reaction and not a result of an unspecific polymerization reaction.

The variety of enzymatic activities within the transcription/repair factor TFIIH led our laboratory to develop a battery of *in vitro* assays that helped us to better understand its crucial role in both processes. XPB and XPD helicase activities are crucial for DNA opening around the start site and for promoter escape. XPB is responsible for the ATP-dependent opening of the transcription initiation complex. The role of XPD in transcription is still unclear, but its presence is required for initiation stimulation and promoter escape. Thanks to the *in vitro* transcription assay and various XPD-mutated TFIIH produced in baculoviruses and purified as described in this chapter, our laboratory was able to relate XPD mutations leading to TTD phenotypes to a defect in the basal transcription activity (Dubaele *et al.*, 2003).

TFIIH is a repair factor as well, whose role is crucial for NER. Magnetic bead technology allowed us to get insight into the NER complex formation at the site of the lesion. Being able to associate and remove the various factors in different buffer conditions, each step of NER was analyzed and the intermediate complexes investigated (Riedl *et al.*, 2003). As an example, we described formation of the XPC-hH23B/TFIIH subcomplex. Interestingly, while the binding of XPC-hHR23B and TFIIH to the lesion was ATP independent, the following recruitment of XPA, RPA, XPG, and XPF strictly relied on the ATP-dependent unwinding of the DNA by TFIIH. Using permanganate footprinting, we showed that NER is a dynamic process in which the DNA is remodeled around the lesion in response to the various NER factors. The binding of XPC/hHR23B and TFIIH to the damaged structure initiates the opening of the helix from position −6 to +6. Thus, it provides the topological structure for recruiting the rest of the NER complex (Tapias *et al.*, 2004). This technology will help precise the role of other repair factors during NER and detail the link between dual incision and DNA resynthesis. It will also be possible to investigate the repair and transcription mechanisms on chromatinized templates: XPE is not necessary for NER and on a naked DNA fragment (Aboussekhra *et al.*, 1995), but data have linked XPE to damage recognition *in vivo* (Moser *et al.*, 2005). The same way that transcriptional activators interact with RNA pol II and/or transcription factors could be approached in a chromatinized environment. It would also be possible to study the mechanism connecting transcription to repair (transcription-coupled repair).

Acknowledgments

This study was supported by funds from La Ligue contre le Cancer (equipe labelisee, contract N°EL2004), the ministere de l'education national et de la recherche for ACI grants (Biologie cellulaire et structurale N°03 2 535). J.P.L. is supported by grants from the Association pour la Recherche contre le Cancer (ARC) and the Fondation pour La

Recherche Medicale (FRM) and V.M. by a grant from INSERM. We are thankful to Ph. Catez, A. Larnicol, and C. Braun for sharing protocols and technical advice.

References

Aboussekhra, A., Biggerstaff, M., Shivji, M. K., Vilpo, J. A., Moncollin, V., Podust, V. N., Protic, M., Hubscher, U., Egly, J. M., and Wood, R. D. (1995). Mammalian DNA nucleotide excision repair reconstituted with purified protein components. *Cell* **80,** 859–868.

Bradsher, J., Coin, F., and Egly, J. M. (2000). Distinct roles for the helicases of TFIIH in transcript initiation and promoter escape. *J. Biol. Chem.* **275,** 2532–2538.

Coin, F., Bergmann, E., Tremeau-Bravard, A., and Egly, J. M. (1999). Mutations in XPB and XPD helicases found in xeroderma pigmentosum patients impair the transcription function of TFIIH. *EMBO J.* **18,** 1357–1366.

Coin, F., Marinoni, J. C., Rodolfo, C., Fribourg, S., Pedrini, A. M., and Egly, J. M. (1998). Mutations in the XPD helicase gene result in XP and TTD phenotypes, preventing interaction between XPD and the p44 subunit of TFIIH. *Nat. Genet.* **20,** 184–188.

Dip, R., Camenisch, U., and Naegeli, H. (2004). Mechanisms of DNA damage recognition and strand discrimination in human nucleotide excision repair. *DNA Repair (Amst)* **3,** 1409–1423.

Dubaele, S., Proietti De Santis, L., Bienstock, R. J., Keriel, A., Stefanini, M., Van Houten, B., and Egly, J. M. (2003). Basal transcription defect discriminates between xeroderma pigmentosum and trichothiodystrophy in XPD patients. *Mol. Cell* **11,** 1635–1646.

Evans, E., Moggs, J. G., Hwang, J. R., Egly, J. M., and Wood, R. D. (1997). Mechanism of open complex and dual incision formation by human nucleotide excision repair factors. *EMBO J.* **16,** 6559–6573.

Frit, P., Bergmann, E., and Egly, J. M. (1999). Transcription factor IIH: A key player in the cellular response to DNA damage. *Biochimie* **81,** 27–38.

Gerard, M., Fischer, L., Moncollin, V., Chipoulet, J. M., Chambon, P., and Egly, J. M. (1991). Purification and interaction properties of the human RNA polymerase B(II) general transcription factor BTF2. *J. Biol. Chem.* **266,** 20940–20945.

Giglia-Mari, G., Coin, F., Ranish, J. A., Hoogstraten, D., Theil, A., Wijgers, N., Jaspers, N. G., Raams, A., Argentini, M., van der Spek, P. J., Botta, E., Stefanini, M., Egly, J. M., Aebersold, R., Hoeijmakers, J. H., and Vermeulen, W. (2004). A new, tenth subunit of TFIIH is responsible for the DNA repair syndrome trichothiodystrophy group A. *Nat. Genet.* **36,** 714–719.

Hansson, J., and Wood, R. D. (1989). Repair synthesis by human cell extracts in DNA damaged by cis- and trans-diamminedichloroplatinum(II). *Nucleic Acids Res.* **17,** 8073–8091.

Holstege, F. C., van der Vliet, P. C., and Timmers, H. T. (1996). Opening of an RNA polymerase II promoter occurs in two distinct steps and requires the basal transcription factors IIE and IIH. *EMBO J.* **15,** 1666–1677.

Humbert, S., van Vuuren, H., Lutz, Y., Hoeijmakers, J. H., Egly, J. M., and Moncollin, V. (1994). p44 and p34 subunits of the BTF2/TFIIH transcription factor have homologies with SSL1, a yeast protein involved in DNA repair. *EMBO J.* **13,** 2393–2398.

Lehmann, A. R. (2003). DNA repair-deficient diseases, xeroderma pigmentosum, Cockayne syndrome and trichothiodystrophy. *Biochimie* **85**, 1101–1111.

Lu, H., Zawel, L., Fisher, L., Egly, J. M., and Reinberg, D. (1992). Human general transcription factor IIH phosphorylates the C-terminal domain of RNA polymerase II. *Nature* **358**, 641–645.

Marinoni, J. C., Rossignol, M., and Egly, J. M. (1997). Purification of the transcription/repair factor TFIIH and evaluation of its associated activities *in vitro*. *Methods* **12**, 235–253.

Moreland, R. J., Tirode, F., Yan, Q., Conaway, J. W., Egly, J. M., and Conaway, R. C. (1999). A role for the TFIIH XPB DNA helicase in promoter escape by RNA polymerase II. *J. Biol. Chem.* **274**, 22127–22130.

Moser, J., Volker, M., Kool, H., Alekseev, S., Vrieling, H., Yasui, A., van Zeeland, A. A., and Mullenders, L. H. (2005). The UV-damaged DNA binding protein mediates efficient targeting of the nucleotide excision repair complex to UV-induced photo lesions. *DNA Repair (Amst)* **4**, 571–582.

Riedl, T., Hanaoka, F., and Egly, J. M. (2003). The comings and goings of nucleotide excision repair factors on damaged DNA. *EMBO J.* **22**, 5293–5303.

Roy, R., Adamczewski, J. P., Seroz, T., Vermeulen, W., Tassan, J. P., Schaeffer, L., Nigg, E. A., Hoeijmakers, J. H., and Egly, J. M. (1994). The MO15 cell cycle kinase is associated with the TFIIH transcription-DNA repair factor. *Cell* **79**, 1093–1101.

Schaeffer, L., Roy, R., Humbert, S., Moncollin, V., Vermeulen, W., Hoeijmakers, J. H., Chambon, P., and Egly, J. M. (1993). DNA repair helicase: A component of BTF2 (TFIIH) basic transcription factor. *Science* **260**, 58–63.

Shivji, M. K., Moggs, J. G., Kuraoka, I., and Wood, R. D. (1999). Dual-incision assays for nucleotide excision repair using DNA with a lesion at a specific site. *Methods Mol. Biol.* **113**, 373–392.

Tapias, A., Auriol, J., Forget, D., Enzlin, J. H., Scharer, O. D., Coin, F., Coulombe, B., and Egly, J. M. (2004). Ordered conformational changes in damaged DNA induced by nucleotide excision repair factors. *J. Biol. Chem.* **279**, 19074–19083.

[16] An Assay for Studying Ubiquitylation of RNA Polymerase II and Other Proteins in Crude Yeast Extracts

By JAMES REID and JESPER Q. SVEJSTRUP

Abstract

   In recent years, much research effort has been invested in understanding the role and importance of protein ubiquitylation. This chapter presents protocols that enable ubiquitylation of RNAPII and other proteins to be reconstituted in crude yeast extracts *in vitro*. These procedures, combined with the ease of genetic manipulations in yeast, provide a new powerful tool for investigating the mechanisms underlying protein ubiquitylation.

Introduction

   Ultraviolet light (UV) irradiation causes bulky lesions such as cyclobutane pyrimidine dimers (CPDs) in DNA that, if left untreated, will obstruct transcription, eventually leading to genomic instability and cell death. To ensure the rapid recovery of mRNA synthesis after UV-induced DNA damage, a repair pathway called transcription-coupled repair (TCR) preferentially repairs the transcribed strand of active genes (Bohr *et al.*, 1985; Mellon *et al.*, 1987; Svejstrup, 2002).
   RNA polymerase II (RNAPII) stalls at UV induced pyrimidine dimers *in vitro* (Selby *et al.*, 1997), and there are several lines of evidence to suggest that such a stalled polymerase is required to trigger TCR (Svejstrup, 2002). In addition, another consequence of UV irradiation is the rapid ubiquitylation and degradation of the large subunit of RNAPII, Rpb1 (Beaudenon *et al.*, 1999; Ratner *et al.*, 1998). Mutations in the TCR factors CSA and CSB cause the hereditary disorder Cockayne's syndrome (CS) and lead to loss of preferential repair in the transcribed strand of active genes. The functionally conserved homolog of CSB in the yeast *Saccharomyces cerevisiae* is Rad26 (van Gool *et al.*, 1994). It has been shown that Rad26 exists in a stable complex with a protein called Def1 in yeast chromatin (Woudstra *et al.*, 2002). Cells lacking Def1 are unable to ubiquitylate and degrade Rpb1 in response to DNA damage, but are still able to support efficient TCR. In contrast, deletion of Rad26 causes accelerated degradation of RNAPII. This separation of repair and RNAPII degradation suggests that degradation of RNAPII represents an alternative

METHODS IN ENZYMOLOGY, VOL. 408   0076-6879/06 $35.00
Copyright 2006, Elsevier Inc. All rights reserved.   DOI: 10.1016/S0076-6879(06)08016-5

to TCR, a last resort. We have proposed a model in which Rad26 displaces RNAPII stalled at DNA damage and recruits nucleotide repair factors to remove the DNA lesion. If, however, the damage cannot be repaired by the TCR pathway, then the polymerase is removed from the damage site by Def1-mediated ubiquitylation and degradation, thereby allowing alternative repair pathways access to the DNA lesion (Svejstrup, 2003; Woudstra *et al.*, 2002).

In vivo data from the study by Woudstra *et al.* (2002), however, did not show conclusively that Def1 is directly involved in ubiquitylation of RNAPII. To address this we needed to reconstitute the ubiquitylation reaction *in vitro*. The ubiquitylation reaction requires three enzymes: the E1 activating enzyme, and E2 conjugating enzyme, and an E3 ubiquitin ligase in addition to ubiquitin, ATP, and the substrate (Hershko and Ciechanover, 1998). Because we did not have the relevant, purified E1, E2, or E3 enzymes available, an assay based on a crude yeast whole cell extract was developed (Reid and Svejstrup, 2004). The advantage of using such a system is that the whole cell extract contains all of the factors required for ubiquitylating not only RNAPII, but also a wide range of other proteins. As a result it is not necessary to purify all of the factors involved in the particular ubiquitylation reaction of interest, and the assay can be modified easily for the study of other substrates. This chapter describes preparation of the ubiquitylation competent whole cell extract and its use in the *in vitro* RNAPII ubiquitylation assay. The procedure can be modified easily for the study of ubiquitylation of other proteins.

Whole Cell Extract Preparation

Preparation of this extract is based on the protocol of Kong and Svejstrup (2002), modified from Schultz *et al.* (1991), and has been shown previously to be competent for studying different aspects of transcription and nucleotide excision repair *in vitro* (Kong and Svejstrup, 2002; Kong *et al.*, 2005; Schultz *et al.*, 1991). The procedure described here is based on a 2-liter yeast culture. However, we have used the protocol successfully with cultures up to 100 liters in size.

Saccharomyces cerevisiae cells of the desired strain are grown in a 2-liter YPD culture to a density of 2×10^7 cells/ml. The cells are harvested by chilling the culture in ice water, followed by centrifugation at 4000g for 10 min. The resulting cell pellet is washed in 1/10 culture volume of ice-cold, sterile water followed by washing in 1/10 volume of extraction buffer (200 mM Tris–Cl, pH 7.5, 0.39 M ammonium sulfate, 10 mM magnesium sulfate, 20% [v/v] glycerol, 1 mM EDTA, 1 mM dithiothreitol [DTT], and protease inhibitors [284 ng/ml leupeptin, 1.37 μg/ml pepstatin A, 170 μg/ml phenylmethysulfonyl fluoride,

and 330 μg/ml benzamindine; protease inhibitors are prepared as a 100×
solution in ethanol]). This is followed by centrifugation at 4000g for 10 min.

A chilled pestle and mortar is prepared: the mortar is placed on a bed of
dry ice, allowed time to cool, and then filled with liquid nitrogen. Once the
mortar has cooled completely, it is refilled with liquid nitrogen. The cell
pellet is scraped into the body of a syringe and, after taking a note of the
final volume in the syringe (yeast pellet volume), the cells are extruded
directly into the liquid nitrogen. Please note that it is not possible to use a
needle with the syringe as it will freeze in the nitrogen vapor and become
blocked.

The resulting strands of frozen yeast are ground to a fine powder in
order to lyse the cells. Ten grams of cells will take approximately 30 min to
grind by hand. As an alternative, we have used a SPEX Certiprep freezer
mill (Glen Creston) to grind the cells mechanically under liquid nitrogen.
The freezer mill also enables the quick and convenient processing of larger
quantities of cells. Once the cells have been lysed, the liquid nitrogen is
allowed to evaporate before adding one pellet volume of cold extraction
buffer. The cells are then allowed to thaw at 4°.

The thawed lysate is centrifuged at 120,000g for 2 h at 4°. Pour the
supernatant into a new tube and immediately repeat centrifugation. This
separates the lysate into a thick cloudy layer of cell debris at the bottom of
the tube, a clear central layer, and a cloudy upper lipid layer. The clear
central layer is carefully removed and transferred to a fresh centrifuge
tube. Be particularly careful not to remove or disturb the upper or lower
cloudy layers. It is often necessary to sacrifice at least 20–30% of the clear
central layer to ensure that it does not become mixed with the other
inhibitory layers.

Ammonium sulfate is then added to the isolated extract to a final
concentration of 2.94 M (60% saturation) by adding 337 mg of solid
ammonium sulfate per milliliter of lysate. Add the ammonium sulfate
slowly over the course of an hour while stirring gently at 4°. Once the
ammonium sulfate has dissolved completely, continue stirring for another
hour. This allows the efficient precipitation of proteins and improves the
final yield.

The precipitated proteins are now collected by centrifugation at 40,000g
for 15 min at 4°. The supernatant is discarded, and the pellet is resuspended
in a minimal amount (approximately 50 μl buffer per milliliter of initial cell
mass) of dialysis buffer (20 mM HEPES-KOH, pH 7.5, 20% glycerol,
10 mM magnesium sulfate, 10 mM EGTA, 5 mM DTT, plus protease
inhibitors [see earlier discussion]). The extract is dialyzed for 12 to 16 h
at 4° against at least two changes of dialysis buffer. Insoluble proteins are

removed by centrifugation at 14,000 rpm, 4° for 10 min in a benchtop centrifuge. The clarified extract is then snap frozen in liquid nitrogen and stored at −80°. The extract should be concentrated, typically at least 25 mg/ml, and will retain significant activity even after several freeze/thaw cycles.

Preparation of myc-Tagged Ubiquitin

Although many tagged and mutant forms of ubiquitin are now available commercially, this section describes a simple method for the purification of recombinant myc-tagged ubiquitin from *Escherichia coli*. This method is adapted from a procedure described by Burch and Haas (1994).

Recombinant myc-tagged ubiquitin is expressed from the plasmid pET myc-Ub in the *E. coli* strain BL21(DE3). After induction, the cells are harvested by centrifugation at 5000 rpm for 10 min and resuspended in 20 mM Tris–Cl, pH 7.5, 150 mM sodium chloride, plus protease inhibitors (see earlier discussion). The cell suspension is lysed by sonication and clarified by centrifugation at 12,000g for 10 min. The lysate is adjusted to pH 4.5 by the slow addition of acetic acid, which causes the majority of proteins to precipitate. Precipitated proteins are removed by centrifugation at 12,000g for 10 min at 4°, and the clear supernatant is transferred to a fresh tube. The pH of the solution is adjusted to 5.1 with sodium hydroxide before applying it to a Sepharose SP fast flow cation-exchange column (Amersham) equilibrated in 50 mM ammonium acetate, pH 5.1. The column is washed with 5 column volumes of the equilibration buffer before eluting bound proteins with a linear gradient to 500 mM ammonium acetate, pH 5.1. Myctagged ubiquitin elutes at approximately 300 mM ammonium acetate. Fractions containing ubiquitin are pooled and dialyzed against 20 mM Tris–Cl, pH 7.5, 50 mM sodium chloride, and stored at −80°. A typical protein concentration is 0.25–0.5 mg/ml. Purified ubiquitin will retain activity even after several freeze/thaw cycles. An example of the highly purified ubiquitin is shown in Fig. 1A.

Purification of RNA Polymerase II

Purification of RNAPII is done essentially as described by Cramer *et al.* (2001). Up to 100 liters of an *S. cerevisiae* strain expressing 3XHA-tagged Rpb1 (Reid and Svejstrup, 2004) is grown in YPD to late log phase, harvested, washed in ice-cold water, and lysed in an equal pellet volume of HSB 150 (50 mM Tris–Cl, pH 7.9, 150 mM potassium chloride, 1 mM

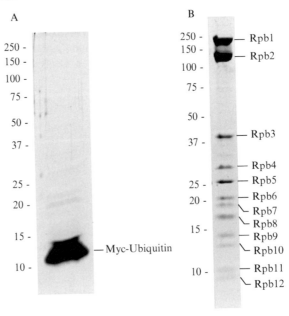

Fig. 1. Purified proteins used in the *in vitro* ubiquitylation assay. (A) Myc-Ubiquitin. (B) 3XHA-tagged RNA polymerase II.

DTT, 10 μM zinc chloride, 10% glycerol, and protease inhibitors). Cells can be lysed as described earlier or by using a mechanical bead beater (see Sayre *et al.*, 1992). The extract is clarified by centrifugation at 4000g for 20 min followed by a second spin at 25,000g for 30 min. A heparin-Sepharose column (Amersham Biosciences) is equilibrated with three column volumes of HSB 150 buffer before loading the extract at 1–2 column volumes per hour. Column capacity is approximately 60 mg extract/ml resin. The column is washed with three further column volumes of HSB 150 before eluting the bound proteins in HSB 600 (HSB buffer + 600 mM potassium chloride). Ammonium sulfate is then added to the eluate to a final concentration of 50% saturation by adding 281 mg of solid ammonium sulfate per milliliter of eluate. Add the ammonium sulfate slowly over the course of an hour while stirring gently at 4°. Once the ammonium sulfate has dissolved completely, continue stirring for another hour. The pre-cipitated protein is now collected by centrifugation at 40,000g for 15 min at 4°. The supernatant is discarded, and the pellet is frozen in liquid nitrogen until use in the next purification step or immediately resuspended in TEZ 0 (TEZ buffer minus ammonium sulfate) until the conductivity is equivalent to that of 400–500 mM ammonium sulfate in the same buffer.

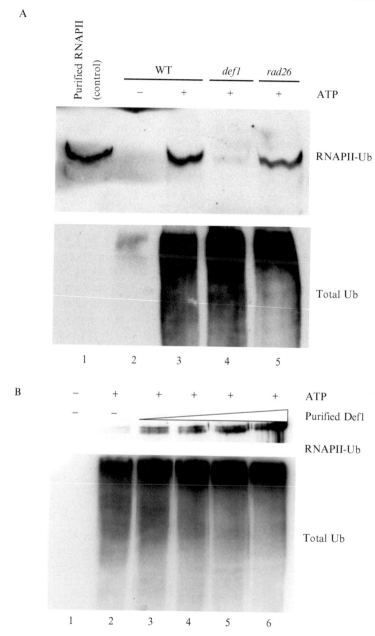

Fig. 2. Ubiquitylation *in vitro* requires Def1. (A) RNAPII ubiquitylation with extracts prepared from wild type (WT), *def1*, and *rad26* strains. (Top) Membrane probed with 12CA5 antibodies (RNAPII detection). (Bottom) Membrane probed with 9E10 antibodies [detection

Anti-Rpb1 affinity beads are prepared using the RNA polymerase II antibody 8WG16 (Covance) by cross-linking the antibody to cyanogen bromide-activated Sepharose (Amersham Biosciences) at a concentration of 5 mg antibody per milliliter of resin according to the manufacturer's instructions. After preparation, the beads are equilibrated with three column volumes of TEZ 250 (50 mM Tris–Cl, pH 7.5, 250 mM ammonium sulfate, 1 mM EDTA, 10 μM zinc chloride, 1 mM DTT, and protease inhibitors). The ammonium sulfate-precipitated, redissolved heparin eluate is loaded onto the antibody beads at 2 to 4 column volumes per hour. After loading, the resin is washed with 10 column volumes (at ~4 column volumes per hour) of TEZ 500 at 4°, followed by a further 10 column volumes at room temperature. The bound polymerase is finally eluted at room temperature using TEZ 500 containing 50% glycerol. Eluates are frozen in liquid nitrogen or are immediately subjected to DEAE chromatography.

To purify and concentrate the polymerase further, the sample is subjected to chromatography on an TSK-Gel DEAE 5PW HPLC column (Supelco) using an Amersham Pharmacia Äkta chromatography system or equivalent. Prior to application, the column is equilibrated with at least three volumes of TEZ 60 plus 10% glycerol. The antibody column eluate is diluted 1:5 with TEZ 0 to reduce both the salt and the glycerol concentration before applying it to the column at a rate of 0.33 ml/min. Unbound material is washed from the column using a further three column volumes of TEZ 60 + 10% glycerol. The purified polymerase is finally eluted with a five column volume linear gradient from TEZ 60 to TEZ 500 (TEZ containing 10% glycerol). The polymerase elutes at approximately 300 mM ammonium sulfate. Fractions containing polymerase are aliquoted, frozen in liquid nitrogen, and stored at −80°. An example of the highly purified RNAPII complex is shown in Fig. 1B.

In Vitro Ubiquitylation

Whole cell extracts prepared using the method described earlier are competent for ubiquitylation of many proteins, including the large subunit of RNA polymerase II (RNAPII) (see Fig. 2). To study RNAPII

---

of total ubiquitylated (*ubi*) protein]. (B) Addition of purified Def1 protein to *def1* extracts recovers RNAPII ubiquitylation activity. (Top) Membrane probed with 12CA5 antibodies (RNAPII detection). (Bottom) Membrane probed with 9E10 antibodies (detection of total ubiquitylated protein).

ubiquitylation, the *in vitro* reactions are performed as follows: for each 100-μl ubiquitylation reaction, 200 μg of extract derived from a strain of interest is incubated with 2 μg myc-ubiquitin, 1 μg 3XHA-tagged RNAPII, 1 mM ATP, and 10 mM lactacystin in ubiquitylation reaction buffer (20 mM HEPES, pH 7.9, 10% glycerol, 60 mM potassium chloride, 7 mM magnesium chloride, plus protease inhibitors). Although the yeast extract contains all factors required for ubiquitylation of RNAPII, we have found that addition of the purified E1-activating enzyme Uba1, which is now available commercially (100 nM E1 activating enzyme per reaction, Boston Biochemicals), can enhance the reaction significantly.

The reactions are incubated for 1 h at 30° before chilling on ice and diluting in 200 μl NP-40 buffer (20 mM Tris, pH 7.5, 150 mM sodium chloride, 10% glycerol, 1% NP-40 plus protease inhibitors). To each reaction mix is added 10 μl protein A-Sepharose beads coupled with the anti-myc antibody 9E10 (available from numerous commercial providers). Myc-tagged ubiquitin conjugates are allowed to bind the resin at 4°, with gentle mixing, for 4 h to overnight. The resin is washed twice with 400 μl NP-40 buffer, then twice in high salt with 400 μl NP-40 buffer containing 1 M sodium chloride, and finally with 400 μl NP-40 buffer again. Extensive washing is required to remove any nonspecifically bound proteins from the beads, especially nonubiquitylated RNAPII.

The resin-associated, ubiquitylated proteins are eluted by the addition of SDS–PAGE sample buffer and boiling for 5 min. The samples are then loaded onto a 6% SDS–PAGE gel, Western blotted, and probed for the presence of RNAPII using the anti-HA antibody 12CA5. Using this system, we were able to show that Def1 is required for ubiquitylation of RNAPII *in vitro* (Reid and Svejstrup, 2004). Ubiquitylated RNAPII can be detected in reactions with the wild type and *rad26* extracts, but not with the *def1* extract [Fig. 2A, top (Reid and Svejstrup, 2004)]. Importantly, the lack of RNAPII ubiquitylation is not due to a general defect in ubiquitylation, as the *def1* extract is still capable of ubiquitylating other proteins as efficiently as the wild type and *rad26* extracts (Fig. 2A, bottom). More importantly, Def1 is directly required for RNAPII ubiquitylation, as addition of the purified protein to the inactive *def1* extract restores ubiquitylation activity (Fig. 2B).

In recent years, interest in studying protein ubiquitylation has increased dramatically. No longer considered to be just a marker for protein degradation, ubiquitylation is rapidly emerging as an important posttranslational modification involved in regulating a wide range of fundamental cellular processes (Conaway *et al.*, 2002; Johnson, 2002; Pickart, 2004; Schnell and Hicke, 2003). The assay described here can potentially be adapted to study the ubiquitylation of many other proteins apart from RNAPII and should be particularly useful when the identity of factors required for ubiquitylation of a target protein remains unknown or when these factors are difficult to purify. The opportunities for genetic manipulation in yeast, combined with the straightforward protocol for preparing extracts presented here, thus make it possible to study the basic mechanisms and regulation of protein ubiquitylation using an easily accessible mixture of *in vitro* and *in vivo* approaches.

References

Beaudenon, S. L., Huacani, M. R., Wang, G., McDonnell, D. P., and Huibregtse, J. M. (1999). Rsp5 ubiquitin-protein ligase mediates DNA damage-induced degradation of the large subunit of RNA polymerase II in *Saccharomyces cerevisiae*. *Mol. Cell. Biol.* **19**, 6972–6979.

Bohr, V. A., Smith, C. A., Okumoto, D. S., and Hanawalt, P. C. (1985). DNA repair in an active gene: Removal of pyrimidine dimers from the DHFR gene of CHO cells is much more efficient than in the genome overall. *Cell* **40**, 359–369.

Burch, T. J., and Haas, A. L. (1994). Site-directed mutagenesis of ubiquitin: Differential roles for arginine in the interaction with ubiquitin-activating enzyme. *Biochemistry* **33**, 7300–7308.

Conaway, R. C., Brower, C. S., and Conaway, J. W. (2002). Emerging roles of ubiquitin in transcriptional regulation. *Science* **296**, 1254–1258.

Cramer, P., Bushnell, D. A., and Kornberg, R. D. (2001). Structural basis of transcription: RNA polymerase II at 2.8 angstrom resolution. *Science* **292**, 1863–1876.

Hershko, A., and Ciechanover, A. (1998). The ubiquitin system. *Annu. Rev. Biochem.* **67**, 425–479.

Johnson, E. S. (2002). Ubiquitin branches out. *Nat. Cell Biol.* **4**, E295–E298.

Kong, S. E., Kobor, M. S., Krogan, N. J., Somesh, B. P., Sogaard, T. M., Greenblatt, J. F., and Svejstrup, J. Q. (2005). Interaction of Fcp1 phosphatase with elongating RNA polymerase II holoenzyme, enzymatic mechanism of action, and genetic interaction with elongator. *J. Biol. Chem.* **280**, 4299–4306.

Kong, S. E., and Svejstrup, J. Q. (2002). Incision of a 1,3-intrastrand d(GpTpG)-cisplatin adduct by nucleotide excision repair proteins from yeast. *DNA Repair (Amst)* **1**, 731–741.

Mellon, I., Spivak, G., and Hanawalt, P. C. (1987). Selective removal of transcription-blocking DNA damage from the transcribed strand of the mammalian DHFR gene. *Cell* **51**, 241–249.

Pickart, C. M. (2004). Back to the future with ubiquitin. *Cell* **116**, 181–190.

Ratner, J. N., Balasubramanian, B., Corden, J., Warren, S. L., and Bregman, D. B. (1998). Ultraviolet radiation-induced ubiquitination and proteasomaldegradation of the large subunit of RNA polymerase II: Implications for transcription-coupled DNA repair. *J. Biol. Chem.* **273**, 5184–5189.

Reid, J., and Svejstrup, J. Q. (2004). DNA damage-induced Def1-RNA polymerase II interaction and Def1 requirement for polymerase ubiquitylation *in vitro*. *J. Biol. Chem.* **279**, 29875–29878.

Sayre, M. H., Tschochner, H., and Kornberg, R. D. (1992). Reconstitution of transcription with five purified initiation factors and RNA polymerase II from *Saccharomyces cerevisiae*. *J. Biol. Chem.* **267**, 23376–23382.

Schnell, J. D., and Hicke, L. (2003). Non-traditional functions of ubiquitin and ubiquitin-binding proteins. *J. Biol. Chem.* **278**, 35857–35860.

Schultz, M. C., Choe, S. Y., and Reeder, R. H. (1991). Specific initiation by RNA polymerase I in a whole-cell extract from yeast. *Proc. Natl. Acad. Sci. USA* **88,** 1004–1008.

Selby, C. P., Drapkin, R., Reinberg, D., and Sancar, A. (1997). RNA polymerase II stalled at a thymine dimer: Footprint and effect on excision repair. *Nucleic Acids Res.* **25,** 787–793.

Svejstrup, J. Q. (2002). Mechanisms of transcription-coupled DNA repair. *Nat. Rev. Mol. Cell. Biol.* **3,** 21–29.

Svejstrup, J. Q. (2003). Rescue of arrested RNA polymerase II complexes. *J. Cell Sci.* **116,** 447–451.

van Gool, A. J., Verhage, R., Swagemakers, S. M., van de Putte, P., Brouwer, J., Troelstra, C., Bootsma, D., and Hoeijmakers, J. H. (1994). RAD26, the functional *S. cerevisiae* homolog of the Cockayne syndrome B gene ERCC6. *EMBO J.* **13,** 5361–5369.

Woudstra, E. C., Gilbert, C., Fellows, J., Jansen, L., Brouwer, J., Erdjument-Bromage, H., Tempst, P., and Svejstrup, J. Q. (2002). A Rad26-Def1 complex coordinates repair and RNA pol II proteolysis in response to DNA damage. *Nature* **415,** 929–933.

[17] Analysis of the Excision Step in Human DNA Mismatch Repair

By JOCHEN GENSCHEL and PAUL MODRICH

Abstract

The reaction responsible for replication error correction by mismatch repair proceeds via several steps: mismatch recognition, mismatch-provoked excision, repair DNA synthesis, and ligation. Key steps in this process are the recognition and subsequent exonucleolytic removal of the mispair. A minimal system comprised of human MutSα (MSH2•MSH6), MutLα (MLH1•PMS2), exonuclease I (EXOI), replication protein A (RPA), proliferating cell nuclear antigen (PCNA), and replication factor C (RFC) is sufficient to support mismatch-provoked excision *in vitro*. This chapter describes methods for analysis of the reconstituted excision reaction.

Introduction

Mismatch repair provides several genetic stabilization functions, but the best understood is its role in the correction of DNA biosynthetic errors. Base–base mismatches can arise as a consequence of DNA polymerase misincorporation, whereas the production of insertion/deletion mispairs is usually attributed to template or primer strand slippage during the process of replication (Kunkel, 1992, 1993). Mismatches produced in this manner are recognized and rectified by mismatch repair (Jiricny, 1998; Kolodner

METHODS IN ENZYMOLOGY, VOL. 408   0076-6879/06 $35.00
                 DOI: 10.1016/S0076-6879(06)08017-7

and Marsischky, 1999; Kunkel and Erie, 2005; Modrich and Lahue, 1996; Schofield and Hsieh, 2003). Inasmuch as inactivation of this pathway confers a 100- to 1000-fold increase in the mutation rate, it is evident that mismatch repair contributes in a major way to genetic stabilization. Inactivation of the human pathway is the cause of hereditary nonpolyposis colon cancer (Kolodner et al., 1999; Lynch and de la Chapelle, 1999; Rowley, 2005), but has also been implicated in the development of a subset of sporadic tumors that occur in a variety of tissues (Eshleman and Markowitz, 1995; Peltomaki, 2003).

Correction of DNA biosynthetic errors relies on secondary signals in the helix that direct mismatch repair to the newly synthesized DNA strand. Although the strand signals that direct eukaryotic mismatch repair have not been elucidated, a strand-specific single strand break is sufficient to direct repair in nuclear extracts of human cells (Fang and Modrich, 1993; Holmes et al., 1990; Thomas et al., 1991), and it has been suggested that DNA termini that occur at the replication fork during the natural course of DNA replication may serve to direct the reaction in vivo (Holmes et al., 1990; Pavlov et al., 2002).

Analysis of human mismatch repair in nuclear extracts has demonstrated that the strand break that directs repair can be located either 3' or 5' to the mismatch on the incised strand (Fang and Modrich, 1993; Wang and Hays, 2002). Mismatch-provoked excision removes DNA spanning the two DNA sites, with excision tracts terminating at a number of sites 90–170 nucleotides beyond the mismatch (Fang and Modrich, 1993). Repair synthesis and ligation restore double helical integrity to the repaired molecule. To date, eight activities have been implicated in human mismatch repair: MutSα (MSH2•MSH6), MutSβ (MSH2•MSH3), MutLα (MLH1•PMS2), exonuclease I (EXOI), replication protein A (RPA), proliferating cell nuclear antigen (PCNA), replication factor C (RFC), and DNA polymerase δ (Dzantiev et al., 2004; Genschel and Modrich, 2003; Genschel et al., 2002; Kolodner and Marsischky, 1999; Longley et al., 1997; Modrich, 1997).

We have shown that near homogeneous MutSα, MutLα, EXOI, RPA, RFC, and PCNA comprise a minimal system that is sufficient to support mismatch-provoked excision directed by a strand break located either 3' or 5' to the mispair. Analysis of this system has demonstrated that RFC and PCNA play key roles in determining the directionality of excision. When the strand break that directs the reaction is located 5' to the mismatch, MutSα activates mismatch-provoked excision by EXOI, which occurs with the 5' to 3' polarity characteristic of EXOI and terminates upon mismatch removal if RPA is present (Genschel and Modrich, 2003). Hydrolysis directed by a 5' strand break can occur in the absence of RFC and PCNA, but excision directed by a 3′ strand break is absolutely dependent on the two proteins, as well as MutSα, MutLα, and EXOI (Dzantiev *et al.*, 2004; Genschel *et al.*, 2002). Although EXOI hydrolyzes DNA with 5′ to 3′ polarity in the absence of other proteins, the EXOI active site is also required for 3′-directed excision. Molecular details of EXOI involvement in the 3′ reaction are not yet clear.

The sections that follow describe procedures for assay of mismatch-provoked excision, as well as isolation of activities required for this reaction.

Methods

Isolation of MutSα, MutLα, and EXOI

Baculovirus constructs expressing MutSα, MutLα, or EXOI have been described (Blackwell *et al.*, 2001; Genschel *et al.*, 2002) and are available upon request. RPA and PCNA are purified from overproducing *Escherichia coli* strains using published protocols (Dzantiev *et al.*, 2004; Fien and Stillman, 1992; Genschel and Modrich, 2003; Henricksen *et al.*, 1994). RFC is isolated from the HeLa nuclear extract using procedures described previously (Dzantiev *et al.*, 2004; Tsurimoto and Stillman, 1989).

Protocol 1. General Fractionation Procedures

REAGENTS. Glycerol (ultrapure, MP Biomedicals, LLC); dithiothreitol (DTT; USB); ATP (USB); phenylmethylsulfonyl fluoride (PMSF; Sigma); leupeptin (Peptides International); E-64 (Peptides International); and aprotinin (USB).

BUFFERS AND GENERAL PROCEDURES

Buffer A: 25 mM HEPES/KOH, pH 7.5, 0.1 mM EDTA
Buffer B: 20 mM KPO$_4$, pH 7.5, 0.1 mM EDTA

Buffers A and B are supplemented with KCl and glycerol as indicated (all glycerol concentrations are v/v) and are then autoclaved; for use in FPLC chromatography, buffers are also sterile filtered using 0.22-μm filter cartridges. Solutions prepared in this manner are then supplemented as indicated with DTT and freshly prepared protease inhibitors [0.1 % (v/v) PMSF relative to saturated solution in isopropanol; 1 mg/l leupeptin; 0.5 mg/l E-64; 0.1 mg/l aprotinin]. DTT and protease inhibitors are routinely omitted from column buffers during FPLC chromatography. They are instead added to fraction collector tubes prior to collection such that their final concentrations in the fractions are the same as those given earlier.

All fractionation steps are done at 4° or on ice. Protein concentrations are determined by the Bradford assay using bovine serum albumin (BSA) as a standard. KCl concentrations of chromatographic fractions are determined by conductivity measurement (radiometer CDM 83) after a 1:400 dilution into distilled H_2O relative to standards of known KCl concentration.

Protocol 2. Isolation of EXOI. Four hundred milliliters of SF9 cells is cultured at 27° in serum-free HyQ-SFX medium (HyClone, Inc.). At a density of 2×10^6 per milliliter, cells are infected (MOI = 2) with baculovirus expressing the b splice variant of human EXOI, and incubation is continued for 60 h (Genschel *et al.*, 2002). Cells are collected by centrifugation (Sorvall H6000A rotor, 3000 rpm, 10 min, 4°). Cell pellets are resuspended (3 ml per gram cells) in buffer A containing 5 mM KCl, 1 mM DTT, and protease inhibitors and are disrupted with 20 strokes using a Dounce B pestle. The salt concentration of the disrupted cell suspension is adjusted to 0.15 M KCl by the addition of 3 M KCl in buffer A, and cellular debris is removed by centrifugation (Sorvall SS-34 rotor, 10,000 rpm, 20 min, 4°).

The clear supernatant is loaded at 1 ml/min onto a 5-ml HiTrap Q FF anion-exchange column (Amersham Biosciences) equilibrated with buffer B containing 0.15 M KCl, 10% glycerol, 1 mM DTT, and protease inhibitors using a peristaltic pump. The column is then attached to an FPLC chromatography system and washed with 100 ml buffer B containing 0.15 M KCl, 10% glycerol at 2 ml/min. The column is developed with a 50-ml gradient of 0.15 to 0.4 M KCl in buffer B containing 10% glycerol at 1 ml/min. One-milliliter fractions are collected into tubes containing DTT and protease inhibitors as described earlier and processed immediately. After removal of small samples (20–50 μl) the remainder of the fractions are quick frozen in liquid N_2 and stored at −70°. The samples are used to determine KCl concentration and to monitor the EXOI elution profile by SDS gel electrophoresis. EXOI, which migrates with an apparent molecular mass of 115 kDa, elutes at approximately 0.25 M KCl.

Fractions containing EXOI are thawed on ice, diluted with buffer B containing 10% glycerol, 1 mM DTT, and protease inhibitors to a conductivity equivalent to 0.2 M KCl and are loaded at 0.5 ml/min onto a 1-ml HiTrap heparin column (Amersham Biosciences) attached to an FPLC chromatography system and equilibrated with buffer B containing 0.2 M KCl and 10% glycerol. After washing with 30 ml buffer B containing 0.2 M KCl, 10% glycerol at 0.5 ml/min, the column is developed at 0.5 ml/min with a 20-ml gradient of 0.2 to 0.6 M KCl in buffer B containing 10% glycerol. Fractions (0.5 ml) are collected into tubes containing DTT and protease inhibitors as described earlier, which are sampled and frozen in liquid N_2 as for HiTrap Q FF eluates. EXOI elutes from HiTrap heparin at approximately 0.35 M KCl.

Fractions containing EXOI are thawed, diluted with buffer B containing 10% glycerol, 1 mM DTT, and protease inhibitors to a conductivity corresponding to 0.2 M KCl and are loaded at 0.5 ml/min onto a 1-ml FPLC MonoS column (HR 5/5, Amersham Biosciences) equilibrated with buffer B containing 0.2 M KCl, 10% glycerol. After washing at 0.5 ml/min with 30 ml of equilibration buffer, the column is developed with a 10-ml gradient of 0.2 to 0.6 M KCl in buffer B containing 10% glycerol at 0.25 ml/min. Fractions (0.25 ml) are collected into tubes containing DTT and protease inhibitors as described earlier and processed as for the HiTrap heparin column. EXOI elutes from Mono S in a broad peak centered about 0.35 M KCl. Peak fractions, which are better than 95% pure, are pooled, distributed into small aliquots, frozen in liquid N_2, and stored at $-70°$.

COMMENTS ON EXOI. In the absence of glycerol and at KCl concentrations below 0.2 M, EXOI tends to precipitate. However, we have typically been able to achieve concentrations of 0.5 mg/ml in the presence of 0.2 M KCl and 10% glycerol.

Protocol 3. Isolation of MutSα. A 500-ml culture of SF9 cells grown to 2×10^6 per milliliter as described earlier is infected at an MOI of 2 with a baculovirus that expresses both MSH2 and MSH6 subunits of MutSα (Blackwell *et al.*, 2001). After further incubation for 60 h, cells are harvested and lysed as described earlier for isolation of EXOI. The salt concentration of the extract is adjusted to 0.2 M KCl and insoluble material is removed by centrifugation (Sorvall SS-34 rotor, 10,000 rpm, 20 min, 4°).

The clear supernatant is loaded with a peristaltic pump at 2 ml/min onto a 10-ml ssDNA-cellulose column (USB) equilibrated with buffer A containing 0.2 M KCl, 1 mM DTT, and protease inhibitors. The column is washed sequentially with 10 column volumes of buffer A containing 0.2 M KCl, 1 mM DTT, and protease inhibitors followed by 10 column volumes of buffer A containing 0.2 M KCl, 2.5 mM MgCl$_2$, 1 mM DTT, and protease inhibitors. MutSα is then eluted with 10 column volumes of buffer A containing 0.2 M KCl, 2.5 mM MgCl$_2$, 1 mM ATP, 1 mM DTT, and protease inhibitors. The eluate is immediately loaded at 2 ml/min onto a 3-ml Q-Sepharose FF column (Amersham Biosciences) equilibrated with buffer A containing 0.2 M KCl, 1 mM DTT, and protease inhibitors. After washing with 3 column volumes of equilibration buffer, the column is step eluted with buffer A containing 0.6 M KCl, 1 mM DTT, and protease inhibitors. One-milliliter fractions are collected and checked for protein content by the Bradford assay. Protein-containing fractions are pooled. At this step, MutSα is approximately 90% pure but contains adenine nucleotides from the elution buffer. The Q-Sepharose eluate is diluted with buffer

B containing 10% glycerol, 1 mM DTT, and protease inhibitors to a conductivity corresponding to 0.2 M KCl and is loaded at 0.5 ml/min onto a 1-ml Mono Q FPLC column (HR 5/5, Amersham Biosciences) equilibrated with buffer B containing 0.2 M KCl, 10% glycerol. After wash at 0.5 ml/min with 30 ml buffer B containing 0.2 M KCl, 10% glycerol, the column is developed at 0.5 ml/min with a 20-ml gradient of 0.2 to 0.6 M KCl in buffer B containing 10% glycerol. Fractions (0.5 ml) are collected into tubes containing DTT and protease inhibitors as described earlier. MutSα fractions, which elute at approximately 0.35 M KCl, are analyzed by SDS gel electrophoresis and pooled. On SDS gels, MSH2 and MSH6 migrate with apparent molecular masses of 105 and 160 kDa, respectively. The protein is dispensed into small aliquots, which are frozen in liquid N$_2$ and stored at $-70°$. If the protein concentration is less than 0.5 mg/ml, BSA (0.25 mg/ml) can be added as a stabilizing agent prior to freezing.

Although MutSα is better than 95% pure after Mono Q chromatography, some preparations are contaminated by significant levels of a single strand-specific exonuclease. This activity can be removed by Mono S chromatography. The Mono Q eluate is diluted with buffer B containing 10% glycerol, 1 mM DTT, and protease inhibitors to a conductivity equivalent to 0.075 M KCl and loaded at 0.5 ml/min onto a 1-ml Mono S FPLC column (HR 5/5, Amersham Biosciences) equilibrated with buffer B containing 0.075 M KCl, 10% glycerol. The column is washed at 0.5 ml/min with 20 ml of buffer B containing 0.075 M KCl, 10% glycerol, and developed with a 20-ml gradient of 0.075 to 0.5 M KCl in buffer B, 10% glycerol. MutSα elutes at approximately 0.175 M KCl. Fractions (0.5 ml) are processed, aliquoted, and frozen as described earlier.

Protocol 4. Isolation of MutLα. MutLα is isolated from SF9 cells using two different baculoviruses that individually express the MLH1 and PMS2 subunits of the heterodimer (Blackwell *et al.*, 2001). Four hundred milliliters of insect cells at 2×10^6 cells/ml is infected at an MOI of 5 with each of the two viruses and incubation continued for 60 h. Cells are collected, disrupted, and adjusted to a final KCl concentration of 0.1 M, and cellular debris is removed as described earlier for EXOI isolation. The clarified supernatant is loaded at 1 ml/min onto a 5-ml HiTrap heparin column (Amersham Biosciences) equilibrated with buffer B containing 0.1 M KCl, 10% glycerol, 1 mM DTT, and protease inhibitors using a peristaltic pump. After attachment to an FPLC system, the column is washed with 100 ml buffer B containing 0.1 M KCl, 10% glycerol at 2 ml/min and is then eluted at 1 ml/min with a 60-ml gradient of 0.1 to 0.5 M KCl in buffer B containing 10% glycerol. Fractions are collected into tubes containing DTT and protease inhibitors. MutLα, which elutes at approximately 0.25 M KCl, is identified by SDS gel electrophoresis (MLH1 and PMS2 migrate with apparent molecular masses of 85 and 110 kDa, respectively), and peak fractions are combined. The HiTrap heparin eluate is diluted to a conductivity corresponding to 0.1 M KCl with buffer B containing 10% glycerol, 1 mM DTT, and protease inhibitors and is loaded at 0.5 ml/min onto a 1-ml Mono Q FPLC column (HR 5/5, Amersham Biosciences) equilibrated with buffer B containing 0.1 M KCl, 10% glycerol. After a wash at 0.5 ml/min with buffer B containing 0.1 M KCl, 10% glycerol, the column is developed at 0.5 ml/min with a 20-ml gradient of 0.1 to 0.5 M KCl in buffer B, 10% glycerol, with 0.5-ml fractions collected. As described earlier, collection tubes contain DTT and protease inhibitors. MutLα which elutes at about 0.28 M KCl, is identified by SDS gel electrophoresis, and peak fractions are pooled. The Mono Q eluate is diluted to a conductivity corresponding to 0.1 M KCl with buffer B containing 10% glycerol, 1 mM DTT, and protease inhibitors and is loaded at 0.5 ml/min onto a 1-ml Mono S FPLC column (HR 5/5, Amersham Biosciences) equilibrated with buffer B containing 0.1 M KCl, 10% glycerol. The column is washed at 0.5 ml/min with 20 ml buffer B containing 0.1 M KCl, 10% glycerol, and developed at 0.5 ml/min with a 20-ml gradient of 0.1 to 0.5 M KCl in buffer B, 10% glycerol. Fractions (0.5 ml) are collected into tubes containing DTT and protease inhibitors. MutLα, which elutes at approximately 0.25 M KCl, is identified as described earlier, and peak fractions are combined and quick frozen as small aliquots in liquid N_2 prior to storage at $-70°$.

COMMENTS ON MutLα. MutLα can precipitate at low salt concentrations. Several column-loading steps involve dilution of the protein to yield a KCl concentration of 0.1 M. To avoid precipitation of the protein, this procedure is performed in a batchwise manner on small aliquots (10 to 20%), which are immediately loaded onto the column.

Mismatch-Provoked Excision

BUFFERS. 10× reaction buffer: 0.2 M Tris/HCl, pH 7.6, 10 mM reduced glutathione (Calbiochem), 15 mM ATP (USB), 50 mM MgCl$_2$, 0.5 mg/ml BSA (Roche); dilution buffer: 8 mM HEPES/KOH, pH 7.5, 0.2 M KCl, 10% (v/v) glycerol, 2 mg/ml BSA, 0.6 mM DTT; stop buffer: 10 mM Tris/HCl, pH 8.0, 0.16% (w/v) SDS, 1 mg/ml glycogen (Roche), 0.12 mg/ml proteinase K (Roche); agarose gel-loading buffer: 0.25 M EDTA, pH 8.0, 1.2 % (w/v) SDS, 25% (w/v) sucrose, 0.1 % (w/v) bromphenol blue; alkaline-loading buffer: 0.3 M NaOH, 50 mM EDTA, 18% (w/v) Ficoll 400 (Amersham Biosciences), 0.2% (w/v) bromocresol green; hybridization solution: 0.5 M NaPO$_4$, pH 7.2, 7% (w/v) SDS, 0.5% (w/v) blocking reagent (Roche); wash solution: 0.04 M NaPO$_4$, pH 7.2, 1% (w/v) SDS, 0.5% (w/v) blocking reagent (Roche)

STANDARD REACTION CONDITIONS. Excision reactions are performed in 20-μl reactions containing 100 ng (24 fmol) of f1 heteroduplex (or control homoduplex) DNA, 100 ng MutSα (400 fmol), 50 to 100 ng MutLα (300–600 fmol), 100 to 200 ng RPA (900–1800 fmol), 25 ng PCNA (300 fmol), 20 to 50 ng RFC (80–200 fmol), 2–5 ng EXOI (20–50 fmol), and KCl at a final concentration of 0.1 to 0.15 M KCl, including salt contributed by DNA and proteins.

Protocol 1. Mismatch-Provoked Excision Reactions. Heteroduplex and control homoduplex DNAs used to study human mismatch repair are derived from 6.4 kbp f1 bacteriophage DNA (Dzantiev *et al.*, 2004; Fang and Modrich, 1993; Su *et al.*, 1988; see Fig. 1). Although space limitations preclude description of heteroduplex preparation in this chapter, we will provide a protocol upon request.

A 10-μl solution containing 2 μl of 10× buffer, DNA, and KCl as necessary is placed at the bottom of a 1.5-ml plastic centrifuge tube on ice. In our standard reactions the final KCl concentration is 0.15 M, conditions that yield maximal mismatch specificity; product yield can be increased by reducing the KCl concentration to 0.1 M, but this occurs at the expense of mismatch specificity. Eight microliters of a solution containing MutSα, MutLα, RPA, PCNA, and RFC in dilution buffer is then placed on the side of the centrifuge tube. Two microliters of EXOI in dilution buffer is added to the droplet containing the other proteins. The tube is quick spun in an Eppendorf microfuge, vortexed, quick spun again, and transferred to a 37° water bath. After 5 min, hydrolysis is terminated by the addition of 80 μl of stop buffer, and the tube is returned to the water bath for an additional 15 min at 37°. The reaction is then extracted twice with 50 μl phenol (equilibrated with 10 mM Tris/HCl, pH 8.0), and the aqueous phase, which is recovered after centrifugation at 13,000 rpm for 1 min in an Eppendorf microfuge, is transferred to a fresh tube containing 10 μl 3 M NaOAc, pH 5.2, and 290 μl 100% ethanol. After 1 h at −20°, DNA is collected by centrifugation at 13,000 rpm for 20 min in an Eppendorf centrifuge at 4°. The supernatant is removed carefully and the DNA pellet is washed twice with 1 ml of 70% ethanol and DNA recovered by centrifugation. The final pellet, which is typically not visible to the eye, is dried under vacuum for 10 min.

Protocol 2. Scoring Mismatch-Provoked Excision by NheI Cleavage. Because our heteroduplex DNAs containing an *Nhe*I site 5 bp from the mismatch, mismatch-provoked excision can be scored routinely by virtue of conversion of this site to an endonuclease-resistant form (Genschel *et al.*, 2002; see Fig. 1). For this method, the DNA pellet obtained earlier is dissolved in 10 μl of restriction buffer (New England Biolabs buffer 2). Three microliters of a restriction enzyme cocktail containing 2 units each of

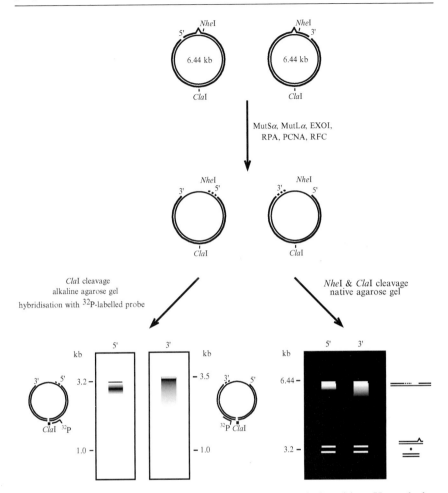

FIG. 1. Schematic flowchart for assay of mismatch-provoked excision. Heteroduplex substrates contain a strand break either 129 bp 5′ to the mismatch or 141 bp 3′ to the mismatch. (Dzantiev *et al.*, 2004). After incubation with MutSα, MutLα, EXOI, RPA, PCNA, and RFC as described in the text, excision products are scored as *Nhe*I-resistant DNA by native agarose gel electrophoresis (lower left, protocol 2). Excision can also be scored by indirect end labeling after electrophoresis through alkaline agarose, a method that also serves to localize excision tract end points (lower right, protocol 3). The latter method employs ^{32}P-labeled oligonucleotides that hybridize to the incised DNA strand on either side of the *Cla*I cleavage site. Hybridization in this manner permits localization of 3′ and 5′ termini in excision products. The oligonucleotides that we use routinely for this purpose correspond to viral strand sequences near the *Cla*I site of f1MR1 (Su *et al.*, 1988) [oligonucleotide V2531 for the 5′ substrate: d(ATGGTTTCATTGGTGACGTTTC), nucleotides 2531–2553; oligonucleotide V2505 for the 3′ substrate: d(CGCTACTGATTACGGTGCTGCT), nucleotides 2505–2527) (Dzantiev *et al.*, 2004; Genschel and Modrich, 2003)].

*Cla*I and *Nhe*I in NEB buffer 2 is added, and the DNA is digested for 1 h at 37°. Two microliters of agarose gel-loading buffer is added, and the DNA is electrophoresed through 1% agarose (14-cm flatbed gel, Owl Scientific Type B3) at 30 V for 15 h in 40 mM Tris/HOAc, pH 8.0, 1 mM EDTA (Maniatis *et al.*, 1982). After staining with 0.5 μg/ml ethidium bromide and destaining, DNA products are photographed using a cooled photometric grade CCD camera (Photometrics). DNA bands in each lane are quantitated using the NIH image 1.63 software package. As summarized in Fig. 1, cleavage of the native substrate with *Nhe*I and *Cla*I yields two DNA fragments. However, because excision products are *Nhe*I resistant, cleavage with the two enzymes yields DNA species that migrate slightly faster than the full-length linear molecule (6.44 kbp).

Protocol 3. Scoring Mismatch-Provoked Excision Tracts by Indirect End Labeling. While the *Nhe*I resistance assay is useful for routine determination of mismatch-provoked excision, it provides no information concerning the nature of excision tract end points. Such questions can be addressed by indirect end labeling (Fang and Modrich, 1993; Fig. 1). To this end, the 10-μl solution obtained by dissolving the DNA excision products in New England Biolabs buffer 2 (see earlier discussion) is split into two 5-μl aliquots. One aliquot is processed for *Nhe*I resistance and native gel electrophoresis as described earlier. The other half is digested with only *Cla*I for 1 h at 37° and is used for alkaline agarose gel electrophoresis. After digestion the reaction is supplemented with 2 μl alkaline-loading buffer, and samples are subjected to electrophoresis through 1% agarose (14-cm gel) in 50 mM NaOH, 1 mM EDTA at room temperature with buffer circulation 25 V for 20 h.

DNA is then transferred to a Hybond N+ membrane (Amersham Biosciences) by capillary transfer in 0.4 M NaOH for approximately 20 h. After two 5-min washes with 50 mM NaPO$_4$, pH 7.2, the membrane is air dried and transferred to a hybridization tube. After the addition of 50 ml of hybridization solution, the tube is incubated with rotation in a hybridization oven (Bellco Glass Inc. Autoblot) for several hours at 45°. After blocking in this manner, the solution is replaced with 10 ml hybridization solution containing a ^{32}P-labeled oligonucleotide probe (2–5 pmol oligonucleotide, 2–5 10^6 cpm) specific for the excised strand of the mismatch containing substrate (Fig. 1). After overnight hybridization at 45°, the probe solution is removed, and the membrane is washed twice with 50 ml wash solution for 20 min at 45°. The membrane is then covered with plastic wrap and exposed to a phosphoimager screen (Molecular Dynamics). Undigested DNA and excision products are then visualized with a Typhoon phosphoimager (Molecular Dynamics).

Discussion

This chapter described methods to analyze a key step in the process of human DNA mismatch repair, the hydrolytic removal of the mismatch. Although the procedures described are presented in the context of a purified system, the analytical methods for scoring mismatch-provoked excision work equally well with nuclear extracts. While MutSα, MutLα, EXOI, RPA, PCNA, and RFC are sufficient to support mismatch-provoked excision directed by a 3′ or 5′ strand break, we regard this set of proteins as a minimal system for several reasons. For example, the behavior of EXOI-depleted extracts and the phenotype of EXOI-deficient mouse cells suggest that alternate excision activities may exist that can function in a manner that is redundant with respect to EXOI (Genschel et al., 2002; Wei et al., 2003). Furthermore, the mismatch dependence of 5′-directed excision in the purified system is significantly less than that observed in nuclear extracts, suggesting that the purified system may be deficient in one or more factors that enhance the specificity of the reaction (Dzantiev et al., 2004). We are currently seeking these other activities.

Acknowledgments

This work was supported in part by Grant GM45190 from the National Institutes of Health. P.M. is an Investigator of the Howard Hughes Medical Institute.

References

Blackwell, L. J., Wang, S., and Modrich, P. (2001). DNA chain length dependence of formation and dynamics of hMutSα•hMutLα• heteroduplex complexes. *J. Biol. Chem.* **276,** 33233–33240.

Dzantiev, L., Constantin, N., Genschel, J., Iyer, R. R., Burgers, P. M., and Modrich, P. (2004). A defined human system that supports bidirectional mismatch-provoked excision. *Mol. Cell* **15,** 31–41.

Eshleman, J. R., and Markowitz, S. D. (1995). Microsatellite instability in inherited and sporadic neoplasms. *Curr. Opin. Oncol.* **7,** 83–89.

Fang, W.-H., and Modrich, P. (1993). Human strand-specific mismatch repair occurs by a bidirectional mechanism similar to that of the bacterial reaction. *J. Biol. Chem.* **268,** 11838–11844.

Fien, K., and Stillman, B. (1992). Identification of replication factor C from *Saccharomyces cerevisiae*: A component of the leading-strand DNA replication complex. *Mol. Cell. Biol.* **12,** 155–163.

Genschel, J., Bazemore, L. R., and Modrich, P. (2002). Human exonuclease I is required for 5′ and 3′ mismatch repair. *J. Biol. Chem.* **277,** 13302–13311.

Genschel, J., and Modrich, P. (2003). Mechanism of 5′-directed excision in human mismatch repair. *Mol. Cell* **12,** 1077–1086.

Henricksen, L. A., Umbricht, C. B., and Wold, M. S. (1994). Recombinant replication protein A: Expression, complex formation, and functional characterization. *J. Biol. Chem.* **269,** 11121–11132.

Holmes, J., Clark, S., and Modrich, P. (1990). Strand-specific mismatch correction in nuclear extracts of human and *Drosophila melanogaster* cell lines. *Proc. Natl. Acad. Sci. USA* **87,** 5837–5841.

Jiricny, J. (1998). Replication errors: Cha(lle)nging the genome. *EMBO J.* **17,** 6427–6436.

Kolodner, R. D., and Marsischky, G. T. (1999). Eukaryotic DNA mismatch repair. *Curr. Opin. Genet. Dev.* **9,** 89–96.

Kolodner, R. D., Tytell, J. D., Schmeits, J. L., Kane, M. F., Gupta, R. D., Weger, J., Wahlberg, S., Fox, E. A., Peel, D., Ziogas, A., Garber, J. E., Syngal, S., Anton-Culver, H., and Li, F. P. (1999). Germ-line msh6 mutations in colorectal cancer families. *Cancer Res.* **59,** 5068–5074.

Kunkel, T. A. (1992). DNA replication fidelity. *J. Biol. Chem.* **267,** 18251–18254.

Kunkel, T. A. (1993). Nucleotide repeats: Slippery DNA and diseases. *Nature* **365,** 207–208.

Kunkel, T. A., and Erie, D. A. (2005). DNA mismatch repair. *Annu. Rev. Biochem.* **74,** 681–710.

Longley, M. J., Pierce, A. J., and Modrich, P. (1997). DNA polymerase δ is required for human mismatch repair *in vitro*. *J. Biol. Chem.* **272,** 10917–10921.

Lynch, H. T., and de la Chapelle, A. (1999). Genetic susceptibility to non-polyposis colorectal cancer. *J. Med. Genet.* **36,** 801–818.

Maniatis, T., Fritsch, E. F., and Sambrook, J. (1982). "Molecular Cloning." Cold Spring Harbor Laboratory, Cold Spring Harbor, NY.

Modrich, P. (1997). Strand-specific mismatch repair in mammalian cells. *J. Biol. Chem.* **272,** 24727–24730.

Modrich, P., and Lahue, R. (1996). Mismatch repair in replication fidelity, genetic recombination, and cancer biology. *Annu. Rev. Biochem.* **65,** 101–133.

Pavlov, Y. I., Newlon, C. S., and Kunkel, T. A. (2002). Yeast origins establish a strand bias for replicational mutagenesis. *Mol. Cell* **10,** 207–213.

Peltomaki, P. (2003). Role of DNA mismatch repair defects in the pathogenesis of human cancer. *J. Clin. Oncol.* **21,** 1174–1179.

Rowley, P. T. (2005). Inherited susceptibility to colorectal cancer. *Annu. Rev. Med.* **56,** 539–554.

Schofield, M. J., and Hsieh, P. (2003). DNA mismatch repair: Molecular mechanisms and biological function. *Annu. Rev. Microbiol.* **57,** 579–608.

Su, S.-S., Lahue, R. S., Au, K. G., and Modrich, P. (1988). Mispair specificity of methyl-directed DNA mismatch correction *in vitro*. *J. Biol. Chem.* **263,** 6829–6835.

Thomas, D. C., Roberts, J. D., and Kunkel, T. A. (1991). Heteroduplex repair in extracts of human HeLa cells. *J. Biol. Chem.* **266,** 3744–3751.

Tsurimoto, T., and Stillman, B. (1989). Purification of a cellular replication factor, RF-C, that is required for coordinated synthesis of leading and lagging strands during simian virus 40 DNA replication *in vitro*. *Mol. Cell. Biol.* **9,** 609–619.

Wang, H., and Hays, J. B. (2002). Mismatch repair in human nuclear extracts: Quantitative analyses of excision of nicked circular mismatched DNA substrates, constructed by a new technique employing synthetic oligonucleotides. *J. Biol. Chem.* **277,** 26136–26142.

Wei, K., Clark, A. B., Wong, E., Kane, M. F., Mazur, D. J., Parris, T., Kolas, N. K., Russell, R., Hou, H., Jr., Kneitz, B., Yang, G., Kunkel, T. A., Kolodner, R. D., Cohen, P. E., and Edelmann, W. (2003). Inactivation of exonuclease 1 in mice results in DNA mismatch repair defects, increased cancer susceptibility, and male and female sterility. *Genes Dev.* **17,** 603–614.

[18]   Characterization of the "Mismatch Repairosome" and Its Role in the Processing of Modified Nucleosides *In Vitro*

By Katja Baerenfaller, Franziska Fischer, and Josef Jiricny

Abstract

The process of postreplicative mismatch repair (MMR) increases the fidelity of DNA replication by eliminating biosynthetic errors from newly synthesized DNA. In addition, MMR proteins are also involved in the processing of intermediates of mitotic and meiotic recombination and, in mammalian cells, play a role in DNA damage signaling. As mismatches cannot be induced in the DNA of living cells, the study of the molecular transactions during MMR is restricted to *in vitro* systems. This chapter describes the construction of heteroduplex substrates that can be used for DNA affinity purification of MMR protein complexes and for the study of the role of eukaryotic MMR proteins in the processing of modified nucleosides.

Introduction

The mismatch repair (MMR) system can increase the fidelity of DNA replication by up to three orders of magnitude. It accomplishes this task by recognizing base/base mismatches and strand misalignments (insertion/deletion loops, IDLs) that have escaped the proofreading function of the replicative polymerase and by catalyzing a regional degradation of the mismatch-containing DNA strand, which results in removal of the non-Watson–Crick structures. Resynthesis of the degraded region restores the integrity of the replicated duplex (Modrich and Lahue, 1996). As it is the newly synthesized DNA strand that carries the erroneous genetic information, the correction process has to be directed to this strand (Modrich, 1997). In eukaryotes, the directionality of MMR is made possible by an intimate coupling of the replication and repair processes, mediated most likely by proliferating cell nuclear antigen, the processivity factor of DNA polymerases (Jiricny and Marra, 2003). Because the eukaryotic MMR system lacks detectable endonuclease activity (Holmes *et al.*, 1990; Thomas *et al.*, 1991), degradation of the error-containing strand must begin at a preexisting strand discontinuity, such as either end of an Okazaki fragment or the 3′ terminus of the primer strand (Modrich, 1997).

METHODS IN ENZYMOLOGY, VOL. 408   0076-6879/06 $35.00
DOI: 10.1016/S0076-6879(06)08018-9

The mammalian MMR reaction works extremely well in cell-free systems. To date, the tested substrates have been heteroduplexes based on fd phage (Holmes *et al.*, 1990) or M13 (Thomas *et al.*, 1991), which were obtained by annealing of a linearized double-stranded DNA of one phage with the single-stranded circular (viral) form of another phage that differed from the first in a defined position. Isolation of circular double-stranded molecules yielded a substrate containing a mismatch at a defined site, together with a nick a given distance 5′ or 3′ from the mispair. The disadvantage of this approach was the large size of the DNA used (fd phage and M13 are 6–7 kb in length), which limits the yields, and the fact that the nick could only be positioned in the complementary strand.

We are interested primarily in isolating the "mismatch repairosome" by DNA affinity purification, which requires considerable amounts of heteroduplex DNA. To this end, we have modified the construction method of the heteroduplex substrate. We make use of phagemid DNA, which is ∼3 kb in size and can be isolated in large amounts in both single- and double-stranded forms. Our second field of study is concerned with the investigation of the role of MMR proteins in the processing of modified nucleosides. In this experimental system we need to introduce site-specific modifications into either strand of the heteroduplex. To achieve this goal, we have modified the pGEM-13Zf(+) vector, such that it carries only a single recognition site for *N.Bst*NBI, an enzyme, which cleaves only a single strand of this asymmetric sequence and thus introduces only a single nick into our duplex DNA molecules (see also Wang and Hays, 2001, 2002).

The following paragraphs describe the preparation of both these substrates and include also some pilot experiments with extracts of human cells.

Construction of Substrates Containing Insertion/Deletion Loops

The heteroduplex substrates containing an insertion/deletion loop (IDL) of a single extrahelical nucleotide are constructed from two different plasmids, one of which is 1 bp longer than the other. To produce these, two oligonucleotides, 5′-GGCCGCTCCATGCAAGCCTAG**AGATCT**CCC-TCAAGTAGTA-3′ and 5′-GGCCGCTCCATGCAAGCCTAGGATC-TCCCTCAAGTAGTA-3′, differing in one position (denoted in bold), are cloned between the *Hind*III and the *Not*I restriction sites of the phagemid pGem-13Zf(+). The first of the resulting plasmids, pGem_IDLinker40, contains a unique *Bgl*II restriction site, whereas the second, pGem_IDLinker39, contains a unique *Avr*II site. By linearizing the two

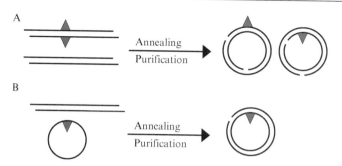

FIG. 1. Schematic representation of the production of (+/−) substrates (A) and 3′ and 5′ substrates (B).

double-stranded DNAs with different restriction enzymes, denaturing and reannealing, a mixture of two circular heteroduplexes can be obtained, 50% of which contain an extrahelical adenosine in one strand and the other 50% an extrahelical thymidine in the other strand (Fig. 1A). This substrate, which is referred to as the (+/−) substrate, is simple to obtain in large amounts and is used for DNA affinity purifications. The second heteroduplex can be obtained by annealing a linearized double-stranded DNA of one phagemid with the circular single-stranded DNA of the other (Fig. 1B). In this way, heteroduplexes containing either a single extrahelical thymidine in the viral strand or a single extrahelical adenosine in the complementary strand can be obtained. By using different phagemids, heteroduplexes carrying any type of base/base mismatch or any size IDL can be obtained.

Construction of the (+/−) Substrate

Linearization. To produce the (+/−) heteroduplex substrate, 100 μg pGem_IDLinker39 is digested with *Bsa*I (New England Biolabs) and 100 μg pGem_IDLinker40 with *Ban*II (1 U enzyme/μg DNA in 500 μl volume) (New England Biolabs). To produce the (+/−) homoduplex substrate, 100 μg pGem_IDLinker39 is digested with *Bsa*I and 100 μg pGem_IDLinker39 with *Ban*II. Completeness of the digest is checked on a 1% agarose gel (Fig. 2, lane 1) and, if complete, the DNA is ethanol precipitated and then dissolved in 500 μl 10 m*M* Tris–HCl, pH 8.0.

Annealing. The annealing is performed in 3 ml total volume containing 10 m*M* NaCl, 1 m*M* EDTA, 50 m*M* Tris–HCl, pH 7.6, 100 μg pGem_IDLinker40 (*Ban*II), and 100 μg pGem_IDLinker39 (*Bsa*I) to produce

FIG. 2. Production of the (+/−) substrate. Aliquots of the various steps in the preparation of this substrate were separated on a 1% agarose gel containing 0.5 μg/ml ethidium bromide and visualized on a UV transilluminator. M, 1-kb DNA ladder; lane 1, 1 μl before annealing; lane 2, 2 μl after annealing; lane 3, 40 ng after DNase V digest and precipitation; lane 4, 40 ng after *Nde*I digest; lane 5, 40 ng after Klenow fill in and Sephadex purification; lane 6, double volume of DNA sup; and lane 7, 40 ng linearized pGemIDLinker40. The mobilities of open circular (oc) and linearized DNA molecules are shown on the right.

the (+/−) heteroduplex substrate or 100 μg pGem_IDLinker39 (*Ban*II) and 100 μg pGem_IDLinker39 (*Bsa*I) to produce the (+/−) homoduplex substrate. Ninety microliters of freshly prepared 10 *N* NaOH (200 mg in 500 μl H$_2$O) is added, and the solution is mixed gently and left at room temperature for 5 min. The following solutions are then added sequentially: 300 μl 2.9 *M* acetic acid, 135 μl 3 *M* KCl, 372 μl 1 *M* potassium phosphate, pH 7.4. The solution is incubated for 30 min at 65°, then for 3 h at 37°, and finally left to stand at room temperature while 2 μl of the reaction mix are loaded onto a 1% agarose gel to check for annealing efficiency (Fig. 2, lane 2). The nicked (+/−) substrates run slightly slower than the linearized DNA.

Purification. The annealing mixture is dialyzed for 2 × 1 h against 1 liter of 10 m*M* Tris–HCl, pH 8, 0.1 m*M* EDTA. The linear side product of the reaction is removed by the addition of DNase V and incubation at 37° overnight in 33 m*M* Tris–acetate (stock solution: 1 *M*, pH 7.8), 66 m*M* potassium acetate (stock solution 3 *M*, pH 7.7), 10 m*M* magnesium acetate (stock solution 1 *M*), 1 m*M* ATP, 0.5 m*M* dithiothreitol (DTT), and 50 U Plasmid-Safe ATP-Dependent DNase (Epicentre). The efficiency of the digest is checked on a 1% agarose gel (Fig. 2, lane 3). If the faster migrating band of linearized DNA has disappeared completely, the DNase V is heat inactivated by incubation at 70° for 30 min. The formerly clear solution now becomes turbid and has to be filtered prior to DNA precipitation. If the filtration step is omitted, the DNA will be only partially soluble after the precipitation. The filtration is performed with PD-10 desalting columns (Amersham Biosciences), which contain Sephadex G-25 (two columns/ reaction). The eluate is precipitated with 0.2 *M* NaCl and 2× total volume of ice-cold ethanol, left at −20° for at least 90 min, centrifuged at 20,000g for 30 min at 4°, washed with 6 ml 70% ethanol, and centrifuged again at 20,000g for 10 min at 4°. The pellets, which may not be visible, are dried at room temperature and then dissolved in about 250 μl 10 m*M* Tris–HCl, pH

8.0. The concentration and quality of the DNA are determined by analytical agarose gels and/or on a Nanodrop. The yield is between 25 and 40 μg (25–40%). If the DNA is to be used in mismatch repair assays, it is purified further by gel extraction.

Biotinylation. If the DNA is to be used as an affinity matrix, it has to be labeled with biotin, such that it will bind onto magnetic, streptavidin-coated beads. To this end, the DNA is first cut with *Nde*I (4 U/μg DNA; New England Biolabs), which creates the required TA-5' overhangs. The digest is checked on an agarose gel (Fig. 2, lane 4) and, if complete, the enzyme is heat inactivated by incubation at 65° for 30 min. The ends are filled in with Klenow polymerase under the following conditions: 10 μg (+/−) substrate (*Nde*I), 10 U Klenow polymerase (Roche), 33 μM biotin-16-2'-deoxyuridine-5'-triphosphate (Biotin-16-dUTP; Roche), 33 μM of each dATP, dGTP, and dCTP. The reaction is allowed to proceed at 25° for 30 min and is then stopped by the addition of EDTA to a final concentration of 10 mM and heating to 75° for 20 min. The biotinylated substrate is then purified from the excess of free biotin-16-dUTP on a Sephadex G-25 or G-50 filtration column (self-packed). The flow through will contain the purified, biotinylated and nicked (+/−) substrate (Fig. 2, lane 5).

Coating of Magnetic Beads with the (+/−) Substrate DNA. To obtain complete binding of the biotinylated DNA to Dynabeads M-280 streptavidin (Dynal), 30 μl of the bead suspension is used per microgram of (+/−) substrate. The beads are agitated and the desired amount is taken out and aliquoted into 1.5-ml Eppendorf tubes. These are put into a magnetic particle concentrator (MPC; Dynal MPC-E-1) and allowed to attach for about 2 min. The storage buffer is removed. An equal volume of wash buffer (10 mM Tris–HCl, pH 7.5, 1 mM EDTA, 1 M NaCl, 0.003% [v/v] NP-40) is added and the beads are resuspended by flicking the side of the tube. They are then allowed to attach to the wall in the MPC for about 2 min. This procedure is repeated twice. The beads are then washed once with B&W buffer (10 mM Tris–HCl, pH 7.5, 1 mM EDTA, 2 M NaCl, 0.003% [v/v] NP40).

The DNA is attached to the beads as follows: the desired amount of DNA and an equal volume of B&W buffer are added to the beads and the mixture is incubated at 43° for 1 h. The supernatant (DNA supernatant) is then removed from the beads using the MPC. To ensure complete binding of the DNA, the DNA supernatant should be analyzed on an agarose gel (Fig. 2, lane 6). The beads are then washed twice with wash buffer and once with B&W buffer by resuspension and immobilization in the MPC at each step and are now ready to be used for DNA affinity purification.

Construction of 3' and 5' Substrates

By definition, the 5' substrate contains a single-strand nick upstream from the mismatch. On this substrate, the mismatch is removed by a 5'→3' exonuclease, which starts at the nick. In contrast, the 3' substrate, which contains the nick downstream from the mispair, has to be processed first by a 3'→5' exonuclease. The key difference between these substrates is that the mismatch in the former can, in principle, be "repaired" by nick translation even in MMR-deficient extracts, whereas the latter cannot.

Production of Bacteriophage M13K07. M13K07 is a helper phage. When host bacteria harbor both phagemid and M13KO7, the gene II product of the phage interacts with the f1 intergenic region present in the phagemid and initiates rolling circle replication to generate copies of one strand of the phagemid DNA (ssDNA). The permissive *Escherichia coli* strains must possess pili to accept the phage. One such suitable strain is *XL1 blue*, which contains a tetracycline selectable F' factor. (When working with phage, caution must be used not to contaminate the environment: sterile tips should be employed, used tips and pipettes should be discarded in 80% ethanol, and flasks and centrifuge bottles should first be washed out with 10 N NaOH and then with 80% ethanol).

PRODUCTION OF RECEPTIVE (MALE) XL1 BLUE CELLS. *XL1 blue* cells from a glycerol stock are streaked out on a 2× YT/Tet (5 μg/ml) plate and incubated overnight at 37°. (To produce 1 liter 2× YT medium, 16 g tryptone, 10 g yeast extract, and 5 g NaCl are dissolved in water and the pH is adjusted to pH 7.0 prior to autoclaving. To produce 2× YT plates, 15 g agar are added to 1 liter 2× YT medium prior to autoclaving.) Five single colonies are picked the next day and inoculated into 5 ml of 2× YT/Tet medium (5 μg/ml) each. They are allowed to grow at 37° overnight. (To prepare a new glycerol stock, 10 μl of the overnight culture is added to 5 ml of 2× YT/Tet medium [5 μg/ml] and incubated at 37° for 6 h. Six hundred microliters is then mixed with 400 μl of 50% glycerol and the stocks are stored at $-80°$.) As the F' factor can be lost, the clones should be tested for its presence as follows.

A thick line of each overnight culture is streaked on a 2× YT/Tet (5 μg/ml) plate in the morning using a sterile inoculation loop. One microliter of M13K07 phage from a stock with a titer of at least 10^6 is spotted at the beginning of each line (make also one control plate without phage). After 6 h of incubation at 37°, the lines are checked for plaque formation at the site of phage release. Clear transparent spots (plaques) indicate that the respective *XL1 blue* clone is receptive to phage infection. A few cells are lifted with a sterilized Pasteur pipette (without touching the plaques), streaked out on a fresh $2\times$ YT/Tet (5 μg/ml) plate, and incubated overnight at 37°.

PRODUCTION OF M13K07 PHAGE. Five milliliters of $2\times$ YT/Tet medium (5 μg/ml) are inoculated with a single male *XL1 blue* colony and incubated at 37° for 6–8 h. Five 4 ml aliquots of liquid top agar (autoclaved, melted, aliquoted in snap-cap tubes, and kept a 44°) are prepared. Two hundred microliters of the bacterial culture are added to 100 μl of five different dilutions of M13K07 (or phosphate-buffered saline [PBS] as a blank; reasonable dilutions of a 1.7 $\times 10^{13}$ pfu/ml stock: 10^{-9}–10^{-12} = 1.7 $\times$ 10^4–10^1 pfu/ml = 17,000–17 plaques) and incubated for 10 min at 37°. Bacteria (blank) and bacteria/phage mixtures are added to the top agar, mixed briefly, and then poured evenly onto prewarmed $2\times$ YT/Tet (5 μg/ml) plates. The plates are allowed to dry at room temperature for 20 min and then incubated overnight at 37°. After 12–15 h of incubation, the plates should show a confluent opaque bacterial layer with plaques (number should be consistent with the corresponding phage dilutions). One liter of $2\times$ YT/Tet (5 μg/ml)/Kan (30 μg/ml) medium in a 5-liter conical flask is inoculated with one single plaque from the bacterial layer by aspirating a plug containing the selected plaque with a sterile Pasteur pipette and flushing it into the culture medium. The phage culture is incubated overnight at 37° with shaking.

The next day, it is centrifuged at 2700g for 30 min at 4°. The pellet is discarded, and the supernatant containing the phage is transferred to new tubes and centrifuged at 10,800g for 30 min at 4°. The pellet is discarded and the supernatant is incubated at 70° for 30 min and then left at room temperature for 10 min. The phage solution is centrifuged at 10,800g for 10 min at 4° and the supernatant is transferred to a 2-liter conical flask, to which one-fourth of the volume of 20% PEG/2.5 M NaCl is added, and the phage particles are precipitated overnight at 4° without shaking.

The next day, the PEG-precipitated phage particles are centrifuged at 10,800g for 30 min at 4°. The supernatant is discarded and all traces of it are removed from the neck of the flask with a paper towel. The pellet containing the phages is resuspended in a minimal volume (6–8 ml) PBS, pH 7.4, with shaking for 1 h at room temperature. The suspension is then transferred to Eppendorf tubes and centrifuged at 20,800g for 10 min at 4° in a tabletop centrifuge. The supernatant containing the purified phages is then aliquoted (the pellet consisting of residual bacterial cells is discarded) and the aliquots are stored at 4°.

DETERMINATION OF PHAGE TITER. Five milliliters $2\times$ YT/Tet (5 μg/ml) medium are inoculated with a single colony of the tested male strain and allowed to grow at 37°. Six hours later, seven dry $2\times$ YT/Tet (5 μg/ml) plates and 7 $\times$ 4 ml liquid sterile top agar aliquots in a 44° water bath are prepared. Assuming that the new phage stock concentration is 10^{13} pfu/ml, serial dilutions in PBS, pH 7.4, are made and 100 μl of phage dilutions 10^{-7}–10^{-12} (or PBS as a blank) are added to 200 μl of the bacterial culture, mixed briefly, incubated for 10 min at 37°, mixed with 4 ml top agar, and plated onto prewarmed 2× YT/Tet (5 μg/ml) plates. The plates are incubated at 37° for at least 15 h. The plaques on the plates with the higher dilutions are counted, the average value is calculated, and the pfu/ml are calculated (e.g., 50 plaques on the plate with the 10^{-10} dilution corresponds to 500 pfu/ml and thus the stock has a titer of 5×10^{12} pfu/ml).

Production of Single-Stranded DNA. When using plasmids derived from pGemZf(+), the sequence of the ssDNA corresponds to the lower strand shown in the Promega catalog.

Day 1. *XL1 blue* are transformed with 20 pg of the desired plasmid (e.g., pGem_IDLinker40 to produce the ΔT substrate).

Day 2. One hundred millilters of 2× YT/Amp (100 μg/ml)/Tet (5 μg/ml) medium is inoculated with a single transformed colony and incubated overnight at 37° with shaking at 200–250 rpm.

Day 3. The OD_{600} of a 1/10 dilution of the overnight culture is measured. Then, 2× 1 liter of 2× YT/Amp (100 μg/ml)/Tet (5 μg/ml) medium in 5-liter conical flasks is inoculated with the overnight culture such that the OD_{600} at time 0 is 0.07. The cultures are grown at 37° with shaking at 220 rpm, and the OD_{600} is checked every 20 min until OD_{600} reaches 0.2–0.3 (after about 80 min). Then, a superinfection with the helper phage M13K07 is made at a multiplicity of infection (MOI) of 20. Assuming that $OD_{600} = 1$ corresponds to 5×10^8 cells/ml ($= 1.25 \times 10^{11}$/liter), 2.5×10^{12} phage particles (250 μl from a 10^{13} phage stock) have to be added per liter. The bacteria are kept for 10 min at 37° without shaking and are then shaken at 220 rpm for 90 min at 37°. After 90 min, 30 μg/ml kanamycin are added to select for M13K07-infected cells. The bacteria are allowed to grow for a further 3 h. The cultures are cooled on ice for 30 min, and the suspension is transferred to centrifuge bottles and centrifuged at 5000g for 30 min at 4°. The supernatant is transferred to fresh centrifuge bottles, heated to 70° for 30 min, equilibrated at room temperature for 15 min, and centrifuged at 10,000g for 15 min. This step kills and removes remaining bacteria in the suspension. The supernatant is transferred to a 5-liter conical flask and 1/4 volume ($\sim$ 500 ml) of 20% (w/v) PEG8'000/2.5 M NaCl is added. The phage particles are precipitated overnight at 4°.

Day 4. The phage suspension is transferred into centrifuge bottles and centrifuged at 10,000g for 30 min at 4°. The supernatant is discarded and the remaining liquid is removed with paper towels. The precipitated phage particles form a whitish film on the wall of the centrifuge bottles, which is hardly visible. These are resuspended in a minimum volume of PBS, pH 7.4 (about 15 ml for 2 liter starting culture), with moderate shaking for 1–2 h at room temperature. The resulting suspension is very dense and brownish and is centrifuged at 24,000g for 15 min at 4°. The supernatant containing the phage particles is transferred to fresh centrifuge bottles, and the pellet of residual bacterial cells is discarded. The phage suspension is now opaque yellow.

If *XL1 blue* cells were used, the phage suspension is usually contaminated with chromosomal DNA. To test for this, 2 μl of the phage suspension is loaded onto a 1% agarose gel. To remove the contamination, $MgCl_2$ (15 mM final concentration) and 120 U DNase I are added, and the suspension is incubated at 37° for 2 h. The phage capsid proteins will protect the single-stranded phage DNA from the DNase I. Completeness of the reaction is checked by loading 2 μl of the phage suspension before the addition of DNase I and after 2 h of incubation. When the contaminating DNA has been removed, EDTA, pH 8, is added (60 mM final concentration) to chelate Mg^{2+} and to block the activity of DNase I. The enzyme is inactivated by heating the solution to 80° for 20 min. The phage particles are resistant to such treatment. When the phage solution has cooled down to 50° or less, proteinase K is added to a final concentration of 100 μg/ml and the solution is incubated at 37° overnight to digest the capsid proteins and release the ssDNA. In the course of this process, the solution should become clear.

Day 5. A 1/10 volume of 5% hexadecyltrimethylammonium bromide (CTAB)/0.5 M NaCl (to prepare a 100-ml stock solution, 2.92 g NaCl are dissolved in 80 ml water, 5 g CTAB are added slowly with stirring, and the solution is then heated to 65°; water is added to a final volume of 100 ml) is added to the proteinase K digestion from day 4, kept for 10 min at room temperature, and centrifuged at 10,000g for 20 min at room temperature. The pellet is dissolved in 5 ml 1.2 M NaCl, 12.5 ml of cold absolute ethanol is added, and the solution is kept at −20° for at least 1 h. It is then centrifuged at 23,000g for 30 min at 4°, and the pellet of ssDNA is rinsed with 5 ml 70% ethanol and centrifuged again for 10 min. This second pellet are significantly smaller and whiter than that from the CTAB/NaCl precipitation. It is air dried and dissolved in about 4 ml 10 mM Tris–HCl, pH 8.0, 0.1 mM EDTA. The DNA concentration is determined by measuring the OD_{260} (the multiplier for ssDNA is 40). The yield is typically about 0.5 mg ssDNA/2 liters of starting culture.

Annealing and Purification of 3′ and 5′ Substrates. For construction of the 3′ substrate, 200 μg of pGem_IDLinker39 (for 3′ heteroduplex) and 200 μg of pGem_IDLinker40 (for 3′ homoduplex) are digested with *Sap*I.

The resulting 3' heteroduplex will contain a nick 291 bp 3' from the deletion site; the 3' homoduplex will contain the nick at the same site.

For construction of the 5' substrate, 200 μg of pGem_IDLinker39 (for 5' heteroduplex) and 200 μg of pGem_IDLinker40 (for 5' homoduplex) are digested with *Ban*II. The resulting 5' heteroduplex will contain a nick 369 bp 5' from the deletion site; the 5' homoduplex will contain the nick at the same site. Completeness of the digests is checked on a 1% agarose gel and, when complete, the DNA is precipitated and then dissolved in 500 μl 10 mM Tris–HCl, pH 8.

Annealing is performed in 3 ml of a buffer containing 10 mM NaCl, 1 mM EDTA, 50 mM Tris–HCl, pH 7.6, 200 μg linearized plasmid DNA, and 100 μg ssDNA (mole ratio dsDNA:ssDNA = 1:1). Subsequent steps are performed exactly the same as for the production of the (+/−) substrate. The circular, nicked DNA is purified further by electrophoresis on 0.8% agarose gels to remove the contaminating ssDNA, which interferes with the MMR assays and with the restriction digest by *Nde*I.

Construction of G/T and G/C Mismatch Repair Substrates

Experimental Strategy

The vector used is a derivative of the pGEM-13Zf(+) vector (Promega). The original vector contained six *Ple*I restriction sites (isoschizomer of *N.Bst*NBI), which were removed by QuickChange site-directed mutagenesis (Stratagene) and appropriate primers. Furthermore, the linker sequence 5'-GGCCGCGATCTGATCAGATCCAGACG TCTGTCAACGTTGG-GA-3' was cloned between its *Not*I and *Hin*dIII restriction sites. Using site-directed mutagenesis, a single recognition site for the nickase *N.Bst*NBI (New England Biolabs) was inserted, which is situated either 352 bp 3' from the mismatch if the complementary (top) strand is nicked or 337 bp 5' from it when the viral (bottom) strand is nicked (Fig. 3).

Fig. 3. Schematic representation of the production of mismatch repair substrates by the primer extension method.

To construct the hetero- and homoduplex substrates, the following PAGE-purified primers were used (Microsynth GmbH, Switzerland) in primer extension reactions:

- for G/T: 5′-CCAGACGTCTGTCGACGTTGGGAAGCTTGAG-3′
- for G/C: 5′-CCAGACGTCTGTCAACGTTGGGAAGCTTGAG-3′

In this method, oligonucleotides can be used that carry modified bases, e.g., 6-methylguanine, 8-oxoguanine, thymine dimers, and cisplatin adducts. However, it is highly advisable to check the quality of oligonucleotides containing these modifications by mass spectrometry. For example, ^{6Me}G is incorporated into oligonucleotides as a 6-*O*-methyl-2-*N*-isobutyryl-2′-deoxyriboside. The isobutyryl protecting group does not come off by ammonia treatment (standard deprotection procedure for synthetic oligonucleotides) and has to be removed with (1,8-diazabicyclo [5.4.0] undec. -7-enc (DBU) (a strong base) in anhydrous methanol over the course of several days in the dark. This procedure is often incomplete and the partially protected primers reduce primer extension yields and interfere with the MMR assay.

Phosphorylation of Primers

The primers are phosphorylated in a 100-μl reaction volume containing 4 nmol of the corresponding primer, 1× polynucleotide kinase (PNK) buffer, 5 mM DTT, 1 mM ATP, 100 μg/ml bovine serum albumin (BSA), and 10 U T4 DNA PNK. The reaction is incubated for 1 h at 37°, followed by 10 min at 70° to inactivate the enzyme. One hundred microliters of 2 M LiClO$_4$ and 800 μl acetone are added, and the mixture is placed on ice for 30 min and then centrifuged at 20,800g for 20 min at room temperature in a tabletop centrifuge. The pellet is washed with 1 ml 80% acetone and centrifuged at 20,800g for 10 min at room temperature. It is dried and dissolved in 100 μl 10 mM Tris–HCl, pH 8.0, 0.1 mM EDTA.

Primer Extension

Sixty micrograms of the ssDNA (57 pmol) is annealed with a 2.1-fold molar excess (120 pmol) of the appropriate primer in a total volume of 240 μl containing 1× T4 DNA polymerase reaction buffer. The solution is heated for 6 min at 85° in a heat block and then allowed to slowly cool down to room temperature. The extension is done in a 600-μl reaction volume in 1× T4 DNA polymerase reaction buffer containing 100 μg/ml BSA, 1 mM ATP, 1 mM dNTPs, 2400 U T4 DNA ligase, and 54 U T4 DNA polymerase. The reaction is incubated for 1 h at 37°, followed by incubation at 70° for 20 min to inactivate the enzymes. The efficiency of the reaction is checked on a 1% agarose gel (Fig. 4, lane 3).

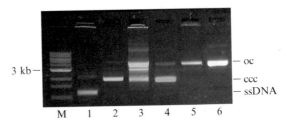

FIG. 4. Production of hetero- and homoduplex MMR substrates by the primer extension method. Aliquots of the various steps in the production of these substrates were separated on a 1% agarose gel containing 0.5 μg/ml ethidium bromide and visualized on a UV transilluminator. M, 1-kb DNA ladder; lane 1, 110 ng ssDNA; lane 2, Miniprep DNA; lane 3, 1 μl primer extension G/T; lane 4, 0.5 μl primer extension G/T after cesium chloride gradient; lane 5, nicked Miniprep DNA; lane 6, nicked G/T substrate. The mobilities of open circular (oc), covalently closed circular (ccc), and single-stranded (ss) DNA molecules are shown on the right.

Cesium Chloride Gradient Purification

Twelve grams of CsCl are dissolved in 10.5 ml TE buffer, pH 8.0, and 240 μl ethidium bromide solution (10 mg/ml) are added. The DNA is nicked in the presence of high ethidium bromide concentrations and light and thus the ethidium bromide- and DNA-containing solutions should be kept in the dark. The primer extension reaction is transferred into a 6-ml ultracentrifuge tube (Sorvall S/L, Sleeve Conical), and the tube is filled to the top with the CsCl/ethidium bromide solution; bubbles are removed and the tube is sealed. The counterweight is prepared in the same way; the weight difference has to be ≤5 mg. The CsCl gradient centrifugation is then performed with a Sorvall Stepsaver 65V13 rotor at 60,000 rpm for 16–18 h at 4° (with slow deceleration below 20,000 rpm). The centrifuge tube, fastened in a clamp in the darkroom, is pierced at the top with a 25-gauge needle to allow air to enter. An ultraviolet (UV) lamp is switched on and the bright orange band that appears about midway of the tube is sucked out with a 18-gauge needle and a 1-ml syringe by puncturing the tube slightly below the substrate band. The needle is removed and the DNA is ejected into a 2-ml Eppendorf tube containing 700 μl water-saturated n-butanol. The solution is mixed by inverting the tube gently, and the upper phase is then removed and the lower phase containing the DNA is transferred to the next Eppendorf tube. This is repeated until the DNA-containing phase is completely colorless (about six times). The DNA is dialyzed for 1 and 3 h against 1 liter of 10 mM Tris–HCl, pH 8.0, 0.1 mM EDTA. Afterward, the DNA is precipitated in 0.3 M NaOAc, pH 5.5, and 2 volumes of ice-cold ethanol, left at −20° for at least 1 h, and centrifuged. After drying, the pellet is redissolved in 90 μl 10 mM Tris–HCl, pH 8.5. The purified substrate is examined on a 1% agarose gel (Fig. 4, lane 4).

Nicking of the Substrate

The purified substrate is nicked with the nicking enzyme N.BstNBI (New England Biolabs) at position 352 in the top (3′) or 337 in the bottom (5′) strand. The nicking reaction is performed in an 80-μl volume containing 45 μl purified substrate, 0.1 mg/ml BSA, 1× N.BstNBI buffer, and 50 U N.BstNBI and incubation is carried out overnight at 16°. One microliter is loaded on a 1% agarose gel to check for completeness of the reaction. If complete, the DNA is purified employing the MinElute reaction cleanup kit (Qiagen). The DNA is eluted in 20 μl 10 mM Tris–HCl, pH 8.5, and the volume is adjusted to 80 μl with the same buffer. The concentration and quality of the final substrate are determined with analytical agarose gels (Fig. 4, lane 6) and/or a Nanodrop.

Preparation of Nuclear Extracts

Nuclei are isolated as described (Iaccarino *et al.*, 1998) and are resuspended in 500 μl cold extraction buffer/1.5 ml packed nuclei (extraction buffer: 25 mM HEPES/KOH, pH 7.5, 10% sucrose, 1 mM phenylmethylsulfonyl fluoride [PMSF], 0.5 mM DTT, 1 μg/ml leupeptine, 1 tablette/5 ml of protease inhibitor cocktail "Complete Mini" [Roche]). The volume of resuspended nuclei is measured, and 0.031 volume of cold 5 M NaCl (final concentration of NaCl is 0.155 M) is slowly (1 drop/min) added with gentle stirring in the cold room. The mixture is then rotated for 1 h at 4° to allow the proteins to leave the nucleus. After this, nuclei are pelleted by centrifugation at 14,500g for 20 min at 2° in a tabletop centrifuge. The supernatant is transferred to a dialysis bag and dialyzed for 2× 1 h against 2 liters cold dialysis buffer (25 mM Hepes/KOH, pH 7.5, 50 mM KCl, 0.1 mM EDTA, 10% sucrose, 1 mM PMSF, 2 mM DTT, 1 μg/ml leupeptine). The extract is clarified by centrifugation at 20,000g for 15 min at 2° in a tabletop centrifuge. The supernatant is aliquoted, snap frozen in liquid nitrogen, and stored at −80°. The concentration of the proteins is determined with a Bradford assay, and the salt concentration is determined using a conductivity meter.

In Vitro Mismatch Repair Assays

Mismatch Repair Assays with G/T and G/C Substrates

Restriction enzymes and positions of the restriction sites used in the interpretation of MMR efficiency are shown in Fig. 5. The G/T substrate is

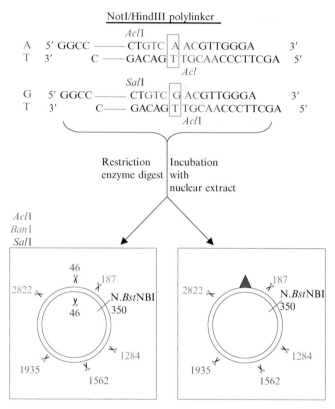

FIG. 5. Schematic representation of the G/T and G/C substrates. The relevant restriction sites and their nucleotide positions are shown. (See color insert.)

refractory to cleavage with *Acl*I and *Sal*I. If the G/T mismatch is repaired to A/T, an *Acl*I recognition sequence is restored (*Acl*I cuts the plasmid now three times instead of twice). When the G/T is repaired to G/C, a unique *Sal*I recognition site is restored.

In vitro mismatch repair assays are carried out in 20-μl volumes containing 20 mM Tris–HCl, pH 7.6, 110 mM KCl, 5 mM MgCl$_2$, 1 mM glutathione, 50 μg/ml BSA, 0.1 mM each dNTP, 1.5 mM ATP, 5 ng/μl substrate (total 100 ng ≅ 48 fmol), and 3.75 μg/μl nuclear extract (total 75 μg). Generally, a 10× MMR buffer containing 200 mM Tris–HCl, pH 7.6, 400 mM KCl, 50 mM MgCl$_2$, 10 mM glutathione, 500 μg/ml BSA, and 1 mM each dNTP (if the assay is to be performed with [α-^{32}P]dATP, dATP has to be omitted) is prepared, and 20-μl aliquots are made and stored at −20°. These aliquots are only thawed once and kept for less than a month.

In the final reaction, the concentration of KCl is adjusted to 110 mM with 1 M KCl. We observed that the reaction works well even if only 1.7 pmol [α-^{32}P]dATP together with dGTP, dTTP, and dCTP are included in the reaction. It should thus be noted that, despite the dialysis, the nuclear extracts still contain a residual pool of dNTPs, which permits DNA synthesis.

For the radioactive MMR assay, 10× MMR buffer (-dATP), 1 M KCl, DNA substrate, 15 mM ATP, 1.7 pmol [α-^{32}P]dATP 10 μCi, and the nuclear extract are mixed to reach the aforementioned concentrations. For the cold MMR assay, 10× MMR buffer containing all four dNTPs is used and the [α-^{32}P]dATP is omitted. The MMR reaction is incubated for 45 min at 37° and is then stopped by adding 30 μl freshly prepared stop solution (1.12% SDS, 41.67 mM EDTA, 83.33 μg/ml proteinase K) and incubated again for 30 min at 37°. The DNA is purified employing the MinElute reaction cleanup kit (Qiagen). The eluted DNA is digested with the appropriate restriction enzymes in a 50-μl volume for 3 h at 37°.

The RNA in the samples is removed by adding 2.5 μg RNase A (taking care to eliminate any contaminating DNase activity of the RNase A preparations before use) and incubation for 15 min at 37°. Then, 0.2% SDS and 3.6 μg proteinase K are added and the mixture is incubated for 15 min at 37°. The DNA is precipitated with 0.3 M NaOAc, pH 5.5, and 2.5× volumes cold 100% ethanol. The pellet is dissolved in 10.5 μl 10 mM Tris–HCl, pH 8.5, and 6× bromphenol blue loading buffer (15%[w/v] Ficoll 400, 0.25% [w/v] bromphenol blue) is added to a final concentration of 1.5-fold. The DNA is loaded onto a 1% agarose gel and run for 45 min at 90 V. The DNA can be visualized in a UV transilluminator (Fig. 6, left); in case of radiolabeled assays, the gel is vacuum dried at 80° for 75–90 min and autoradiographed using a PhosphoScreen (Molecular Dynamics) (Fig. 6, right).

Mismatch Repair Assay with Insertion/Deletion Substrates

In vitro MMR assays are carried out basically as described earlier except for monitoring with the restriction enzymes. The efficiency of repair of the (+/−) substrates is checked by subjecting the DNA to a restriction digest with *Bsa*I (New England Biolabs), which linearizes the DNA substrates, and with the sensor enzymes *Avr*II and *Bgl*II (both New England Biolabs), which cut only when deletion (in the case of *Avr*II) or insertion (in the case of *Bgl*II) occurs. The two fragments (1833 and 1361 bp) indicative of repair are visualized on 0.8% agarose gels.

The efficiency of repair of 3' and 5' substrates is visualized by subjecting the DNA to restriction digest with the linearizing enzyme *Bsa*I and the

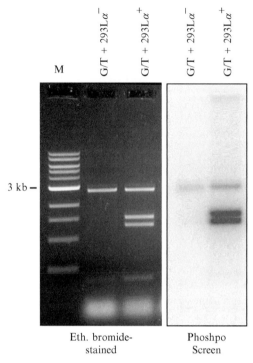

FIG. 6. Repair of the G/T substrate *in vitro*. The substrate containing a G/T mispair in the *AclI/SalI* site of the phagemid substrate was incubated with MMR-proficient or -deficient nuclear extracts in the presence of [α-^{32}P]dATP and digested with the appropriate restriction enzymes to test for efficiency of the repair reaction and incorporation of radioactivity into the different fragments. After the MMR assay, the restriction fragments were separated on a 1% agarose gel containing 0.5 μg/ml ethidium bromide and visualized on a UV transilluminator (left). The dried gel was subsequently exposed to a PhosphoScreen (right).

single sensor enzyme *BglII*. As the 3′ and 5′ substrates contain the extra-helical ΔT on the continuous strand, insertion occurs almost exclusively.

DNA Affinity Purification

DNA-coated beads are washed twice with buffer W110 (20 mM Tris–HCl, pH 7.6, 110 mM KCl, 5 mM MgCl$_2$) and once with equilibration buffer (1× MMR buffer adjusted to 110 mM KCl) while the MPC is standing on ice. The conditions to allow binding of the proteins to DNA are 1× MMR buffer, total of 110 mM KCl, 1.5 mM ATP, and 360 μg of nuclear extract per microgram of DNA. First, the components without the nuclear extract are added to the beads and incubated for 5 min at the

STEP 1: Coating of magnetic beads with DNA substrate

STEP 2: Incubation of the coated beads with nuclear extract

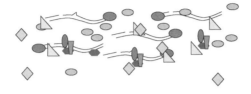

STEP 3: Isolation of the specifically-bound proteins in a MPC

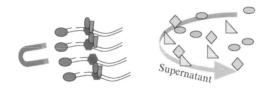

STEP 4: Elution of the specifically bound proteins

STEP 5: Analysis of the eluted proteins

FIG. 7. Scheme of the DNA affinity purification of the "mismatch repairosome." (See color insert.)

desired temperature. The nuclear extract is then added and incubated with the DNA for the desired time at the desired temperature. After this, the tubes are put into the cold MPC and the beads are allowed to attach. The supernatant is withdrawn and the beads are washed twice with equilibration buffer and once with buffer W110, buffer W200 (20 mM Tris–HCl, pH 7.6, 200 mM KCl, 5 mM MgCl$_2$), and DNase buffer (20 mM Tris–HCl, pH 7.6, 110 mM KCl, 10 mM MgCl$_2$). Proteins are eluted by cleavage of the

DNA with DNase I by adding 0.9 U DNase I (Roche) in 15 μl DNase buffer per microgram of DNA and incubation at 37°. The DNase I solution is usually added in two portions with incubation at 37° for 10 min after each addition. The eluate is then put into the MPC once more, as some beads are usually cotransferred. The supernatant of this step constitutes the eluate, which is either subjected directly to PAGE or precipitated with 20% TCA (Fig. 7).

Conclusions

The eukaryotic MMR is a complex interplay of numerous polypeptides. Given that MMR malfunction is linked to human cancers, both inherited and sporadic (Jiricny and Marra, 2003; Truninger et al., 2005), detailed understanding of the molecular transactions that take place during the recognition and metabolism of the various types of MMR substrates may help identify new genetic loci that predispose to malignancy. Moreover, identification of new interaction partners of the individual MMR factors may lead to the discovery of new metabolic pathways that involve these versatile polypeptides. Thus, in addition to mismatch correction, MMR proteins and their homologs are involved in genetic recombination, both meiotic and mitotic (Harfe and Jinks-Robertson, 2000; Lipkin et al., 2002; Snowden et al., 2004), and in somatic hypermutation and class switch recombination of immunoglobulin loci (Neuberger et al., 2005). The role of MMR proteins in DNA damage signaling is another process that is highly relevant to human health, as MMR malfunction can result in substantial resistance to cancer chemotherapy (Stojic et al., 2004). The methods described in this chapter permit the study of the interactions of the MMR system with DNA modifications such as those generated in genomic DNA by chemotherapeutics.

Acknowledgments

The authors thank Claudia Perrera and Richard Brun for their excellent contributions during the initial phase of this work. We also gratefully acknowledge the generous financial support of the Swiss National Science Foundation (K.B., Grant 3100/068182.02 to JJ) and the UBS Stiftung (F.F.).

References

Harfe, B. D., and Jinks-Robertson, S. (2000). DNA mismatch repair and genetic instability. Annu. Rev. Genet. **34,** 359–399.
Holmes, J. J., Clark, S., and Modrich, P. (1990). Strand-specific mismatch correction in nuclear extracts of human and Drosophila melanogaster cell lines. Proc. Natl. Acad. Sci. USA **87,** 5837–5841.

Iaccarino, I., Marra, G., Palombo, F., and Jiricny, J. (1998). hMSH2 and hMSH6 play distinct roles in mismatch binding and contribute differently to the ATPase activity of hMutSalpha. *EMBO J.* **17,** 2677–2686.

Jiricny, J., and Marra, G. (2003). DNA repair defects in colon cancer. *Curr. Opin. Genet. Dev.* **13,** 61–69.

Lipkin, S. M., Moens, P. B., Wang, V., Lenzi, M., Shanmugarajah, D., Gilgeous, A., Thomas, J., Cheng, J., Touchman, J. W., Green, E. D., Schwartzberg, P., Collins, F. S., and Cohen, P. E. (2002). Meiotic arrest and aneuploidy in MLH3-deficient mice. *Nat. Genet.* **31,** 385–390.

Modrich, P. (1997). Strand-specific mismatch repair in mammalian cells. *J. Biol. Chem.* **272,** 24727–24730.

Modrich, P., and Lahue, R. (1996). Mismatch repair in replication fidelity, genetic recombination, and cancer biology. *Annu. Rev. Biochem.* **65,** 101–133.

Neuberger, M. S., Di Noia, J. M., Beale, R. C., Williams, G. T., Yang, Z., and Rada, C. (2005). Somatic hypermutation at A.T pairs: Polymerase error versus dUTP incorporation. *Nat. Rev. Immunol.* **5,** 171–178.

Snowden, T., Acharya, S., Butz, C., Berardini, M., and Fishel, R. (2004). hMSH4-hMSH5 recognizes Holliday junctions and forms a meiosis-specific sliding clamp that embraces homologous chromosomes. *Mol. Cell* **15,** 437–451.

Stojic, L., Brun, R., and Jiricny, J. (2004). Mismatch repair and DNA damage signalling. *DNA Repair (Amst.)* **3,** 1091–101.

Thomas, D. C., Roberts, J. D., and Kunkel, T. A. (1991). Heteroduplex repair in extracts of human HeLa cells. *J. Biol. Chem.* **266,** 3744–3751.

Truninger, K., Menigatti, M., Luz, J., Russell, A., Haider, R., Gebbers, J. O., Bannwart, F., Yurtsever, H., Neuweiler, J., Riehle, H. M., Cattaruzza, M. S., Heinimann, K., Schar, P., Jiricny, J., and Marra, G. (2005). Immunohistochemical analysis reveals high frequency of PMS2 defects in colorectal cancer. *Gastroenterology* **128,** 1160–1171.

Wang, H., and Hays, J. B. (2001). Simple and rapid preparation of gapped plasmid DNA for incorporation of oligomers containing specific DNA lesions. *Mol. Biotechnol.* **19,** 133–140.

Wang, H., and Hays, J. B. (2002). Mismatch repair in human nuclear extracts: Quantitative analyses of excision of nicked circular mismatched DNA substrates, constructed by a new technique employing synthetic oligonucleotides. *J. Biol. Chem.* **277,** 26136–26142.

[19] Analysis of DNA Mismatch Repair in Cellular Response to DNA Damage

By LIYA GU and GUO-MIN LI

Abstract

Significant advances have been made in identifying and characterizing the roles of DNA mismatch repair (MMR) proteins in cellular response to DNA damage. Insights into this process have been obtained by performing interactions of mismatch recognition proteins (e.g., MutSα) with DNA adduct-containing duplexes and by analyzing cellular responses (including

METHODS IN ENZYMOLOGY, VOL. 408
0076-6879/06 $35.00
DOI: 10.1016/S0076-6879(06)08019-0

cell cycle checkpoints and apoptosis) of cell lines and animals with various MMR capacities. This chapter presents detailed methods for gel-shift analysis to determine the interaction between MutSα and oligonucleotide duplex containing a single DNA adduct and for apoptotic assays in cell lines and experimental animals. In addition, a step-by-step protocol is also provided for the purification of MutSα from human cells, the preparation of DNA substrates containing a defined DNA adduct, and the treatment of MMR-proficient and deficient cell lines as well as MMR knockout mice.

Introduction

The importance of DNA mismatch repair (MMR) in the maintenance of genetic integrity is underscored by the fact that deficiency in MMR causes an increased rate of genomic mutations and susceptibility to certain types of cancer, including hereditary nonpolyposis colorectal cancer (Kolodner and Marsischky, 1999; Kunkel and Erie, 2005; Li, 2003; Modrich and Lahue, 1996). It is well known that MMR promotes genomic stability by correcting mismatches that arise during DNA replication (Kolodner and Marsischky, 1999; Kunkel and Erie, 2005; Li, 2003; Modrich and Lahue, 1996) and by blocking homologous recombination (Harfe and Jinks-Robertson, 2000; Myung et al., 2001; Schofield and Hsieh, 2003). However, evidence suggests that the genome maintenance function of MMR may also involve its role in mediating a cellular response to DNA damage, which triggers damaged cells to arrest at certain cell cycle checkpoints and/or undergo programmed cell death or apoptosis (Li, 1999; Stojic et al., 2004a), thereby preventing tumorigenesis.

Since misincorporation occurs in the newly synthesized strand, mismatch correction has to target exclusively to the strand. However, as both parent and daughter strands contain normal Watson–Crick bases at the site of the mispair, a strand discrimination signal is necessary to direct the strand-specific repair. Although the strand discrimination signal in mammalian cells has not been identified, the in vitro MMR reaction can be directed by a preexisting strand break (Holmes et al., 1990; Thomas et al., 1991). Thus, it is possible that the strand discrimination signals for mammalian MMR could be gaps between Okazaki fragments on the lagging strand or the free 3′ terminus on the leading strand. It is well accepted that the strand-specific MMR reaction in mammalian cells is initiated by key MMR components MutSα (the heterodimer of MSH2 and MSH6) and MutLα (the heterodimer of MLH1 and PMS2). MutSα specifically recognizes the mismatch and undergoes an ATP-dependent conformational change, which promotes formation of the MutSα–MutLα ternary complex. This ternary complex has been proposed to either stay at the mismatch

(Junop *et al.*, 2001) or translocate bidirectionally (Allen *et al.*, 1997; Gradia *et al.*, 1999) in search for the strand discrimination signal. Localization of the strand discrimination signal by the MutSα–MutLα complex triggers mismatch-provoked excision, which starts at the strand break and terminates at a location shortly past the mismatch, generating a single-stranded DNA gap. The single strand gap is filled by DNA polymerase to leave a nick that can be sealed by a DNA ligase (Kunkel and Erie, 2005; Modrich and Lahue, 1996).

Like the repair function, the DNA damage-signaling function of MMR also relies on MutSα and MutLα. Cells defective in MSH2 (a subunit of MutSα) or MLH1 (a subunit of MutLα) exhibit an elevated hypermutable phenotype when treated with alkylating and other DNA damage agents, including many chemotherapeutic drugs (Li, 1999). Nevertheless, the mutant cells are highly tolerant to the cytotoxicity of these agents. In contrast, these treatments induce cell cycle arrests in MMR-proficient cells (Hawn *et al.*, 1995; Stojic *et al.*, 2004b) and eventually lead to apoptotic cell death (Li, 1999; Wu *et al.*, 1999). Although the molecular mechanism by which the MMR system mediates DNA damage-induced cellular response is not clear, two models have been proposed to account for this process. The futile recycling model (Karran, 2001; Kat *et al.*, 1993) suggests that a repetitive attempt by MMR to remove a DNA adduct in the template strand induces ssDNA gaps. This is because DNA adducts in the template strand can lead to misincorporation, which can be recognized by the MutSα–MutLα complex, provoking strand-specific MMR. Because MMR is always targeted to the newly synthesized strand and adducts in the template strand cannot be removed, thus misincorporation reoccurs upon resynthesis during repair. As a result, the repair event reinitiates. Such a futile cycle could be the reason why cells undergo cell cycle arrest and apoptosis. A second model suggests that the response signal could come from the binding of the MutSα–MutLα complexes to DNA adducts in the replication fork and/or unwound DNA helix. These stable protein–DNA complexes may block DNA replication and repair, which serve as a signal for cell cycle arrest and apoptosis (Li, 1999). Strong support for these models is provided by the fact that MutSα specifically recognizes a variety forms of DNA adduct (reviewed in Li, 1999). Studies have shown that treating MMR-proficient but not -deficient cells with alkylating agents leads to the formation of ssDNA breaks/gaps and activates the ATR-ATRIP signaling pathway (Stojic *et al.*, 2004b). Evidence for the involvement of apoptotic transducers p53 and p73 in MMR-mediated DNA damage response has also been demonstrated (Gong *et al.*, 1999; Li, 1999). Despite great progress made in the past several years, the molecular basis for MMR in cellular response to DNA damage is not fully understood. This chapter describes the methods used in the authors' laboratory to analyze MMR-mediated DNA damage signaling.

In Vitro Biochemistry Studies

A direct piece of evidence showing the involvement of MMR proteins in cellular response to DNA damage is the demonstration by gel-shift analyses that mismatch recognition proteins specifically recognize a variety of DNA adducts (Li, 1999). To carry out the gel-shift analysis successfully, purified MutSα and DNA substrates containing a defined DNA adduct are required.

Purification of MutSα

Procedures to purify both the native and the recombinant MutSα have been documented (Drummond et al., 1995, Iaccarino et al., 1998). HeLa cells are normally used as a source for the purification of the native MutSα. Nuclear extracts prepared from 2×10^9 HeLa cells (Holmes et al., 1990) are precipitated with 0.215 g/ml of ammonium sulfate, and the supernatant is recovered by centrifugation (20 min at 15,000g) and precipitated with 0.192 g/ml of ammonium sulfate. Precipitates are dissolved in buffer A [25 mM HEPES-KOH, pH 7.6, 0.1 mM EDTA, 2 mM dithiothreitol (DTT), 1 μg/ml leupeptin and 0.1% phenylmethylsulfonyl fluoride] containing 5 mM KCl and dialyzed against the same buffer until the conductivity reaches that of buffer A containing 110 mM KCl. The dialyzed nuclear extracts (50 mg) are diluted to a final protein concentration of 5 mg/ml with buffer A containing 0.2 M KCl and applied onto a ssDNA cellulose column (1.8 cm^2 × 1.6 cm; Sigma: 3.6 mg of DNA/g of cellulose) equilibrated with the dilution buffer. After washing with 10 ml of the same buffer, the column is eluted stepwise with 50 ml of buffer A containing 0.2 M KCl and 2.5 mM MgCl$_2$ followed by the same buffer containing 1 mM ATP, which elutes MutSα. Peak fractions are pooled and loaded onto a Pharmacia HR 5/5 Mono Q column equilibrated with buffer A containing 0.1 M KCl. After washing with 10 ml of the equilibration buffer, the column is developed with a 20-ml linear KCl gradient (from 0.1 to 0.65 M) in buffer A. MutSα, which elutes at ∼0.35 M KCl, is loaded onto a Pharmacia HR Superdex-200 column (25 ml) equilibrated with buffer A containing 0.1 M KCl. The column is eluted with the same buffer, and a near homogenous MutSα peak is eluted at a volume of 24–26 ml. To stabilize the protein, bovine serum albumin and sucrose are added to the purified protein at a final concentration of 1 mg/ml and 10% (w:v), respectively. The protein is frozen in small aliquots in liquid N$_2$ and stored at $-80°$.

Recombinant MutSα protein can be obtained by coexpressing MSH2 and MSH6 in insect cells through the baculovirus system and purifying the heterodimer as described elsewhere (Iaccarino *et al.*, 1998).

Construction of DNA Substrates Containing a Defined DNA Adduct

An ideal oligonucleotide duplex for MutSα binding usually has a length of 30–50 bp and a lesion at the central location of the duplex. Because a DNA damage agent usually interacts with several bases or several positions of a particular base, it is not so easy to obtain an oligonucleotide duplex containing a defined DNA adduct by directly reacting DNA duplexes with a chemical or physical agent. In many cases, only a short oligomer (7–10 nucleotides) containing the defined adduct is available. Thus, one has to perform ligation reactions to obtain desired DNA substrates. Figure 1 shows a couple of ways that we used to construct a single lesion-containing DNA substrate.

In the first method, three small oligonucleotides are annealed to larger oligonucleotides (50-mer), with the lesion-containing oligomer in the middle, as indicated in Fig. 1A. The oligomers are then ligated together with *Escherichia coli* DNA ligase. In the second method, only the adduct-containing oligomer and the 5' end oligomer are annealed to the 50-mer and ligated by DNA ligase. The ligated product is elongated by using the 50-mer as a template in the presence of dNTPs and the Klenow fragment of DNA polymerase I (Fig. 1B). The resulting 50-mer duplexes containing the desired site-specific DNA adduct are purified by polyacrylamide gel

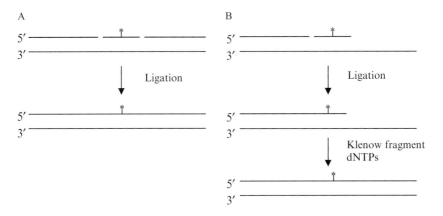

FIG. 1. Construction of 50-mer deoxyoligonucleotide containing a defined DNA adduct. Asterisk represents the desired DNA adduct. See text for description.

electrophoresis as described (Li *et al.*, 1996; Zou *et al.*, 1995). Similarly, a homoduplex without the DNA adduct and a heteroduplex containing a mismatch at the adduct site are normally constructed for negative and positive controls, respectively.

Gel-Shift Analysis

The gel-shift assay is usually performed in a 25-μl reaction containing 0.1–0.5 pmol of ^{32}P-labeled adducted oligonucleotide duplexes, 0.25–1.0 pmol of MutSα, 10 mM HEPES-KOH (pH 7.5), 110 mM KCl, 1 mM EDTA, 1 mM DTT, and 10–30 pmol nonlabeled oligonucleotides (noncompetitive DNA). The reaction is incubated on ice for 20 min, followed by the addition of 5 μl 50% (w/v) of sucrose. The samples are then fractionated at room temperature through a 6% nondenaturing polyacrylamide gel in a buffer containing 6.7 mM Tris–acetate (pH 7.5) and 1 mM EDTA. The buffer is recirculated with a pump during electrophoresis. Bands are detected by autoradiography or a phosphor imager. Figure 2 show a typical gel-shift analysis demonstrating specific binding of MutSα to a chemical carcinogen-adducted DNA.

Interactions of MutSα with adducted DNA can also be measured by real-time total internal reflectance using surface plasmon resonance as described (Tajima *et al.*, 2004).

Studies at Cellular Level

Whereas biochemical studies are important to elucidate the roles of MMR proteins in the DNA damage response at the molecular level, understanding how the repair pathway regulates DNA damage signaling requires detailed investigations at the cellular level. Such investigations usually utilize isogenic cell lines that are only different in their MMR capability. In our laboratory, MMR-deficient cell lines used are MutSα-deficient MT1 and MutLα-deficient HCT116, and the corresponding wild-type cell lines used are TK6 and HCT116-Chr.3 (Gu *et al.*, 2002; Wu *et al.*, 1999, 2003). MT1 was derived from the TK6 lymphoblastoid cell line by frameshift mutagenesis and selected by its tolerance to *N*-methyl-*N'*-*N*-nitrosoguanidine (MNNG) (Goldmacher *et al.*, 1986); HCT116-Chr.3 was derived from colorectal tumor cell line HCT116 by transferring chromosome 3, which carries the wild-type *MLH1* gene (Koi *et al.*, 1994).

Clonogenic Survival Analysis

Clonogenic survival analysis is an assay designed to measure the sensitivity of cells of interest to a particular chemical or physical agent. To determine how MMR-proficient and -deficient cells respond to a DNA

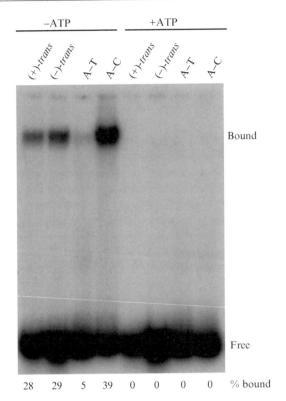

FIG. 2. Binding of MutSα to DNA containing an adenine adduct induced by an environmental chemical carcinogen benzo[c]phenanthrene 3,4-dihydrodiol-1,2 epoxide (B[c] PhED). Gel-shift assays were performed as described in the text in the presence or absence of 1 mM ATP, as indicated. ATP has been shown to inhibit the interaction between MutSα and its substrates. (+)-*trans*, (+)-*trans-anti*-B[c]PhDE; (−)-*trans*, (−)-*trans-anti*-B[c]PhDE; A–T, oligonucleotide substrate containing no mismatches; A–C, oligonucleotide substrate containing an A–C mismatch. Reproduced from Wu *et al.* (2003) with permission.

damage agent, an MMR-proficient cell line and its corresponding mutant cell line, e.g., TK6 and MT1 or HCT116-Chr.3 and HCT116 lines, can be treated with various concentrations of the agent in culture conditions (i.e., 37° and 5% CO_2) for 1 h, and the cells are removed from the DNA damage agent by centrifugation (5 min at 500g) and resuspended in fresh medium. Treated cells are placed in 96-well plates at a density of 1–2 cells/well in 0.2 ml culture medium and are cultured at 37° in a humidified atmosphere containing 5% CO_2. Cell growth is monitored daily under a microscope to check the clone formation, a process usually taking 2–4 weeks depending on the clone-forming ability of the cells. For some cells growing

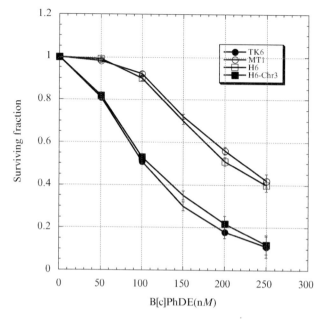

FIG. 3. Clonogenic survival analysis. MMR-proficient (TK6 and HCT116-3-6) and MMR-deficient (MT1 and HCT116) cells were treated with various concentrations of B[c]PhDE, as indicated, for 1 h and treated cells were subjected to clonogenic analysis as described in the text. Reproduced from Wu *et al.* (2003) with permission.

in suspension, e.g., TK6 and MT1 cells, 1×10^5 feeder cells/well may be used to facilitate clone formation, and these feeder cells can be obtained by treating TK6 cells with a γ irradiator (^{137}Cs) for a total dose of 2.1 krad as described (Goldmacher *et al.*, 1986). With these treatments, the feeder cells will not proliferate. Clones are counted, and surviving fraction (dividing the number of clones with the total number of cells used for the clonogenic assay) can be plotted out as a function of the concentration of the DNA damage agent used. An example of the clonogenic survival assay is shown in Fig. 3.

Apoptotic Analysis

As described earlier, among the cellular events associated with the MMR-mediated DNA damage response are cell cycle arrests (or checkpoints) and apoptotic cell death. Readers who are interested in studying cell cycle checkpoints are encouraged to follow the protocols described previously (Hawn *et al.*, 1995; Stojic *et al.*, 2004b). Methods for detecting MMR-mediated apoptosis in response to DNA damage are described next.

Classic DNA Fragmentation Analysis. MMR-proficient and -deficient cells that are growing in exponential phase are treated with a DNA damage agent for 1 h in a 5% CO_2/37° incubator. After the agent is removed by centrifugation, cells should be cultured continuously under the normal culture conditions until harvesting at the time points required, e.g., 0, 6, 12, 24, 48, and 72 h. For a control, cells receiving a mock treatment should also be harvested at the desired time points.

Cells (~1 × 10^7 cells) harvested at each time point are washed with ice-cold phosphate-buffered saline (PBS) twice and recovered by centrifugation (5 min, 500g) before being suspended in 0.3 ml of a digestion buffer that contains 10 mM Tris–HCl (pH 8), 25 mM EDTA, 0.5% SDS, 100 mM NaCl, and 0.1 mg/ml proteinase K. Samples are incubated at 55° with shaking overnight in tightly capped tubes and are then vigorously extracted at room temperature with an equal volume of phenol/chloroform/isoamyl alcohol (24:24:1). The aqueous phase is collected after separation from the organic phase by centrifugation (10 min, 17,000g in a swing bucket) and reextracted with phenol/chloroform/isoamyl alcohol for at least two more times or until no more white material is present at the interface. Genomic DNA (in aqueous phase) is precipitated with ethanol and suspended in 100 μl TE buffer. To visualize DNA fragmentations, 2–5 μg of genomic DNA is fractionated through agarose gel (1.5%) electrophoresis and detected by UV illumination in the presence of ethidium bromide. Typical results of MMR-mediated apoptosis in response to DNA damage are shown in Fig. 4A.

Flow Cytometry Analysis. Apoptotic cell death can also be measured by terminal deoxynucleotidyl transferase (TdT)-mediated dUTP nick end labeling (TUNEL) analysis using flow cytometry, based on the fact that DNA fragmentation is a characteristic of apoptosis. Several kits have been developed to detect apoptosis by enzymatically labeling free 3′ OH termini of DNA breaks with fluorescein-containing dUTP using TdT. The following methods discuss TUNEL analysis using the APO-DIRECT Kit (PharMingen, San Diego, CA), which include cell fixation, staining, and flow cytometric analysis.

FIXATION. After washing with PBS, treated cells (1–2 × 10^6) are suspended in 0.5 ml of PBS and fixed by incubating with 5 ml of 1% paraformaldehyde in PBS on ice for 15 min. After removing the paraformaldehyde solution by centrifugation (5 min at 300g), cells are washed with 5 ml of PBS twice, suspended in 0.5 ml of PBS, and mixed with 5 ml of ice-cold 70% ethanol, followed by incubation at −20° for at least 12 h prior to staining.

STAINING. Fixed cells are collected by centrifugation and the 70% ethanol is removed by aspiration. After washing twice with 1.0 ml of wash buffer (a part of the APO-DIRECT kit), cells are recovered by centrifugation

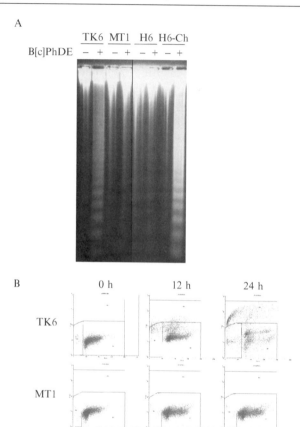

FIG. 4. MMR-mediated apoptosis in response to DNA damage. (A) DNA fragmentation analysis. Cells were treated with B[c]PhDE (0.25 μM) for 1 h and cultured in fresh medium for 24 h. Genomic DNA was isolated and fractionated through 1.5% agarose gels and visualized under UV light in the presence of ethidium bromide. (B) TUNEL analysis. Cells were treated with 0.25 μM B[c]PhDE at 37° for 1 h, cultured in fresh medium for various time as indicated, and subjected to TUNEL analysis as described in the text. Reproduced from Wu *et al.* (2003) with permission. (See color insert.)

and resuspended in 50 μl of the staining solution (a part of the apoptosis kit), which contains reaction buffer, TdT, and fluorescein isothiocyanate (FITC)-dUTP. Cells are stained for 60 min at 37° and the staining reaction is terminated by rinsing twice with 1.0 ml of rinse buffer, followed by centrifugation (300*g* for 5 min). The supernatant in each case is removed by aspiration, and the cell pellet is suspended and incubated in 0.5 ml of the propidium iodide (PI)/RNase A solution in the dark for 30 min at room temperature.

FLOW CYTOMETRY ANALYSIS. Stained cells can be analyzed using a Becton-Dickinson flow cytometer and Cell Quest software. A typical TUNEL assay is shown in Fig. 4B.

Animal Models

Thanks to the availability of MMR gene knockout mice, the MMR-mediated DNA damage response can be analyzed in an animal model. TUNEL assay kits have been developed to measure *in situ* apoptosis in animal organs or tissues. Because intestinal neoplasia is associated with loss of MMR function (Kolodner and Marsischky, 1999; Li, 2003; Modrich and Lahue, 1996) and because apoptosis within the murine small intestine has been well characterized (Clarke *et al.*, 1994), mouse small intestine is an ideal target for analyzing MMR-mediated apoptosis in response to DNA damage. The following methods outline *in situ* apoptosis in mouse small intestine using the ApopTag kit (Intergen Co., Purchase, NY). Like the APOP DIRECT kit, the ApopTag kit also relies on TdT to catalyze a template-independent addition of modified nucleotide triphosphates to the 3' OH ends of DNA breaks. The incorporated nucleotides form an oligomer composed of digoxigenin, which can be recognized by an antidigoxigenin antibody that is conjugated to a peroxidase reporter molecule. The bound peroxidase antibody conjugate enzymatically produces a permanent, intense, and localized stain for chromogenic substrates, thereby providing sensitive detection of 3' OH ends localized in apoptotic bodies.

Treatment of Mice with DNA Damage Agents

MMR knockout mice ($Msh2^{-/-}$ or $Mlh1^{-/-}$) and wild-type mice with an age of 8–12 weeks are given an intraperitoneal injection of the desired amount of a DNA damage agent (see Table I). Chemicals should be made freshly by first dissolving in an organic solvent such as dimethyl sulfoxide and then diluting to a final concentration in PBS. The total injection volume should be no more than 0.25 ml. At least three mice are sacrificed at each time point (e.g., at hour 0, 3, 6, 12, or 24) after injection. The small intestine is removed, flushed with water, and fixed in 10% (v:v) neutral-buffered formalin (mixing 10 ml of commercial formalin solution with 90 ml of PBS, pH 7.4). The fixed small intestine can be stored at room temperature for a period of time.

Paraffin Embedding and Sectioning

The formalin-fixed samples are subjected to paraffin embedding and sectioning. The embedded samples should be sectioned at 5 μm. In our

TABLE I
DOSE OF DNA DAMAGE REAGENTS USED IN MOUSE MODEL

Agent	Dose	Reference
Chemical		
MNNG	50 mg/kg	Toft *et al.* (1999)
Cisplatin	10 mg/kg	Toft *et al.* (1999)
B[c]PhDE	1.0 mg/kg	Wu *et al.* (2003)
O^6-Benzylguanine	60 mg/kg	Toft *et al.* (1999)
Temozolomide	100 mg/kg	Toft *et al.* (1999)
Physical		
UVB	500–2000 J/m^2	van Oosten *et al.* (2005)
γ irradiation (^{137}Crs)	4 Gy	Toft *et al.* (1999)

studies, these steps were performed by the Histology Laboratory of the Department of Pathology, University of Kentucky Medical Center. For those who are interested in performing these procedures, a step-by-step protocol is available (Carson, 1990).

Apoptosis Detection Using the ApopTag Kit

Pretreatments of Slides. Paraffin section slides should be treated in a coplin jar with three changes of xylene, with each treatment for 5 min, to remove paraffin from the tissue specimen, followed by washing with absolute ethanol (twice, 5 min each), 95% ethanol (once for 3 min), 75% ethanol (once for 3 min), and PBS (once for 5 min). The specimens are then treated with proteinase K (20 μg/ml) for 15 min at room temperature and washed with two changes (2 min each) of dH$_2$O to remove proteins, followed by treating with 3.0% hydrogen peroxide in PBS for 5 min at room temperature and rinsing twice (each for 5 min) with water to quench endogenous peroxidase.

Peroxidase Staining. Pretreated slides are incubated with the equilibration buffer provided by the ApopTag kit for 30 s by directly applying 75 μl of the buffer on the specimen. Immediately after removing the equilibration buffer by aspiration, the slides are incubated with 55 μl/5 cm^2 of the working strength TdT enzyme at 37° for 1 h. The reaction is terminated by placing the slides in a coplin jar containing the working strength stop/wash buffer (a part of the ApopTag kit) for 10 min at room temperature and washing with PBS twice (each for 1 min). The samples are then incubated with the antidigoxigenin peroxidase conjugate at room temperature for 30 min by directly applying 65 μl/5 cm^2 of the conjugate to the specimen

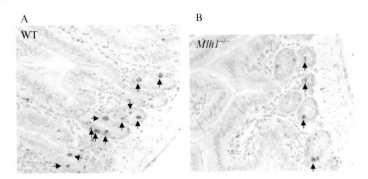

FIG. 5. MMR-mediated apoptosis in mouse small intestines. Mice (wild type, or $Mlh1^{-/-}$) were injected with B[c]PhDE (1 mg/kg) and sacrificed at hour 6 after injection. Paraffin sections of small intestines were analyzed for apoptosis by ApopTag peroxidase kits. Apoptotic cells are indicated by arrows. Reproduced from Wu *et al.* (2003) with permission.

surface. After washing with four changes (each for 2 min) of PBS in a coplin jar, the specimens are treated for 3 to 6 min with the peroxidase substrate (a part of the ApopTag kit) by completely covering the specimen area with the substrate. To determine the optimal staining time, individual slides should be checked for color development under a microscope. Upon washing with three changes (each for 2 min) of dH_2O, the slides are counterstained with 0.5% (w:v) of methyl green in a coplin jar for 10 min at room temperature. After washing with three changes (each for 30 s) of dH_2O and three changes (each for 30 s.) of 100% *n*-butanol, the slides are dehydrated by moving through three jars of xylene and incubating for 2 min in each jar, followed by mounting with Permount under a glass coverslip.

Apoptotic Score under Microscope

Apoptosis in mouse small intestine can be scored within the crypts of small intestines under microscope. Nuclei in apoptotic cells are stained brown. A typical apoptotic analysis by the ApopTag kit is shown in Fig. 5. For each time point, a minimum of 50 half-crypts per animal should be scored.

Acknowledgments

Work was supported by NIH Grants CA72956 and ES013193 to GML. GML is a James-Gardner Chair in Cancer Research.

References

Allen, D. J., Makhov, A., Grilley, M., Taylor, J., Thresher, R., Modrich, P., and Griffith, J. D. (1997). MutS mediates heteroduplex loop formation by a translocation mechanism. *EMBO J.* **16,** 4467–4476.

Carson, F. L. (1990). *In* "Histotechnology, a Self-instructional Text," pp. 31–55. American Society of Clinical Pathologists Press, Chicago.

Clarke, A. R., Gledhill, S., Hooper, M. L., Bird, C. C., and Wyllie, A. H. (1994). p53 dependence of early apoptotic and proliferative responses within the mouse intestinal epithelium following gamma-irradiation. *Oncogene* **9,** 1767–1773.

Drummond, J. T., Li, G. M., Longley, M. J., and Modrich, P. (1995). Isolation of an hMSH2-p160 heterodimer that restores DNA mismatch repair to tumor cells. *Science* **268,** 1909–1912.

Goldmacher, V. S., Cuzick, R. A., Jr., and Thilly, W. G. (1986). Isolation and partial characterization of human cell mutants differing in sensitivity to killing and mutation by methylnitrosourea and N-methyl-N'-nitro-N-nitrosoguanidine. *J. Biol. Chem.* **261,** 12462–12471.

Gong, J. G., Costanzo, A., Yang, H. Q., Melino, G., Kaelin, W. G., Jr., Levrero, M., and Wang, J. Y. (1999). The tyrosine kinase c-Abl regulates p73 in apoptotic response to cisplatin-induced DNA damage. *Nature* **399,** 806–809.

Gradia, S., Subramanian, D., Wilson, T., Acharya, S., Makhov, A., Griffith, J., and Fishel, R. (1999). hMSH2-hMSH6 forms a hydrolysis-independent sliding clamp on mismatched DNA. *Mol. Cell* **3,** 255–261.

Gu, L., Wu, J., Qiu, L., Jennings, C. D., and Li, G. M. (2002). Involvement of DNA mismatch repair in folate deficiency-induced apoptosis. *J. Nutr. Biochem.* **13,** 355–363.

Harfe, B. D., and Jinks-Robertson, S. (2000). DNA mismatch repair and genetic instability. *Annu. Rev. Genet.* **34,** 359–399.

Hawn, M. T., Umar, A., Carethers, J. M., Marra, G., Kunkel, T. A., Boland, C. R., and Koi, M. (1995). Evidence for a connection between the mismatch repair system and the G2 cell cycle checkpoint. *Cancer Res.* **55,** 3721–3725.

Holmes, J., Jr., Clark, S., and Modrich, P. (1990). Strand-specific mismatch correction in nuclear extracts of human and Drosophila melanogaster cell lines. *Proc. Natl. Acad. Sci. USA* **87,** 5837–5841.

Iaccarino, I., Marra, G., Palombo, F., and Jiricny, J. (1998). hMSH2 and hMSH6 play distinct roles in mismatch binding and contribute differently to the ATPase activity of MutSα. *EMBO J.* **17,** 2677–2686.

Junop, M. S., Obmolova, G., Rausch, K., Hsieh, P., and Yang, W. (2001). Composite active site of an ABC ATPase: MutS uses ATP to verify mismatch recognition and authorize DNA repair. *Mol. Cell* **7,** 1–12.

Karran, P. (2001). Mechanisms of tolerance to DNA damaging therapeutic drugs. *Carcinogenesis* **22,** 1931–1937.

Kat, A., Thilly, W. G., Fang, W. H., Longley, M. J., Li, G. M., and Modrich, P. (1993). An alkylation-tolerant, mutator human cell line is deficient in strand-specific mismatch repair. *Proc. Natl. Acad. Sci. USA* **90,** 6424–6428.

Koi, M., Umar, A., Chauhan, D. P., Cherian, S. P., Carethers, J. M., Kunkel, T. A., and Boland, C. R. (1994). Human chromosome 3 corrects mismatch repair deficiency and microsatellite instability and reduces N-methyl-N'-nitro-N-nitrosoguanidine tolerance in colon tumor cells with homozygous hMLH1 mutation. *Cancer Res.* **54,** 4308–4312.

Kolodner, R. D., and Marsischky, G. T. (1999). Eukaryotic DNA mismatch repair. *Curr. Opin. Genet. Dev.* **9,** 89–96.

Kunkel, T. A., and Erie, D. A. (2005). DNA mismatch repair. *Annu. Rev. Biochem.* **74,** 681–710.

Li, G. M. (1999). The role of mismatch repair in DNA damage-induced apoptosis. *Oncol. Res.* **11,** 393–400.

Li, G. M. (2003). DNA mismatch repair and cancer. *Front. Biosci.* **8,** d997–d1017.

Li, G. M., Wang, H., and Romano, L. J. (1996). Human MutSα specifically binds to DNA containing aminofluorene and acetylaminofluorene adducts. *J. Biol. Chem.* **271,** 24084–24088.

Modrich, P., and Lahue, R. (1996). Mismatch repair in replication fidelity, genetic recombination, and cancer biology. *Annu. Rev. Biochem.* **65,** 101–133.

Myung, K., Datta, A., Chen, C., and Kolodner, R. D. (2001). SGS1, the *Saccharomyces cerevisiae* homologue of BLM and WRN, suppresses genome instability and homeologous recombination. *Nature Genet.* **27,** 113–116.

Schofield, M. J., and Hsieh, P. (2003). DNA mismatch repair: Molecular mechanisms and biological function. *Annu. Rev. Microbiol.* **57,** 579–608.

Stojic, L., Brun, R., and Jiricny, J. (2004a). Mismatch repair and DNA damage signalling. *DNA Repair (Amst.)* **3,** 1091–1101.

Stojic, L., Mojas, N., Cejka, P., Di Pietro, M., Ferrari, S., Marra, G., and Jiricny, J. (2004b). Mismatch repair-dependent G2 checkpoint induced by low doses of SN1 type methylating agents requires the ATR kinase. *Genes Dev.* **18,** 1331–1344.

Tajima, A., Hess, M. T., Cabrera, B. L., Kolodner, R. D., and Carethers, J. M. (2004). The mismatch repair complex hMutS alpha recognizes 5-fluorouracil-modified DNA: Iimplications for chemosensitivity and resistance. *Gastroenterology* **127,** 1678–1684.

Thomas, D. C., Roberts, J. D., and Kunkel, T. A. (1991). Heteroduplex repair in extracts of human HeLa cells. *J. Biol. Chem.* **266,** 3744–3751.

Toft, N. J., Winton, D. J., Kelly, J., Howard, L. A., Dekker, M., te Riele, H., Arends, M. J., Wyllie, A. H., Margison, G. P., and Clarke, A. R. (1999). Msh2 status modulates both apoptosis and mutation frequency in the murine small intestine. *Proc. Natl. Acad. Sci. USA* **96,** 3911–3915.

van Oosten, M., Stout, G. J., Backendorf, C., Rebel, H., de Wind, N., Darroudi, F., van Kranen, H. J., de Gruijl, F. R., and Mullenders, L. H. (2005). Mismatch repair protein Msh2 contributes to UVB-induced cell cycle arrest in epidermal and cultured mouse keratinocytes. *DNA Repair (Amst.)* **4,** 81–89.

Wu, J., Gu, L., Wang, H., Geacintov, N. E., and Li, G. M. (1999). Mismatch repair processing of carcinogen-DNA adducts triggers apoptosis. *Mol. Cell. Biol.* **19,** 8292–8301.

Wu, J., Zhu, B. B., Yu, J., Zhu, H., Qiu, L., Kindy, M. S., Gu, L., Seidel, A., and Li, G. M. (2003). *In vitro* and *in vivo* modulations of benzo[c]phenanthrene-DNA adducts by DNA mismatch repair system. *Nucleic Acids Res.* **31,** 6428–6434.

Zou, Y., Liu, T. M., Geacintov, N. E., and Van Houten, B. (1995). Interaction of the UvrABC nuclease system with a DNA duplex containing a single stereoisomer of dG-(+)- or dG-(−)-anti-BPDE. *Biochemistry* **34,** 13582–13593.

[20] Characterization of *Escherichia coli* Translesion Synthesis Polymerases and Their Accessory Factors

By PENNY J. BEUNING, SHAROTKA M. SIMON, VERONICA G. GODOY, DANIEL F. JAROSZ, and GRAHAM C. WALKER

Abstract

Members of the Y family of DNA polymerases are specialized to replicate lesion-containing DNA. However, they lack $3'$-$5'$ exonuclease activity and have reduced fidelity compared to replicative polymerases when copying undamaged templates, and thus are potentially mutagenic. Y family polymerases must be tightly regulated to prevent aberrant mutations on undamaged DNA while permitting replication only under conditions of DNA damage. These polymerases provide a mechanism of DNA damage tolerance, confer cellular resistance to a variety of DNA-damaging agents, and have been implicated in bacterial persistence. The Y family polymerases are represented in all domains of life. *Escherichia coli* possesses two members of the Y family, DNA pol IV (DinB) and DNA pol V (UmuD'$_2$C), and several regulatory factors, including those encoded by the *umuD* gene that influence the activity of UmuC. This chapter outlines procedures for *in vivo* and *in vitro* analysis of these proteins. Study of the *E. coli* Y family polymerases and their accessory factors is important for understanding the broad principles of DNA damage tolerance and mechanisms of mutagenesis throughout evolution. Furthermore, study of these enzymes and their role in stress-induced mutagenesis may also give insight into a variety of phenomena, including the growing problem of bacterial antibiotic resistance.

Introduction

The Y family of DNA polymerases is characterized by its ability to bypass noninstructive template sites or lesions in DNA that block replicative DNA polymerases (Ohmori *et al.*, 2001). This specialized ability comes at a potential mutagenic cost, as the Y family polymerases replicate undamaged DNA with reduced fidelity compared to replicative polymerases (Ohmori *et al.*, 2001). *Escherichia coli* has two such polymerases: DNA pol IV (DinB) and DNA pol V (UmuD'$_2$C). These two polymerases are among the genes induced as part of the SOS response to DNA damage (Courcelle

METHODS IN ENZYMOLOGY, VOL. 408   0076-6879/06 $35.00
     DOI: 10.1016/S0076-6879(06)08020-7

et al., 2001; Friedberg *et al.*, 2005b). In the absence of DNA damage, the LexA repressor binds to operator sites and to various extents represses the transcription of genes under its control. The SOS response is induced when RecA polymerizes on regions of single-stranded DNA (ssDNA), which are formed in response to DNA damage and/or replication fork stalling (Fig. 1), creating a nucleoprotein filament. The RecA/ssDNA nucleoprotein filament binds the LexA repressor and stimulates the latent ability of LexA to cleave itself, leading to a decrease in the intracellular concentration of LexA$_2$ and the derepression of at least 1% of the *E. coli* genome (Courcelle *et al.*, 2001). Each Y family polymerase in *E. coli* is encoded in an SOS-inducible operon (Fig. 1) (Courcelle *et al.*, 2001; Friedberg *et al.*, 2005b).

The *umuC* gene is organized in an operon with the the *umuD* gene, which is tightly regulated by the LexA repressor (Kitagawa *et al.*, 1985; Sommer *et al.*, 1993). After induction, full-length UmuD is expressed and then slowly undergoes RecA/ssDNA nucleoprotein filament-mediated

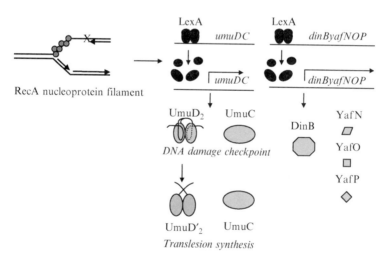

FIG. 1. Induction of the SOS response is initiated when a region of ssDNA forms due to failed attempts of a cell to replicate damaged DNA. RecA polymerizes on the ssDNA, forming the RecA/ssDNA nucleoprotein filament (RecA*). LexA undergoes autocatalytic cleavage in the presence of RecA*, leading to the expression of at least 40 genes, including the polymerase operons shown. UmuD cleavage is also facilitated by interaction with the RecA/ssDNA filament, which provides a temporal switch between its action with UmuC in a DNA damage checkpoint and its facilitation of the translesion synthesis of UmuC.

proteolysis to remove its N-terminal 24 amino acids to generate UmuD'. A complicated set of interactions results in UmuD being the predominant species for the first 30 min, approximately, after the initial DNA damaging treatment, and then UmuD' becomes the predominant form (Friedberg et al., 2005b; Opperman et al., 1999). Whereas there are approximately 180 UmuD molecules and fewer than 20 UmuC molecules in an uninduced cell, UmuD' is not detectable (Woodgate and Ennis, 1991). After full induction of the SOS response with mitomycin C, these levels rise to approximately 930 UmuD, 1900 UmuD', and 200 UmuC proteins per cell (Woodgate and Ennis, 1991). This compares with a maximal increase in mRNA levels of about 40-fold for UmuC and 30-fold for UmuD after ultraviolet (UV)–light induction of the SOS response (Courcelle et al., 2001).

The temporal ordering of the appearance of UmuD followed by UmuD' has important implications throughout the SOS response. UmuD$_2$, together with UmuC, plays a role in affecting a DNA damage checkpoint (Opperman et al., 1999; Sutton and Walker, 2001). Elevated levels of the umuDC gene products specifically cause a decrease in the rate of DNA replication compared to the rates of RNA or protein synthesis (Marsh and Walker, 1985). In contrast, UmuD'$_2$ is required for UmuC to act as a translesion polymerase (Reuven et al., 1999; Tang et al., 1999). Additionally, RecA is required for UmuD'$_2$C to carry out translesion synthesis; this complex is referred to as the pol V mutasome (Fujii et al., 2004; Schlacher et al., 2005; Tang et al., 1998). UmuD is also important for protein turnover, as it targets its dimeric partner, whether UmuD or UmuD', for proteolysis by the ClpXP protease (Gonzalez et al., 2000; Neher et al., 2003).

The DinB subfamily is the only branch of Y family polymerases that is conserved throughout all domains of life (Ohmori et al., 2001). In E. coli, dinB is organized in an operon with three genes of unknown function, yafNOP. DinB protein is present at considerably higher levels than UmuC in the absence of DNA damage—about 250 molecules per cell (Kim et al., 2001). Upon induction of the SOS response with mitomycin C, this number increases approximately 10-fold (Kim et al., 2001). After UV induction, the DinB mRNA levels are induced 7-fold, and the yafNOP mRNA levels are each induced from 3- to 5-fold (Courcelle et al., 2001). The role, if any, of yafNOP genes in modulating DinB activity is not well understood (McKenzie et al., 2003). Indeed, while much is known about the cellular mechanisms that regulate UmuC activity (Friedberg et al., 2005a,b; Goodman, 2002; Sutton et al., 2000), DinB function and regulation have remained largely enigmatic (Fuchs et al., 2004).

The two Y family polymerases of E. coli appear to bypass distinct template lesions. UmuD'$_2$C is able to perform translesion synthesis opposite abasic sites, N^6-dA- and N^2-dG-benzo[a]pyrene, thymine–thymine cyclobutane dimers, and [6-4] photoproducts (Shen *et al.*, 2002; Tang *et al.*, 2000). Although UmuD'$_2$C bypasses T–T cyclobutane dimers accurately, it causes both lesion-targeted and untargeted mutagenesis (Tang *et al.*, 2000). The misincorporation frequency of UmuD'$_2$C on undamaged templates is estimated to be 10^{-3} to 10^{-4} (Tang *et al.*, 2000). UmuD'$_2$C tends to cause base pair transition mutations, as it inserts guanine opposite the 3'-thymine of [6-4] T–T photoproducts, while it bypasses T–T cyclobutane dimers relatively accurately (Tang *et al.*, 2000). UmuD'$_2$C is also responsible for G → T transversion mutations when bypassing N^2-dG-(+)-*trans-anti*-benzo[a]pyrene *in vivo* in a specific sequence context (Lenne-Samuel *et al.*, 2000; Yin *et al.*, 2004). In contrast, *dinB* causes −1 frameshifting *in vivo* when present at high concentrations (Kim *et al.*, 1997, 2001). DinB is able to extend from bulged primer/template termini (Kobayashi *et al.*, 2002; Wagner *et al.*, 1999), partly explaining its ability to generate frameshift mutations. DinB has also been shown to weakly, although accurately, bypass the major N^2-dG adduct of benzo[a]pyrene (Shen *et al.*, 2002). Abasic sites can be weakly bypassed *in vitro* by DinB, although DinB activity is not required *in vivo* for abasic site bypass (Maor-Shoshani *et al.*, 2003). The error rate of DinB on undamaged DNA is 10^{-3} to 10^{-5}, which compares with the error rate of the pol III α catalytic subunit without proofreading of about 10^{-5} (Goodman, 2002; Kornberg and Baker, 1992). The Y family polymerases are characterized by their low processivity, although this is stimulated to varying extents by addition of the β processivity clamp (Fujii *et al.*, 2004; Wagner *et al.*, 2000).

Thus, the two Y family polymerases in *E. coli* are regulated at the level of transcription by the SOS response and at least UmuD is also regulated posttranslationally with implications for UmuC activity. Furthermore, the two *E. coli* Y family polymerases have largely nonoverlapping substrate specificities. A major unresolved question in the field is how their access to primer termini is governed (Friedberg *et al.*, 2005a; Pages and Fuchs, 2002). There is growing evidence that the β processivity clamp and/or the manager proteins UmuD and UmuD' play important roles in permitting access of either of the *E. coli* Y family polymerases to the replication fork (Becherel *et al.*, 2002; Beuning *et al.*, 2006; Bunting *et al.*, 2003; Sutton *et al.*, 1999, 2001b; Wagner *et al.*, 2000). The understanding of the mechanisms regulating these enzymes will require dissecting the known genetic pathways *in vitro* and *in vivo*, with special attention to regulatory protein–protein interactions. Moreover, discovering new phenotypes *in vivo* will give insights into the global role of these polymerases in response to DNA damage and environmental stress.

Purification and Characterization of umuD Gene Products

Expression and Purification of UmuD and UmuD'

Expression and purification of the *umuD* gene products have been typically accomplished using *umuD*[+] *E. coli* strain BL21(DE3) harboring a *umuD* expression plasmid. Unfortunately, a drawback of the use of this strain for the preparation of UmuD variants is that some contaminating wild-type protein will be present due to expression from the chromosomal copy of the gene. Therefore, a chloramphenicol-resistant *ΔumuDC* derivative of *E. coli* BL21 was constructed by P1 transduction from GW8017 (Guzzo *et al.*, 1996; Silhavy *et al.*, 1984; Woodgate, 1992). UmuD and UmuD' expression plasmids, pSG5 and pSG4, respectively, are based on pET11T (Nguyen *et al.*, 1993) and were constructed from pMAD (UmuD) or pMADp (UmuD'), respectively (Reuven *et al.*, 1998). An *NdeI* site was added 5' of the *umuD* gene in pMAD or *umuD'* gene in pMADp, by site-directed mutagenesis, and the *NdeI–BamHI* fragment was ligated into the *NdeI–BamHI* linearized pET11T. The integrity of the construct was confirmed by sequencing. Colony-dependent variation in expression has been observed in our laboratory when using the pET11T expression plasmid so several colonies are routinely screened for efficient overexpression.

The major advantage of this purification protocol over those published previously (Burckhardt *et al.*, 1988; Ferentz *et al.*, 1997; Lee *et al.*, 1994) is the use of fast flow FPLC columns, reducing purification time greatly. Using a single FPLC, two UmuD variants can be purified simultaneously, with one preparation on the instrument while the other is assayed. One should allow two consecutive days after growth, induction, and harvesting for the preparation of UmuD proteins.

Transformed BL21(DE3) *ΔumuDC* are grown at $37°$ in LB medium or M9 medium (Sambrook *et al.*, 1989) supplemented with 100 μg/ml ampicillin in baffled flasks until an OD_{600} of 0.6 to 0.8 is reached. Isopropyl-β-D-thiogalactoside (IPTG) is added to 1 mM, at which time the culture is transferred to $30°$ for 3–4 h. Cells are harvested by centrifugation at 5000g for 30 min and resuspended in 15 ml lysis buffer (50 mM Tris–HCl, pH 8.0, 2 mM EDTA) per liter of culture. After resuspension, cells can be frozen dropwise in liquid nitrogen and stored at $-80°$ indefinitely or can be lysed immediately. Typically, protein from 2 to 4 liters of culture is purified at once.

Frozen cells are thawed on ice overnight at $4°$. All subsequent steps are performed at $4°$. Cell lysis is accomplished by two passages at 10 kpsi through a French press (Thermo Spectronic) or via sonication at 50% output with alternating 15-s bursts followed by 15-s rest periods for 3–4 min.

After lysis, cell debris is pelleted by centrifugation at 14,000g for 30 min. The first ammonium sulfate precipitation of the supernatant is to 20% saturation (0.121 g/ml). The precipitate is removed by centrifugation at 14,000g for 30 min, and the supernatant is subjected to a second ammonium sulfate precipitation, this time adding ammonium sulfate to 40% saturation (0.151 g/ml). Centrifugation is the same as for the first ammonium sulfate precipitation, but the pellet is retained. The pellet itself may be stored on ice overnight. However, once chromatography is started, it must be completed within the same day, as either extensive storage at 4° or freezing and thawing results in a loss of yield.

The pellet is resuspended in 50% PS_A (10 mM sodium phosphate, pH 6.8, 0.1 mM EDTA, 1 mM dithiothreitol [DTT], 1 M [NH_4]$_2SO_4$) and 50% PS_B (10 mM sodium phosphate, pH 6.8, 0.1 mM EDTA, 1 mM DTT) plus one complete, EDTA-free protease inhibitor cocktail tablet (Roche Molecular Biochemicals), until there is no visible debris (about 10 ml per liter original culture). The sample is centrifuged at 14,000g for 30 min to pellet remaining debris and then filtered through a 0.25-μm Millex GV syringe filter. The filtered sample is loaded onto a HiTrap fast flow, low substitution phenyl-Sepharose column (GE Healthcare) preequilibrated with PS_A. The protein is eluted with a gradient of 0–100% PS_B over 5 column volumes at a flow rate of 5 ml/min. Fractions are collected at the end of the gradient, after 80% PS_B is reached, and during a 2 column volume wash with PS_B. UmuD proteins will elute as the last peak, starting at 100% PS_B. Aliquots (10 μl) of each fraction are assayed by 14% SDS–PAGE and stained with Coomassie blue.

Fractions that contain UmuD are pooled and applied to a HiTrap fast flow Q- Sepharose column (GE Healthcare) equilibrated with Q_A (10 mM sodium phosphate, pH 6.8, 0.1 mM EDTA, 1 mM DTT, 100 mM NaCl). The protein is eluted with a gradient of 0–100% Q_B (10 mM sodium phosphate, pH 6.8, 0.1 mM EDTA, 1 mM DTT, 1 M NaCl) over 10 column volumes at a flow rate of 5 ml/min. UmuD elutes at approximately 400 mM NaCl. Fractions are analyzed by SDS–PAGE.

Fractions chosen for further purification are concentrated using spin concentrators with a 5-kDa molecular weight cutoff membrane (VivaSpin) to 1–3% of the volume of a Superdex 75 gel filtration column (GE Healthcare). After concentration, the sample is filtered through a Millex GV syringe filter before injection onto the column. The column is run isocratically at 1.5 ml/min with buffer Q_A. UmuD elutes as the last peak. Fractions are analyzed for purity by SDS–PAGE. Fractions with pure UmuD are pooled, concentrated to 0.1–4 mM, and flash frozen in liquid N_2 in conveniently sized aliquots. The purified proteins are stored at −80°, and once an aliquot is thawed, it should be used the same day. The protein is more stable when concentrated than when left dilute after purification. Each liter of induced BL21(DE3) will yield 10–20 mg protein. Both UmuD$_2$ and UmuD'$_2$ can be purified with this protocol, although purifying UmuD'$_2$ will tend to result in higher yields than UmuD$_2$.

RecA/ssDNA Coprotease-Facilitated Cleavage of UmuD In Vitro and In Vivo

A RecA/ssDNA nucleoprotein filament (RecA*) formed during the SOS response acts as a coprotease to facilitate the processing of UmuD$_2$ to UmuD'$_2$. In this self-catalyzed cleavage reaction, the UmuD active site Ser60-Lys97 catalytic dyad cleaves the peptide backbone between Cys24 and Gly25, removing the N-terminal 24 amino acids (Friedberg et al., 2005b). UmuD binding to the RecA/ssDNA nucleoprotein filament seems to play an important role in bringing the active site functional groups of the Ser-Lys dyad into the proper conformation for catalysis (Ferentz et al., 2001). The cleavage reaction can be assayed in vitro with purified proteins or the products of the in vivo reaction can be detected by immunoblotting.

In vitro UmuD cleavage reactions are carried out in buffer LG (40 mM Tris–HCl, pH 8.0, 6.8 mM MgCl$_2$, 30 mM NaCl, 0.3 mM DTT) with 0.68 mM ATPγS (Sigma-Aldrich), 0.35 μM 24-mer DNA oligonucleotide, 3.15 μM RecA (New England Biolabs), 10 μM UmuD (monomer concentration) in a reaction volume of 20 μl (Lee et al., 1994). The reaction is incubated at 37° for at least 30 min and quenched in 4× SDS–PAGE loading buffer [1× = 25 mM Tris–HCl, pH 6.8, 2% SDS, 5% glycerol, 0.1% bromphenol blue, and 1.25% β-mercaptoethanol (β-ME)]. UmuD and UmuD' can be resolved by 14% or 4–20% gradient SDS–PAGE (Fig. 2). Several protocols for this cleavage reaction have been published by our laboratory and others (Battista et al., 1990; Burckhardt et al., 1988; Lee et al., 1996; Sutton et al., 2001a). The conditions given earlier result in efficient UmuD cleavage, although cleavage to varying extents is also observed with the addition of 10% glycerol, salts up to 10 mM MgCl$_2$ or 100 mM NaCl, up to 2.5 mM ATPγS, 2 μM DNA oligonucleotide, or buffered at pH 7.5. DNA oligonucleotides used as RecA substrates to form RecA nucleoprotein filaments are at least 20 nucleotides in length and should not contain any recognizable secondary structure such as hairpins.

Products of UmuD cleavage in vivo are detected directly from cultures by immunoblotting (McDonald et al., 1998; Sutton et al., 2001a). Saturated overnight cultures are subcultured 1:50 into fresh LB medium (with

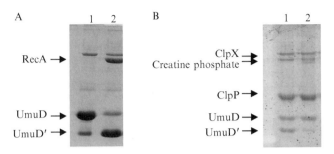

FIG. 2. (A) *In vitro* RecA/ssDNA filament-facilitated cleavage of UmuD. Lane 1, −RecA; lane 2, +RecA $t = 1$ h. A small amount of autocleavage is observed in lane 1. Unlabeled bands are impurities in the UmuD preparation. (B) *In vitro* ClpXP proteolysis of wild-type UmuD' with wild-type UmuD as adaptor protein. Lane 1, $t = 0$; lane 2, $t = 2$ h.

appropriate antibiotics). After approximately 1.5 h of growth, at an OD_{600} of 0.2–0.3, approximately 2.5×10^{10} cells are harvested by centrifugation at $5000g$ for 10 min. Cells are washed twice with 5 ml of 0.85% saline and are resuspended in 5 ml 0.85% saline. The resuspended cells are transferred to a 15×100-mm petri dish and irradiated with varying amounts (typically 25 J/m^2) of UV (254 nm; from a GE 15 W G15T8 germicidal lamp). Cells are collected by centrifugation, resuspended in 5 ml LB, and incubated at 37° for 1–3 h to allow expression and cleavage of UmuD. After harvesting by centrifugation, cells can either be processed for immunoblot analysis immediately or stored as pellets at −80° for several days.

To analyze samples by immunoblotting, pellets are resuspended in 100 μl 0.85% saline, to which 50 μl 4× SDS–PAGE loading buffer is added. Samples are boiled 15 min, and cell debris is pelleted by centrifugation at $12,000g$ for 1 min. UmuD and UmuD' in the supernatant can be resolved by SDS–PAGE as described earlier. Typically a 10- or 20-μl aliquot is analyzed per lane of the gel. The proteins are then transferred to polyvinylidene fluoride (PVDF) membrane (Millipore) in CAPS transfer buffer (9 mM CAPS, adjusted to pH 11.5 with a final concentration of 8.33 mM NaOH, 10% methanol). The blot is processed according to the protocol supplied by Pierce for the SuperSignal West Dura Extended Duration Substrate.

Alkaline Cleavage of UmuD

UmuD and other enzymes of its class also possess the ability to cleave themselves under alkaline conditions, without the RecA coprotease, although this autoproteolysis of UmuD is inefficient compared to that of

LexA or phage λ repressor (Kulaeva *et al.*, 1995; Little, 1984; McDonald *et al.*, 1998). To determine whether specific noncleavable variants of UmuD are defective in interaction with RecA or in the cleavage mechanism, UmuD autoproteolysis can be assayed under alkaline conditions. UmuD ($10–30 \mu M$) is incubated in buffer G [100 mM glycine, pH 10, 10 mM CaCl$_2$, 50 mM NaCl, 10 mM DTT, 0.1–0.25 mg/ml bovine serum albumin (BSA)] at 37° for 24–48 h or longer (Kulaeva *et al.*, 1995). The two UmuD species can be analyzed by SDS–PAGE as described earlier.

ClpXP Degradation Assay

UmuD and UmuD′ exist in three dimeric species: UmuD$_2$ or UmuD′$_2$ homodimers or the most thermodynamically stable UmuD′D heterodimer (Battista *et al.*, 1990; Jonczyk and Nowicka, 1996; Lee *et al.*, 1994). When UmuD$_2$ or UmuD′$_2$ homodimers are coincubated *in vitro* at 25° in equal concentration, the only dimeric species detectable after 30 min is the UmuD′D heterodimer (Battista *et al.*, 1990). In this context, the purified ClpXP protease will specifically target the UmuD′ partner of the heterodimer for degradation in the presence of ATP (Gonzalez *et al.*, 2000).

This assay allows for evaluation of the ability of a protein to act as either a substrate of ClpXP degradation, as does UmuD′ and UmuD, or an adaptor protein delivering substrates to ClpXP, as does UmuD (Neher *et al.*, 2003). It is important when making these comparisons that a wild-type UmuD′D control is included in each assay, as the extent of UmuD′ degradation can vary with ClpXP preparation and length of storage.

ClpX and ClpP were a gift from Professor Tania Baker at MIT (Kim *et al.*, 2000; Neher *et al.*, 2003). UmuD′D is formed by coincubating 4.5 μM UmuD$_2$ and 4.5 μM UmuD′$_2$ with 0.8 μM ClpP$_{14}$ in 50 mM Tris–HCl, pH 8.0, 100 mM KCl, 10 mM MgCl$_2$, 1 mM DTT at 30° in 38-μl reactions for 30 min. A 2-μl aliquot of 20× ATP regeneration mix [50 mM creatine phosphate, 1 mg/ml rabbit muscle creatine kinase (both from Sigma-Aldrich), and 80 mM ATP] is added. The reaction is initiated with addition of ClpX$_6$ to 0.3 μM. At this point, a 20-μl aliquot of the reaction is removed and quenched with 5 μl 4× SDS–PAGE loading buffer as a $t = 0$ control. The quenched reaction is flash frozen in liquid N$_2$ and stored at −20° until analysis. The remaining reaction is incubated at 30° for a given time, at which point it is quenched and frozen as described earlier. Typically, the reaction is complete after 2 h.

Samples are boiled 2 to 3 min and analyzed by 14% SDS–PAGE. Due to the small molecular weight difference between UmuD and UmuD′, the gel must be run to completion to resolve both species. The gel is stained with SYPRO Orange (Molecular Probes), and UmuD/UmuD' are visualized and quantitated by phosphorimager and ImageQuant software (GE Healthcare). It is possible that UmuD bands will become fainter over the time course of the reaction as well, as ClpXP will degrade one UmuD in the context of the homodimer in the absence of the UmuD'D heterodimer (Fig. 2) (Neher et al., 2003).

Using Intrinsic Tryptophan Fluorescence to Determine Binding Constants between UmuD Proteins and Their Interaction Partners

Dimeric forms of the UmuD proteins interact with a variety of other E. coli proteins (Gonzalez et al., 2000; Jonczyk and Nowicka, 1996; Mustard and Little, 2000; Schlacher et al., 2005; Sutton et al., 1999, 2001b,c; Tang et al., 1998, 1999; Woodgate et al., 1989). Although the interactions made with $UmuD_2$ or with $UmuD'_2$ largely overlap, some differ in magnitude (Sutton et al., 1999, 2002) or result in different effects on the interacting protein. Namely, $UmuD'_2$ activates UmuC as a translesion polymerase (Reuven et al., 1998, 1999; Tang et al., 1998, 1999), while $UmuD_2C$ is involved in a DNA damage checkpoint (Opperman et al., 1999; Sutton and Walker, 2001). In an effort to gain insight into the myriad roles that the different forms of the UmuD protein play in regulation of the E. coli SOS response, it is important to quantitate the affinity of the dimeric UmuD proteins for their interactors.

Because UmuD has no tryptophan, binding of UmuD proteins to most other proteins will not contribute significantly to the overall emission spectrum of the complex. Samples are prepared in 50 mM HEPES, pH 7.5, 0.1 mM EDTA, 100 mM NaCl. Unless there is danger of aberrant disulfide bond formation, no reductant is added due to the intrinsic fluorescence of DTT and the high volatility of β-ME. To measure the effect of UmuD binding on the intrinsic tryptophan fluorescence of other proteins, the tryptophan-containing protein should be kept at a convenient concentration (on the order of 5 μM), while the concentration of the interacting UmuD protein can be varied from 0 to 200 μM or more, if necessary. A separate sample of UmuD at the same concentration is made in the absence of an interactor. The fluorescence of UmuD alone is subtracted from the spectrum of the combined proteins to eliminate noise from the aromatic residues in UmuD. For each binding curve, the fluorescence from a sample of the interacting partner in the absence of UmuD is made as a starting point.

Samples with more than one protein present are preincubated at room temperature for 2 h prior to data collection in order to allow the protein complex to reach equilibrium. The samples are excited at either 278 nm for measurement of all aromatic residues or at 295 nm to selectively excite tryptophan. Emission is measured from 300 to 400 nm. To ensure that the complex has reached equilibrium, the emission spectrum is acquired at additional 1- to 2-h intervals until no change in the spectrum is observed.

The center of spectral mass $\Sigma(\lambda \times I)_i / \Sigma I_i$ is calculated for each sample, where λ is the wavelength in nanometers and I is the intensity in arbitrary units for each step i. The center of spectral mass is plotted vs [UmuD], and the plot is fit to $\lambda_{obs} = \lambda_0 \pm [\lambda_1 [(C_0 + D + K_D) - ((C_0 + D + K_D)^2 - 4C_0 \times D)^{0.5}]]/(2C_0)$ (Segel, 1975), where λ_{obs} is the variable center of spectral mass at UmuD protein concentration D, with λ_0 as λ_{obs} at $D = 0$, and C_0 is the fixed concentration of interacting protein.

If UmuD'D is tested, an additional step is required to generate UmuD'D from purified UmuD and UmuD'. The two proteins are mixed in equimolar concentrations and are allowed to equilibrate for 30 min at room temperature before addition of the interacting protein.

Characterization of *Escherichia coli* Y Family Polymerases *In Vitro* and *In Vivo*

Preparation of Native DinB

A DinB expression plasmid was constructed by cloning the *dinB* gene into pET11T (Nguyen *et al.*, 1993) using standard cloning procedures (Sambrook *et al.*, 1989). The protein can be purified from *E. coli* BL21 (DE3) pLysS (Novagen). BL21(DE3) cells (1–3 liters of culture) harboring the DinB expression plasmid are grown at 37° in LB medium to an OD_{600} of 0.6. Cultures are then transferred to 30° and grown to an OD_{600} of 0.8 before adding IPTG to a final concentration of 1 mM. After 3–4 h of induction, cells are harvested by centrifugation at 6000g for 10 min. The cell pellets can be frozen directly and stored at $-80°$ for up to several months.

To purify DinB, the cell pellets are thawed on ice and resuspended in 3–5 ml/g wet cell paste of lysis buffer [50 mM HEPES, pH 7.5, 300 mM NaCl, 7.5 mM MgCl$_2$, 2 mM β-ME, 10 μg/ml phenylmethylsulfonyl fluoride (PMSF), and 5% glycerol] and one protease inhibitor cocktail tablet (Roche Molecular Biochemicals) and then homogenized by one to two brief sonication pulses of 15 s each at 50% power. Lysis is accomplished by treatment with lysozyme at a final concentration of 0.3 mg/ml for 30 min on ice, followed by one freeze–thaw cycle. The lysate is cleared by centrifugation at 14,000g for 1 h at 4° and then treated with deoxyribonuclease I (1 μg/ml; Sigma-Aldrich) for 30 min. Following filtration through a 0.2-μm membrane, the cleared lysate is diluted twofold in S$_A$ (50 mM HEPES, pH

7.5, 2 mM β-ME, 10 μg/ml PMSF) and loaded onto a HiTrap fast flow Mono S column (GE Healthcare). After washing the column with 5 column volumes of S_A, DinB is eluted with a gradient of 0–100% S_B (50 mM HEPES, pH 7.5, 1 M NaCl, 2 mM β-ME, 10 μg/ml PMSF) over 10 column volumes. DinB-containing fractions are identified by SDS–PAGE, pooled, diluted 1:1 with P_A [50 mM HEPES, pH 7.5, 2 mM β-ME, 1 M (NH$_4$)$_2$SO$_4$], and loaded onto a HiTrap fast flow low substitution phenyl-Sepharose column (GE Healthcare). The column is washed with 5 column volumes of P_A, and DinB is eluted with a gradient of 0–100% S_A over 10 column volumes. DinB-containing fractions are again identified by SDS–PAGE, pooled, and dialyzed against a 1000-fold excess of S_A containing 10% glycerol for 4 h, with at least one exchange of buffer. Each liter of induced BL21(DE3) will yield approximately 30 mg of protein.

DinB-Dependent Bypass of N^2-dG Adducts In Vitro

DinB has been shown to bypass an N^2-dG-benzo[a]pyrene adduct *in vitro* with modest efficiency (Shen *et al.*, 2002). Lesion bypass assays are performed with purified wild-type or mutant DinB using established procedures (Creighton *et al.*, 1995) with some minor modifications. Running start reactions are used to address the intrinsic ability of DinB to bypass a given lesion *in vitro*, whereas standing start reactions are used to address the fidelity of dNTP insertion opposite to the lesion. Extension from the lesion is addressed using standing start experiments in which the base at the 3′ terminus of the primer is varied to assess the ability of DinB to insert the next nucleotide correctly. Running start reactions typically contain 2 nM DinB and 10 nM ^{32}P-labeled primer/template in 30 mM HEPES, pH 7.5, 100 mM KCl, 7.5 mM MgSO$_4$, 2 mM β-ME, 1% BSA, and 4% glycerol and are initiated with the addition of dNTPs to a final concentration of 250 $\mu$$M$ each. Reactions are performed in a 10-μl final volume and incubated at 37° for 10 min prior to quenching with 95% formamide, 50 mM EDTA, 0.025% xylene cyanol, and 0.025% bromphenol blue. Time points should be taken throughout the time course to ensure that the reaction remains linear. Standing start reactions are carried out in a similar fashion, except that they are initiated with the addition of various concentrations (0–2 mM) of a single nucleotide. Again, linearity of product formation must be established over the course of the reaction. Products from the quenched reactions are analyzed on a denaturing polyacrylamide gel (12–15%). The gel is then dried and imaged on a storage phosphor screen (Molecular Dynamics) from which the products are quantified on a phosphorimager using ImageQuant software (GE Healthcare). V_{max} and K_m values are determined by plotting V_0 vs [dNTP], and only reactions in which there is no more than 20% loss of oligonucleotide substrate are included in the analysis to ensure that experiments are conducted under saturating conditions.

Assays of Mutagenesis and Survival Phenotypes Due to Y Family Polymerases In Vivo

UV-Induced Mutagenesis and Survival Attributed to umuDC. E. coli DNA pol V causes both lesion-targeted and untargeted mutagenesis (Tang et al., 2000). Because pol V inserts guanine opposite the 3′-thymine of [6-4] T–T photoproducts, the mutagenic signature of *umuC* function is usually considered transition mutations (Tang et al., 2000). Thus, *umuC* function is responsible for most UV-induced mutagenesis. G→T transversions at template N^2-dG-(+)-*trans-anti*-benzo[a]pyrene sites in a specific sequence context are also *umuDC* dependent (Yin et al., 2004). Therefore, *umuC* function can be inferred by the detection of mutations in specific reporter genes. The reversion of nonfunctional *argE3* and *hisG4* alleles from auxotrophy to prototrophy is commonly used in such an assay. Arginine auxotrophy of strains harboring the *argE3* allele is due to the Gln388ochre (CAA → TAA) mutation (Kato and Shinoura, 1977; Meinnel et al., 1992). The reversion of the *argE3* allele can occur via a number of mutations, such as by direct back mutation of the TAA codon to CAA or by transition or transversion mutations in any of several tRNA genes that would lead to suppressor tRNAs (Mironova et al., 2005; Sledziewska-Gojska et al., 1992). *umuC*-dependent reversion is likely to occur in the *argE3* allele, as the sequence context of the ochre mutation is ATT-TAA, and UmuD′$_2$C inserts guanine opposite the 3′-thymine of T–T [6-4] photoproducts (Meinnel et al., 1992; Tang et al., 2000). The *hisG4* allele also harbors an ochre mutation, but fewer of the possible suppressor tRNAs are found to actually suppress the mutation (Eggertsson and Adelberg, 1965; Sledziewska-Gojska et al., 1992). Thus, the *hisG4* allele is a more selective monitor of mutagenesis, whereas the *argE3* allele provides more sensitivity due to the higher number of tRNA gene mutations that can suppress the ochre codon (Sledziewska-Gojska et al., 1992). The reversion of rifampicin[S] to Rif[R] can also be assayed in Rif[S] strains, such as the widely used AB1157 and related strains (Jin and Gross, 1988). Rif[R] mutations can arise from base substitutions at various sites in the *rpoB* gene (Jin and Gross, 1988). Typically, complementation of the Δ*umuC* phenotype is assayed by expressing plasmid-borne UmuC, UmuD, or UmuD′ variants in a Δ*umuDC* strain (Elledge and Walker, 1983), although chromosomal *umuD* and *umuC* variants have also been assayed (Elledge and Walker, 1983).

To assay the *umuC*-dependent mutagenesis phenotype using reversion of the *argE3* allele, freshly saturated overnight cultures are subcultured 1:50 into fresh LB medium (with appropriate antibiotics when using plasmid-bearing strains). After the culture reaches an OD_{600} of 0.2–0.3, 7.5×10^9 cells are harvested by centrifugation at 5000g for 10 min. Cells are washed twice with 1 ml of 0.85% saline and resuspended in 1.5 ml 0.85% saline. A 700-μl aliquot of the resuspended cells is transferred to each of two 15 × 60-mm petri dishes and irradiated with 25 J/m^2 of UV from a 254-nm lamp. To assay *umuC*-dependent mutagenesis or cell survival as a function of UV dose, all 1.5 ml of cells are transferred to one petri dish and 300-μl aliquots are removed at designated UV exposures, including time $t=0$. Induced mutagenesis is determined by plating 100- to 200-μl aliquots of irradiated cells on M9 minimal plates with trace arginine (1 μg/ml) (1× M9 salts [Sambrook *et al.*, 1989], 0.2% glucose, 2 mM MgSO$_4$, 0.1 mM CaCl$_2$, 0.0005% thiamine, 40 μg/ml of each Thr, Leu, Ile, Val, Pro, His, and 1.5% agar, and appropriate antibiotics). The trace amount of arginine is necessary to allow the bacteria to grow enough (one generation) to accumulate the mutations that lead to the Arg^+ phenotype. Spontaneous mutagenesis is determined by plating an equivalent amount of untreated cells. Colony-forming units are scored after 48 h of growth at 37°. Survival is determined by plating either 10-μl spots of serial dilutions made in 0.85% saline or a suitable dilution of the cells to give 100–200 colonies per plate on equivalent M9 minimal plates with 40 μg/ml arginine. Non-UV-irradiated cultures should be treated identically to assess the spontaneous mutation frequency. *hisG4* reversion is assayed similarly, using trace histidine (1 μg/ml) plates. Typically, multiple trials (two to three) of each of three independent cultures are performed.

To determine mutation frequencies by reversion to Rif^R in liquid cultures, cells are centrifuged at 5000g for 10 min and resuspended in 1× M9 salts (Miller, 1972) before plating on LB versus LB supplemented with 100 μg/ml rifampicin. Frequency of mutation to rifampicin resistance ranges from 10^{-7} to 10^{-9}. Because rifampicin resistance can arise from a variety of specific mutations in the *rpoB* gene (Jin and Gross, 1988), sequencing the *rpoB* gene of the Rif^R mutants can be employed to investigate the spectrum of induced or spontaneous mutagenesis (Wolff *et al.*, 2004). The use of specific *lac*⁻ alleles that become Lac$^+$ each by a particular reversion allows for more detailed insights into *dinB*- or *umuC*-dependent mutation spectra (Table I) (Cupples and Miller, 1989; Cupples *et al.*, 1990).

Chemical-Induced Mutagenesis and Survival Attributed to DinB and UmuC. The role of Y family polymerases in mutagenesis and killing due to treatment with DNA damaging agents can be assayed by comparing the sensitivities of wild-type and deletion strains to these agents. For example,

TABLE I
COLLECTION OF STRAINS BEARING REVERTIBLE *LAC*⁻ ALLELES[a]

Strain name	Reversion
CC101	AT→CG[b]
CC102	GC→AT[b]
CC103	GC→CG[b]
CC104	GC→TA[b]
CC105	AT→TA[b]
CC106	AT→GC[b]
CC107	+1G[c]
CC108	−1G[c]
CC109	−2CG[c]
CC110	+1A[c]
CC111	−1A[c]

[a] The specific mutation needed to become Lac⁺ is shown in the right-hand column.
[b] Information about this strain and the reversion specificities was compiled from Cupples *et al.* (1990).
[c] Information about this strain and reversion specificities was compiled from Cupples and Miller (1989).

the bypass of N^2-dG adducts *in vivo* can be assayed using nitrofurazone (NFZ) or 4-nitroquinoline-1-oxide (4-NQO) (Jarosz *et al.*, 2006). The *E. coli* strain AB1157 (DeWitt and Adelberg, 1962) and its Δ*dinB* or Δ*umuC* derivatives are grown overnight, subcultured with a 1:1000 dilution, and grown to fresh saturation. Treatment with DNA-damaging agents can be performed on plates or in liquid. Because NFZ and 4-NQO decompose in solution, stocks (1 mg/ml in *N,N*-dimethylformamide) are made immediately prior to use and can be stored at −20°, shielded from light, for not more than 24 h. For treatment on plates, the NFZ or 4-NQO stock is diluted directly into media to an appropriate concentration (0–5 μg/ml final concentration for each agent). A number of cells of each strain is then deposited on the plate such that approximately 200 colonies can be observed the following day. Complementation of the sensitivity can be observed by the addition of *dinB* or *umuC* under its native promoter on low-copy number plasmids (Kim *et al.*, 1997; Sommer *et al.*, 1993), with appropriate antibiotics for plasmid maintenance. Through complementation, various mutants of *dinB* or *umuC* can be tested for their ability to restore parental-level resistance to various treatments (Kim *et al.*, 1997; Sommer *et al.*, 1993). Typically, survival will vary by one to three orders of magnitude, depending on the dose of chemical used. For treatment in liquid, freshly saturated cultures are diluted 1:1000 into LB containing NFZ or 4-NQO (0–5 μg/ml final concentration) and treated in the dark at 37° with shaking at 200 rpm for 6 h. The cultures are then diluted and plated on LB agar for determination of colony-forming units. Complementation in liquid is also performed with a low-copy plasmid harboring the *dinB* or *umuC* gene. Chemically induced mutagenesis can then be assayed as described earlier with *argE3*, *hisG4*, Rif[S], or the *lac*[−] alleles.

Assay for the Rate of Nascent DNA Synthesis by Thymine Incorporation

The Y family polymerases in *E. coli* function both in translesion synthesis and in a DNA replication checkpoint. Additionally, a decrease in the rate of DNA replication can be caused by treatment with specific chemicals or DNA-damaging agents. The rate of DNA synthesis *in vivo* is assayed by measuring the incorporation of [³H] thymidine (Dupont, stable for approximately 3 months) or [³H] thymine (Moravek, also stable for approximately 3 months). This assay takes advantage of the fact that thymine does not arise exclusively through the action of ribonucleotide reductase, but can also be formed from reductive methylation of dUMP at C5 by thymidylate synthase (*thyA*) to give dTMP (Neuhard and Kelln, 1996). The DNA of wild-type strains (*thyA*[+]) can be labeled with [³H]thymidine by the addition of deoxyadenosine or deoxyguanine (50 µg/ml) (O'Donovan, 1978). To facilitate incorporation of thymidine, a P90C derivative (Cairns and Foster, 1991) has been made *thyA*[−], thus requiring thymidine for growth, by transduction of the allele from the strain EGSC#6827 (*E. coli* Genetic Stock Center). In general, thymidine is a better choice than thymine for labeling nascent DNA because cells uptake it well and it is readily incorporated into DNA. Thymine, however, is better suited for detection of global DNA synthesis because its incorporation requires accumulation of deoxyribose-1-phosphate in order to drive the generation of thymidine and thereby dTTP (O'Donovan, 1978). This is accomplished in cells lacking the *thyA* gene by accumulation of dUMP (Neuhard and Kelln, 1996). Care should be taken when using strain AB1157 and its derivatives because they are *tsx* (nucleoside transporter) mutants (Neuhard and Kelln, 1996), which uptake thymidine poorly at concentrations lower than 1 µg/ml even in a *thyA*[−] strain (O'Donovan, 1978).

Typically, 1 µCi/ml of [³H] thymidine is used for labeling cells with and without treatment to investigate whether there is still active DNA replication in wild-type cells that are sensitive to treatment. Cells are grown to saturation in M9GC [M9 medium with 0.1–0.2% glucose, 0.3% casaminoacids (Difco)] plus 50 µg/ml thymidine. The cultures are then diluted 1:1000 in 50% M9GC, 50% LB, plus 10 µg/ml of thymidine. Immediately following inoculation, cultures are treated with the DNA-damaging agent. Aliquots of 0.5 ml diluted cells (in triplicate) are labeled with 1 µCi/ml of thymidine for 10 min at 37° before precipitation with an equal volume of 20% cold trichloroacetic acid (TCA) for at least 20 min on ice. After TCA precipitation, samples are filtered through a glass filter (Millipore) and washed thoroughly with approximately 30 ml 5% TCA followed by a final wash with about 10 ml of 100% ethanol. The dried filters are deposited in scintillation vials to which 5 ml of scintillation fluid (National Diagnostics) are added. It is important for the filters to be completely dry before counting because detection of the weak γ radiation of ^{3}H can easily be hindered by water. The vials are well shaken to optimize detection by scintillation counting (Beckman). For this experiment, time points are typically taken every 30 min for 5 h. Concomitantly, samples are also taken to determine viability.

Role of DinB in Damage-Independent DNA Replication Stalling

Hydroxyurea (HU) stalls DNA replication by inhibiting class I ribonucleotide reductases (Gerez *et al.*, 1997). Ribonucleotide reductases are the only known enzymes capable of reducing ribonucleotides to deoxyribonucleotides (Stubbe and Riggs-Gelasco, 1998), which represents the rate-determining step in DNA biosynthesis in most organisms. It is thought that DNA replication stalling mediated by HU is brought about by dNTP starvation. The Y family polymerases may play a role in resistance to HU treatment related to their roles in a DNA damage checkpoint.

The HU treatment is carried out by preparing a fresh 1 *M* stock solution of HU (Calbiochem) in double-distilled sterile water, which is filter sterilized using a 0.2-μm filter (Millipore). The strains to be tested are grown to saturation in LB (Miller, 1974) at 37° and then diluted 1:1000 in LB with and without HU (100 m*M* final concentration). These cultures are then incubated for 6 h at 37°, making sure that there is sufficient aeration for aerobic growth at all times. To obtain a time course of colony-forming units during HU treatment, samples are taken every hour, diluted serially, and plated on rich medium. Colonies are counted after overnight incubation at 37°.

A similar procedure involving the addition of HU to solid LB medium is used to determine the sensitivity of numerous derivatives in a high-throughput fashion. This is a particularly useful procedure when screening large numbers of *dinB* or other mutants. Twelve strains can be tested at the same time by serial dilution in a 96-well plate, plating 10 μl of each dilution on large (15 × 150 mm) LB agar plates with HU. The kinetics of survival or killing under HU treatment are then obtained as described earlier. The advantage of using the solid media approach is that many different HU concentrations can be assessed for up to 12 different strains or isolates whose sensitivity to HU may vary. It is also helpful to spot plate the serially diluted derivatives to determine the proper dilution to measure the kinetics of survival or killing.

Assay for umuDC-Dependent Cold-Sensitive Phenotype and Its Use in Genetic Selections

Elevated levels of *umuDC* gene products cause a cold-sensitive growth phenotype that correlates with the inhibition of the rate of DNA synthesis (Opperman *et al.*, 1999; Sutton and Walker, 2001). Thus, the cold-sensitive phenotype can be used to infer the presence of the DNA damage checkpoint. This is assayed most commonly by a quantitative transformation assay, but also can be assayed by growth curves (Murli *et al.*, 2000; Opperman *et al.*, 1996). The cold-sensitive phenotype requires elevated levels of the *umuDC* gene products. Chromosomal levels are insufficient to observe the phenotype; however, constitutive expression of *umuDC* genes from low-copy plasmids is sufficient (Opperman *et al.*, 1996). In the quantitative transformation assay, equal amounts of each plasmid to be tested (0.1 μg) are added to a 25-μl aliquot of competent cells (Sambrook *et al.*, 1989) and incubated on ice for 10 min. After a 5-min heat shock at 37 ° and a further 10-min incubation on ice, transformation mixtures are recovered in 750 μl LB at 37° for 1.5–2 h with gentle shaking. Equal volumes are plated on LB plates containing the appropriate antibiotics for incubation under different temperatures, typically 30 versus 37 or 42°. After 24–36 h, colonies are counted to determine the surviving fraction at the nonpermissive temperature. The cold-sensitive phenotype is typically manifested as a 100- to 1000-fold or greater difference in the transformation efficiency of *umuDC*-expressing plasmids at 30° versus 37°, depending on the plasmid copy number and expression level. Higher expression levels of the *umuDC* gene products lead to a more extreme cold-sensitive phenotype (Opperman *et al.*, 1996). An empty vector control should yield a ratio of colony-forming units (cfus) at 30° versus 37° of approximately one.

The cold-sensitive phenotype can be used to select for *umuDC* variants that fail to exhibit it and therefore presumably fail to function in the DNA damage checkpoint. Moreover, strains bearing these variants sometimes fail to act in induced mutagenesis or may separate these two functions of the *umuDC* gene products (Marsh *et al.*, 1991). Because the cold-sensitive phenotype seems to require an interaction of UmuC with the *umuD* gene products, as well as with other proteins, this phenotype provides an efficient genetic selection to isolate mutants and map protein–protein interactions (Sutton *et al.*, 2001b).

Conclusions

We have illustrated common techniques for *in vivo* and *in vitro* analysis of Y family polymerases from *E. coli*. The experiments described here lend insight into the functions of these specialized polymerases and their accessory factors in cellular survival and mutagenesis. Major unanswered questions about these polymerases include detailed methods of regulation, especially temporal control of activity and polymerase switching according to specific lesions. Further characterization will undoubtedly continue to address these questions and reveal the properties of the Y family polymerases and their accessory factors.

Acknowledgments

This work was supported by Grant CA21615 from the National Cancer Institute to G.C.W. Additional support was received from the Center for Environmental Health Sciences at MIT, which is supported by NIEHS Center Grant P30ES02109. P.J.B. was supported by a fellowship from the Damon Runyon Cancer Research Foundation. S.M.S. was supported by a Cleo and Paul Schimmel Fellowship. Additional support was received from the American Cancer Society and the Center for Environmental Health Studies at MIT, which is supported by NIEHS Center Grant Number P30ES02109. We thank members of the Walker Lab for helpful comments on the manuscript.

References

Battista, J. R., Ohta, T., Nohmi, T., Sun, W., and Walker, G. C. (1990). Dominant negative *umuD* mutations decreasing RecA-mediated cleavage suggest roles for intact UmuD in modulation of SOS mutagenesis. *Proc. Natl. Acad. Sci. USA* **87,** 7190–7194.

Becherel, O. J., Fuchs, R. P. P., and Wagner, J. (2002). Pivotal role of the β-clamp in translesion DNA synthesis and mutagenesis in *E. coli* cells. *DNA Repair* **1,** 703–708.

Beuning, P. J., Barsky, D., Sawicka, D., and Walker, G. C. (2006). Two processivity clamp interactions differentially alter the dual activities of UmuC. *Molecular Microbiolog.* **59,** 460–474.

Bunting, K. A., Roe, S. M., and Pearl, L. H. (2003). Structural basis for recruitment of translesion DNA polymerase Pol IV/DinB to the β-clamp. *EMBO J.* **22,** 5883–5892.

Burckhardt, S. E., Woodgate, R., Scheuermann, R. H., and Echols, H. (1988). UmuD mutagenesis protein of *Escherichia coli*: Overproduction, purification, and cleavage by RecA. *Proc. Natl. Acad. Sci. USA* **85,** 1811–1815.

Cairns, J., and Foster, P. L. (1991). Adaptive reversion of a frameshift mutation in *Escherichia coli*. *Genetics* **128,** 695–701.

Courcelle, J., Khodursky, A., Peter, B., Brown, P. O., and Hanawalt, P. C. (2001). Comparative gene expression profiles following UV exposure in wild-type and SOS-deficient *Escherichia coli*. *Genetics* **158,** 41–64.

Creighton, S., Bloom, L. B., and Goodman, M. F. (1995). Gel fidelity assay measuring nucleotide misinsertion, exonucleolytic proofreading, and lesion bypass efficiencies. *Methods Enzymol.* **262,** 232–256.

Cupples, C. G., Cabrera, M., Cruz, C., and Miller, J. H. (1990). A set of *lacZ* mutations in *Escherichia coli* that allow rapid detection of specific frameshift mutations. *Genetics* **125,** 275–280.

Cupples, C. G., and Miller, J. H. (1989). A set of *lacZ* mutations in *Escherichia coli* that allow rapid detection of each of the six base substitutions. *Proc. Natl. Acad. Sci. USA* **86,** 5345–5349.

DeWitt, S. K., and Adelberg, E. A. (1962). The occurrence of a genetic transposition in a strain of *Escherichia coli. Genetics* **47,** 577–585.

Eggertsson, G., and Adelberg, E. A. (1965). Map positions and specificities of suppressor mutations in *Escherichia coli* K-12. *Genetics* **52,** 319–340.

Elledge, S. J., and Walker, G. C. (1983). Proteins required for ultraviolet light and chemical mutagenesis: Identification of the products of the *umuC* locus of *Escherichia coli. J. Mol. Biol.* **164,** 175–192.

Ferentz, A. E., Opperman, T., Walker, G. C., and Wagner, G. (1997). Dimerization of the UmuD′ protein in solution and its implications for regulation of SOS mutagenesis. *Nature Struct. Biol.* **4,** 979–983.

Ferentz, A. E., Walker, G. C., and Wagner, G. (2001). Converting a DNA damage checkpoint effector (UmuD$_2$C) into a lesion bypass polymerase (UmuD′$_2$C). *EMBO J.* **20,** 4287–4298.

Friedberg, E. C., Lehmann, A. R., and Fuchs, R. P. (2005a). Trading places: How do DNA polymerases switch during translesion DNA synthesis? *Mol. Cell* **18,** 499–505.

Friedberg, E. C., Walker, G. C., Siede, W., Wood, R. D., Schultz, R. A., and Ellenberger, T. (2005b). "DNA Repair and Mutagenesis." ASM Press, Washington, DC.

Fuchs, R. P., Fujii, S., and Wagner, J. (2004). Properties and functions of *Escherichia coli*: pol IV and pol V. *Adv. Protein Chem.* **69,** 229–264.

Fujii, S., Gasser, V., and Fuchs, R. P. (2004). The biochemical requirements of DNA polymerase v-mediated translesion synthesis revisited. *J. Mol. Biol.* **341,** 405–417.

Gerez, C., Elleingand, E., Kauppi, B., Eklund, H., and Fontecave, M. (1997). Reactivity of the tyrosyl radical of *Escherichia coli* ribonucleotide reductase: Control by the protein. *Eur. J. Biochem.* **249,** 401–407.

Gonzalez, M., Rasulova, F., Maurizi, M. R., and Woodgate, R. (2000). Subunit-specific degradation of the UmuD/D′ heterodimer by the ClpXP protease: The role of trans recognition in UmuD′ stability. *EMBO J.* **19,** 5251–5258.

Goodman, M. F. (2002). Error-prone repair DNA polymerases in prokaryotes and eukaryotes. *Annu. Rev. Biochem.* **71,** 17–50.

Guzzo, A., Lee, M. H., Oda, K., and Walker, G. C. (1996). Analysis of the region between amino acids 30 and 42 of intact UmuD by a monocysteine approach. *J. Bacteriol.* **178,** 7295–7303.

Jarosz, D. F., Godoy, V. G., Delaney, J. C., Essigmann, J. M., and Walker, G. C. (2005). A single residue in DNA polymerase IV effects enhanced activity on N2-dG damaged template. *Nature,* **439,** 225–228.

Jin, D. J., and Gross, C. A. (1988). Mapping and sequencing of mutations in the *Escherichia coli rpoB* gene that lead to rifampicin resistance. *J. Mol. Biol.* **202,** 45–58.

Jonczyk, P., and Nowicka, A. (1996). Specific *in vivo* protein–protein interactions between *Escherichia coli* SOS mutagenesis proteins. *J. Bacteriol.* **178,** 2580–2585.

Kato, T., and Shinoura, Y. (1977). Isolation and characterization of mutants of *Escherichia coli* deficient in induction of mutations by ultraviolet light. *Mol. Gen. Genet.* **156,** 121–131.

Kim, S.-R., Maenhaut-Michel, G., Yamada, M., Yamamoto, Y., Matsui, K., Sofuni, T., Nohmi, T., and Ohmori, H. (1997). Multiple pathways for SOS-induced mutagenesis in *Escherichia coli*: An overexpression of *dinB/dinP* results in strongly enhancing mutagenesis in the absence of any exogenous treatment to damage DNA. *Proc. Natl. Acad. Sci. USA* **94,** 13792–13797.

Kim, S.-R., Matsui, K., Yamada, M., Gruz, P., and Nohmi, T. (2001). Roles of chromosomal and episomal *dinB* genes encoding DNA pol IV in targeted and untargeted mutagenesis in *Escherichia coli. Mol. Genet. Genom.* **266,** 207–215.

Kim, Y. I., Burton, R. E., Burton, B. M., Sauer, R. T., and Baker, T. A. (2000). Dynamics of substrate denaturation and translocation by the ClpXP degradation machine. *Mol. Cell* **5,** 639–648.

Kitagawa, Y., Akaboshi, E., Shinagawa, H., Horii, T., Ogawa, H., and Kato, T. (1985). Structural analysis of the *umu* operon required for inducible mutagenesis in *Escherichia coli. Proc. Natl. Acad. Sci. USA* **82**, 4336–4340.

Kobayashi, S., Valentine, M. R., Pham, P., O'Donnell, M., and Goodman, M. F. (2002). Fidelity of *Escherichia coli* DNA polymerase IV: Preferential generation of small deletion mutations by dNTP-stabilized misalignment. *J. Biol. Chem.* **277**, 34198–34207.

Kornberg, A., and Baker, T. A. (1992). "DNA Replication." Freeman, New York.

Kulaeva, O. I., Wootton, J. C., Levine, A. S., and Woodgate, R. (1995). Characterization of the *umu*-complementing operon from R391. *J. Bacteriol.* **177**, 2737–2743.

Lee, M. H., Guzzo, A., and Walker, G. C. (1996). Inhibition of RecA-mediated cleavage in covalent dimers of UmuD. *J. Bacteriol.* **178**, 7304–7307.

Lee, M. H., Ohta, T., and Walker, G. C. (1994). A monocysteine approach for probing the structure and interactions of the UmuD protein. *J. Bacteriol.* **176**, 4825–4837.

Lenne-Samuel, N., Janel-Bintz, R., Kolbanovsky, A., Geacintov, N. E., and Fuchs, R. P. P. (2000). The processing of benzo(a)pyrene adduct into a frameshift or a base substitution mutation requires a different set of genes in *Escherichia coli. Mol. Microbiol.* **38**, 299–307.

Little, J. W. (1984). Autodigestion of LexA and phage lambda repressors. *Proc. Natl. Acad. Sci. USA* **81**, 1375–1379.

Maor-Shoshani, A., Hayashi, K., Ohmori, H., and Livneh, Z. (2003). Analysis of translesion replication across an abasic site by DNA polymerase IV of *Escherichia coli. DNA Repair* **2**, 1227–1238.

Marsh, L., and Walker, G. C. (1985). Cold sensitivity induced by overproduction of UmuDC in *Escherichia coli. J. Bacteriol.* **162**, 155–161.

Marsh, L., Nohmi, T., Hinton, S., and Walker, G. C. (1991). New mutations in cloned *Escherichia coli umuDC* genes: Novel phenotypes of strains carrying a *umuC125* plasmid. *Mutat. Res.* **250**, 183–197.

McDonald, J. P., Maury, E. E., Levine, A. S., and Woodgate, R. (1998). Regulation of UmuD cleavage: Role of the amino-terminal tail. *J. Mol. Biol.* **282**, 721–730.

McKenzie, G. J., Magner, D. B., Lee, P. L., and Rosenberg, S. M. (2003). The *dinB* operon and spontaneous mutagenesis in *Escherichia coli. J. Bacteriol.* **185**, 3972–3977.

Meinnel, T., Schmitt, E., Mechulam, Y., and Blanquet, S. (1992). Structural and biochemical characterization of the *Escherichia coli argE* gene product. *J. Bacteriol.* **174**, 2323–2331.

Miller, J. H. (1972). "Short Course in Bacterial Genetics: A Laboratory Manual and Handbook for *Escherichia coli* and Related Bacteria." Cold Spring Harbor Laboratory Press, Cold Spring Harbor, NY.

Miller, J. H. (1974). "Experiments in Molecular Genetics." Cold Spring Harbor Laboratory Press, Cold Spring Harbor, NY.

Mironova, R., Niwa, T., Handzhiyski, Y., Sredovska, A., and Ivanov, I. (2005). Evidence for non-enzymatic glycosylation of *Escherichia coli* chromosomal DNA. *Mol. Microbiol.* **55**, 1801–1811.

Murli, S., Opperman, T., Smith, B. T., and Walker, G. C. (2000). A role for the *umuDC* gene products of *Escherichia coli* in increasing resistance to DNA damage in stationary phase by inhibiting the transition to exponential growth. *J. Bacteriol.* **182**, 1127–1135.

Mustard, J. A., and Little, J. W. (2000). Analysis of *Escherichia coli* RecA interactions with LexA, lambda cI, and UmuD by site-directed mutagenesis of *recA. J. Bacteriol.* **182**, 1659–1670.

Neher, S. B., Sauer, R. T., and Baker, T. A. (2003). Distinct peptide signals in the UmuD and UmuD' subunits of UmuD/D' mediate tethering and substrate processing by the ClpXP protease. *Proc. Natl. Acad. Sci. USA* **100**, 13219–13224.

Neuhard, J., and Kelln, R. A. (1996). Biosynthesis and vonversion of purines. In "*Escherichia coli* and *Salmonella*" (F. C. Neidhardt, ed.), pp. 580–599. ASM Press, Washington, DC.

Nguyen, L. H., Jensen, D. B., and Burgess, R. R. (1993). Overproduction and purification of σ^{32}, the *Escherichia coli* heat shock transcription factor. *Protein Expr. Purif.* **4,** 425–433.

O'Donovan, G. A. (1978). Thymidine metabolism in bacteria (and how, or how not, to label DNA). In "DNA Synthesis: Present and Future" (I. Molineux and M. Kohiyama, eds.), pp. 219–253. Plenum, New York.

Ohmori, H., Friedberg, E. C., Fuchs, R. P. P., Goodman, M. F., Hanaoka, F., Hinkle, D., Kunkel, T. A., Lawrence, C. W., Livneh, Z., Nohmi, T., Prakash, L., Prakash, S., Todo, T., Walker, G. C., Wang, Z., and Woodgate, R. (2001). The Y-family of DNA polymerases. *Mol. Cell* **8,** 7–8.

Opperman, T., Murli, S., and Walker, G. C. (1996). The genetic requirements for UmuDC-mediated cold sensitivity are distinct from those for SOS mutagenesis. *J. Bacteriol.* **178,** 4400–4411.

Opperman, T., Murli, S., Smith, B. T., and Walker, G. C. (1999). A model for a *umuDC*-dependent prokaryotic DNA damage checkpoint. *Proc. Natl. Acad. Sci. USA* **96,** 9218–9223.

Pages, V., and Fuchs, R. P. (2002). How DNA lesions are turned into mutations within cells? *Oncogene* **21,** 8957–8966.

Reuven, N. B., Tomer, G., and Livneh, Z. (1998). The mutagenesis proteins UmuD' and UmuC prevent lethal frameshifts while increasing base substitution mutations. *Mol. Cell* **2,** 191–199.

Reuven, N. B., Arad, G., Maor-Shoshani, A., and Livneh, Z. (1999). The mutagenesis protein UmuC is a DNA polymerase activated by UmuD', RecA, and SSB and is specialized for translesion replication. *J. Biol. Chem.* **274,** 31763–31766.

Sambrook, J., Fritsch, E. F., and Maniatis, T. (1989). "Molecular Cloning: A Laboratory Manual." Cold Spring Harbor Laboratory Press, Cold Spring Harbor, NY.

Schlacher, K., Leslie, K., Wyman, C., Woodgate, R., Cox, M., and Goodman, M. (2005). DNA polymerase V and RecA protein, a minimal mutasome. *Mol. Cell* **17,** 561–572.

Segel, E. H. (1975). "Enzyme Kinetics: Behavior and Analysis of Rapid Equilibrium and Steady State Enzyme Systems." Wiley-Interscience, New York.

Shen, X., Sayer, J. M., Kroth, H., Ponten, I., O'Donnell, M., Woodgate, R., Jerina, D. M., and Goodman, M. F. (2002). Efficiency and accuracy of SOS-induced DNA polymerases replicating benzo[a]pyrene-7,8-diol 9,10-epoxide A and G adducts. *J. Biol. Chem.* **277,** 5265–5274.

Silhavy, T. J., Berman, M. L., and Enquist, L. W. (1984). "Experiments with Gene Fusions." Cold Spring Harbor Laboratory Press, Cold Spring Harbor, NY.

Sledziewska-Gojska, E., Grzesiuk, E., Plachta, A., and Janion, C. (1992). Mutagenesis of *Escherichia coli*: A method for determining mutagenic specificity by analysis of tRNA suppressors. *Mutagenesis* **7,** 41–46.

Sommer, S., Knezevic, J., Bailone, A., and Devoret, R. (1993). Induction of only one SOS operon, *umuDC*, is required for SOS mutagenesis in *Escherichia coli*. *Mol. Gen. Genet.* **239,** 137–144.

Stubbe, J., and Riggs-Gelasco, P. (1998). Harnessing free radicals: Formation and function of the tyrosyl radical in ribonucleotide reductase. *Trends Biochem. Sci.* **23,** 438–443.

Sutton, M. D., and Walker, G. C. (2001). *umuDC*-mediated cold sensitivity is a manifestation of functions of the UmuD$_2$C complex involved in a DNA damage checkpoint control. *J. Bacteriol.* **183,** 1215–1224.

Sutton, M. D., Opperman, T., and Walker, G. C. (1999). The *Escherichia coli* SOS mutagenesis proteins UmuD and UmuD′ interact physically with the replicative DNA polymerase. *Proc. Natl. Acad. Sci. USA* **96**, 12373–12378.

Sutton, M., Kim, M., and Walker, G. C. (2001a). Genetic and biochemical characterization of a novel *umuD* mutation: Insights into a mechanism for UmuD self-cleavage. *J. Bacteriol.* **183**, 347–357.

Sutton, M. D., Narumi, I., and Walker, G. C. (2002). Posttranslational modification of the *umuD*-encoded subunit of *Escherichia coli* DNA polymerase V regulates its interactions with the β processivity clamp. *Proc. Natl. Acad. Sci. USA* **99**, 5307–5312.

Sutton, M. D., Smith, B. T., Godoy, V. G., and Walker, G. C. (2000). The SOS response: Recent insights into *umuDC*-dependent mutagenesis and DNA damage tolerance. *Annu. Rev. Genet.* **34**, 479–497.

Sutton, M. D., Farrow, M. F., Burton, B. M., and Walker, G. C. (2001b). Genetic interactions between the *Escherichia coli umuDC* gene products and the β processivity clamp of the replicative DNA polymerase. *J. Bacteriol.* **183**, 2897–2909.

Sutton, M. D., Murli, S., Opperman, T., Klein, C., and Walker, G. C. (2001c). *umuDC-dnaQ* interaction and its implications for cell cycle regulation and SOS mutagenesis in *Escherichia coli*. *J. Bacteriol.* **183**, 1085–1089.

Tang, M., Shen, X., Frank, E. G., O'Donnell, M., Woodgate, R., and Goodman, M. F. (1999). UmuD′(2)C is an error-prone DNA polymerase, *Escherichia coli* pol V. *Proc. Natl. Acad. Sci. USA* **96**, 8919–8924.

Tang, M., Pham, P., Shen, X., Taylor, J. S., O'Donnell, M., Woodgate, R., and Goodman, M. F. (2000). Roles of *E. coli* DNA polymerases IV and V in lesion-targeted and untargeted SOS mutagenesis. *Nature* **404**, 1014–1018.

Tang, M., Bruck, I., Eritja, R., Turner, J., Frank, E. G., Woodgate, R., O'Donnell, M., and Goodman, M. F. (1998). Biochemical basis of SOS-induced mutagenesis in *Escherichia coli*: Reconstitution of *in vitro* lesion bypass dependent on the UmuD′$_2$C mutagenic complex and RecA protein. *Proc. Natl. Acad. Sci. USA* **95**, 9755–9760.

Wagner, J., Fujii, S., Gruz, P., Nohmi, T., and Fuchs, R. P. (2000). The β clamp targets DNA polymerase IV to DNA and strongly increases its processivity. *EMBO Rep.* **1**, 484–488.

Wagner, J., Gruz, P., Kim, S. R., Yamada, M., Matsui, K., Fuchs, R. P., and Nohmi, T. (1999). The *dinB* gene encodes a novel *E. coli* DNA polymerase, DNA pol IV, involved in mutagenesis. *Mol. Cell* **4**, 281–286.

Wolff, E., Kim, M., Hu, K., Yang, H., and Miller, J. H. (2004). Polymerases leave fingerprints: Analysis of the mutational spectrum in *Escherichia coli rpoB* to assess the role of polymerase IV in spontaneous mutation. *J. Bacteriol.* **186**, 2900–2905.

Woodgate, R. (1992). Construction of a *umuDC* operon substitution mutation in *Escherichia coli*. *Mutat. Res.* **281**, 221–225.

Woodgate, R., and Ennis, D. G. (1991). Levels of chromosomally encoded Umu proteins and requirements for *in vivo* UmuD cleavage. *Mol. Gen. Genet.* **229**, 10–16.

Woodgate, R., Rajagopalan, M., Lu, C., and Echols, H. (1989). UmuC mutagenesis protein of *Escherichia coli*: Purification and interaction with UmuD and UmuD′. *Proc. Natl. Acad. Sci. USA* **86**, 7301–7305.

Yin, J., Seo, K. Y., and Loechler, E. L. (2004). A role for DNA polymerase V in G->T mutations from the major benzo[a]pyrene N^2-dG adduct when studied in a 5′-TGT sequence in *E. coli*. *DNA Repair* **3**, 323–334.

[21] Measuring the Fidelity of Translesion DNA Synthesis

By Scott D. McCulloch and Thomas A. Kunkel

Abstract

A method is described to measure the fidelity of copying past a DNA lesion in a defined sequence on a synthetic oligonucleotide primer-template. The DNA product is the result of a complete lesion bypass reaction, i.e., containing all four deoxynucleotide triphosphates and requiring both insertion opposite the lesion and multiple extensions from the resulting primer termini containing the lesion. The nascent strand is recovered and hybridized to a gapped region of the *lacZα* complementation gene of the M13mp2 genome. When this DNA is introduced into *Escherichia coli*, errors made during translesion DNA synthesis are detected by M13 plaque colors. Sequencing of DNA from mutant plaques defines the types of errors and permits calculation of error rates for base substitutions, insertions, and deletions. The method is illustrated here for bypass of a *cis–syn* thymine–thymine dimer by human DNA polymerase η. The assay can be used with other lesions in various sequence contexts and with other polymerases with or without accessory proteins.

Introduction

With the relatively recent discovery of numerous highly specialized DNA polymerases (reviewed in Bebenek and Kunkel, 2004) came the realization that the fidelity of DNA synthesis can vary over an extraordinary range (Kunkel, 2004). For example, Y family polymerases implicated in translesion DNA synthesis and X family polymerases implicated in nonhomologous end joining of double strand DNA breaks are much less accurate when copying undamaged DNA *in vitro* than the major replicative DNA polymerases. This variation in the fidelity of DNA synthesis has been established primarily using two methods. One involves monitoring insertion of individual dNTP into oligonucleotide primer-templates to establish the extent of discrimination between correct and incorrect dNTP insertions at individual template bases and the efficiency of correct dNTP insertion from matched or mismatched primer termini. The second method involves copying a region of the wild-type *lacZα* complementation gene in bacteriophage M13mp2 and scoring DNA synthesis errors by reduced blue plaque

METHODS IN ENZYMOLOGY, VOL. 408   0076-6879/06 $35.00
                                                   DOI: 10.1016/S0076-6879(06)08021-9

color phenotype upon transfection of the copied DNA into *Escherichia coli* (Kunkel, 1985). Because the *lacZα* complementation gene is not essential for M13 plaque formation, this forward mutation assay can score hundreds of different errors in numerous sequence contexts, thereby providing a comprehensive view of polymerase error rates and error specificity. The assay has also been adapted to include several "reversion assays" that involve copying templates containing preexisting *lacZ* mutations encoding colorless or light blue plaque phenotypes. With these reversion substrates, subsets of base substitution, insertion, and deletion errors in specific sequence contexts are scored as dark blue plaque revertants. These M13-based assays have been used to determine the fidelity of copying undamaged DNA by DNA-dependent DNA polymerases in families A, B, X, and Y (Kunkel, 2004), the fidelity of bidirectional replication of duplex DNA in human cell extracts (Roberts *et al.*, 1993), the fidelity of copying DNA and RNA templates by reverse transcriptases (Boyer *et al.*, 1996), and the fidelity of the base excision repair reaction acting upon DNA containing uracil using purified proteins (Matsuda *et al.*, 2003). This chapter describes a further adaptation of this methodology to measure the fidelity of translesion DNA synthesis (Kokoska *et al.*, 2003). This approach to measure fidelity with damaged DNA is distinct from kinetic determinations of single nucleotide insertion or mismatch extension in that it permits quantitative determination of polymerase error rates during DNA synthesis to completely bypass defined, site-specific lesions in DNA templates in reactions containing all four dNTPs (Kokoska *et al.*, 2003; McCulloch *et al.*, 2004).

Experimental Outline of Translesion Synthesis Fidelity Assays

The approach (Fig. 1) uses an oligonucleotide primer template with a defined, site-specific lesion. The primer is extended to a full-length product by the polymerase of choice in the presence of all four dNTPs, one of which is radiolabeled. The template is a modified sequence of the N-terminal region of the *lacZα* complementation gene of bacteriophage M13mp2. It contains a TAG amber codon and a *Pst*I recognition site in the primer template duplex region (Matsuda *et al.*, 2003). Following DNA synthesis, the now fully duplex 45-mer is digested with *Pst*I and the resulting 35-mer of the newly synthesized strand, along with fragments slightly longer and shorter to score insertion and deletion errors, is separated from the unlabeled 39-mer template strand by denaturing polyacrylamide gel electrophoresis (dPAGE). The newly synthesized DNA products are then recovered from gel slices by a "crush-and-soak" method (Sambrook *et al.*, 1989). The gel-purified products are annealed to an M13 DNA molecule containing

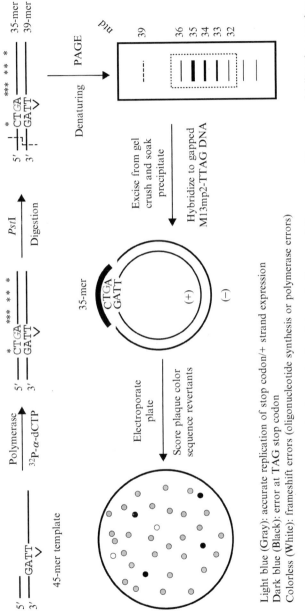

FIG. 1. Measuring the fidelity of lesion bypass *in vitro*. Schematic diagram of the assay used to measure fidelity of a complete lesion bypass reaction. See text for description.

Light blue (Gray): accurate replication of stop codon/+ strand expression
Dark blue (Black): error at TAG stop codon
Colorless (White): frameshift errors (oligonucleotide synthesis or polymerase errors)

a 35 nucleotide gap corresponding to 35 nucleotides of the starting template. The hybridized DNA is introduced into competent *E. coli* cells and plated onto minimal media plates in the presence of the chromogenic indicator X-gal (5-bromo-4-chloro-3-indolyl-β-D-galactoside) and a lawn of *E. coli* α complementation host cells (strain CSH50). Error-free synthesis across the TAG stop codon yields a light blue plaque phenotype due to a lack of α complementation. Eight of nine possible single base substitutions at this codon result in a dark blue plaque phenotype (misinsertion of T opposite the G results in the TAA ochre codon; Matsuda *et al.*, 2003). Sequence analysis of revertant (i.e., dark blue) plaques identifies the responsible substitutions, allowing error rates to be calculated for each of the eight different errors. Fortuitously, slight translational read through of the amber codon results in the light blue plaque phenotype, thus permitting detection of frameshift errors as true colorless plaques. Thus, this approach measures error rates for a complete bypass reaction, encompassing both insertion opposite the lesion and extension past the lesion; during both steps the polymerase has available all four dNTPs in direct competition with each other. Parallel assays with undamaged DNA templates allow comparisons of fidelity when bypassing undamaged versus damaged templates. This method is illustrated using a specific example of human Pol η bypassing a *cis–syn* cyclobutane thymine–thymine dimer (TTD).

Materials

Stock Solutions

> Crush and soak buffer: 0.5 *M* ammonium acetate, 10 m*M* magnesium acetate, 1 m*M* EDTA (pH 8.0), 0.1% SDS. Filter sterilize at 0.2 μm.
> All other stock solutions used have been described in detail previously (Bebenek and Kunkel, 1995; Boyer *et al.*, 1996; Kokoska *et al.*, 2003; Matsuda *et al.*, 2003).

Enzymes and Reagents

Restriction endonucleases *Pst*I and *Eco*RI are obtained from New England Biolabs (Beverly, MA). HPLC-purified deoxyribonucleoside triphosphates (dNTPs) are 100 m*M* solutions from Pharmacia LKB (Piscataway, NJ). Qiagen columns for plasmid purification are from Qiagen Inc. (Chatsworth, CA). Agar, tryptone, and yeast extract are from Difco Labs (Detroit, MI), PEG-8000 and deionized formamide are from Sigma (St. Louis, MO), isopropyl-β-D-thiogalactoside (IPTG) is from BRL (Gaithersburg, MD), X-gal is from Biosynth AG (Switzerland), and phenol-chloroform is from United States Biochemical (Cleveland, OH). Oligonucleotides are from Invitrogen (Carlsbad, CA), and all radioisotopes are from Amersham Corporation (Arlington Heights, IL). The TT dimer containing 45-mer template is a kind gift of Dr. S. Iwai. Human polymerase η is purified as described previously (Masutani *et al.*, 2000).

Bacterial Strains

Escherichia coli strains used for DNA preparation, electroporation, and lawn formation have been described extensively (Bebenek and Kunkel, 1995; Boyer *et al.*, 1996).

Bacteriophage: M13mp2, developed by Messing *et al.* (1977) and later modified by Matsuda *et al.* (2003), is modified further by site-directed mutagenesis using a commercial kit from Stratagene (La Jolla, CA) to create a local sequence context of 5'-...GGT<u>TAG</u>CCT...-3' surrounding the TAG amber stop codon (underlined). The new bacteriophage is named M13mp2-TTAG.

Experimental Procedures

Lesion Bypass DNA Polymerase Assays

Reaction conditions (pH, salt composition and concentration, dNTP concentration, temperature, time) to obtain full-length lesion bypass products should be determined for each polymerase/lesion combination. We have successfully used conditions of large polymerase excess (up to 25-fold) and reaction times up to 60 min. For polymerases with $3' \rightarrow 5'$ exonuclease activity, conditions must be chosen to prevent (if possible) digestion of the DNA. We generally perform "test" reactions with a substrate prepared (as described later) using a 5'-^{32}P end-labeled primer and analyze the products by 12% dPAGE (Fig. 2A). The goal is to find conditions that give full-length products in quantities sufficient for subsequent fidelity determinations. If a full-length product is produced in low quantities, a larger amount of starting substrate can be used. The following reactions are for human polymerase η that efficiently bypasses the lesion of interest, in this case a TTD.

Substrates are constructed by mixing 200 pmol of the 25-mer primer LBP-25 (5'-AATTTCTGCAGGTCGACTCCAAAGG), 300 pmol template DNA 45-mer (5'-CCAGCTCGGTACCGGG*TT*AGCCTTTGGA-GTCGACCTGCAGAAATT; either undamaged or containing a TTD at the italicized site), 1 μl of 1 M Tris–HCl (pH 7.5), 2.5 μl of 20× SSC and dH$_2$O in a total volume of 50 μl, heating to 75° for 5 min, and cooled to

A End-labelled reaction phospho image

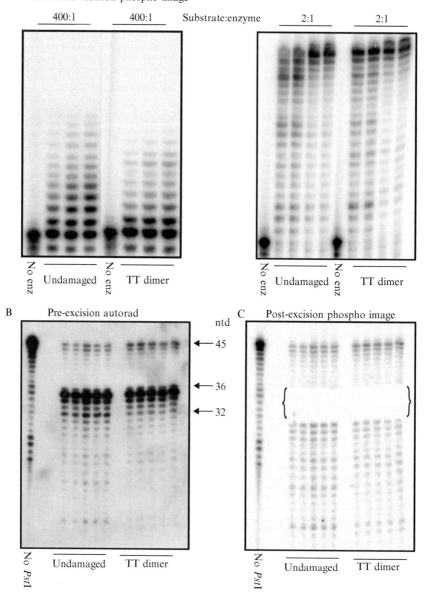

FIG. 2. Denaturing polyacrylamide gel of lesion bypass reaction products. (A) Phospho-image of 5′ ³²P end-labeled reactions products of single interaction (left, 3-, 6-, and 9-min time points; from McCulloch *et al.*, 2004) and complete synthesis (right, 5-, 10-, 15-, and 20-min time points) reactions by human Pol η at the indicated ratios of substrate to enzyme.

room temperature over 3–4 h. The annealing reaction can be left overnight and should be protected from light. The annealed substrate (4 μM) is stored at $-20°$. Lesion bypass assays with Pol η include 4 pmol substrate, 2 pmol polymerase, 40 mM Tris–HCl (pH 8.0), 10 mM MgCl$_2$, 60 mM KCl, 2.5% glycerol, 100 μg/ml bovine serum albumin (BSA), 100 μM each of dATP, dCTP, dGTP, and dTTP, and $\sim$50 μCi ^{32}P-α-dCTP in a total volume of 30 μl. The reaction is incubated for 20 min at $37°$ and is then brought to 25 mM EDTA (2.5$\times$ Mg^{2+} concentration) to stop the reaction. Unincorporated ^{32}P-α-dCTP is removed by passing the reaction mixture through a MicroSpin G25-Sephadex column (Amersham Biosciences, England). The volume is brought to 50 μl with TE, extracted once with 25:24:1 phenol:chloroform:isoamyl alcohol, the organic phase back extracted with 50 μl TE, and the combined aqueous phases ethanol precipitated.

The reaction products are resuspended in 10 μl water, and 10 μl of 2$\times$ PstI digestion mix is added. The digestion mix contains 2 μl of 10$\times$ buffer (as recommended by the manufacturer), 2 μl of 1 mg/ml BSA, 1 μl of 20 U/μl PstI, and 5 μl dH$_2$O. Reactions are incubated at $37°$ for 2 h and are stopped by the addition of an equal volume of formamide loading buffer (95% deionized formamide, 25 mM EDTA, 0.1% bromphenol blue, 0.1% xylene cyanol). Digestion products are heated to $95°$ for 3 min and then separated by 12% dPAGE. We generally use four to six lanes loaded with as much volume as possible. If desired, size markers can be loaded in the gel, although we have found that the PstI digest is never 100% complete (possibly due to the proximity of the recognition sequence to the end of the DNA) so that an internal size marker is present (Fig. 2B). The gel is run until the bromphenol blue is near the end, transferred to support (either Cambrex Gel-Bond PAG or a piece of developed film), covered in plastic wrap, and exposed to X-ray film. We use 1–3 μl of one of the "test" reactions to spot the gel in several places. The formamide loading solution provides a visual marker to facilitate alignment of the developed film and gel. If relatively fresh ^{32}P-α-dCTP is used in the reaction, a 1- to 3-h exposure at $-80°$ is long enough to see the digested reaction products.

The developed film and the gel are lined up, and the bands of interest are excised with a clean razor blade. The full-length product should be 35 nucleotides, although we recover bands of 32–36 nucleotides in length to account for possible insertion/deletion errors. The unlabeled template DNA strand is 39 nucleotides long, and care must be taken not to recover

(B) Internally labeled reactions (2:1 substrate:enzyme) used for fragment recovery. The left-most lane shows the full-length undigested product (45-mer). PstI digestion produces fragments of 32–36 bases in length on the synthesized strand visible after radiography. (C) Phospho image of dried gel after excision of bands of interest.

these fragments, as they will interfere with the downstream annealing step. After excision, it is advisable to dry the gel and visualize the products to verify that the correct region was cut (Fig. 2C). The excised gel slices are placed into a 1.5-ml tube and crushed into a paste against the side of the tube with a disposable pipette tip. Crush and soak buffer is added (500–750 μl) and the tube is rotated or shaken for at least 4 h (overnight is best) at 25–42°. The sample is then centrifuged for 1 min, and the supernatant is transferred to a new tube, taking care not to transfer any pieces of acrylamide. We have also recovered the crush and soak buffer by inverting the tube over a 0.45-μm spin filter (Amicon Bioseparations UFC40HV25) and centrifuging at 500g for 5 min. While the recovered DNA fragments can be precipitated with ethanol, we have found that recovery yields vary greatly, possibly due to the low efficiency of precipitation in the presence of ammonium (Sambrook *et al.*, 1989). A more time-consuming but also more reliable method is to concentrate and desalt the recovered oligonucleotide using a Centricon YM-3 spin filter. We generally try to obtain a final volume of purified oligonucleotide between 10 and 20 μl.

Quantitation of Oligonucleotide Recovery

Conversion factors, decay tables, and specific activity (SA) formulas can be found in most common laboratory manuals. A description of how the calculations are made follows, with an example specific to Pol η given in italics.

a. Count radioactivity in sample. An aliquot (1 μl) of the recovered oligonucleotide is placed onto a Whatman 24-mm GF/C filter, dried under a heat lamp, placed into scintillation fluid, and counted. *The value obtained is 6214 cpm.*

b. Determine the Curies (Ci) in sample. Assuming a 100% counting efficiency, counts per minute (cpm) of a sample are taken as the disintegrations per minute (dpm). To convert from dpm to Ci, divide the value by 2.22×10^{12}. *6214 dpm = 2.8×10^{-9} Ci.*

c. Determine the amount of dCTP that originated from isotope. To determine the μCi of dCTP added to the reaction, first calculate the SA (in μCi/μl) of the dCTP used based on the reference activity and date and then multiply by the volume used: (μCi/μl)*(vol used) = μCi used. To calculate the amount of dCTP originating from isotope ("hot" dCTP), the μCi in the reaction is converted to moles of dCTP using the SA (in Ci/mmol) of the isotope on the date used: (μCi dCTP)/(Ci/mmol) = mmol dCTP. The percentage of dCTP in the reaction that is "hot" is calculated by dividing this value by the total dCTP present: (hot dCTP/total dCTP). *If a 3000-Ci/mmol (10 μCi/μl) stock of ^{32}P-α-dCTP is used 2 days after the reference* *date, the SA on the day of the reaction will be 2805 Ci/mmol (9.3 μCi/μl). The addition of 5 μl of this stock adds 46.5 μCi to the reaction. Therefore, the amount of hot dCTP added to the reaction is (46.5 μCi)/(2805 × 10^6 μCi/mmol) = 1.66 × 10^{-8} mmol, or 16.6 pmol. In a 30-μl reaction with 100 μM of each cold dNTP, 3000 pmol of cold dCTP is also present, making the percentage of "hot" dCTP: 16.6/(3000 + 16.6) = 0.55%.*

d. Determine specific activity of incorporated dCTP. The SA of incorporated dCTP is calculated as follows: (SA of isotope on date counted)* (% hot dCTP). This calculation "corrects" for the cold dCTP in the reaction. *Assuming the polymerase assays and PstI digestion are performed in 1 day, the gel, excision, "crush and soak" steps the next day, and then recovery and quantitation the following day, the SA of the isotope (4 days post reference date) is 2624 Ci/mmol. Therefore, the SA of the incorporated dCTP is (2624 Ci/mmol)*(0.55/100) = 14.4 Ci/mmol.*

e. Calculate amount recovered. The concentration and yield of recovered oligonucleotide are calculated as (Ci in aliquot)/(SA of incorporated dCTP) = mmol dCTP incorporated. Because there are 7 Gs in the template, this value is ?tlsb?> then divided by 7 to give mmol recovered oligonucleotide. If 1 μl is counted, this value is also the per μl concentration of oligonucleotide. To calculate the total amount recovered, multiply by the total volume recovered. Percentage (%) recovery is calculated by dividing by the starting amount of substrate. *In step A, there was 2.8 × 10^{-9} Ci/μl sample. So the calculation is (2.8 × 10^{-9} Ci)/(14.4 Ci/mmol) = 1.9 × 10^{-10} mmol dCTP incorporated, and therefore 2.7 × 10^{-11} mmol oligonucleotide, or 27 fmol (and also 27 fmol/μl). With a total volume of 20 μl, there is 540 fmol oligo recovered, and therefore 540/4000 = 13.5% recovery.*

Recovery is typically between 10 and 30% of starting substrate (400–1200 fmol). This is more than needed to achieve insert excess during gap hybridization (see later). However, due to the low recovery yield and potentially small amounts of full-length product (from low efficiency bypass reactions), the time between the initial polymerase reaction and the final recovery step should be minimized in order to give a maximal scintillation signal during quantitation. Other ways to increase the signal include the addition of more radioactive label to the reaction, lowering the amount of cold dCTP in the reaction (both of which will increase the percentage of hot dCTP in the reaction and therefore the SA of incorporated dCTP), or adding two or more radiolabeled dNTPs to the reaction instead of only dCTP (the number of positions that labeled dNTP can be inserted must be taken into account). The recovered oligonucleotide is stored at −20° until further use.

Construction of Gapped M13 DNA Substrate

Single-stranded and double-stranded DNA of the M13mp2-TTAG bacteriophage are purified as described previously (Bebenek and Kunkel, 1995). Details of the digestion, purification, and gap production protocol have been published previously (Bebenek and Kunkel, 1995) so only the specifics for this system are given. Double-stranded DNA is digested to completion at 37° with the restriction endonucleases *Eco*RI and *Pst*I, each at a concentration of 0.2 U/μg DNA using the manufacturer's recommended conditions. After confirming complete digestion by visualization on an agarose gel, the DNA is ethanol precipitated, resuspended in TE (pH 8.0), and adjusted to 5.5% PEG-8000 and 0.55 *M* NaCl, with the final DNA concentration no more than 0.1 mg/ml (we assume 100% recovery from starting amounts). The DNA is incubated at 45° for 20 min and the large digestion fragment is recovered by centrifugation at 20,800*g* for 10 min. The supernatant is carefully removed, the tube is briefly centrifuged again, and a fine tipped pipette is used to remove trace liquid. The pellet is resuspended in water and residual PEG/NaCl is removed by ethanol precipitation. The PEG precipitation protocol and supernatant removal steps are critical because even though a 35-mer fragment should remain soluble under the PEG/NaCl concentrations used, *Pst*I and *Eco*RI both contain four base overhangs that can allow for the small fragment to be "carried along" during the precipitation step under less stringent conditions. If this occurs, the final substrate preparation can potentially contain a significant fraction of molecules that already have a 35-mer oligonucleotide hybridized to the gapped region, a situation that can affect both the reproducibility of experiments over time and the magnitude of the values calculated.

To form the gapped DNA, the digested large fragment is diluted and hybridized to single-stranded circular, viral (+) DNA as described previously (Bebenek and Kunkel, 1995; Kokoska *et al.*, 2003; Matsuda *et al.*, 2003), followed by ethanol precipitation and resuspension in TE buffer. An aliquot is then analyzed by agarose gel electrophoresis (see Bebenek and Kunkel, 1995) to verify gap production. The gapped product is separated from other DNA species using a 0.8% agarose gel, electroeluted from gel slices, ethanol precipitated, and resuspended in TE. The concentration of the gap DNA is determined by UV spectroscopy. The resulting molecule contains a 35 nucleotide gap between the *Eco*RI and the *Pst*I sites of the *lacZ*α gene in the M13mp2 genome.

Hybridization of Recovered Oligonucleotide to Gapped DNA

Excess nascent strand 32- to 36-mer DNA is mixed with gapped M13mp2-TTAG DNA. We typically use a 5- to 10-fold excess of recovered oligonucleotide over gapped DNA to ensure complete annealing, although we have observed similar results with anywhere from a 2- to 50-fold excess. Tube A, containing 9 fmol (40 ng) of gapped DNA, 2 μl of 20× SSC, and ddH$_2$O, is incubated at 48° for 3 min. Simultaneously, tube B, containing 90 fmol of oligonucleotide (10-fold excess over gapped DNA) is incubated at 75° for 3 mins. The contents of tube B are added to tube A to achieve a final volume of 20 μl (thus, 2× SSC). This mixture is further incubated at 48° for 10 min, then at 42° for 10 min, and finally cooled slowly to room temperature over several hours. Once annealed, the DNA is stored at −20° until further use. This example assumes a 20-μl final volume, but depending on the concentration of the recovered oligonucleotide, the final volume of the hybridization mixture and the amount of SSC and ddH$_2$O needed can be varied accordingly.

Electroporation and Plating

A portion of the annealed DNA solution is introduced into competent cells to score mutant frequencies (for a protocol to prepare competent cells, see Bebenek and Kunkel, 1995). Typically, 1 μl of a 1:5 dilution in dH$_2$O is used. The DNA is mixed with 50 μl of competent cells on ice, which are then transferred to an ice-cold 0.2-cm cuvette. Electroporation is with a Bio-Rad gene pulser set at 2.01 kV, 400 W, 25 mF. Time constants should be 9.0 to 9.4 ms. Immediately following electroporation, 1–2 ml of SOC medium (Bebenek and Kunkel, 1995) is added to the cells, which are kept at room temperature. Cells are plated onto minimal agar plates within 20 min of electroporation to avoid the release of progeny M13 viral particles. An aliquot (usually 5 to 50 μl) of the electroporated cells is added to a tube containing 2.5 ml of soft agar, 20 μl of 25 mg/ml IPTG, and 50 μl of 50 mg/ml X-gal at 45–49°. Finally, 0.5 ml of a log phase culture of CSH50 cells (the α complementation strain) is added and the tubes are vortexed gently. The soft agar is poured onto the minimal plate and allowed to solidify. Plates are inverted and incubated overnight (∼16–18 h) at 37° and then scored for plaque color. In theory, the TAG codon in the lacZα coding sequence should give colorless plaques. In practice, we have observed this sequence to instead give rise to a light blue phenotype, presumably due to some translational read through of the codon. This is fortuitous, as it allows detection of both base substitution errors at the T, A, and G positions (by reversion to a dark blue phenotype), and also frameshift errors (colorless plaque) on the same plates. An appropriate amount of the electroporated cells is used to yield 200 to 400 plaques per plate. The total number of plaques is determined by counting all plaques for which the color phenotype is obvious. Problematic plaques (those that are too small, on the edge of the plate, or in occasional regions of smearing) are not counted. Plaques with a light blue phenotype result from error-free bypass

TABLE I
Mutation Frequency and Error Rates at the 3′ T for Pol η Bypass of a TT Dimer

	Plaque numbers[a] (mutation frequency)			Sequencing results[b]				Error rate ($\times 10^{-4}$)		
					3′ T to			3′ T to		
	Dark blue	Colorless	Total	Total	A	C	G	A	C	G
Undamaged	284 (3.6%)	285 (3.6%)	7993	101	5	67	1	30	398	6
TT dimer	184 (2.8%)	230 (3.5%)	6486	97	4	75	2	19	361	11

[a] Data from (McCulloch *et al.*, 2004). Dark blue plaques score for base substitutions at the TAG stop codon. Colorless plaques score for frameshift errors anywhere in the synthesized strand.

[b] Changes at the A and G residues not shown.

or from expression of the viral, (+) strand template, dark blue plaques result from base substitution errors at the T, A, and G positions of the amber codon, and colorless plaques result largely from frameshift errors made by the polymerase or during chemical synthesis of the starting DNA template. Table I shows the results (from McCulloch *et al.*, 2004) of reactions with human Pol η copying either an undamaged or TT dimer containing template. The counts for each type of plaque are given. The mutation frequencies (described in more detail later) for dark blue and colorless plaques are an indication of the base substitution and frameshift fidelity of the lesion bypass reaction.

Plaque Purification and DNA Sequencing

The procedure used to purify mutant plaques to confirm plaque color has been described previously (Bebenek and Kunkel, 1995). Confirmed mutants are processed to prepare DNA for sequencing using either the TempliPhi system (Amersham Biosciences, England) or a previously described method (Bebenek and Kunkel, 1995). We use BigDye or dRhodamine cycle sequencing, using 5 pmol of primer F256 (5′-GATCGCACT-CCAGCCAGC), following the instructions given by the manufacturer (Applied Biosystems, Foster City, CA). Samples are analyzed by capillary sequencing by the NIEHS core sequencing group.

Mutant Frequency Data and Calculation of Error Rates

Mutation frequencies for dark blue and colorless plaques are calculated by dividing the number of each plaque type by the total number of plaques counted. DNA sequence analysis of independent mutants of each type allows calculation of the error rates as follows: $((N_x/T)*MF)/0.6$, where N_x is the number of observed errors of a given type, T is the total number of sequenced plaques of that particular phenotype, and MF is the mutation frequency for that particular phenotype. This value is then divided by 0.6 to correct for the average $(-)$ strand expression value $(\sim60\%)$ for several different mismatches (Kokoska et al., 2003). For example, 75 of 97 sequenced dark blue mutants recovered from TT dimer bypass reactions had a T→C mutation at the 3′ T of the dimer (Table I). Because the dark blue mutation frequency for these reactions is 2.8%, the calculation is $[(75/97) \times 0.028]/0.6 = 3.6 \times 10^{-2}$, or 360×10^{-4}. For frameshift error rates, only events involving the dimer site (defined as either T or either flanking base) are counted. It is important to note that colorless plaques may result from frameshift mutations anywhere in the 35 nucleotide fragment, including polymerase errors and errors made during chemical synthesis of the starting oligonucleotides.

Plaque Hybridization to Detect Silent Mutations

Due to a lack of colorimetric detection of errors at the 5′ T of the dimer or the G→A error in the amber codon, a plaque hybridization assay is used instead to monitor fidelity. Briefly, plates containing between 200 and 500 plaques are prepared as described earlier, and membrane lifts are made using a circular nylon membrane (PerkinElmer Life Sciences, Boston). Blotting with 5′ ^{32}P end-labeled probe oligonucleotides is then performed as described previously (McCulloch et al., 2004). The probe should be 14–16 nucleotides long and contain the target base in the middle of the sequence. Positive plaques will show up as intense dark spots against a background of light gray spots. The total number of plaques on each plate is counted after the lifts are made. Error rates are calculated by $(N/P_t)/E$, where N is the number of mutant plaques observed, P_t is the total number of plaques screened, and E is 0.6, the efficiency (60%) of expressing the $(-)$ strand (see earlier discussion). Control reactions to screen for 3′ T→C changes using this method give similar results to DNA sequencing (McCulloch and Kunkel, unpublished data).

Applications of the Method

The approach described here is applicable to any polymerase that is capable of performing TLS. It can also be applied to any lesion that can be stably placed in a defined oligonucleotide. The assay depicted in Fig. 1 can assess bypass of lesions of template T, A, or G and is currently being adapted to monitor bypass fidelity for lesions on template C as well. The approach is adaptable to studying lesions in multiple sequence contexts (Kokoska *et al.*, 2003) and can be used to study single (Brieba *et al.*, 2004; Kokoska *et al.*, 2003) or compound lesions, e.g., those involving two bases such as *cis–syn* cyclobutane pyrimidine dimers (McCulloch *et al.*, 2004), 6-4 photoproducts, or cisplatin adducts. Parallel assays with undamaged DNA templates allow direct comparisons of fidelity when bypassing undamaged versus damaged templates. To date, we have used this assay successfully to analyze TLS by polymerases that have either high or low lesion bypass ability and either high or low fidelity with undamaged DNA (Brieba *et al.*, 2004; Kokoska *et al.*, 2003; McCulloch *et al.*, 2004). Experiments currently in progress demonstrate that the role of accessory proteins on TLS fidelity can also be measured using longer template primers.

Acknowledgments

The authors thank Drs. Mercedes Arana and John Fortune for their critical comments on the manuscript.

References

Bebenek, K., and Kunkel, T. A. (1995). Analyzing the fidelity of DNA polymerases. *Methods Enzymol.* **262**, 217–232.

Bebenek, K., and Kunkel, T. A. (2004). Functions of DNA polymerases in DNA repair and replication. *In* "Advances in Protein Chemistry" (W. Yang, ed.).

Boyer, J. C., Bebenek, K., and Kunkel, T. A. (1996). Analyzing the fidelity of reverse transcription and transcription. *Methods Enzymol.* **275**, 523–537.

Brieba, L. G., Eichman, B. F., Kokoska, R. J., Doublie, S., Kunkel, T. A., and Ellenberger, T. (2004). Structural basis for the dual coding potential of 8-oxoguanosine by a high-fidelity DNA polymerase. *EMBO J.* **23**(17), 3452–3461.

Kokoska, R. J., McCulloch, S. D., and Kunkel, T. A. (2003). The efficiency and specificity of apurinic/apyrimidinic site bypass by human DNA polymerase η and *Sulfolobus solfataricus* Dpo4. *J. Biol. Chem.* **278**, 50537–50545.

Kunkel, T. A. (1985). The mutational specificity of DNA polymerase-β during *in vitro* DNA synthesis: Production of frameshift, base substitution, and deletion mutations. *J. Biol. Chem.* **260**, 5787–5796.

Kunkel, T. A. (2004). DNA replication fidelity. *J. Biol. Chem.* **279**, 16895–16898.

Masutani, C., Kusumoto, R., Iwai, S., and Hanaoka, F. (2000). Mechanisms of accurate translesion synthesis by human DNA polymerase η. *EMBO J.* **19**, 3100–3109.

Matsuda, T., Vande Berg, B. J., Bebenek, K., Osheroff, W. P., Wilson, S. H., and Kunkel, T. A. (2003). The base substitution fidelity of DNA polymerase beta-dependent single nucleotide base excision repair. *J. Biol. Chem.* **278**, 25947–25951.

McCulloch, S. D., Kokoska, R. J., Masutani, C., Iwai, S., Hanaoka, F., and Kunkel, T. A. (2004). Preferential *cis-syn* thymine dimer bypass by DNA polymerase η occurs with biased fidelity. *Nature* **428**, 97–100.

Messing, J., Gronenborn, B., Muller-Hill, B., and Hans Hopschneider, P. (1977). Filamentous coliphage M13 as a cloning vehicle: Insertion of a HindII fragment of the lac regulatory region in M13 replicative form *in vitro. Proc. Natl. Acad. Sci. USA* **74**, 3642–3646.

Roberts, J. D., Nguyen, D., and Kunkel, T. A. (1993). Frameshift fidelity during replication of double-stranded DNA in HeLa cell extracts. *Biochemistry* **32**, 4083–4089.

Sambrook, J., Fritsch, E. F., and Maniatis, J. (1989). "Molecular Cloning. A Laboratory Manual." Cold Spring Harbor Laboratory, Cold Spring Harbor, NY.

[22]   DNA Polymerases for Translesion DNA Synthesis: Enzyme Purification and Mouse Models for Studying Their Function

By PAULA L. FISCHHABER, LISA D. MCDANIEL, and ERROL C. FRIEDBERG

Abstract

This chapter discusses experimental methods and protocols for the purification and preliminary characterization of DNA polymerases that are specialized for the replicative bypass (translesion DNA synthesis) of base or other types of DNA damage that typically arrest high-fidelity DNA synthesis, with particular emphasis on DNA polymerase κ (Polκ from mouse cells). It also describes some of the methods employed in the evaluation of mouse strains defective in genes that encode these enzymes.

Introduction

Translesion DNA synthesis (TLS) is a fundamental DNA damage tolerance mechanism that both prokaryotic and eukaryotic cells are able to deploy in the face of arrested or stalled DNA replication caused by base damage (Fischhaber and Friedberg, 2005; Friedberg *et al.*, 2005). The process utilizes a class of novel DNA polymerase that are specialized for incorporating deoxyribonucleotides opposite and beyond sites of base damage. At least 10 such specialized polymerases have been identified in higher eukaryotes, a number of which belong to a new protein family called the Y family (Fischhaber and Friedberg, 2005; Friedberg *et al.*, 2005). Structurally, Y family polymerases are endowed with a more open configuration at their active sites, and in addition to palm, thumb, and fingers domains they carry an extra little finger or PAD (polymerase accessory domain) (Bunting *et al.*, 2003). The mammalian Y family (but not members of the other polymerases families) comprises four DNA polymerases designated Rev1, Polη, Polκ, and Polι, all of which have ubiquitin-binding

METHODS IN ENZYMOLOGY, VOL. 408
0076-6879/06 $35.00
DOI: 10.1016/S0076-6879(06)08022-0

motifs. Such domains are not present in the other specialized DNA polymerases, suggesting that the Y family is functionally distinct in a manner that remains to be determined. All known specialized DNA polymerases are devoid of $3' \rightarrow 5'$ exonucleolytic proofreading ability and have markedly reduced processivity and fidelity compared to replicative DNA polymerases (Fischhaber and Friedberg, 2005; Friedberg et al., 2005).

Purification of Specialized DNA Polymerases as Tagged Proteins Following Expression of Cloned Fusion Genes

Specialized DNA polymerases are present in cells in very small amounts and little is known about the endogenous proteins. Thus, most, if not all, reported studies deal with recombinant proteins. Additionally, most studies employ fusions of specialized polymerases with tags that render them suitable for affinity purification. The coding region for any specialized DNA polymerase of interest can be cloned into a variety of plasmids that introduce affinity tags in frame with the polymerase coding sequence and that can be expressed in appropriate host cells, usually *Escherichia coli*. Cells are harvested and disrupted by standard methods and purification is based primarily on the use of matrices that specifically bind the affinity tag. Details of specific experimental conditions for the routine handling and storage of affinity-purified polymerases must be determined empirically, but these enzymes are typically soluble, stable, and active in a variety of buffer conditions. Our experience with DNA polymerase κ (Polκ) is that achieving substantial overexpression presents the greatest limitation to success.

Expression Constructs

A variety of affinity tags facilitate extensive protein purification in a single chromatographic step. Tags are expressed in frame at either the N-terminal or the C-terminal end of the coding region of the polymerase of interest, generating a fusion protein upon expression. Glutathione *S*-transferase (GST) (Smith and Johnson, 1988) and histidine tags (His tag) (Gentz *et al.*, 1989; Hochuli, 1988) are two that we have employed with the greatest success, but others to be considered include a chitin-binding domain (Chong *et al.*, 1997), calmodulin-binding peptide (Stofko-Hahn *et al.*, 1992), maltose-binding protein (Bedouelle and Duplay, 1988; di Guan *et al.*, 1988), and the S tag (Karpeisky *et al.*, 1994; Kim and Raines, 1994). Many epitope tags, such as HA (Chen *et al.*, 1993; Wilson, 1984), Myc (Evan *et al.*, 1985) and Flag (Hopp *et al.*, 1988), readily facilitate the detection of proteins by Western analysis.

The presence of a GST affinity domain frequently increases the solubility of the fusion protein compared to expression of the polymerase of interest alone. Most plasmid vectors currently available also carry recognition sites for enzymatic cleavage incorporated into their coding regions that allow the removal of the affinity tag following protein purification. The removal of these tags can be helpful if one is concerned about their presence altering the polymerase activity *in vitro*.

When preparing plasmid constructs for expressing specialized DNA polymerases it is useful to prepare a control construct in parallel in which the catalytic domain of the polymerase has been inactivated by site-specific mutagenesis (Gerlach *et al.*, 2001). All DNA polymerases, including those specialized for translesion DNA synthesis, carry a conserved catalytically required aspartate-glutamate (DE) dipeptide in the polymerase active site. These can be mutated to alanine-alanine (DE → AA), generating a polypeptide that expresses and purifies like the wild-type construct but that is devoid of DNA polymerase activity. This ensures that the activity demonstrated truly derives from the specialized polymerase in question as opposed to a spuriously isolated host cell DNA polymerase(s).

Protocols for plasmid construction and manipulation involve routine molecular biological techniques. Commercial plasmid vectors are typically shipped with a complete set of instructions. Hence, specific protocols are not reiterated here. We routinely propagate plasmid vectors in *E. coli* DH5α (Invitrogen "subcloning efficiency" competent cells") using a rapid DNA ligation kit (Roche "rapid DNA ligation" or Promega "TA cloning kit") to ligate polymerase chain reaction (PCR)-generated cDNAs into commercially obtained *E. coli* expression vectors [pET system vectors (Stratagene and Novagen); we employ pGEX system vectors (Pharmacia)] and site-directed mutagenesis kits (QuickChange or ExSite, Stratagene) for generating mutation or deletion constructs. For PCR, our experience is that platinum *Taq* polymerase (Invitrogen) works well for a wide variety of amplifications and can be particularly useful for difficult polymerase chain reactions.

Expression in E. coli, *Yeast, or SF9 Cells Infected with Baculovirus*

When the desired expression construct has been generated and confirmed by sequencing, the plasmid is transformed into an *E. coli* expression strain. Numerous BLR and BL21 strains have been optimized for protein expression. Enhanced BL21 strains such as BL21 Rosetta (Novagen) or BL21-Codon-Plus(DE3)-RIL (Stratagene) are available commercially. The latter two are useful for handling proteins from higher organisms, such as mouse and human, which utilize the genetic code differently from

E. coli. These strains carry exogenous plasmids that encode tRNAs required to express underutilized codons in *E. coli.*

When selecting a host strain for expression it is prudent to select one compatible with the promoter of the plasmid expression construct. For instance, if the plasmid promoter is from bacteriophage T7, it is necessary to transform an expression host with the DE3 lysogen carrying a copy of the T7 RNA polymerase gene to ensure appropriate promoter induction. However, if the promoter is P_{tac}, for example (composed of the -35 region of the trp promoter and the -10 region of the lacUV5 promoter/operator), the DE3 lysogen feature is not necessary. Such details are generally provided in instructions shipped with commercial plasmids.

Protein expression is typically induced by adding a specific chemical agent to the growth medium, such as isopropyl-β-D-thiogalactoside if one selects a plasmid vector with the P_{tac} or T7 promoter. The time of induction, *E. coli* culture density at induction, concentration of inducing agent, and temperature of the culture during induction can all bear heavily on recovery of the highest levels of soluble protein. We have observed that slower induction with lower concentrations of inducing agent and at lower temperatures (16° or room temperature) facilitates improved yields of soluble protein. The addition of glycylglycine sometimes improves protein solubility as well.

If expression trials in *E. coli* are not successful, other potential host organisms are available. Our experience is that two other hosts are worth considering: the yeast *Schizosaccharomyces pombe* and insect SF9 cells infected with baculoviruses. For such purposes one must obviously switch to appropriate expression vectors. Commercially available kits (such as Gateway Invitrogen/LifeTechnologies) facilitate the preparation of expression constructs that can be employed with either of these host organisms, as well as with *E. coli.*

Expression in Baculovirus-Infected SF9 Cells. Baculoviruses are double-stranded DNA viruses that infect many different species of insects. The baculovirus genome is replicated in the nuclei of infected host cells, transcription of the viral genes occurs, and newly replicated DNA is packaged into rod-shaped nucleocapsids (Burgess, 1977). Infectious viral particles enter susceptible cells and proteins, including that expressed by the cloned gene of interest, are generated, often under control of the polyhedron promotor. The gene of interest is cloned into the baculovirus genome by homologous recombination either *in vitro* (BaculoDirect Baculovirus Expression System, Invitrogen) or *in vivo* (Baculovirus Expression Vector System, BD Pharmingen). Both have been optimized to near 100% recombination efficiency. Both companies offer vectors that include epitopes tags for purification and identification of the authentic protein.

Insect cells, unlike bacterium, are capable of performing many of the processing events that are required for biologically active mammalian proteins, including disulfide bond formation and oligomerization (Kidd and Emery, 1993), *N*- and *O*-linked glycosylation, phosphorylation, acylation, amidation, carboxymethylation, isoprenylation, signal peptide cleavage, and proteolytic cleavage (Hoss *et al.*, 1990; Kloc *et al.*, 1991; Kuroda *et al.*, 1991). The posttranslational protein modifications generated in baculovirus-infected cells are not identical to those affected in mammalian cells. Nonetheless, if such modifications are essential for polymerase activity, those generated in insect SF9 cells are more likely to be active than those produced by a bacterial expression systems (reviewed in Kost *et al.*, 2005). In addition, there are no size limits of the protein to be expressed and multiple proteins can be expressed in the same cell. In some cases, expression of a cloned gene can be higher than in other expression systems (Baixeras *et al.*, 1990; Brandt-Carlson and Butel, 1991; Caroni *et al.*, 1991; Christensen *et al.*, 1993; Hsu *et al.*, 1991; Mattion *et al.*, 1991), achieving as much as 50% of the total cellular protein of infected insect cells. The information provided earlier notwithstanding, there are no reports in the literature using baculovirus expression systems for the purification of specialized DNA polymerases.

Expression in Yeast. Both *S. pombe* and *Saccharomyces cerevisiae* can be used for eukaryotic gene expression, and vectors for both are available commercially. Both types of yeast support posttranslational modifications that increase the probability that the cloned protein will be active (Herscovics and Orlean, 1993; Sambucetti *et al.*, 1986). The Spectra system (Invitrogen) uses their TOPO cloning strategy to insert the gene of interest into vectors designed for *S. pombe* expression. The inserted gene is under control of the *nmt1* promoter, allowing thiamine-regulated expression (Maundrell, 1990). Vectors also contain distinct *nmt1* promoters differing in the TATA box (Basi *et al.*, 1993) for varying expression levels and epitope tags for purification and Western blot analysis. Vectors also contain the *S. pombe ars1* origin of replication for high-copy, nonintegrative maintenance of plasmids in *S. pombe* cells (Heyer *et al.*, 1986).

The pESC vectors (Stratagene) are epitope-tagging vectors for expression in *S. cerevisiae* using the *GAL1* and *GAL10* promoters in opposite directions. This allows for cloning of one or two genes with repressible promoters. Upon introduction into yeast the two genes can be coexpressed and investigated for *in vivo* interactions. These vectors also include the yeast 2μ origin for autonomous replication, auxotrophic markers, and epitope tags.

Expression in Mammalian Cells. COS-7 (CRL-1651, ATCC) cells can provide a mammalian cell alternative for transient or stable transfection of expression plasmids. COS-7 cells are SV-40-transformed African green monkey kidney cells that retain complete permissiveness for lytic growth of the SV-40 virus. Mammalian cell culture systems are not useful for large-scale production and purification of protein. However, they are used widely for analysis and interactions of posttranslationally modified overexpressed protein in mammalian cells (Bross *et al.*, 1994; Kouzuki *et al.*, 1998; Pray and Laimins, 1995; Shin *et al.*, 1993).

Purification by Affinity Chromatography

Escherichia coli cells containing an expressed specialized DNA polymerase of interest are disrupted in a buffer system in which the polymerase is soluble and to which sufficient quantities of a protease inhibitor cocktail have been added so that affinity chromatography can be executed before the protein is degraded. All purification steps from cell disruption onward are performed at 4°. The Complete Tablet (Roche) offers a convenient method of introducing protease inhibitors directly to the cell disruption buffer immediately prior to use. We routinely supplement the Complete Tablet with 1 mM (final concentration in the disruption buffer) phenylmethylsulfonyl fluoride (PMSF) diluted fresh from a 100 mM stock in ethanol; 20° storage. Following cell disruption by sonication the lysate is cleared by centrifugation and loaded onto an equilibrated affinity column.

When using GST as an affinity tag we employ glutathione cross-linked agarose as the affinity matrix. Although GST binding to the resin is rapid, some studies claim to achieve better binding efficiency by incubating the lysate "in batch" with the resin for 30 min and then pouring the lysate and resin together into a drip column to wash out unbound protein. The wash buffer is typically identical to the cellular disruption buffer, and at least 5 column volumes should be allowed to flow through the column. Following the wash step, additional resolution can sometimes be achieved with a second wash in which some feature (such as the ionic strength) of the wash buffer has been modified. Adding salt to the wash buffer can disrupt nonspecific protein–protein interactions. The exact salt concentration used (if any) must be determined empirically, but specialized DNA polymerases are typically soluble to a high concentration of salt (600 mM or more). Hence, adding 150 mM NaCl or more to a wash buffer can remove spuriously bound proteins without bringing the polymerase out of solution.

When using GST affinity tags the GroEL chaperonin (60 kDa) frequently binds to the GST fusion protein (Badcoe *et al.*, 1991) (which can be a sign of misfolded protein). GroEL can be removed on the GST column with a wash buffer containing ATP (approximately 1 mM)

(Thain *et al.*, 1996). Following column washing, elution buffer is added and fractions containing the specialized DNA polymerase of interest are collected. For a GST column the elution buffer contains glutathione in the reduced (thiol) form at a concentration of 1–10 mM. This competes with resin cross-linked glutathione moieties for binding to GST, bringing the resin-bound GST fusion protein back into the mobile phase and hence elutable from the column. Thiols readily oxidize in air. It is thus helpful to add glutathione in the dry powder form to the elution buffer immediately prior to use. Thiols are acidic and depending on the buffering agent used and its concentration, adding powdered glutathione may lower the pH significantly, thereby interfering with protein solubility or stability when the buffer comes in contact with the protein-bound resin. It is therefore important to check the pH of the elution buffer after adding glutathione.

If the affinity tag is His, the required affinity resin is Ni^{2+}- or Co^{2+}-chelated resin. Co^{2+}-chelated resin (Bush *et al.*, 1991; Chaga *et al.*, 1999) (Talon, BD Clontech) usually provides better resolution because it offers fewer coordination sites for the tag and hence fewer non-His-tagged proteins exhibit background binding to the resin. However, extracts from eukaryotic cells such as yeasts, SF9 insect cells, or COS7 cells are replete with numerous small histidine-rich proteins that can compete with the His-tagged protein for resin-binding sites by mass action. Some studies have observed that preincubation of eukaryotic cell lysates with a gel filtration resin harboring a low molecular weight cutoff pore size sequesters many of these small histidine-rich competitors, enabling more efficient binding of the His-tagged protein to the Ni^{2+} or Co^{2+} resin.

Subsequent Purification Steps and Storage

Following careful execution of the affinity purification step, enzyme preparations are generally sufficiently pure for many studies. If further purification is required, gel filtration chromatography is often appropriate. Gel filtration columns in frequent use include Sepharose and Superdex (Pharmacia), both of which provide reliable resolution. Gel filtration chromatography can be efficiently executed with commercially prepared columns in which resins have been packed to tight specification. Gel filtration chromatography is readily achieved with an FPLC or SMART system liquid chromatography instrument that delivers a consistent mobile phase flow. It is, of course, possible to perform gel filtration chromatography manually, but pouring and using the column require care.

Mono Q and Mono S columns (Pharmacia) are also used frequently in later or final purification steps. These anion-exchange resins have proven to be reliable "clean up" steps and appear frequently in the Materials and

Methods sections of contemporary manuscripts. These require the use of FPLC or SMART system liquid chromatography instruments.

Storage of purified specialized DNA polymerases is facilitated by the addition of 10% (or more) glycerol to the buffer (often added to the buffers from the cell lysis step forward), in order to stabilize or solubilize the fusion protein, and flash freezing of sample vials in liquid nitrogen prior to storing at $-80°$. Following purification we typically freeze samples in "single-use aliquots," as freeze–thaw cycles can compromise enzymatic activity.

Testing Enzyme Purity and Activity

Examining the purity of a specialized DNA polymerase preparation is accomplished using SDS–PAGE and staining the gel with either Coomassie brilliant blue-R or silver stain. Specialized DNA polymerases are typically ~100 kDa in size when expressed as full-length polypeptides. Hence, 4–15 or 4–20% gradient SDS–PAGE gels are convenient because they afford good separation in the 100-kDa size range, but retain smaller proteins so that one can assess the extent of polymerase degradation.

DNA polymerase activity is typically tested in primer extension assays. One anneals a $5'$-^{32}P-radiolabeled DNA primer to a complementary oligonucleotide strand bearing additional bases downstream from the $3'$ end of the primer. The radiolabeled primer–template is incubated with the purified polymerase in a buffer near physiological pH containing approximately 1 mM Mg^{2+} (a necessary cofactor) and the four deoxynucleotide $5'$ triphosphates (dATP, dCTP, dGTP, and TTP). Analysis of the reaction products on a denaturing polyacrylamide gel reveals a ladder of radiolabeled bands, indicating extension of the primer toward the $5'$ end of the template strand. If the polymerase is sufficiently processive, one recovers only a single band corresponding to primers extended to the end of the template strand.

Each specialized DNA polymerase bypasses distinctive forms of DNA base damage. Hence, conducting primer extensions with primers annealed to oligonucleotides containing site-specifically incorporated forms of chemically modified bases is used routinely to assess the lesion bypass capability of a polymerase *in vitro*. Some forms of specific DNA damage can be obtained commercially, either embedded in a site-specific manner in DNA oligonucleotides or as phosphoramidite precursors that can be incorporated into a DNA sequence of choice using standard phosphoramidite DNA synthesis techniques. Other forms of base damage can be prepared by direct adduction to DNA oligonucleotides.

Mouse Strains Defective in Specialized DNA Polymerases

Methods for generating mutant mouse strains by targeted disruption of selected genes are well documented elsewhere (Fan and Skoultchi, 2004; Huang *et al.*, 2002; Kojima *et al.*, 2004; Lee *et al.*, 2002; Trifunovic and Larsson, 2002; Wong and Storm, 2002). Table I lists the mutations reported in mouse genes that encode known specialized DNA polymerases. (For a list of targeted alleles in genes affecting all known biological responses to DNA damage, see http://pathcuric1.swmed.edu/research/research.htm [version 6 (Friedberg and Meira, 2004); version 7 in press]).

Targeted disruption of the *Rev3l* gene generates an embryo lethal phenotype. However, embryo lethality has not been reported for any other mouse strains listed in Table I. Cells from *Polh* and *Polk* mutant mice are abnormally sensitive to killing by various DNA-damaging agents. Additionally, *Poli*, *Polh,* and *Poll* mutant mice have been implicated in defective somatic hypermutation of immunoglobulin genes. The following sections detail techniques relevant to the evaluation of these particular phenotypes.

Measuring Cellular Sensitivity to DNA-Damaging Agents

Assays to determine the sensitivity of cells to DNA-damaging agents can be performed with either mouse embryonic fibroblast (MEF) cultures or cultures of embryonic stem (ES) cells. This section describes the use of MEFs for testing sensitivity of cells to ultraviolet (UV) radiation and the use of ES cells for determining sensitivity to ionizing radiation. However, in principle, either cell type can be used to evaluate sensitivity to killing by various DNA-damaging agents.

Measuring the Sensitivity of Mouse Embryonic Fibroblasts to UV Radiation. MEFs are isolated from day 13.5 embryos as described (Meira *et al.*, 2001). With the euthanized pregnant mouse on its back, soak the abdomen with 70% alcohol. Make an incision down the midsection using scissors and expose the uterine horns. Remove the uterus and place it in a dish of phosphate-buffered saline (PBS). Open the uterus, releasing the embryos into the PBS solution. Remove the embryos from their amniotic sacs and place them in a culture dish containing Dulbecco's modified Eagles medium (DMEM) (Gibco) supplemented with 10–15% fetal bovine serum. Mince the embryos into small pieces in the culture dishes and allow cells to attach and grow for several days. Aspirating the minced pieces repeatedly into a syringe with a 16-gauge syringe needle is effective in disassociating cells further if needed. MEFs can often be generated from mice carrying genes that result in embryonic lethality (Chester *et al.*, 1998; Renshaw *et al.*, 1999). However, in the case of *Rev3l* mouse models, no

TABLE I
MOUSE MODELS FOR TLS POLYMERASES

Allele	Type	Target	Selection	Counter selection	Selection marker removed	Background	Phenotype	Reference
Rev1l	Targeted (k-o)	BRCT domain exon 2 and 3 deleted	puro, hyg	Gangciclovir	Yes	129/SvEv	None detected-19 months	Jansen et al. (2005)
Rev3^{tm1Esp}	Targeted (k-o)	Exon 4 disrupted by the insertion of a floxed neo cassette	neo	No	No	C57BL/6	Embryonic lethal	Esposito et al. (2000)
Rev3^{tm1Kaji}	Targeted (k-o)	AA 1327–1336 in exon 12 replaced by a PGK-neo-PolyA. A stop codon placed in the 5' end of the neo cassette results in loss of the DNA polymerase domain	neo	Diptheria toxin	No	129P2/OlaHsd	Embryonic lethal 7.5–10.5 dpc	Kajiwara et al. (2001)
Rev3^{tm1Msn}	Targeted (k-o)	Insertion of a neomycin selection cassette disrupted an exon encoding a portion of the carboxy-terminal DNA polymerase domain. Exon 29	neo	Gangciclovir	No	129	Embryonic lethal 10.5–12.5 dpc	Bemark et al. (2000)
Rev3^{tm1Ndew}	Targeted (k-o)	A 10-kb region of the gene encoding DNA polymerase domains in exons 22–26 was replaced with a PGK-hygro cassette	Hyg	Gangciclovir	No	129P2/OlaHsd	Embryonic lethal 10.5–11.5 dpc	Van Sloun et al. (2002)
Rev3^{tm1Rwd}	Targeted (reporter)	Exons 27 and 28, encoding a portion of polymerase motifs, were deleted by the insertion of a promoterless lacZ and neo fusion gene	neo	None	No	129P2/OlaHsd	Embryonic lethal 10.5–11.5 dpc	Wittschieben et al. (2000)

Allele	Type	Description	Marker			Strain	Phenotype	Reference
$Polk^{tm1.1Esp}$	Targeted (k-o)	This allele is a derivative of Polktm1Esp. The loxP-flanked exon 6 and neo cassette were removed	neo	Gangciclovir	Yes	129P2/OlaHsd	None detected	Schenten et al. (2002)
$Polk^{tm1Esp}$	Targeted (Floxed/Frt)	Exon 6 was floxed by insertion of a single upstream loxP site and a downstream floxed neo resistance cassette	neo	Gangciclovir	No	129P2/OlaHsd	None detected	Schenten et al. (2002)
$Polk^{tm1Taz}$	Targeted (k-o)	Deletion exons 5 and 6 with a PGK-neo cassette	neo	Gangciclovir	No	(C57BL/6 × CBA)F1 129/J, 129/SvJ, 129/ReJ 129/OlaHsd	None detected	Shimizu et al. (2003)
$Poli^{d}$	Spontaneous	A nonsense mutation changes codon 27 from serine (TCG) to an amber stop codon (TAG), resulting in a truncated protein lacking any catalytic function	None	None	None		None detected	McDonald et al. (2003)
$Polh^{tm1Crey}$	Targeted (k-o)	Cre mediated excision of exon 4 in ES cells, resulting in the direct splicing of exons 3–5	neo	None	Yes	129P2/OlaHsd	Ig sequences from mutant mice PNA high B cells displayed the 85% G/C-biased pattern mutation	Delbos et al. (2005)
$Poll^{tm1Crey}$	Targeted (k-o)	Exons 5–7 were replaced with a neomycin resistance gene	neo	Gangciclovir	No	129P2/OlaHsd	None detected	Bertocci et al. (2002)
$Poll^{tm1Nmt}$	Targeted (k-o)	Targeted replacement of exons 1–6 with a PGK-neo cassette also disrupted expression of the Dpcd gene	neo	Gangciclovir	No	129P2/OlaHsd	Hydrocephalus, situs inversus, chronic sinusitis, and male infertility	Kobayashi et al. (2002)
$Polm^{tm1Crey}$	Targeted (k-o)	Exons 7–9 were replaced with a neomycin resistance gene	neo	Gangciclovir	No	129P2/OlaHsd	None detected	Bertocci et al. (2002)

(continued)

TABLE I (*continued*)

Allele	Type	Target	Selection	Counter selection	Selection marker removed	Background	Phenotype	Reference
Polq^{chaos1}	Chemically induced (ENU)	The mutation in the chaos1 mouse was identified as a C-to-T substitution at nucleotide 5794 in the coding region of exon 19. The mutation results in a serine-to-proline change at amino acid residue 1932	None	None	None	C57BL/6J	Marked B-cell deficiency in the spleen and Peyer's patches in about 50% of homozygotes. Lower cellularity of peripheral lymphoid tissues. Maturation of B cells from IgM− to IgM+ is slowed down and light chain rearrangements are impaired	Shima *et al.* (2003)

MEF cultures have been reported for the mutant mouse except in the absence of functional p53 protein (Zander and Bemark, 2004).

The survival of MEFs following exposure to UV radiation can be determined by two methods. The first examines cell survival directly by measuring colony formation from individual cells grown on plates. This technique is the more accurate and sensitive of the two methods described here. The second technique employs a colorimetric assay that measures the amount of dye retained by living cells stained with crystal violet or Giemsa stain. In either protocol cells are placed at a distance from a UVC bulb (G15T8) in order to deliver 254 nm light. The bulb should be enclosed in a box to limit the exposure of the user (Fig. 1). UV fluence is measured by a UV meter (International Light, Inc.). Cells used for these experiments come from asynchronous cultures.

METHOD 1. Early passage cells are plated at a density of 100–200 cells/ 100-mm culture dishes. If cells are expected to be sensitive to killing by UV radiation, proportionally higher numbers should be plated for higher UV dose exposures, such that statistically significant numbers of colonies

FIG. 1. UV light irradiation cabinet. Cabinet configuration is similar to a kitchen upper cabinet. It is 12 in. deep and 36 in. tall. (A) The UV light source is recessed and contains one G15T8 bulb. (B) Slider blocks the UV light while the cabinet outer doors are open, protecting the user from exposure to UV light. When the sample is in place, outer doors are closed and the slider is pulled out to allow UV light to reach the sample for the appropriate time.

can be enumerated in individual plates. Cell survival for each dose of UV light should be evaluated in triplicate or quadruplicate. Four to 6 h after the cells are plated the growth medium is removed by aspiration and the cells are irradiated in the dishes with increasing doses of UVC radiation (e.g., 0, 2, 4, 6, and 8 J/m^2) at a fluence of 0.8–1.0 $J/m^2/s$. Growth medium is replaced quickly and the dishes are incubated in 5% CO_2 at 37°. It is important to replace the medium in the zero dose plates to prevent unattached cells from attaching and thus skewing results. Plates are incubated for 10 to 21 days or until colonies are large enough to observe with the naked eye once stained.

Plates are washed with PBS and fixed with 100% ethanol or methanol. Cell colonies are stained with 0.1% crystal violet in water or PBS or with 6% Giemsa (Invitrogen) for 10 min, rinsed with deionized water, and air dried. Colonies are counted with the naked eye or with low-power microscopic assistance. Colonies at the edges of the plates are not counted because they are shielded from the UV light by the sides of the dishes (shadow effect). Experimental results should be confirmed in at least two separate experiments. Colony numbers are plotted as a percentage of the survival of unirradiated cells (zero UV radiation) on a log scale.

METHOD 2. Early passage cells are plated at a density of 3×10^5 cells/60-mm culture dish. The following day the medium is aspirated and the dishes are irradiated with UVC light at a fluence of 0.8–1.0 J/m2/s. The growth medium is replaced and dishes are incubated for 4 days. The dishes are washed with PBS, fixed (as described earlier), and stained with 0.1% crystal violet. The extent of cell growth and survival in individual dishes is determined by measuring the amount of crystal violet in the attached (viable) cells. To achieve this cells are destained with a standardized volume of 70% ethanol and the amount of crystal violet stain released into the ethanol is determined by reading the optical density at 575 nm. Each dose point is performed in triplicate and confirmed in independent experiments. Survival is plotted as the ratio of the mean of the OD_{575} nm relative to that of unirradiated controls for each dose. The results of this technique are often skewed by the shadow effect described earlier, the dead cells stained by crystal violet (a protein stain), and staining of the plastic growth surface.

Measuring the Sensitivity of ES Cells to Ionizing Radiation. ES cells are grown on feeder layers of mouse STO cells (CRL-1503m, ATCC) rendered nonviable by treatment with 20 rad of γ radiation or 10 μg/ml of mitomycin C (Sigma) for 1.5 h. STO cells can be frozen after treatment and removed from the freezer as needed. STO cells are plated to confluence on gelatin-treated dishes or flasks. ES cells are added to the cultures and fed ES medium (DMEM, 15% ES qualified fetal bovine serum, and 1:100 volume β-ME stock and 1:100 antibiotic–antimycotic, all from GIBCO). Cultures are maintained in logarithmic growth by passage every third day to new feeder layers. Trypsinized control and test ES cells are suspended in ES medium, exposed to 0, 2, 4, 6, 8, and 10 Gy of ionizing radiation and replated on feeder layers. Cultures are maintained in triplicate for each dose and colonies are grown for 10–14 days, washed, fixed, and stained as described earlier.

Measuring Immunological Responses at the Cellular Level

Response to the T-Cell-Dependent Antigen NP-CG. To determine whether B cells are able to mount a normal T-cell-dependent immune response and differentiate into plasma cells secreting antigen-specific Ig, groups of age-matched mice (8–10 weeks of age) are immunized intraperitoneally with 100 μg alum-precipitated (4-hydroxy-3-nitrophenyl)-acetyl-chicken globulin (NP-CG). Add 1 ml of a 50-g/liter aluminum potassium sulfate solution to 1 ml of NP-CG solution (1 mg). Adjust the pH to 6.5 by adding 4 N NaOH in 25-μl aliquots, monitoring pH with indicator strips. Centrifuge for 5 mins in a microfuge (12,000 rpm). Resuspend the pellet in NaCl (9 g/liter) at 1 mg/ml and store at 4° for up to a day. Inject mice intraperitoneally with 0.1 ml of the resuspension (Schenten *et al.*, 2002). Mice can be boosted weekly with the same dose.

After 14 days, place the cage of mice under a heat lamp to induce vasodilation. After several minutes, place the mice to be bled in a restraint with the tail protruding. Using a sharp scalpel blade or a syringe needle, puncture the lateral vein on the underside of the tail. Collect blood into an Eppendorf tube ($\sim$3–5 drops). Stop the bleeding by applying pressure to the vein. Spin down the blood in a microcentrifuge for 5 min (12,000 rpm) and collect the serum.

NP-specific IgG1 titers are determined 24 h later by ELISA. Plates are coated with 50 μl/well of a 10-μg/ml solution of NP15-BSA (Biosearch Technologies) at 4° overnight and are washed three times with PBS-Tween (0.05% Tween 20). Block wells with (200 μl/well) 3% bovine serum albumin in PBS at room temperature for 2 h and wash wells. Serially diluted preimmune and immune sera are added to the wells and incubated for 2 h at room temperature. Wash the wells. Bound IgG1 antibodies are detected in two steps, first by adding 50 μl of a 200-μg/ml stock biotinylated goat antimouse IgG1 antibody (SBA, Birmingham, Alabama) in PBS-BSA solution and incubating at room temperature for 30 min. Wash the wells. The secondary detection step uses 50 μl of 1:5000 to 1:20,000 dilution of the stock streptavidin-alkaline phosphatase (Roche) in PBS-BSA. Final detection of the colorimetric assay uses 100–200 μl of ATBS (Roche) incubated for 1 h at 22–37° in a plate shaker. Measure at 405 nm against an ATBS blank in a plate reader.

Evaluation of Somatic Hypermutation (SH). Evaluation of SH requires enrichment of cells that have undergone SH from the mass of cells with unchanged genomic DNA. If one has a flow sorter available then this is the simplest way to accomplish the separation. If this is not an option, isolation of B cells can be carried out using biotinylated antibodies recognizing the cells you wish to isolate and binding them to streptavidin-coated magnetic beads. Detailed protocols of these techniques are beyond the scope of this chapter and can be worked out with an immunologist.

ISOLATING CELLS BY FLOW CYTOMETRY. Single-cell suspensions from spleens of NP immunized mice are depleted of erythrocytes by treatment with 0.83% NH$_4$Cl. T cells can also be depleted by treatment with anti-Thy1 antibodies (T24/40 and HO13.4) followed by treatment with rabbit complement. Cells are stained with PE-conjugated anti-CD45R/B220 or with FITC-conjugated antimouse IgM. CD45R(B220)+IgM−, CD45R (B220)+IgM+, or CD5+CD45(B220)+ cells are fractionated by flow cytometry on a FACSVantage. For control experiments, CD4+ and/or CD8+ cells can be obtained from the thymus after staining and sorting (Terauchi *et al.*, 2001).

ISOLATING NAÏVE AND GERMINAL CENTER (GC) B CELLS. Splenic GC B cells are enriched 14 days after NP immunization by magnetic cell separation using the MACS system (Miltenyi Biotech, Bergisch Gladbach, Germany). Following erythrocyte lysis, splenocytes are first incubated with an anti-IgD-biotin mAb (clone 1.3–5 [Roes *et al.*, 1995]) or a combination of anti-IgDa-biotin (Ig[5a]7.2 [Oi and Herzenberg, 1979]) and anti-IgDb-biotin (4/4D7, Takeshi and Tokuhisa, unpublished result) mAb and are subsequently stained with streptavidin-CD43 (S7) microbeads (Miltenyi-Biotech). The labeled cells are then applied to a LD separation column (Miltenyi Biotech). The eluted fraction containing naive B220+IgD+ B cells is incubated with anti-B220-phycoerythrin (PE) (clone R33–24.12 [Leptin *et al.*, 1984]) and anti-IgD fluorescein isothiocyanate (FITC) (clone 1.3–5). An unstained aliquot of the eluted fraction is analyzed with anti-B220-FITC and PE-conjugated peanut agglutinin (PNA) (Vector Laboratories, Burlingame, CA) to exclude the presence of contaminating GC B cells. The flow through, enriched for GC B cells, is stained with anti-B220-FITC and PNA-PE, and B220+PNA+ GC B cells are isolated on a FACS 440 cell sorter (Becton-Dickinson). Naïve cells are isolated according to B220 and IgD expression (Schenten *et al.*, 2002).

Analysis of SH by Polymerase Chain Reaction. Sorted cell populations (50 μl/10^5 cells) are lysed in 10 mM Tris–HCl and 0.5 mg/ml proteinase K

(Roche) for 2.5 h at 50°, followed by the denaturation of proteinase K at 95° for 10 min.

PCR Reactions. Primer J558Fr3 anneals in the framework 3′ region of most J558 V genes. and primer JHCHint hybridizes in the intron 3′ of exon JH4 (Jolly *et al.*, 1997).

J558Fr3–CAGCCTGACATCTGAGGACTCTGC
JHCHint–CTCCACCAGACCTCTCTAGACAGC

Reactions contain 20 μl cell lysate in a 50-μl reaction for 30 s at 95°-, 30 s at 65°, and 2 min at 72° for 32 cycles. Products are cloned into the vector in the pGEM-Teasy kit (Stratagene) and sequenced using the JHCHint primer. Amplification of JH4 intronic sequence-flanking rearranged VH genes can also be performed using a mixture of primers amplifying most VH gene families described (Delbos *et al.*, 2005):

V1-FR3–GAGGACTCTGCRGTCTATTWCTGTGC-3′;
V5-FR3–GAGGACACRGCCATGTATTACTGTGC-3′;
V3-FR3–GAGGACACACCCACATATTACTGTGC-3′;
V7-FR3–GAGGACAGTGCCACTTATTACTGTGC-3′;
V9-FR3–ATGAGGACATGGCTACATATTTCTGT-3′, respectively, in a 6:3:1:1:1 ratio, and
JH3′–TGAGACCGAGGCTAGATGCC-3′.

PCR uses 500 or 1000 cells [15 s, 98°, 30 s, 64°, 30 s, 72° for 50 cycles using Phusion DNA polymerase (NEB)]. Products are cloned and sequenced using the JH3′ primer. The Sμ core (class switching) repeat sequence is amplified using primers Sμ-5′-GTTGAGGTACTGATGC-TGTC-3′ and Sμ-3′-CCAGCCTAGTTTAGCTTAGC-3′ RCR cycles are (45 s at 94°, 30 s at 58°, 2 min at 72°, 40 cycles with Pfu Turbo) and the sequence is determined using primer 5′-CTATTCTGGCTCTTCT TAAGC-3′.

Results are often presented in tabular form (Table II). Sequences with one or more mutation are included in the study. The percentage of each type of mutation is entered in the appropriate boxes in the table. The number of mutations per sequence and the average mutation frequency are similar between $Polk^{-/-}$ mice and wild-type animals, with the transition to transversion ratio (Ts/Tv) of ~1 (Schenten *et al.*, 2002). Although data are not presented in the same format, $Poll^{-/-}$ and $Polm^{-/-}$ mice appear to have normal hypermutation rates as well (Bertocci *et al.*, 2002). Only disruption of the *Polh* gene has an effect on SH, reducing the number of mutations observed in Ig genes in the J_H4 region from 1.4 in controls to 0.8 in mutants (Delbos *et al.*, 2005). The Ts/Tv ratio is not presented. However, when calculated from data presented, the ratio in control animals is 1.16

TABLE II
Example of Table Format Used to Report Mutation in SH Studies

n^a	To				Sum	Ts/Tv[b]
	A	G	C	T		
From						
A	—					
G		—				
C			—			
T				—		

[a] Number of mutations found in $V_H D_H J_H 4$ joints.
[b] Ts/Tv ratio of transitions (Ts) to transversions (Tv).

and in mutant animals it is 1.13. This ratio has been reported at 2.5 in $Msh2^{-/-}$ mice (Delbos *et al.*, 2005; Frey *et al.*, 1998; Martin *et al.*, 2003; Martomo *et al.*, 2004). A more useful measure in this case is the number of C or G changes to any other nucleotide. In control animals the proportion of C/G changes is 44.3% and in $Polh^{-/-}$ animals it is 86.1%, similar to the levels observed in XPV patients (Faili *et al.*, 2004). In $Polk^{-/-}$ mice the C/G mutation rates are unchanged: 38.5% versus 40.5% for controls (Schenten *et al.*, 2002).

Intact Animal Studies

Studies on intact animals are obviously predicated on expected or observed phenotypes. Here we restrict ourselves to the measurement of skin tumors in mice exposed to UV radiation, a phenotype observed in *Polh* mice and possibly in *Polk* mice.

Skin Cancer in Mice Exposed to UV Radiation. Twenty mice, 8–12 weeks of age, from each genotype are tested. The dorsal skin is shaved each week using a pet clipper (A5, Oster) with a #40 blade (surgical). Mice are exposed to UVB light at 14 J/m2, 5 days a week, for 18 weeks as described previously (Cheo, 2000; Nahari *et al.*, 2004). Mice are irradiated in a wood cabinet containing two FS20 bulbs (UVB, 270–315 nm with a peak at 313 nm, Westinghouse) filtered by Kodacel sheeting (Kodak, Rochester, NY). Mice are irradiated until either skin tumors are visible to the naked eye or for a maximum of 18 weeks. The mice are monitored for the presence of skin tumors visible to the naked eye at least twice a week for up to 100 weeks. When a skin tumor is visible the mouse is euthanized, tumors are removed, and a portion of each tumor is fixed in 10% neutral-buffered formalin and prepared for routine histology. Skin tumors are graded and classified by standard histopathological criteria. The remaining portion of the tumor is flash frozen in liquid nitrogen and stored at $-80°$. Tumor incidence curves are generated (see next section) and compared to age-matched wild-type mice.

Previous studies with mice defective in genes required for nucleotide excision repair have demonstrated unique mutational spectra in the *Trp53* gene in cells from skin cancers (Nahari *et al.*, 2004; Reis *et al.*, 2000). RNA is isolated from the skin tumors that arise using any RNA isolation kit and are reverse transcribed with one of the many enzymes available commercially using poly-T or random hexamer primers. cDNA from the *Trp53* gene is amplified and sequenced for mutations. The spectrum of mutations is compared to those reported previously for mice defective in nucleotide excision repair (Nahari *et al.*, 2004; Reis *et al.*, 2000) and compared to the mutations in the human *TP53* gene (http://p53.free.fr/ or http://www-p53. iarc.fr/index.html).

Tumor Incidence. There are no published studies of spontaneous or induced tumor incidence in mice defective in a specialized DNA polymerase, although such studies are in progress (L. McDaniel and E. C. Friedberg, unpublished result). In mouse models without obvious tumor susceptibility cohorts of 60 mice are followed for a 2-year period. In general, animal care committees discourage studies that use death as an end point. Two-year studies allow ample time to discern mild-to-moderate tumor susceptibilities (McDaniel *et al.*, 2003). Mice are checked twice weekly for tumor development and are sacrificed when tumors become visible, or at the end of the period of observation if the mice are tumor free. Animals that appear sick while under observation but have no visible tumors must be sacrificed and are not included in the final tumor incidence, as they could not be followed for the full 2 years. All animals are examined internally and externally for tumors. This includes examining superficial and deep cervical lymph nodes, mediastinal, axillary, brachial and inguinal lymph nodes, and the thymus gland. Tissues are fixed in formalin for histological studies. After histopathological studies are completed, the percentage of surviving animals is plotted against the time point at which the animals were sacrificed (Cheo, 2000).

References

Badcoe, I. G., Smith, C. J., Wood, S., Halsall, D. J., Holbrook, J. J., Lund, P., and Clarke, A. R. (1991). Binding of a chaperonin to the folding intermediates of lactate dehydrogenase. *Biochemistry* **30**, 9195–9200.

Baixeras, E., Roman-Roman, S., Jitsukawa, S., Genevee, C., Mechiche, S., Viegas-Pequignot, E., Hercend, T., and Triebel, F. (1990). Cloning and expression of a lymphocyte activation gene (LAG-1). *Mol. Immunol.* **27**, 1091–1102.

Basi, G., Schmid, E., and Maundrell, K. (1993). TATA box mutations in the *Schizosaccharomyces pombe* nmt1 promoter affect transcription efficiency but not the transcription start point or thiamine repressibility. *Gene* **123,** 131–136.

Bedouelle, H., and Duplay, P. (1988). Production in *Escherichia coli* and one-step purification of bifunctional hybrid proteins which bind maltose: Export of the Klenow polymerase into the periplasmic space. *Eur. J. Biochem.* **171,** 541–549.

Bemark, M., Khamlichi, A. A., Davies, S. L., and Neuberger, M. S. (2000). Disruption of mouse polymerase zeta (Rev3) leads to embryonic lethality and impairs blastocyst development *in vitro. Curr. Biol.* **10,** 1213–1216.

Bertocci, B., De Smet, A., Flatter, E., Dahan, A., Bories, J. C., Landreau, C., Weill, J. C., and Reynaud, C. A. (2002). Cutting edge: DNA polymerases mu and lambda are dispensable for Ig gene hypermutation. *J. Immunol.* **168,** 3702–3706.

Brandt-Carlson, C., and Butel, J. S. (1991). Detection and characterization of a glycoprotein encoded by the mouse mammary tumor virus long terminal repeat gene. *J. Virol.* **65,** 6051–6060.

Bross, P., Jensen, T. G., Andresen, B. S., Kjeldsen, M., Nandy, A., Kolvraa, S., Ghisla, S., Rasched, I., Bolund, L., and Gregersen, N. (1994). Characterization of wild-type human medium-chain acyl-CoA dehydrogenase (MCAD) and mutant enzymes present in MCAD-deficient patients by two-dimensional gel electrophoresis: Evidence for post-translational modification of the enzyme. *Biochem. Med. Metab. Biol.* **52,** 36–44.

Bunting, K. A., Roe, S. M., and Pearl, L. H. (2003). Structural basis for recruitment of translesion DNA polymerase Pol IV/DinB to the beta-clamp. *EMBO J.* **22,** 5883–5892.

Burgess, S. (1977). Molecular-weights of lepidopteran baculovirus dnas: Derivation by electron-microscopy. *J. Gen. Virol.* **37,** 501–510.

Bush, G. L., Tassin, A. M., Friden, H., and Meyer, D. I. (1991). Secretion in yeast: Purification and *in vitro* translocation of chemical amounts of prepro-alpha-factor. *J. Biol. Chem.* **266,** 13811–13814.

Caroni, P., Rothenfluh, A., McGlynn, E., and Schneider, C. (1991). S-cyclophilin: New member of the cyclophilin family associated with the secretory pathway. *J. Biol. Chem.* **266,** 10739–10742.

Chaga, G., Bochkariov, D. E., Jokhadze, G. G., Hopp, J., and Nelson, P. (1999). Natural poly-histidine affinity tag for purification of recombinant proteins on cobalt(II)-carboxymethy-laspartate crosslinked agarose. *J. Chromatogr. A* **864,** 247–256.

Chen, Y. T., Holcomb, C., and Moore, H. P. (1993). Expression and localization of two low molecular weight GTP-binding proteins, Rab8 and Rab10, by epitope tag. *Proc. Natl. Acad. Sci. USA* **90,** 6508–6512.

Cheo, D. L., Meira, L. B., Burns, D. K., Reis, A. M., Issac, T., and Friedberg, E. C. (2000). Ultraviolet B radiation-induced skin cancer in mice defective in the Xpc, Trp53, and Apex (HAP1) genes: Genotype-specific effects on cancer predisposition and pathology of tumors. *Cancer Res.* **60,** 1580–1584.

Chester, N., Kuo, F., Kozak, C., O'Hara, C. D., and Leder, P. (1998). Stage-specific apoptosis, developmental delay, and embryonic lethality in mice homozygous for a targeted disruption in the murine Bloom's syndrome gene. *Genes Dev.* **12,** 3382–3393.

Chong, S., Mersha, F. B., Comb, D. G., Scott, M. E., Landry, D., Vence, L. M., Perler, F. B., Benner, J., Kucera, R. B., Hirvonen, C. A., Pelletier, J. J., Paulus, H., and Xu, M. Q. (1997). Single-column purification of free recombinant proteins using a self-cleavable affinity tag derived from a protein splicing element. *Gene* **192,** 271–281.

Christensen, J., Storgaard, T., Bloch, B., Alexandersen, S., and Aasted, B. (1993). Expression of Aleutian mink disease parvovirus proteins in a baculovirus vector system. *J. Virol.* **67,** 229–238.

Delbos, F., De Smet, A., Faili, A., Aoufouchi, S., Weill, J. C., and Reynaud, C. A. (2005). Contribution of DNA polymerase eta to immunoglobulin gene hypermutation in the mouse. *J. Exp. Med.* **201,** 1191–1196.

di Guan, C., Li, P., Riggs, P. D., and Inouye, H. (1988). Vectors that facilitate the expression and purification of foreign peptides in *Escherichia coli* by fusion to maltose-binding protein. *Gene* **67,** 21–30.

Esposito, G., Godindagger, I., Klein, U., Yaspo, M. L., Cumano, A., and Rajewsky, K. (2000). Disruption of the Rev3l-encoded catalytic subunit of polymerase zeta in mice results in early embryonic lethality. *Curr. Biol.* **10,** 1221–1224.

Evan, G. I., Lewis, G. K., Ramsay, G., and Bishop, J. M. (1985). Isolation of monoclonal antibodies specific for human c-myc proto-oncogene product. *Mol. Cell. Biol.* **5,** 3610–3616.

Faili, A., Aoufouchi, S., Weller, S., Vuillier, F., Stary, A., Sarasin, A., Reynaud, C. A., and Weill, J. C. (2004). DNA polymerase eta is involved in hypermutation occurring during immunoglobulin class switch recombination. *J. Exp. Med.* **199,** 265–270.

Fan, Y., and Skoultchi, A. I. (2004). Genetic analysis of H1 linker histone subtypes and their functions in mice. *Methods Enzymol.* **377,** 85–107.

Fischhaber, P. L., and Friedberg, E. C. (2005). How are specialized (low-fidelity) eukaryotic polymerases selected and switched with high-fidelity polymerases during translesion DNA synthesis? *DNA Repair (Amst.)* **4,** 279–283.

Frey, S., Bertocci, B., Delbos, F., Quint, L., Weill, J. C., and Reynaud, C. A. (1998). Mismatch repair deficiency interferes with the accumulation of mutations in chronically stimulated B cells and not with the hypermutation process. *Immunity* **9,** 127–134.

Friedberg, E. C., and Meira, L. B. (2004). Database of mouse strains carrying targeted mutations in genes affecting biological responses to DNA damage (version 6). *DNA Repair (Amst.)* **3,** 1617–1638.

Friedberg, E. C., Lehmann, A. R., and Fuchs, R. P. (2005). Trading places: How do DNA polymerases switch during translesion DNA synthesis? *Mol. Cell* **18,** 499–505.

Gentz, R., Chen, C. H., and Rosen, C. A. (1989). Bioassay for trans-activation using purified human immunodeficiency virus tat-encoded protein: Trans-activation requires mRNA synthesis. *Proc. Natl. Acad. Sci. USA* **86,** 821–824.

Gerlach, V. L., Feaver, W. J., Fischhaber, P. L., and Friedberg, E. C. (2001). Purification and characterization of pol kappa, a DNA polymerase encoded by the human DINB1 gene. *J. Biol. Chem.* **276,** 92–98.

Herscovics, A., and Orlean, P. (1993). Glycoprotein biosynthesis in yeast. *FASEB J.* **7,** 540–550.

Heyer, W. D., Sipiczki, M., and Kohli, J. (1986). Replicating plasmids in *Schizosaccharomyces pombe*: Improvement of symmetric segregation by a new genetic element. *Mol. Cell. Biol.* **6,** 80–89.

Hochuli, E. (1988). Large-scale chromatography of recombinant proteins. *J. Chromatogr.* **444,** 293–302.

Hopp, T. P., Prickett, K. S., Price, V. L., Libby, R. T., March, C. J., Cerretti, D. P., Urdal, D. L., and Conlon, P. J. (1988). A short polypeptide marker sequence useful for recombinant protein identification and purification. *Biotechnology* **6,** 1204–1210.

Hoss, A., Moarefi, I., Scheidtmann, K. H., Cisek, L. J., Corden, J. L., Dornreiter, I., Arthur, A. K., and Fanning, E. (1990). Altered phosphorylation pattern of simian virus 40 T antigen expressed in insect cells by using a baculovirus vector. *J. Virol.* **64,** 4799–4807.

Hsu, C. Y., Hurwitz, D. R., Mervic, M., and Zilberstein, A. (1991). Autophosphorylation of the intracellular domain of the epidermal growth factor receptor results in different effects on its tyrosine kinase activity with various peptide substrates: Phosphorylation of peptides representing Tyr(P) sites of phospholipase C-gamma. *J. Biol. Chem.* **266,** 603–608.

Huang, T. T., Raineri, I., Eggerding, F., and Epstein, C. J. (2002). Transgenic and mutant mice for oxygen free radical studies. *Methods Enzymol.* **349**, 191–213.

Jansen, J. G., Tsaalbi-Shtylik, A., Langerak, P., Calleja, F., Meijers, C. M., Jacobs, H., and de Wind, N. (2005). The BRCT domain of mammalian Rev1 is involved in regulating DNA translesion synthesis. *Nucleic Acids Res.* **33**, 356–365.

Jolly, C. J., Klix, N., and Neuberger, M. S. (1997). Rapid methods for the analysis of immunoglobulin gene hypermutation: Application to transgenic and gene targeted mice. *Nucleic Acids Res.* **25**, 1913–1919.

Kajiwara, K., J, O. W., Sakurai, T., Yamashita, S., Tanaka, M., Sato, M., Tagawa, M., Sugaya, E., Nakamura, K., Nakao, K., Katsuki, M., and Kimura, M. (2001). Sez4 gene encoding an elongation subunit of DNA polymerase zeta is required for normal embryogenesis. *Genes Cells* **6**, 99–106.

Karpeisky, M., Senchenko, V. N., Dianova, M. V., and Kanevsky, V. (1994). Formation and properties of S-protein complex with S-peptide-containing fusion protein. *FEBS Lett.* **339**, 209–212.

Kidd, I. M., and Emery, V. C. (1993). The use of baculoviruses as expression vectors. *Appl. Biochem. Biotechnol.* **42**, 137–159.

Kim, J. S., and Raines, R. T. (1994). A misfolded but active dimer of bovine seminal ribonuclease. *Eur. J. Biochem.* **224**, 109–114.

Kloc, M., Reddy, B., Crawford, S., and Etkin, L. D. (1991). A novel 110-kDa maternal CAAX box-containing protein from Xenopus is palmitoylated and isoprenylated when expressed in baculovirus. *J. Biol. Chem.* **266**, 8206–8212.

Kobayashi, Y., Watanabe, M., Okada, Y., Sawa, H., Takai, H., Nakanishi, M., Kawase, Y., Suzuki, H., Nagashima, K., Ikeda, K., and Motoyama, N. (2002). Hydrocephalus, situs inversus, chronic sinusitis, and male infertility in DNA polymerase lambda-deficient mice: Possible implication for the pathogenesis of immotile cilia syndrome. *Mol. Cell. Biol.* **22**, 2769–2776.

Kojima, H., Jones, B. T., Chen, J., Cascalho, M., and Sitkovsky, M. V. (2004). Hypoxia-inducible factor 1alpha-deficient chimeric mice as a model to study abnormal B lymphocyte development and autoimmunity. *Methods Enzymol.* **381**, 218–229.

Kost, T. A., Condreay, J. P., and Jarvis, D. L. (2005). Baculovirus as versatile vectors for protein expression in insect and mammalian cells. *Nat. Biotechnol.* **23**, 567–575.

Kouzuki, H., Suzuki, H., Ito, K., Ohashi, R., and Sugiyama, Y. (1998). Contribution of sodium taurocholate co-transporting polypeptide to the uptake of its possible substrates into rat hepatocytes. *J. Pharmacol. Exp. Ther.* **286**, 1043–1050.

Kuroda, K., Veit, M., and Klenk, H. D. (1991). Retarded processing of influenza virus hemagglutinin in insect cells. *Virology* **180**, 159–165.

Lee, S. S., Tian, L., Lee, W. S., and Cheung, W. T. (2002). Application of fluorescent differential display and peroxisome proliferator-activated receptor (PPAR) alpha-null mice to analyze PPAR target genes. *Methods Enzymol.* **357**, 214–240.

Leptin, M., Potash, M. J., Grutzmann, R., Heusser, C., Shulman, M., Kohler, G., and Melchers, F. (1984). Monoclonal antibodies specific for murine IgM. I. Characterization of antigenic determinants on the four constant domains of the mu heavy chain. *Eur. J. Immunol.* **14**, 534–542.

Martin, A., Li, Z., Lin, D. P., Bardwell, P. D., Iglesias-Ussel, M. D., Edelmann, W., and Scharff, M. D. (2003). Msh2 ATPase activity is essential for somatic hypermutation at a-T basepairs and for efficient class switch recombination. *J. Exp. Med.* **198**, 1171–1178.

Martomo, S. A., Yang, W. W., and Gearhart, P. J. (2004). A role for Msh6 but not Msh3 in somatic hypermutation and class switch recombination. *J. Exp. Med.* **200**, 61–68.

Mattion, N. M., Mitchell, D. B., Both, G. W., and Estes, M. K. (1991). Expression of rotavirus proteins encoded by alternative open reading frames of genome segment 11. *Virology* **181,** 295–304.

Maundrell, K. (1990). nmt1 of fission yeast: A highly transcribed gene completely repressed by thiamine. *J. Biol. Chem.* **265,** 10857–10864.

McDaniel, L. D., Chester, N., Watson, M., Borowsky, A. D., Leder, P., and Schultz, R. A. (2003). Chromosome instability and tumor predisposition inversely correlate with BLM protein levels. *DNA Repair (Amst.)* **2,** 1387–1404.

McDonald, J. P., Frank, E. G., Plosky, B. S., Rogozin, I. B., Masutani, C., Hanaoka, F., Woodgate, R., and Gearhart, P. J. (2003). 129-derived strains of mice are deficient in DNA polymerase iota and have normal immunoglobulin hypermutation. *J. Exp. Med.* **198,** 635–643.

Meira, L. B., Devaraj, S., Kisby, G. E., Burns, D. K., Daniel, R. L., Hammer, R. E., Grundy, S., Jialal, I., and Friedberg, E. C. (2001). Heterozygosity for the mouse Apex gene results in phenotypes associated with oxidative stress. *Cancer Res.* **61,** 5552–5557.

Nahari, D., McDaniel, L. D., Task, L. B., Daniel, R. L., Velasco-Miguel, S., and Friedberg, E. C. (2004). Mutations in the Trp53 gene of UV-irradiated Xpc mutant mice suggest a novel Xpc-dependent DNA repair process. *DNA Repair (Amst.)* **3,** 379–386.

Oi, V. T., and Herzenberg, L. A. (1979). Localization of murine Ig-1b and Ig-1a (IgG 2a) allotypic determinants detected with monoclonal antibodies. *Mol. Immunol.* **16,** 1005–1017.

Pray, T. R., and Laimins, L. A. (1995). Differentiation-dependent expression of E1–E4 proteins in cell lines maintaining episomes of human papillomavirus type 31b. *Virology* **206,** 679–685.

Reis, A. M., Cheo, D. L., Meira, L. B., Greenblatt, M. S., Bond, J. P., Nahari, D., and Friedberg, E. C. (2000). Genotype-specific Trp53 mutational analysis in ultraviolet B radiation-induced skin cancers in Xpc and Xpc Trp53 mutant mice. *Cancer Res.* **60,** 1571–1579.

Renshaw, M. W., Price, L. S., and Schwartz, M. A. (1999). Focal adhesion kinase mediates the integrin signaling requirement for growth factor activation of MAP kinase. *J. Cell Biol.* **147,** 611–618.

Roes, J., Muller, W., and Rajewsky, K. (1995). Mouse anti-mouse IgD monoclonal antibodies generated in IgD-deficient mice. *J. Immunol. Methods* **183,** 231–237.

Sambucetti, L. C., Schaber, M., Kramer, R., Crowl, R., and Curran, T. (1986). The fos gene product undergoes extensive post-translational modification in eukaryotic but not in prokaryotic cells. *Gene* **43,** 69–77.

Schenten, D., Gerlach, V. L., Guo, C., Velasco-Miguel, S., Hladik, C. L., White, C. L., Friedberg, E. C., Rajewsky, K., and Esposito, G. (2002). DNA polymerase kappa deficiency does not affect somatic hypermutation in mice. *Eur. J. Immunol.* **32,** 3152–3160.

Shima, N., Hartford, S. A., Duffy, T., Wilson, L. A., Schimenti, K. J., and Schimenti, J. C. (2003). Phenotype-based identification of mouse chromosome instability mutants. *Genetics* **163,** 1031–1040.

Shimizu, T., Shinkai, Y., Ogi, T., Ohmori, H., and Azuma, T. (2003). The absence of DNA polymerase kappa does not affect somatic hypermutation of the mouse immunoglobulin heavy chain gene. *Immunol. Lett.* **86,** 265–270.

Shin, Y. S., Tohya, Y., Oshikamo, R., Kawaguchi, Y., Tomonaga, K., Miyazawa, T., Kai, C., and Mikami, T. (1993). Antigenic analysis of feline calicivirus capsid precursor protein and its deleted polypeptides produced in a mammalian cDNA expression system. *Virus Res.* **30,** 17–26.

Smith, D. B., and Johnson, K. S. (1988). Single-step purification of polypeptides expressed in *Escherichia coli* as fusions with glutathione S-transferase. *Gene* **67,** 31–40.

Stofko-Hahn, R. E., Carr, D. W., and Scott, J. D. (1992). A single step purification for recombinant proteins: Characterization of a microtubule associated protein (MAP 2) fragment which associates with the type II cAMP-dependent protein kinase. *FEBS Lett.* **302,** 274–278.

Terauchi, A., Hayashi, K., Kitamura, D., Kozono, Y., Motoyama, N., and Azuma, T. (2001). A pivotal role for DNase I-sensitive regions 3b and/or 4 in the induction of somatic hypermutation of IgH genes. *J. Immunol.* **167,** 811–820.

Thain, A., Gaston, K., Jenkins, O., and Clarke, A. R. (1996). A method for the separation of GST fusion proteins from co-purifying GroEL. *Trends Genet.* **12,** 209–210.

Trifunovic, A., and Larsson, N. G. (2002). Tissue-specific knockout model for study of mitochondrial DNA mutation disorders. *Methods Enzymol.* **353,** 409–421.

Van Sloun, P. P., Varlet, I., Sonneveld, E., Boei, J. J., Romeijn, R. J., Eeken, J. C., and De Wind, N. (2002). Involvement of mouse Rev3 in tolerance of endogenous and exogenous DNA damage. *Mol. Cell. Biol.* **22,** 2159–2169.

Wilson, I. A. (1984). Structure of antigenic and immunogenic determinants in proteins and synthetic peptides. *Ann. Sclavo Collana Monogr.* **1,** 129–138.

Wittschieben, J., Shivji, M. K., Lalani, E., Jacobs, M. A., Marini, F., Gearhart, P. J., Rosewell, I., Stamp, G., and Wood, R. D. (2000). Disruption of the developmentally regulated Rev3l gene causes embryonic lethality. *Curr. Biol.* **10,** 1217–1220.

Wong, S. T., and Storm, D. R. (2002). Generation of adenylyl cyclase knockout mice. *Methods Enzymol.* **345,** 206–231.

Zander, L., and Bemark, M. (2004). Immortalized mouse cell lines that lack a functional Rev3 gene are hypersensitive to UV irradiation and cisplatin treatment. *DNA Repair (Amst.)* **3,** 743–752.

[23]   Purification and Characterization of *Escherichia coli* DNA Polymerase V

By Katharina Schlacher, Qingfei Jiang, Roger Woodgate, and Myron F. Goodman

Abstract

Cell survival and genome rescue after UV irradiation in *Escherichia coli* depends on DNA repair mechanisms induced in response to DNA damage as part of the SOS regulon. SOS occurs in two phases. The first phase is dominated by accurate repair processes such as excision and recombinational DNA repair, while the second phase is characterized by a large ~100-fold increase in mutations caused by an error-prone replication of damaged DNA templates. SOS mutagenesis occurs as a direct result of the action of

0076-6879/06 $35.00
DOI: 10.1016/S0076-6879(06)08023-2

the UmuDC gene-products, which form the low fidelity *Escherichia coli* DNA polymerase V, a heterotrimeric complex composed of $UmuD'_2C$. This chapter describes the preparation of highly purified native pol V that is suitable for a wide range of biochemical studies of protein-protein, protein-DNA interactions and translesion-synthesis (TLS) mechanisms.

Introduction

DNA polymerase V (pol V) is a low-fidelity enzyme that is primarily responsible for DNA damage-induced chromosomal mutations in *Escherichia coli*. The bacteria replicates DNA accurately using pol III holoenzyme, but the fidelity decreases dramatically when cells are exposed to DNA-damaging agents, such as UV light (Friedberg *et al.*, 1995). DNA damage triggers the formation of an activated RecA filament (Friedberg *et al.*, 1995; Kuzminov, 1999) that stimulates the self-cleavage of LexA, the transcriptional repressor of the SOS regulon (Little, 1984; Luo *et al.*, 2001). Following repressor inactivation, more than 40 genes in the SOS regulon are induced to facilitate DNA damage repair (Courcelle *et al.*, 2001; Friedberg *et al.*, 1995).

Pol V, which is induced late in the SOS response, is an error-prone polymerase with the ability to copy and bypass damaged DNA template bases, a process referred to as translesion synthesis (TLS) (Reuven *et al.*, 1998; Tang *et al.*, 1998, 1999). Pol III presumably stalls when encountering a template lesion resulting in collapse of the replication fork (Cox, 2003; Lusetti and Cox, 2002), whereupon pol III is replaced on the β-processivity clamp by one of three SOS-induced polymerases, pol II, pol IV or pol V (Goodman, 2002). These polymerases are involved in both error-free lesion avoidance repair mechanisms (e.g., replication fork regression) involving recombinational pathways (Cox *et al.*, 2000), or error-prone TLS mechanisms (Goodman, 2002).

Pol V is encoded by the damage-inducible *umuDC* operon (Kato and Shinoura, 1977; Steinborn, 1978). The UmuD protein is inactive in SOS mutagenesis until it is processed to UmuD' via a self-cleavage reaction (McDonald *et al.*, 1998) analogous to the cleavage of the LexA repressor protein (Burckhardt *et al.*, 1988; Nohmi *et al.*, 1988; Shinagawa *et al.*, 1988). Cleavage of LexA induces the global SOS response, while cleavage of UmuD to form UmuD' causes SOS mutagenesis. Once generated, UmuD' homodimers bind to UmuC protein to form pol V, a heterotrimeric $UmuD'_2C$ complex (Bruck *et al.*, 1996; Woodgate *et al.*, 1989). Pol V is soluble in aqueous solution (Bruck *et al.*, 1996), whereas UmuC appears to be essentially insoluble (Bruck *et al.*, 1996; Woodgate *et al.*, 1989) unless bound to at least 1 molecule of UmuD' (Shen *et al.*, 2003).

The biochemical properties of pol V have been characterized (Fuchs *et al.*, 2004; Fujii *et al.*, 2004; Pham *et al.*, 2001; Schlacher *et al.*, 2005; Shen *et al.*, 2002, 2003; Tang *et al.*, 1999, 2000). It lacks $3'-5'$ exonuclease proofreading activity and is able to incorporate nucleotides opposite abasic template lesion, *cis-syn* T-T pyrimidine dimers, and (6-4) T-T photoproducts with *in vitro* specificities (Shen *et al.*, 2002; Tang *et al.*, 2000) that accurately reflect SOS mutagenic signatures *in vivo* (Lawrence *et al.*, 1990a,b; LeClerc *et al.*, 1991; Smith *et al.*, 1996). Pol V activity is stimulated by RecA protein and β sliding clamp by 350- and 3-fold, respectively (Pham *et al.*, 2001). The β sliding clamp which binds to and increases the processivity of all five *E. coli* polymerases, is required for TLS *in vivo* (Lenne-Samuel *et al.*, 2002; Sutton *et al.*, 2005) but not *in vitro* (Pham *et al.*, 2002; Schlacher *et al.*, 2005). RecA is essential for pol V-catalyzed TLS *in vivo* and *in vitro*, but not in the form of a RecA nucleoprotein filament (Schlacher *et al.*, 2005). Instead, we have proposed that RecA may serve as an integral part of a pol V holoenzyme, perhaps as UmuD$'_2$C-RecA$_2$ complex (Schlacher *et al.*, 2005). This protein complex, referred to as a pol V-mutasome, is necessary and sufficient for TLS *in vitro* (Schlacher *et al.*, 2005).

In this paper, we describe a procedure to obtain highly purified native pol V (UmuD$'_2$C) from *E. coli*, suitable for a wide variety of biochemical experiments to study protein-protein, protein-DNA interactions and TLS mechanisms.

Assay Methods

Principle

DNA polymerase catalyzes the addition of deoxyribonucleotides to the $3'$-OH termini of a DNA primer annealed to a template. The chemical reaction is

$$DNA_n + dNTP \rightarrow DNA_{n+1} + PP_i$$

Procedure

The polymerase activity of pol V can be measured on DNA substrates containing both undamaged, and damaged templates for trans lesion synthesis (TLS) studies. The assay includes either slowly hydrolysable Adenosine $5'$-O-(3-thio)triphosphate (ATPγS) or continuously regenerated ATP to serve as a nucleotide co-factor for RecA protein. The polymerase activity is examined by extending a P^{32}-labeled primer annealed to a linear template oligomer, where the template overhang region varies, for example, between 6 to 24 nt, and may contain a variety of site-specific damaged or modified bases, e.g., abasic moiety, pyrimidine dimer, bulky lesion, etc. By limiting the length of the template overhang we can distinguish between bona fide pol V activity and a low level of pol III contamination that appears in early (low salt) phophocellulose fractions. Virtually no pol V activity is detected on short template overhangs in the absence of RecA protein whereas pol III activity can be discerned, even in the absence of β clamp.

ATPγS System

The reaction buffer is composed of 20 mM Tris-HCl (pH 7.5), 0.1 mM EDTA (pH 8.0), 25 mM sodium glutamate, 4% (vol/vol) glycerol, 8 mM MgCl$_2$, and 5 mM dithiothreitol (DTT). The reaction mixture (10 μl) contains standard reaction buffer and 20nM primer template DNA, which is preincubated for 3 min at 37° with RecA and ATPγS (0.5 mM). Since pol V is virtually undetectable in the absence of RecA, the concentration of RecA is varied between 0 and 2000 nM to observe pol V activity as a function of RecA protein. The reaction is initiated by the addition of pol V (240 nM) and dNTP's (0.5 mM) and is terminated after 10 min by adding 20 μL of 20 mM EDTA in 95% formamide. The extension products are resolved on a 12% denaturing polyacrylamide gel and the integrated intensities of radiolabelled extension products are quantified by phosphorimaging.

ATP-Regenerating System

The reactions are performed as above, with the exception that instead of ATPγS (0.5 mM), the reaction mixture is preincubated with ATP (1 mM), phosphocreatine (3 mM) and creatine phosphokinase (1unit/reaction).

Purification of DNA Polymerase V

Expression Vectors and Strains

As noted above, UmuD has to undergo a posttranslational cleavage reaction *in vivo* before it can interact with UmuC. To circumvent cleavage of UmuD, UmuD'$_2$ can be overexpressed as a recombinant protein (Frank *et al.*, 1993). To simultaneously overproduce UmuD'$_2$ and UmuC, the two proteins were expressed in an operon (similar to their arrangement on the *E.coli* chromosome) under the control of various inducible promoters. Although greatest expression can be achieved when the *umuD'C* operon is expressed from the thermo-inducible λP_L promoter, or the phage T7 gene *10* promoter, most of the overproduced UmuC protein accumulates in insoluble inclusion bodies. As a consequence, we have found that the greatest amount of soluble UmuD'$_2$C can be obtained when they are expressed from the weaker, IPTG-inducible, pTac promoter in pKK223-3 (Pharmacia Biotech Inc.).

We have used a variety of *E. coli* strains, with different genetic backgrounds, e.g., *uvrA* (Bruck *et al.,* 1996); $\Delta polB$ *dnaE*ts (Tang *et al.,* 1999); $\Delta polB$ $\Delta dinB$ (Shen *et al.,* 2003) to overexpress UmuD'$_2$C. The key, is that all the strains carry a deletion of the chromosomal *umuDC* operon (Woodgate, 1992), so as to avoid possible heterodimer formation between chromosomally expressed UmuD and plasmid expressed UmuD' (Battista *et al.,* 1990) that leads to rapid ClpXP-dependent proteolysis of UmuD/D' heterodimers (Frank *et al.,* 1996; Gonzalez *et al.,* 2000) and the subsequent inactivation of UmuC (Shen *et al.,* 2003).

Cell Growth

Large-scale bacterial growth was conducted in Luria-Bertani (LB) media (10g/L Tryptone, 5g/L Yeast extract, 10g/L NaCl) supplemented with 10 g/L of glucose, in a 300L fermentor (New Brunswick Scientific). The fermentor was inoculated with an overnight culture (1% v/v). The bacteria were grown at 37° at pH 7.0, in the presence of 30% dissolved oxygen. When the culture reaches an OD$_{600}$ ~5.0, 1 mM isopropyl-beta-D-thiogalactoside (IPTG) is added for 2–3 hours to induce pol V. The culture is then cooled to 10° during centrifugation and harvesting of cells. The collected wet cell paste can be frozen at −80° for long-term storage purposes. Depending upon the strain used for protein expression, the yields were ~1.5–3.0 Kg of wet cell paste from a 300L culture.

Cell Lysis

The standard procedure, summarized in Table I and Fig. 1, involves cell lysis with lysozyme, PEI (Polyethyleneimine) precipitation, ammonium sulfate precipitation followed by a second ammonium sulfate precipitation and separation by gel filtration and ion-exchange chromatography. A preparative scale purification starts with about 1500 g of frozen cells and yields about 30 mg of purified pol V. Lysis buffer (100 mM NaCl, 50 mM Tris-HCl pH 7.5, 10% sucrose) is added to frozen cells to a final concentration of 150 g cells/L. This step is typically carried out by partitioning an equal volume of cell slurry into three 4L beakers. Cells are thawed at 4° and lysozyme (10 g/L dissolved in lysis buffer) is added to the slurry of cells to achieve a final concentration of 0.7 mg/mL. The cell slurry is gently stirred for 1 hr at 4°, followed by incubation at 37° for an additional 10 min.

TABLE I
PURIFICATION OF DNA POLYMERASE V FROM *ESCHERICHIA COLI*

Fraction	Volume (mL)	Protein concentration (mg/mL)[a]	Final protein yield (mg)
(A) Lysate[b]	12,000	2.98	35,770
(B) PEI precipitation[b]	3,800	4.00	15,200
(C) 1st ammonium sulfate precipitation[b]	405	14.6	5,913
(D) 2nd ammonium sulfate precipitation (= gel filtration load)	29	17.6	510
(E) Gel filtration column (= PC – load)	79	0.96	75.8
(F) Phosphocellulose column total pol V	24	~1.5[c]	33.4

[a] Protein concentrations were determined by the method of Bradford.
[b] Represents total amounts of supernatant and pellet branch.
[c] Fractions vary in protein concentration.

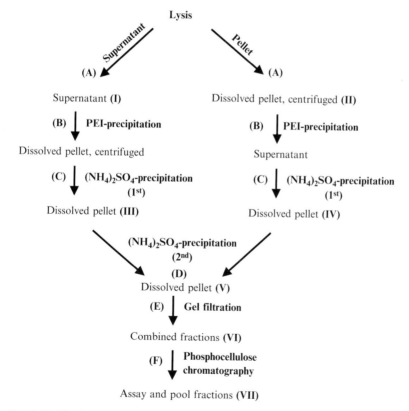

FIG. 1. Purification outline. The Arabic letters (A–F) next to the purification steps refer to the fractions in Table I. The roman numbers (I–VII) refer to the lanes in Fig. 2.

Centrifugation is performed in a GSA rotor at 11,000 rpm and 4° for 1 hr. Both, supernatant (Fig. 2, lane I) and the pellet are saved and handled separately for the PEI and the first $(NH_4)_2SO_4$ precipitation. All subsequent steps are performed at 4°. All enzyme fractions prior to the gel filtration chromatography step (Fig. 2, lane V) are monitored by Western blotting with antibodies against UmuD' and UmuC to identify pol V, along with separations carried out by 12% SDS PAGE and Comassie staining. The additional Western blotting step is useful in distinguishing UmuC from nearby protein bands (see Fig. 2, closed arrows). In our experience, the key steps of the purification scheme are the precipitations prior to the chromatography steps, with emphasis on the two ammonium sulfate cuts. Care taken in these early steps helps to ensure optimal recovery and purification in the later steps.

Supernatant Branch

10% (w/v) Poly(ethyleneimine) solution (PEI, adjusted to pH 7.6 with HCl, Sigma-Aldrich) is added slowly with gentle stirring to the supernatant to a final concentration of 1.1%. The suspension is allowed to sit on ice overnight without stirring. Centrifugation is performed in a GSA rotor at 11,000 rpm for 20 min at 4°. The supernatant is discarded. The pellet is allowed to dissolve overnight in R-buffer (20 mM Tris-HCl pH 7.5, 0.1 mM EDTA, 1 mM DTT, 20% glycerol) with 1M NaCl at 4° by gentle stirring, followed by centrifugation in a GSA rotor at 11,000 rpm for 25 min. The pellet is discarded. $(NH_4)_2SO_4$ is added slowly to the supernatant to a final concentration of 326 g/L until completely dissolved, and the suspension is left on ice for 3 h. The precipitate is collected by centrifugation in a GSA rotor at 11,000 rpm for 35 min. The pellet is resuspended in R-buffer with 1M NaCl (Fig. 2, lane III).

Pellet Branch

The pellet after cell lysis is resuspended into 3L of R-buffer containing 1M NaCl. The solution is stirred overnight at 4° and spun down at 11,000 rpm for 1 ~ 1.5 hr in a SS-34 rotor. The pellet is discarded. 10% PEI (pH 7.6) is added to the supernatant (Fig. 2, lane II) to a final concentration of 1.1%, and the suspension is allowed to sit on ice for 1 hr. Centrifugation is performed at 11,000 rpm for 20 min. The pellet is discarded.

$(NH_4)_2SO_4$ is added slowly to the supernatant to a final concentration of 326 g/L. The suspension is allowed to sit overnight on ice and the ammonium sulfate precipitate is collected by centrifugation at 11,000 rpm for 35 min. The pellet is resuspended in R-buffer containing 1 M NaCl (Fig. 2, IV).

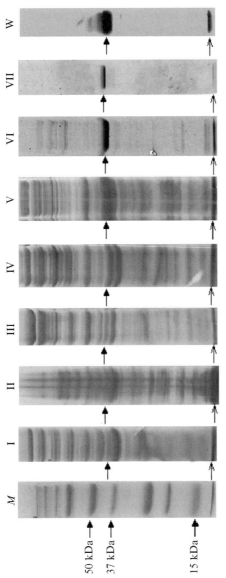

FIG. 2. Comassie stained SDS polyacrylamide gel (12%) showing protein bands during the purification steps of *E. coli* native DNA polymerase V (UmuD'₂C) and a Western blot with antibodies against UmuD' and UmuC. The order of the gel bands are from left to right: (M) Marker (Biorad precision protein standard) (I) supernatant after cell lysis, (II) dissolved cell lysis pellet after centrifugation to clarity, (III) first (NH₄)₂SO₄ redisolved precipitate from the supernatant branch, (IV) first (NH₄)₂SO₄ redisolved precipitate from the pellet branch, (V) Gel filtration chromatography load, (VII) Phosphocellulose chromatography load, (VIII) purified pooled pol V fractions. 30–40 µg protein was loaded in each lane except for lane (VII) (15µg) and lane (VIII) (1µg). UmuD' (12.3 kDa) runs close to the front of the protein gel and is visualized in lane (W), where antibodies against UmuD' and UmuC where used in a Western blot for the pooled fractions. In all lanes except (M), the closed arrow indicates UmuC (47.7 kDa) and the open arrow indicates UmuD' (12.3 kDa).

2nd Ammonium Sulfate Precipitation

The dissolved $(NH_4)_2SO_4$ precipitates from both the supernatant-(Fig. 2, lane III) and the pellet – branches (Fig. 2, lane IV) are combined and dialyzed overnight against R buffer containing 1 M NaCl at 4°. $(NH_4)_2SO_4$ is added to the combined fractions to a final concentration of 164 g/L. The suspension is kept on ice for 3 h and the precipitate is collected at 11,000 rpm for 35 min in a SS-34 rotor. The pellet is resuspended in R-buffer containing 1M NaCl (the buffer volume is kept as small as possible) and dialyzed against the same buffer overnight at 4° (Fig. 2, lane V).

Gel-Filtration Chromatography

A Superdex^{TM}200 gel-filtration HiLoad^{Tm} 26/60 (Pharmacia Biotech) column is equilibrated in R-buffer with 1M NaCl. To avoid pressure problems, glycerol is omitted from the R-buffer for the gel-filtration column but is added to each fraction eluted off the column. The best resolution is obtained when no more than 5 mL of sample is loaded on a column of this size. We have typically carried out this step in 5 separate runs. The dialyzed sample (Fig. 2, lane V) is clarified by centrifugation at 11,000 rpm for 20 min in a SS-34 rotor. Recentrifugation may be necessary to eliminate any residual turbidity. The clear supernatant is loaded on the column at a flow rate of 1 mL/min. Pol V is eluted with a total of 400 mL R-buffer (containing 1 M NaCl) at a flow rate of 1.5 mL/min. Fractions (3ml) are collected between 150–222 mL and imediately supplemented with glycerol to a final concentration of 20%. The pol V content of every other fraction is visualized by Comassie-stained 12% SDS polyacrylamide gels. Pol V-rich fractions from all runs are combined and dialyzed overnight at 4° against R-buffer containing 300 mM NaCl (Fig. 2, lane VI).

Phosphocellulose Chromatography

Whatman cellulose phosphate ion-exchange resin P11 is used. The resin is prepared as recommended by the manufacturer, decanted into a 20 mL disposable column (Poly-Prep, Bio-Rad) such that a packed column volume is about 5mL. The column is equilibrated with 4 column volumes R-buffer containing 200 mM NaCl, washed with 4 column volumes of R buffer containing 1 M NaCl and again washed with 10 column volumes of R-buffer with 200 mM NaCl at a flow rate of 0.2 mL/min.

Immediately prior to loading, the dialyzed sample (Fig. 2, lane VI) is diluted in R-buffer in the absence of NaCl to reach a conductivity that is approximately equal to the column buffer conductivity. The sample is loaded very slowly at 0.15 mL/min. Depending on the sample volume, the loading process may be performed overnight at 4°.

The column is washed with six column volumes of R-buffer containing 200 mM NaCl and four column volumes of R-buffer containing 250 mM NaCl, followed by another gradient wash between 250–270 mM over 3 column volumes (flow rate of 0.2 mL/min). Pol V is not expected to elute at these salt concentrations. However it is advisable to begin collecting fractions (2 mL) at 250 mM NaCl onwards. Pol V protein is eluted between a salt gradient of about 270 mM to 800 mM NaCl in R-buffer, 7 column volumes at a flow rate of 0.2 mL/min. The main peak of pol V is usually observed at NaCl concentrations between 460 and 760 mM.

The pol V fractions are analyzed for purity using Comassie-stained SDS polyacrylamide gels and for polymerase activity as described above. Fractions with robust TLS activity in the presence of optimal levels of RecA protein, but with very little polymerase activity in the absence of RecA (Fig. 3), can then be pooled if desired (Fig. 2, lane VII and lane W).

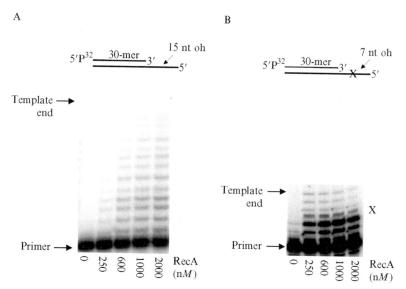

Fig. 3. Assay for pol V activity. Pol V activity was measured using an undamaged P^{32}-labeled primer/template DNA (A). TLS was detected using a primer/template DNA containing an abasic (tetrahydrofuran) moiety X on a 7 nt template overhang (oh) (B). The polymerase activity is dependent on the presence of RecA protein and a nucleotide cofactor in form of either ATP (A) or ATPγS (B).

Individual and pooled fractions are stable when stored at −80° for longer than 6 months, without a measurable loss of activity.

Acknowledgments

This work was supported by NIH grants ESO12259 and R37GM21422 to MFG and by the NIH Intramural Research Program to RW.

References

Battista, J. R., Ohta, T., Nohmi, T., Sun, W., and Walker, G. C. (1990). Dominant negative *umuD* mutations decreasing RecA-mediated cleavage suggest roles for intact UmuD in modulation of SOS mutagenesis. *Proc. Natl. Acad. Sci. USA* **87,** 7190–7194.

Bruck, I., Woodgate, R., McEntee, K., and Goodman, M. F. (1996). Purification of a Soluble UmuD'C Complex from *Escherichia coli:* Cooperative Binding of UmuD'C to Single-Stranded DNA. *J. Biol. Chem.* **271,** 10767–10774.

Burckhardt, S. E., Woodgate, R., Scheuremann, R. H., and Echols, H. (1988). UmuD mutagenesis protein of *Escherichia coli:* Overproduction, purification, and cleavage by RecA. *Proc. Natl. Acad. Sci. USA* **85,** 1811–1815.

Courcelle, J. A., Khodursky, A., Peter, B., Brown, P. O., and Hanawalt, P. C. (2001). Comparative gene expression profiles following UV exposure in wild-type and SOS deficient *Escherichia coli. Genetics* **158,** 41–64.

Cox, M. M. (2003). The bacterial RecA protein as a motor protein. *Annu. Rev. Microbiol.* **57,** 551–577.

Cox, M. M., Goodman, M. F., Kreuzer, K. N., Sherattt, D. J., Sandler, S. J., and Marians, K. J. (2000). The importance of repairing stalled replication forks. *Nature* **404,** 37–41.

Frank, E. G., Gonzalez, M., Ennis, D. G., Levine, A. S., and Woodgate, R. (1996). Regulation of SOS mutagenesis by proteolysis. *Proc. Natl. Acad. Sci. USA* **93,** 10291–10296.

Frank, E. G., Hauser, J., Levine, A. S., and Woodgate, R. (1993). Targeting of the UmuD, UmuD', and MucA' mutagenesis proteins to DNA by RecA protein. *Proc. Natl. Acad. Sci. USA* **90,** 8169–8173.

Friedberg, E. C., Walker, G. C., and Siede, W. (1995). DNA Repair and Mutagenesis ASM Press, Washington, DC.

Fuchs, R. P., Fujii, S., and Wagner, J. (2004). Properties and functions of *Escherichia coli:* Pol IV and Pol V. *Dna Repair And Replication* **69,** 229–264.

Fujii, S., Gasser, V., and Fuchs, R. P. (2004). The biochemical requirements of DNA polymerase V-mediated translesion synthesis revisited. *J. Mol. Biol.* **341,** 405–417.

Gonzalez, M., Rasulova, F., Maurizi, M. R., and Woodgate, R. (2000). Subunit-specific degradation of the UmuD/D' heterodimer by the ClpXP protease: The role of *trans* recognition in UmuD' stability. *EMBO J.* **19,** 5251–5258.

Goodman, M. F. (2002). Error-prone repair DNA polymerases in prokaryotes and eukaryotes. *Annu. Rev. Biochem.* **70,** 17–50.

Kato, T., and Shinoura, Y. (1977). Isolation and characterization of mutants of *Escherichia coli* deficient in induction of mutagenesis by ultraviolet light. *Mol. Gen. Genet.* **156,** 121–131.

Kuzminov, A. (1999). Recombinational Repair of DNA Damage in *Escherichia coli* and Bacteriophage λ. *Microbiol. Mol. Biol. Rev.* **63,** 751–813.

Lawrence, C. W., Banerjee, S. K., Borden, A., and LeClerc, J. E. (1990a). T-T cyclobutane dimers are misinstructive, rather than non-instructive, mutagenic lesions. *Mol. Gen. Genet.* **222**, 166–168.

Lawrence, C. W., Borden, A., Banerjee, S. K., and LeClerc, J. E. (1990b). Mutation frequency and spectrum resulting from a single abasic site in a single-stranded vector. *Nucleic. Acids. Res.* **18**, 2153–2157.

LeClerc, J. E., Borden, A., and Lawrence, C. W. (1991). The thymine-thymine pyrimidine-pyrimidone(6-4) ultraviolet light photoproduct is highly mutagenic and specifically induces 3' thymine-to-cytosine transitions in *Escherichia coli*. *Proc. Natl. Acad. Sci. USA* **88**, 9685–9689.

Lenne-Samuel, N., Wagner, J., Etienne, H., and Fuchs, R. P. (2002). The processivity factor beta controls DNA polymerase IV traffic during spontaneous mutagenesis and translesion synthesis *in vivo*. *EMBO Rep.* **3**, 45–49.

Little, J. W. (1984). Autodigeston of lexA and phage λ repressors. *Proc. Natl. Acad. Sci. USA* **81**, 1375–1379.

Luo, Y., Pfuetzner, R. A., Mosimann, S., Paetzel, M., Frey, E. A., Cherney, M., Kim, B., Little, J. W., and Strynadka, N. C. (2001). Crystal structure of LexA: A conformational switch for regulation of self-cleavage. *Cell* **106**, 585–594.

Lusetti, S. L., and Cox, M. M. (2002). The bacterial RecA protein and the recombinational DNA repair of stalled replication forks. *Annu. Rev. Biochem.* **71**, 71–100.

McDonald, J. P., Frank, E. G., Levine, A. S., and Woodgate, R. (1998). Intermolecular cleavage by UmuD-like mutagenesis proteins. *Proc. Natl. Acad. Sci. USA* **95**, 1478–1483.

Nohmi, T., Battista, J. R., Dodson, L. A., and Walker, G. C. (1988). RecA-mediated cleavage activates UmuD for mutagenesis: Mechanistic relationship between transcriptional derepression and posttranslational activation. *Proc. Natl. Acad. Sci. USA* **85**, 1816–1820.

Pham, P., Bertram, J. G., O'Donnell, M., Woodgate, R., and Goodman, M. F. (2001). A model for SOS-lesion targeted mutations in *E. coli* involving pol V, RecA, SSB and β sliding clamp. *Nature* **409**, 366–370.

Pham, P., Seitz, E. M., Saveliev, S., Shen, X., Woodgate, R., Cox, M. M., and Goodman, M. F. (2002). Two distinct modes of RecA action are required for DNA polymerase V-catalyzed translesion synthesis. *Proc. Natl. Acad. Sci. USA* **99**, 11061–11066.

Reuven, N. B., Tomer, G., and Livneh, Z. (1998). The Mutagenesis Proteins UmuD' and UmuC Prevent Lethal Frameshifts While Increasing Base Substitution Mutations. *Mol. Cell* **2**, 191–199.

Schlacher, K., Leslie, K., Wyman, C., Woodgate, R., Cox, M. M., and Goodman, M. F. (2005). DNA polymerase V and RecA protein, a minimal mutasome. *Mol. Cell* **17**, 561–572.

Shen, X., Sayer, J. M., Kroth, H., Pontén, I., O'Donnell, M., Woodgate, R., Jerina, D., and Goodman, M. F. (2002). Efficiency and Accuracy of SOS-induced DNA Polymerases Replicating Benzo[*a*]pyrene-7,8-diol 9,10-Epoxide A and G Adducts. *J. Biol. Chem.* **277**, 5265–5274.

Shen, X., Woodgate, R., and Goodman, M. F. (2003). *Escherichia coli* DNA Polymerase V Subunit Exchange: A Post-SOS Mechanism To Curtail Error-Prone DNA Synthesis. *J. Biol. Chem.* **278**, 52546–52550.

Shinagawa, H., Iwasaki, T., Kato, T., and Nakata, A. (1988). RecA protein-dependent cleavage of UmuD protein and SOS mutagenesis. *Proc. Natl. Acad. Sci. USA* **85**, 1806–1810.

Smith, C. A., Wang, M., Jiang, N., Che, L., Zhao, X., and Taylor, J. S. (1996). Mutation spectra of M13 vectors containing site-specific Cis-Syn, Trans-Syn-I, (6-4), and Dewar pyrimidone photoproducts of thymidylyl-(3'—>5')-thymidine in *Escherichia coli* under SOS conditions. *Biochemistry* **35**, 4146–4154.

Steinborn, G. (1978). Uvm mutants of *Escherichia coli* K12 deficient in UV mutagenesis. I. Isolation of *uvm* mutants and their phenotypical characterization in DNA repair and mutagenesis. *Mol. Gen. Genet.* **165**, 87–93.

Sutton, M. D., Duzen, J. M., and Maul, R. W. (2005). Mutant forms of the *Escherichia coli* β sliding clamp that distinguish between its roles in replication and DNA polymerase V-dependent translesion DNA synthesis. *Mol. Microbiol.* **55**, 1751–1766.

Tang, M., Bruck, I., Eritja, R., Turner, J., Frank, E. G., Woodgate, R., O'Donnell, M., and Goodman, M. F. (1998). Biochemical Basis of SOS Mutagenesis in *Escherichia coli:* Reconstitution of *in vitro* lesion bypass dependent on the UmuD$_2$'C mutagenic complex and RecA protein. *Proc. Natl. Acad. Sci. USA* **95**, 9755–9760.

Tang, M., Pham, P., Shen, X., Taylor, J.-S., O'Donnell, M., Woodgate, R., and Goodman, M. F. (2000). Roles of *E. coli* DNA polymerases IV and V in lesion-targeted and untargeted SOS mutagenesis. *Nature* **404**, 1014–1018.

Tang, M., Shen, X., Frank, E. G., O'Donnell, M., Woodgate, R., and Goodman, M. F. (1999). UmuD'$_2$C is an error-prone DNA polymerase, *Escherichia coli* pol V. *Proc. Natl. Acad. Sci. USA* **96**, 8919–8924.

Woodgate, R. (1992). Construction of a *UmuDC* operon substitution mutation in *Escherichia coli*. *Mutation Res.* **281**, 221–225.

Woodgate, R., Rajagopalan, M., Lu, C., and Echols, H. (1989). UmuC mutagenesis protein of *Escherichia coli:* Purification and interaction with UmuD and UmuD'. *Proc. Natl. Acad. Sci. USA* **86**, 7301–7305.

[24] Yeast and Human Translesion DNA Synthesis Polymerases: Expression, Purification, and Biochemical Characterization

By Robert E. Johnson, Louise Prakash, and Satya Prakash

Abstract

The emergence of translesion DNA synthesis (TLS) as a primary mechanism by which eukaryotic cells tolerate DNA damage has led to a large effort to characterize the biochemical properties of the individual DNA polymerases and their roles in promoting replication past DNA lesions. The low-fidelity Y family DNA polymerases constitute a large proportion of TLS polymerases, and four of the five subfamilies of this class of polymerases are represented in eukaryotes. The eukaryotic B family DNA polymerase Polζ also functions in TLS. We have had success in expressing and purifying these TLS polymerases from yeast cells, sometimes in milligram quantities. The purified proteins have been used to determine their ability to synthesize DNA on various modified templates

METHODS IN ENZYMOLOGY, VOL. 408
0076-6879/06 $35.00
DOI: 10.1016/S0076-6879(06)08024-4

and to analyze the kinetic efficiencies with which bypass occurs. Purified proteins have also been used to determine the X-ray crystal structures of several Y-family DNA polymerases. This chapter describes a general outline of methods used in our laboratory for the expression and purification of these TLS DNA polymerases from yeast cells and for assaying some of their biochemical properties.

Introduction

Eukaryotes harbor four of the five subfamilies of Y-family DNA polymerases. These include Rad30A (Polη), Rad30B (Polι), DinB (Polκ), and Rev1, and each is distinguished by markedly different structural features and lesion bypass properties (Prakash *et al.*, 2005). For instance, only Polη has the ability to replicate through a cyclobutane pyrimidine dimer (CPD) efficiently and accurately (Johnson *et al.*, 1999b, 2000b; Washington *et al.*, 2000, 2003). The significance of this polymerase in promoting replication through sunlight induced CPDs is attested by the fact that defects in human Polη cause the variant form of xeroderma pigmentosum (XP-V) (Johnson *et al.*, 1999a; Masutani *et al.*, 1999). Polι displays marked template specificity, being more efficient at synthesizing DNA opposite template purines than pyrimidines (Haracska *et al.*, 2001b; Johnson *et al.*, 2000a; Tissier *et al.*, 2000; Washington *et al.*, 2004a; Zhang *et al.*, 2000). This specificity derives from its unique ability to orient the template purine nucleotide in the *syn* conformation, utilizing Hoogsteen base pairing to direct deoxynucleotide incorporation (Johnson *et al.*, 2005a; Nair *et al.*, 2004). Human Polκ is adept at extending from base pair mismatches (Washington *et al.*, 2002) and from misaligned primer–templates (Wolfle *et al.*, 2003), and its proficiency at extending from a C opposite template G containing N^2 minor groove adducts provides for a mechanism to replicate through such lesions (Washington *et al.*, 2004c). The Rev1 protein is a G-template-specific DNA polymerase, but also has a strong bias for incorporating dCTP opposite abasic sites and other template residues (Haracska *et al.*, 2002a; Nelson *et al.*, 1996a). In addition, eukaryotes harbor a B-family TLS polymerase, Polζ, which exists as a complex of the Rev3 and Rev7 proteins (Nelson *et al.*, 1996b). Polζ is very efficient at extending from mispaired and damage containing template–primer termini (Haracska *et al.*, 2001d; Johnson *et al.*, 2000a, 2003; Washington *et al.*, 2004b)

Although many laboratories routinely use *Escherichia coli*-based expression systems to express eukaryotic proteins, we have had much success in overexpressing TLS polymerases in the yeast *Saccharomyces cerevisiae*. For some TLS polymerases, we find that the proteins are more soluble in yeast and host toxicity problems are reduced and, unlike other eukaryotic hosts, such as insect cells, large quantities of yeast cells can be grown relatively quickly. Not all the problems are avoided, however, as some proteins, such as Polζ, remain largely insoluble. Still, sufficient quantities of Polζ for enzymatic and interaction studies can be obtained. At the other end of the spectrum, though, milligram quantities of several TLS polymerases have been purified from yeast for determining their X-ray crystal structures (Nair et al., 2004; Trincao et al., 2001; Uljon et al., 2004). Because each TLS polymerase often presents its own unique expression and purification characteristics, this chapter describes a general outline of the methods used for their expression and purification and for assaying their biochemical properties. Any variations that apply to individual polymerases are indicated.

Expression of GST-Tagged TLS Polymerases in Yeast

Expression Vector

Our yeast expression system is based on the galactose-inducible (GAL1–10 UAS) phosphoglycerate kinase (PGK) promoter developed previously (Kingsman et al., 1990). The glutathione S-transferase (GST) gene has been cloned downstream of the GAL-PGK promoter, maintaining a unique BglII site used for cloning the genes of interest. Although earlier versions of the vector used for expressing TLS polymerases contained a factor Xa or a thrombin cleavage site (Johnson et al., 1999b, 2000a, 2001) that allowed for the cleavage of the GST tag from the purified protein, these proteases are rather inefficient and can cause nonspecific cleavage of the protein. The vector we routinely use now, pBJ842 (Fig. 1), contains a Prescission protease (Amersham Pharmacia) cleavage site near the BglII cloning site. pBJ842 is a 2-μm-based vector that carries the yeast leu2-d allele, which lacks a complete promoter. Yeast cells harboring pBJ842 acquire an increased copy number of the plasmid in order to maintain growth on media lacking leucine (Kingsman et al., 1990), thereby leading to increasing expression of the TLS polymerase. We have also generated a series of GAL-PGK:GST(prescission) expression vectors that harbor the wild-type yeast TRP1, URA3, or LEU2 genes, but these do not typically result in expression levels as high as those observed with the leu2-d containing plasmid pBJ842. Fusion proteins are expressed by cloning the genes of interest in frame with GST(prescission) at the unique BglII site, in which the codon 5′-GAT-3′ codes for aspartic acid (see inset in Fig. 1).

To coexpress two different proteins in yeast cells, we clone one of the genes in pBJ842, which allows for purification of a GST-tagged protein,

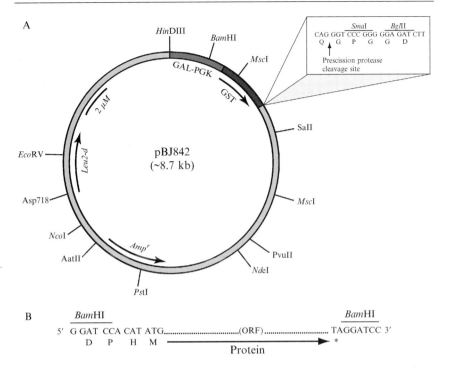

FIG. 1. Cloning of TLS polymerases. (A) Map of GST expression vector pBJ842. Positions of several restriction endonuclease sites, the *Leu2-d* allele, and the *Amp*[r] gene are indicated. (Inset) Sequence of the cloning site used to generate fusions with GST, along with the location of the prescission protease cleavage site (indicated by an arrow) and the *Bgl*II site used for cloning. (B) Example of a PCR-generated TLS polymerase open reading frame for expression. The leader sequence contains a *Bam*HI restriction site for cloning the ORF in frame with GST at the *Bgl*II site in pBJ842. The TAG translation termination codon, as well as a *Bam*HI site, is generated at the 3′ end of the ORF. The ATG is the initiating methionine (M) codon of the ORF.

while the other gene is expressed from the *GAL-PGK* promoter in a vector that contains either the *URA3* or the *TRP1* gene and produces the native protein.

Cloning

The open reading frames (ORFs) of TLS polymerases and other proteins are amplified by standard polymerase chain reaction (PCR) techniques using genomic DNA or cloned genes, or from mRNA by RT-PCR. PCR primers are designed to generate restriction endonuclease sites just upstream and downstream of the ORF for cloning into the BglII site in pBJ842. In all cases, the integrity of the amplified DNAs is verified by sequencing. Alternatively, previously cloned wild-type genes are used to replace internal regions of the PCR-generated fragments. We have identified a leader sequence that gives high levels of protein expression when used as the linker between the GST and the fused protein. This sequence, 5'-GGATCCACATATG-3', where the ATG encodes for the initiating methionine, contains a BamHI site (Fig. 1B) used for cloning the gene in frame with GST at the BglII site in pBJ842 (Fig. 1A, inset). Substitution of the BamHI site in this sequence with BglII has resulted in significantly lower expression for the same protein. For this reason, we have often removed internal BamHI sites within genes by silent mutation so that the ORF could be easily cloned using the aforementioned sequence to maximize their expression. Use of the aforementioned sequence will leave a seven amino acid leader peptide (GPGGDPH) attached to the N terminus of the purified protein when the GST tag is cleaved off by prescission protease (Fig. 1). At the 3' end of the ORF, the termination codon TAG, followed by either a BamHI site or a BglII site, is used, eliminating the 3'-untranslated region (Fig. 1B). The ORF is then cloned in frame with GST in pBJ842 at the BglII site (Fig. 1A, inset) or in the BglII site of the native expression vector. Although available, we have not utilized the SmaI site in pBJ842 (Fig. 1A, inset) and thus do not know the expression levels or the stability of fusion proteins when using this site for cloning.

Growth of Yeast Cells and Induction of Protein Expression

Expression constructs are transformed into the protease-deficient yeast strain BJ5464 (MATα ura3-52 trp1 leu2Δ1 his3Δ200 pep4::HIS3 prb1Δ1.6R GAL) (Jones, 1991) or its derivatives. Although not confirmed genetically, the trp1 allele in BJ5464 is likely to be trp1–289 because it exhibits a high rate of reversion to TRP^+, characteristic of this mutation. To facilitate the use of plasmids harboring the TRP1 gene, we have made a trp1Δ derivative of BJ5464, named YRP654. Yeast cells transformed with vectors containing the Leu2-d allele will not grow if they are plated onto SC-leu media directly because the initial low plasmid copy number is not sufficient to complement the leu2Δ1 mutation. To circumvent this problem, Leu2d containing vectors are cotransformed with a CEN/ARS URA3 vector and cells are plated on SC-ura medium. The resulting Ura^+ transformants can then be replica plated onto SC-Leu media to isolate the Leu2-d transformants. Alternatively, the transformed cells can be grown in liquid SC media for 8–24 h at 30° before plating on SC-Leu. For achieving high levels of expression, the use of freshly transformed cells is recommended. The protocol for growing *Leu2-d* yeast cells and for inducing protein is as follows.

1. Inoculate cells using a population of transformants at $\sim$1–2 $\times$ 10^6 cells/ml in SC-leu medium containing 2% dextrose, 2% glycerol, and 1.8% sodium DL-lactate (60% stock, 30 ml/liter) and grow for 16 h at 30°.
2. Dilute the overnight (O/N) culture 1:20 in SC-leu medium containing 2% glycerol and 1.8% sodium DL-lactate, but lacking dextrose. Grow for an additional 16–24 h at 30°.
3. Induce protein by the addition of 2% galactose, 1% peptone, and 0.5% yeast extract and grow for 6–7 h. Solid constituents can be added. Harvest cells by centrifugation and store the yeast paste at −70°.

Typical yields are about 3–10 g yeast pellet per liter of culture. Cells should not be washed with distilled water. Although individual transformants can be prescreened to identify the highest expressers, the slight increase in expression over that observed in the general population does not typically warrant the extra time and effort.

Protein Purification

Preparation of Yeast Cell Extract

All steps are done on ice. Frozen yeast paste is resuspended in a total of 2.5 volumes 1× cell breakage buffer (CBB) (50 mM Tris–HCl, pH 7.5, 10% sucrose, 1 mM EDTA), keeping in mind that 1 g of yeast occupies approximately 1 ml volume. Add 1.25 ml 2× CBB per gram yeast paste, and 4 M NaCl can be added to a final concentration of 0–500 mM, depending on the protein(s) being purified. For a GST-fusion protein, 500 mM is typically used. Water is then added to bring the final suspension volume up to 2.5 ml per gram. Also added to a final concentration are 10 mM β-mercaptoethanol, 0.5 mM benzamidine–HCl, 0.5 mM phenylmethylsulfonyl fluoride, and 5 μg/ml of protease inhibitors leupeptin, aprotinin, chymostatin, and pepstatin A. Complete protease inhibitor tablets (Roche) are a good alternative to using individual protease inhibitors.

We have used two methods for the breakage of yeast cells. For volumes of 15 to 200 ml, samples are passed through a 40K French press cell (Spectronic Instruments) twice at 20,000 psi at 4°. For larger volumes, a mechanical bead beater (Biospec Products) is used with precooled 0.5-mm zirconia/silica beads and a cooling chamber. The cooling chamber is filled with an ice/ethanol/dry ice mixture because ice alone does not keep the sample near 4°. Cells are disrupted five times in 45-s bursts, at 45-s intervals.

Cell debris is first removed by centrifugation at 10,000 rpm for 10 min, and the clarified extract is subsequently prepared from the resulting supernatant by ultracentrifugation at ~100,000g (e.g., 60,000 rpm for 30 min or 40,000 rpm for 45 min in Beckman type 70 Ti or type 45 Ti rotors, respectively). Extract is harvested from the clear, middle portion of the sample, avoiding any white floating lipid and the very bottom layer. Figure 2 (lanes 1 and 3) shows SDS–PAGE analysis of 1 μl each of extract from yeast strains expressing the catalytic cores of Polκ (residues 19–526) and Polι (residues 1–420), respectively. The extracts are estimated to contain about 1–2 mg/ml of GST-tagged polymerase.

Purification by Binding to Glutathione-Sepharose Affinity Column

Glutathione-Sepharose, in the form of free matrix or prepacked columns, is available from a variety of sources. We use a glutahione-Sepharose fast flow matrix packed in HR5 type columns (Amersham Pharmacia) because the prefilters prohibit any insoluble material from entering the column and because the matrix can be removed from the column prior to batch elution.

GST-fusion proteins can be bound to glutathione-Sepharose directly from yeast extract, but the use of an ammonium sulfate precipitation step first is often desirable. We have found that the majority of GST-tagged TLS polymerases precipitate with 0.208 g/ml (0–35% fraction) ammonium sulfate (Fig. 2, lanes 2 and 4). The ammonium sulfate precipitate is resuspended in 1× GST binding buffer (GBB) containing 10 mM β-mercaptoethanol and protease inhibitors in a volume equal to that of the extract prior to precipitation. GBB is equivalent to CBB except that 10% glycerol is substituted for 10% sucrose. The sample is then dialyzed against 10 to 100 volumes 1× GBB for several hours prior to loading on the column.

The binding capacity of glutathione-Sepharose varies from 5 to 15 mg protein/ml matrix, and the amount of matrix used is proportional to the expected total GST-tagged protein in the extract as predicted from SDS–PAGE analysis. Samples are passed over the column slowly, taking approximately 2–3 h to load. The method of rocking the extract and the matrix is not recommended, as this binding procedure typically leads to aggregation/precipitation of the bound protein. The column is washed with 10 column volumes 1× CBB, or 1× GBB, containing 1 M NaCl (or less as needed), 10 mM β-mercaptoethanol, and 0.01% NP-40. Protease inhibitors are no longer needed, and their use at this stage inhibits the subsequent

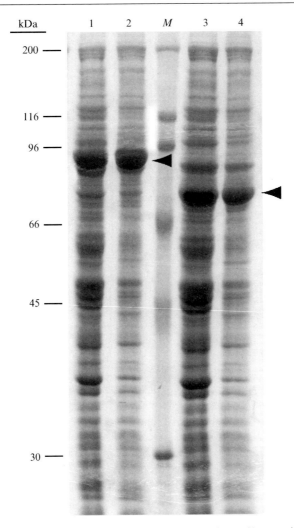

FIG. 2. SDS–PAGE analysis of yeast extract and ammonium sulfate precipitations from yeast cells expressing GST fusion proteins. Yeast cells harboring GST-Polκ (residues 19–526) or GST-Polι (residues 1–420) were grown, and extract was prepared as described. Lanes 1 and 2, 1 μl of yeast extract or 1 μl of the 0–35% ammonium sulfate pellet from cells expressing GST-Polκ (residues 19–526) protein, respectively. Lanes 3 and 4, 1 μl of yeast extract or 1 μl of the 0–35% ammonium sulfate pellet from cells expressing GST-Polι (residues 1–420) protein, respectively. Lane M, protein markers. Molecular masses (kDa) are indicated on the left. Arrowheads indicate positions of the GST-tagged polymerases. The gel was stained with Coomassie R250.

proteolytic cleavage by prescission protease. After equilibrating the matrix with 5 column volumes of elution buffer [50 mM Tris–HCl, pH 7.5, 150 mM NaCl, 10% glycerol, 5 mM dithiothreitol (DTT), 0.01% NP-40], the protein can be eluted as either the GST-tagged or the -untagged protein containing the seven amino acid leader peptide. Although the following steps can be performed on the column, we prefer to remove the matrix from the column and transfer to an appropriately sized tube.

The GST-tagged polymerase is eluted by incubating the matrix with an equal volume of elution buffer containing 40 mM glutathione for 10 min on ice (the pH of elution buffer should be adjusted to pH 7.5 with NaOH after the addition of reduced glutathione). The untagged form of the polymerase (with the seven amino acid leader peptide) can be cleaved directly off the beads by treatment with prescission protease, which is active at 4° and very specific for cleaving the recognition peptide sequence LEVLFQGP. The matrix is resuspended in an equal volume of elution buffer containing 50 units protease/ml and is rocked slowly for 15 h at 4°. The prescission protease is GST tagged so it will bind the matrix and will not elute with the cleaved protein. In both cases, the elution mixture is transferred to a precooled, empty spin column, placed in a 15 Sarstedt tube, and spun briefly in a tabletop centrifuge at 4° to collect the protein.

At this point, the proteins are typically 75–95% pure, as analyzed by SDS–PAGE and Coomassie R250 staining. Regardless of whether proteins are eluted by glutathione or by prescission protease, the presence of two contaminating proteins with molecular masses of ∼70 and ∼75 kDa may be evident. We think they are heat shock proteins that bind the fusion protein and do not bind the glutathione-Sepharose directly. The ammonium sulfate precipitation step after extract preparation often reduces the copurification of these contaminants. Further purification by standard ion-exchange column chromatography is done to obtain a nearly homogeneous preparation of the TLS polymerase. Most of the TLS polymerases can be effectively concentrated by centrifugation in a microconcentrator at 4°, and buffer constituents such as salt or glutathione can be removed by standard dialysis against an appropriate buffer.

Determination of accurate enzyme concentrations is a prerequisite for comparing the efficiencies of wild-type and mutant versions of these TLS polymerases. Enzyme concentrations are determined by a combination of Coomassie-stained gel analysis and by absorbance using the Bradford (1976) and BCA (Smith *et al.*, 1985) assays using bovine serum albumin (BSA) to create a standard curve. Because the Bradford assay is based on Coomassie stain binding and because not all proteins bind Coomassie equally, the results should be consistent with results from the BCA assay and with the comparison of the Coomassie-stained protein with a known amount of molecular weight standard. Aliquots (5–10 μl) of purified proteins are flash frozen in liquid nitrogen and stored at $-70°$.

DNA Synthesis Assays

DNA Substrates

The standard DNA substrate consists of an undamaged or damage-containing template annealed to a 5′ ^{32}P-labeled primer (Fig. 3). Oligonucleotide primers can be designed for either a running start assay (Fig. 3A), in which

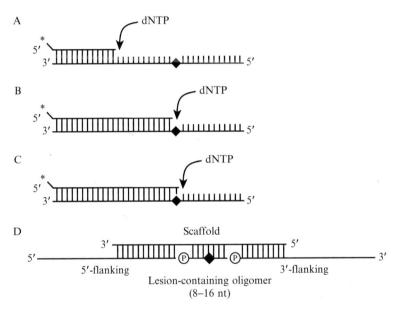

FIG. 3. DNA substrates used for lesion bypass. (A) A running start substrate. The primer anneals several bases 3′ to the lesion, depicted by a diamond in the template. The polymerase must synthesize several nucleotides before encountering the lesion. (B) The standing start substrate for examining dNTP incorporation opposite the lesion. The primer anneals such that the 3′ terminus pairs with the template base 3′ to the lesion, and incorporation by the polymerase occurs opposite the lesion. (C) Standing start lesion substrate for examining extension. The primer anneals such that its 3′ nucleotide is opposite the lesion. The assay measures the ability of the polymerase to extend from the nucleotide opposite the lesion. (D) Generation of a long DNA template from a short damage-containing oligomer. The 5′- and 3′-flanking oligomers and the lesion-containing oligomer are annealed to a complementary scaffold oligomer, which holds the components of the template together for the ligation reaction to occur. The lesion-containing oligomer and the 3′-flanking oligomer are first phosphorylated by T4 kinase and ATP, as indicated by the circled P.

the primer terminus is located at a distance 3′ to the lesion and requires the polymerase to synthesize a length of DNA before encountering the lesion, or a standing start assay, where the primer terminus is situated one base 3′ to the lesion. Standing start primers can be designed to assay for nucleotide incorporation opposite the lesion site (Fig. 3B) or to examine subsequent extension of the primer terminus paired with a lesion (Fig. 3C). While a variety of DNA lesions and degenerate bases are available as phosphoramidites and can be incorporated into a DNA template by standard oligonucleotide synthesis, many interesting lesions are yet unavailable in this form. For our studies, several lesions have been obtained by ultraviolet (UV) or chemical treatment of a short oligonucleotide, followed by HPLC purification. The *cis–syn* TT dimer and the 6-4 (TT) photoproduct, for instance, were obtained by UV treatment of a 10 nucleotide oligomer, and the adducted oligomers were then purified by HPLC and identified by their unique absorption spectrum. Although now available as a phosphoramidite, for our studies, thymine glycol was prepared by treatment of a short oligomer with $KMnO_4$. The damaged short oligomer is then incorporated into a longer sequence context by ligation to flanking oligonucleotides of ∼20–40 nucleotides in length using a scaffold oligonucleotide (∼ 40 nucleotides) to hold the template together (Fig. 3D). Briefly, the adducted oligomer (1000 pmol) and the 3′-flanking oligonucleotide (1500 pmol) are phosphorylated in 50 μl using T4 polynucleotide kinase (Roche) and 1 mM ATP in 1× ligase buffer for 1 h at 37°. After heat inactivation of the kinase, the 5′-flanking oligomer (1500 pmol), the scaffold oligomer (2500 pmol), and 1 μl 10× ligase buffer are added and the volume is increased to 60 μl. The mixture is heated to 95° and allowed to cool slowly over several hours to anneal the oligomers. Fresh 1 mM ATP and 5 U of ligase are added and incubated for 8 h at 4°. Five units of fresh DNA ligase are then added, and the reaction is incubated for 16 h at 16° before finally shifting to room temperature for an additional 4 h. The full-length adducted template is PAGE purified.

Oligonucleotides templates can also be made with biotin attached to the 5′ and 3′ ends, and subsequent to annealing the primer, streptavidin molecules can be bound to the biotin ends. These biotin–streptavidin-bound DNA substrates are needed for assaying the stimulation of TLS polymerases by proliferating cell nuclear antigen (PCNA), as streptavidin prevents the sliding of the PCNA ring off the linear DNA, after it has been loaded onto the DNA substrate by the eukaryotic clamp loader RF-C (Haracska *et al.*, 2001b, 2002b).

Primers are 5′ ^{32}P labeled using T4 polynucleotide kinase (Roche) and [γ-^{32}P]ATP (6000 Ci/mmol, Amersham Pharmacia), followed by gel filtration to remove the unincorporated label. The template and the labeled primer oligonucleotides are combined in a 1.5:1 molar ratio such that the primer concentration is 100 nM in 10 mM Tris–HCl containing 20 mM NaCl and annealed by heating the mixture in a beaker of 95° water followed by cooling to room temperature over a period of several hours by placing it in a covered ice bucket.

Although oligonucleotide stock concentrations are calculated from OD$_{260}$ in 8 M urea and from their extinction coefficients, it is important to ensure that when comparing the kinetics of insertion opposite a DNA lesion versus the undamaged template, the two templates are of equal concentration. This can be determined by testing DNA synthesis of a polymerase from a running start primer on the two templates. Primer usage should be the same on both templates, thus ensuring that any difference observed in synthesis opposite the lesion is not due to differences in template concentration. Additionally, the integrity of any DNA lesion in the template should be verified by the inability of a non-TLS, replicative DNA polymerase, such as Polδ, to bypass it. For example, a *cis–syn* TT dimer presents a complete block to synthesis by Polδ (Johnson *et al.*, 1999b).

DNA Polymerase Assays

The standard DNA polymerase assay contains 25 mM Tris–HCl, pH 7.5, 1 mM DTT, 0.1 mg/ml BSA, 10% glycerol, 5 mM MgCl$_2$, and 10 nM DNA substrate. For assays involving PCNA stimulation, 150 mM NaCl should be added (Haracska *et al.*, 2001a,b,c, 2002b). Deoxynucleoside 5′ triphosphates are included at various concentrations depending on the specific assay. For a typical DNA synthesis assay, 10–100 μM each of dGTP, dATP, dTTP, and dCTP is sufficient. A 5× solution of DNA polymerase is made in ice-cold 5× polymerase buffer (125 mM Tris–HCl, pH 7.5, 5 mM DTT, 0.5 mg/ml BSA) and stored on ice prior to use. We find that when stored on ice, these TLS polymerases are stable in 5× polymerase buffer for several days to a week without any appreciable loss in activity, even at polymerase concentrations as low as 0.5 nM. Reactions are carried out in a 96-well PCR (Fisher) plate using a standard reaction volume of 5 μl consisting of 2.5 μl 20% glycerol, 0.5 μl 50 mM MgCl$_2$, 0.5 μl DNA substrate (0.1 μM), and 0.5 μl 10× dNTP and 1 μl 5× buffer containing the polymerase. The 96-well PCR plate can be cut with scissors to the desired number of wells. Reactions are assembled on ice by mixing the glycerol, DNA substrate, MgCl$_2$, and dNTP to a final volume of 4 μl and then equilibrated to the reaction temperature in a thermocycler for 1 min. We assay yeast TLS polymerases at 30° and human TLS polymerases at 37°. The archeal DPO4 polymerase can be assayed at 60–70° (Johnson *et al.*, 2005b). Temperatures higher than that result in denaturation of the substrate. Reactions are initiated by the addition of 1 μl 5× polymerase/buffer and are terminated at various times by the addition of 30 μl stop solution (90% formamide, 1× TBE, 0.3% bromphenol blue, 0.3% xylene cyanole). After terminating the reactions, the PCR plate is covered with sealing tape and cycled to 95° for 3 min, followed by rapid cooling to 4°.

Reaction products are separated on thin (0.4 mm) TBE polyacrylamide gels containing 8 M urea. Each sample (2.5 μl) is loaded into a 3-mm well (64-well comb), and the gel is run at a constant 75 W until there is sufficient separation of the reaction products from the primer. Products can be visualized by autoradiography or phosphor imager analysis.

Efficiency and Fidelity Assays

Because the Y-family TLS polymerases and Polζ have no 3′→5′ proofreading exonuclease activity, their efficiencies and fidelities on various DNA templates can be determined from steady-state kinetic analysis using the primer extension gel assay as described by Creighton et al. (1995). We stress the requirement for using appropriate enzyme concentrations in these assays, particularly for examining the synthesis opposite DNA lesions. TLS polymerases need to be assayed under conditions in which nucleotide incorporation is linear with time and protein concentration, and for single nucleotide incorporation studies, under conditions in which less than 20% of the primers are utilized. Synthesis by TLS polymerases opposite a lesion should always be compared to that on an undamaged DNA template.

Standard polymerase reaction conditions are used, except that reactions contain a single dNTP, ranging in concentration from 0 to 4 mM. dNTP concentrations should range from 10-fold lower to 10-fold higher than the K_m. The 10× dNTP stock solutions containing equal amounts of dNTP and MgCl$_2$ are made from appropriate dilutions of individual 0.1 M dNTPs (Roche) and 0.1 M MgCl$_2$, and aliquots are stored at −70°. Because each efficiency experiment requires multiple reactions in which only the dNTP concentration varies, we have developed a manual, high-throughput procedure utilizing a thermocycler and a 96-well PCR plate. The various 10× dNTPs (0.5 μl) are added to the bottom of individual wells of the PCR plate, and then 3.5 μl of a stock solution of template (100 nM, 0.5 μl per reaction), glycerol (20%, 2.5 μl per reaction), and MgCl$_2$ (50 mM, 0.5 μl per reaction) is added to the side of the well. The substrate/MgCl$_2$/glycerol mixture can be coaxed to the bottom of the wells by gently tapping the plate on the bench top. The PCR plate containing the reaction mixtures is then equilibrated to the reaction temperature in the thermocycler for

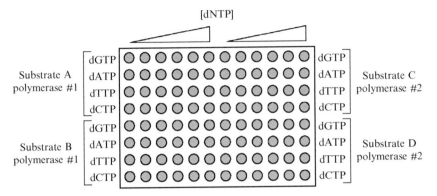

FIG. 4. A typical kinetic assay layout using a 96-well PCR plate. Each well contains a 4-μl reaction mixture containing DNA substrate, glycerol, MgCl$_2$, and dNTP. The plate is prewarmed to the reaction temperature in a thermocycler for 1 min, and reactions are initiated by the addition of 1 μl 5× polymerase buffer containing the TLS polymerase. Each column of eight reactions can be initiated at 10- to 20-s intervals using an eight-channel multipipetter and subsequently terminated at the designated time using the same 10- to 20-sec intervals, thereby keeping the time constant.

1 min. Using an eight-channel micropipetter, each row or column of reactions is initiated by the addition of 1 μl DNA polymerase in 5× reaction buffer (Fig. 4). At the appropriate time, a second eight-channel pipetter can be used to terminate the reactions with 30 μl stop solution. Multiple sets of reactions can be performed in the same PCR plate, e.g., those with different dNTPs, templates, or TLS polymerases; by initiating different sets of reaction at 20-s intervals and subsequently terminating at similar 20-s intervals, a large array of assays can be performed in a short period. All assays are repeated in triplicate. The samples are then denatured and resolved by denaturing TBE–PAGE as described earlier.

Determination of Kinetic Parameters

The products of nucleotide insertion reaction are analyzed by a phosporimager (Molecular Dynamics) using the ImageQuant software (Amersham-Pharmacia). To quantitate the band intensities, two boxes are used: one box encompasses both the primer and the incorporation product(s) and the second box contains only the reaction product(s) (or the equivalent position in the no dNTP control lane). The background correction for all boxes is set to histogram peak and the volumes of each box are then determined (Fig. 5). Dividing the product(s) band volume by the total volume for each lane and multiplying by the DNA substrate

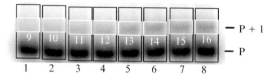

FIG. 5. ImageQuant analysis of a phosphor image scan. The primer (P) and the reaction products of nucleotide incorporation (P + 1) are indicated. Black boxes encompass both the primer and the extension product, and white boxes encompass only the nucleotide insertion product. The background correction for all boxes is set for the histogram peak, and the volumes are determined.

concentration (10 nM) give the amount of product formed. This is then divided by the reaction time and the enzyme concentration to give the rate of nucleotide incorporation per enzyme molecule. The rate of nucleotide incorporation (y) is then plotted as a function of the nucleotide concentration (x) and then fit using a program such as sigma plot to the Michaelis–Menten equation:

$$y = k_{cat} * x/(k_m + x)$$

The efficiency of nucleotide incorporation is defined as k_{cat}/K_m. The relative efficiencies for the incorporation of different nucleotides opposite various template residues can then be compared. The frequency of misincorporation (f_{inc}) is defined as $k_{cat}/k_m^{(incorrect)}/k_{cat}/k_m^{(correct)}$. The fidelity (efficiency of correct dNTP incorporation relative to incorrect dNTP incorporation) is then presented as $1/f_{inc}$. The efficiency dNTP inorporation opposite a lesion site can be determined as $k_{cat}/k_m^{(damaged)}/k_{cat}/k_m^{(undamaged)}$.

Special Considerations for Individual TLS Polymerases

*Yeast Pol*η. Yeast Polη precipitates in the 35–45% ammonium sulfate fraction. It purifies well on Q Sepharose with a NaCl gradient.

*Human Pol*η. hPolη should only be induced for 3 h to prevent proteolytic degradation.

*Human Pol*κ. hPolκ should only be induced for 3 h to prevent proteolytic degradation.

*Human Pol*ι. This polymerase purifies well on Q Sepharose with a NaCl gradient. It is very stable in solution. Polι displays marked template specificity and exhibits high K_m values for dNTP incorporation opposite templates T and C.

Yeast and Human Rev1. These highly purified proteins tend to precipitate from solution and microconcentration or dialysis will result in loss of protein.

Yeast Polζ. To increase solubility in extract, 0.1–0.2 % Triton X-100 can be added to the breaking buffer (Nelson *et al.*, 1996b). Polζ has a maximal solubility of ∼75–100 ng/μl.

Acknowledgment

This work was supported by National Institutes of Health Grants CA107650, CA094006, and ES012411.

References

Bradford, M. M. (1976). A rapid and sensitive method for the quantitation of microgram quantities of protein utilizing the principle of protein-dye binding. *Anal. Biochem.* **72**, 248–254.

Creighton, S., Bloom, L. B., and Goodman, M. F. (1995). Gel fidelity assay measuring nucleotide misinsertion, exonucleolytic proofreading, and lesion bypass efficiencies. *Methods Enzymol.* **262**, 232–256.

Haracska, L., Prakash, S., and Prakash, L. (2002a). Yeast Rev1 protein is a G template-specific DNA polymerase. *J. Biol. Chem.* **277**, 15546–15551.

Haracska, L., Kondratick, C. M., Unk, I., Prakash, S., and Prakash, L. (2001c). Interaction with PCNA is essential for yeast DNA polymerase η function. *Mol. Cell* **8**, 407–415.

Haracska, L., Johnson, R. E., Unk, I., Phillips, B., Hurwitz, J., Prakash, L., and Prakash, S. (2001a). Physical and functional interactions of human DNA polymerase η with PCNA. *Mol. Cell. Biol.* **21**, 7199–7206.

Haracska, L., Johnson, R. E., Unk, I., Phillips, B. B., Hurwitz, J., Prakash, L., and Prakash, S. (2001b). Targeting of human DNA polymerase ι to the replication machinery via interaction with PCNA. *Proc. Natl. Acad. Sci. USA* **98**, 14256–14261.

Haracska, L., Unk, I., Johnson, R. E., Johansson, E., Burgers, P. M. J., Prakash, S., and Prakash, L. (2001d). Roles of yeast DNA polymerases δ and ζ and of Rev1 in the bypass of abasic sites. *Genes Dev.* **15**, 945–954.

Haracska, L., Unk, I., Johnson, R. E., Phillips, B. B., Hurwitz, J., Prakash, L., and Prakash, S. (2002b). Stimulation of DNA synthesis activity of human DNA polymerase κ by PCNA. *Mol. Cell. Biol.* **22**, 784–791.

Johnson, R. E., Prakash, S., and Prakash, L. (1999b). Efficient bypass of a thymine-thymine dimer by yeast DNA polymerase, Polη. *Science* **283**, 1001–1004.

Johnson, R. E., Prakash, L., and Prakash, S. (2005a). Biochemical evidence for the requirement of Hoogsteen base pairing for replication by human DNA polymerase ι. *Proc. Natl. Acad. Sci. USA* **102**, 10466–10471.

Johnson, R. E., Prakash, L., and Prakash, S. (2005b). Distinct mechanisms of *cis-syn* thymine dimer bypass by Dpo4 and DNA polymerase η. *Proc. Natl. Acad. Sci. USA* **102**, 12359–12364.

Johnson, R. E., Kondratick, C. M., Prakash, S., and Prakash, L. (1999a). *hRAD30* mutations in the variant form of xeroderma pigmentosum. *Science* **285**, 263–265.

Johnson, R. E., Washington, M. T., Prakash, S., and Prakash, L. (2000b). Fidelity of human DNA polymerase η. *J. Biol. Chem.* **275**, 7447–7450.

Johnson, R. E., Haracska, L., Prakash, S., and Prakash, L. (2001). Role of DNA polymerase η in the bypass of a (6-4) TT photoproduct. *Mol. Cell. Biol.* **21**, 3558–3563.

Johnson, R. E., Yu, S.-L., Prakash, S., and Prakash, L. (2003). Yeast DNA polymerase zeta (ζ) is essential for error-free replication past thymine glycol. *Genes Dev.* **17,** 77–87.

Johnson, R. E., Washington, M. T., Haracska, L., Prakash, S., and Prakash, L. (2000a). Eukaryotic polymerases ι and ζ act sequentially to bypass DNA lesions. *Nature* **406,** 1015–1019.

Jones, E. W. (1991). Tackling the protease problem in *Saccharomyces cerevisiae*. *Methods Enzymol.* **194,** 428–453.

Kingsman, S. M., Cousens, D., Stanway, C. A., Chambers, A., Wilson, M., and Kingsman, A. J. (1990). High-efficiency yeast expression vectors based on the promoter of the phosphoglycerate kinase gene. *Methods in Enzymology* **185,** 329–341.

Masutani, C., Kusumoto, R., Yamada, A., Dohmae, N., Yokoi, M., Yuasa, M., Araki, M., Iwai, S., Takio, K., and Hanaoka, F. (1999). The *XPV* (xeroderma pigmentosum variant) gene encodes human DNA polymerase η. *Nature* **399,** 700–704.

Nair, D. T., Johnson, R. E., Prakash, S., Prakash, L., and Aggarwal, A. K. (2004). Replication by human DNA polymerase ι occurs via Hoogsteen base-pairing. *Nature* **430,** 377–380.

Nelson, J. R., Lawrence, C. W., and Hinkle, D. C. (1996a). Deoxycytidyl transferase activity of yeast *REV1* protein. *Nature* **382,** 729–731.

Nelson, J. R., Lawrence, C. W., and Hinkle, D. C. (1996b). Thymine-thymine dimer bypass by yeast DNA polymerase ζ. *Science* **272,** 1646–1649.

Prakash, S., Johnson, R. E., and Prakash, L. (2005). Eukaryotic translesion synthesis DNA polymerases: Specificity of structure and function. *Annu. Rev. Biochem.* **74,** 317–353.

Smith, P. K., Korohn, R. I., Hermanson, G. T., Mallia, A. K., Gartner, F. H., Provenzano, M. D., Fujimoto, E. K., Goeke, N. M., Oldson, R. J., and Klenk, D. C. (1985). Measurement of protein using bicinchoninic acid. *Anal. Biochem.* **150,** 76–85.

Tissier, A., McDonald, J. P., Frank, E. G., and Woodgate, R. (2000). Polι, a remarkably error-prone human DNA polymerase. *Genes Dev.* **14,** 1642–1650.

Trincao, J., Johnson, R. E., Escalante, C. R., Prakash, S., Prakash, L., and Aggarwal, A. K. (2001). Structure of the catalytic core of *S. cerevisiae* DNA polymerase η: Implications for translesion DNA synthesis. *Mol. Cell* **8,** 417–426.

Uljon, S. N., Johnson, R. E., Edwards, T. A., Prakash, S., Prakash, L., and Aggarwal, A. K. (2004). Crystal structure of the catlytic core of human DNA polymerase kappa. *Structure* **12,** 1395–1404.

Washington, M. T., Prakash, L., and Prakash, S. (2003). Mechanism of nucleotide incorporation opposite a thymine-thymine dimer by yeast DNA polymerase η. *Proc. Natl. Acad. Sci. USA* **100,** 12093–12098.

Washington, M. T., Johnson, R. E., Prakash, S., and Prakash, L. (2000). Accuracy of thymine-thymine dimer bypass by *Saccharomyces cerevisiae* DNA polymerase η. *Proc. Natl. Acad. Sci. USA* **97,** 3094–3099.

Washington, M. T., Johnson, R. E., Prakash, L., and Prakash, S. (2002). Human *DINB1*-encoded DNA polymerase κ is a promiscuous extender of mispaired primer termini. *Proc. Natl. Acad. Sci. USA* **99,** 1910–1914.

Washington, M. T., Johnson, R. E., Prakash, L., and Prakash, S. (2004a). Human DNA polymerase ι utilizes different nucleotide incorporation mechanisms dependent upon the template base. *Mol. Cell. Biol.* **24,** 936–943.

Washington, M. T., Minko, I. G., Johnson, R. E., Haracska, L., Harris, T. M., Lloyd, R. S., Prakash, S., and Prakash, L. (2004b). Efficient and error-free replication past a minor-groove N^2-guanine adduct by the sequential action of yeast Rev1 and DNA polymerase ζ. *Mol. Cell. Biol.* **24,** 6900–6906.

Washington, M. T., Minko, I. G., Johnson, R. E., Wolfle, W. T., Harris, T. M., Lloyd, R. S., Prakash, S., and Prakash, L. (2004c). Efficient and error-free replication past a minor groove DNA adduct by the sequential action of human DNA polymerases ι and κ. Mol. Cell. Biol. **24,** 5687–5693.

Wolfle, W. T., Washington, M. T., Prakash, L., and Prakash, S. (2003). Human DNA polymerase κ uses template-primer misalignment as a novel means for extending mispaired termini and for generating single-base deletions. Genes Dev. **17,** 2191–2199.

Zhang, Y., Yuan, F., Wu, X., and Wang, Z. (2000). Preferential incorporation of G opposite template T by the low-fidelity human DNA polymerase ι. Mol. Cell. Biol. **20,** 7009–7108.

[25]   Localization of Y-Family Polymerases and the DNA Polymerase Switch in Mammalian Cells

By PATRICIA KANNOUCHE and ALAN LEHMANN

Abstract

During translesion synthesis past sites of damaged DNA, specialized Y-family polymerases are employed by the cell to replace the high stringency replicative polymerases and synthesize DNA past the damaged site. These polymerases are localized in replication factories during the S phase of the cell cycle. When progress of the replication fork is blocked, the polymerase accessory protein, proliferating cell nuclear antigen (PCNA), becomes ubiquitinated and the monoubiquitinated PCNA has an increased affinity for Y-family DNA polymerase η (polη). This chapter describes methods for visualizing the polymerases in replication factories, for analyzing the ubiquitination status of PCNA, and for measuring its interaction with polη in chromatin extracts.

Introduction

When cellular DNA is damaged, most types of damage act as complete blocks to replicative DNA polymerases. In order to overcome the blocks, the cell has to employ either damage avoidance processes by means of recombinational exchanges or carry out translesion synthesis (TLS) directly past the lesion (Lehmann, 2002, 2005). The replicative polymerases cannot carry out TLS past DNA lesions, but in mammals there are at least four specialized DNA polymerases able to carry out TLS past different types of damage (Lehmann, 2002, 2005). These recently discovered polymerases belong to the Y family and have an open conformation that

METHODS IN ENZYMOLOGY, VOL. 408
0076-6879/06 $35.00
DOI: 10.1016/S0076-6879(06)08025-6

enables them to accommodate different types of DNA damage in their active sites.

In order to engage in TLS, these specialized polymerases need to be localized in the subnuclear complexes, designated replication factories, in which the DNA is replicated. Second, they need to be switched into the replication machinery at the sites of the blocked forks. This chapter describes methods for (1) observing the localization of TLS polymerases using fluorescence microscopy and (2) measuring the interaction with the sliding clamp proliferating cell nuclear antigen (PCNA) at the site of the blocked fork.

Cellular Localization of Translesion Synthesis Polymerases

Ideally, localization of a protein in the cell is detected at endogenous levels using a specific antibody of high sensitivity. Unfortunately, because none of the antibodies raised against TLS polymerases has sufficient sensitivity to meet this requirement, the protein has to be overexpressed and either detected immunologically using a specific antibody or fused to enhanced green fluorescent protein (eGFP) or one of its derivatives and detected using autofluorescence of the eGFP.

Transfection

Materials

Coverslips sterilized
35-mm dishes
Fugene 6 (Clontech) (Keep capped except when using)
Eagle's MEM (Invitrogen)

We use vigorously growing SV40-transformed fibroblasts cultured in Eagle's MEM supplemented with 10% fetal calf serum. DNA encoding the whole of a specialized polymerase open reading frame or a fragment of it is cloned either into pCDNA3zeo (Invitrogen) or into peGFP-C3 (Clontech). In both cases, expression of the cloned gene is under control of the strong cytomegalovirus (CMV) promoter. In the peGFP plasmid the GFP tag is N-terminal to the expressed protein, and we delete the initiating ATG codon from the cDNA to be cloned.

Methodology

1. Day 0: 2×10^5 cells are plated in 2 ml complete medium in a series of 35-mm dishes containing a sterile coverslip.

2. Day 1: Transfection. Per dish 2 μg DNA and 3 μl Fugene 6 are mixed with 97 μl serum-free medium (SFM) as follows. A Fugene 6–SFM mix can be made for up to six dishes. The Fugene is added dropwise to the medium in an Eppendorf tube, avoiding touching the walls of tube. The contents are mixed well by flicking and are left for 1–5 min at room temperature.

3. Fugene 6–SFM (100 μl) is added to the DNA in a separate tube, again dropwise, avoiding the sides of tube, followed again by mixing by flicking. The Fugene–DNA–SFM mix is incubated for 30–45 min at room temperature.

4. One hundred microliters of mix is added to each dish dropwise. The contents of the dish are mixed carefully, and the dishes are incubated at 37° for 20–24 h.

DNA Damage

On the day after transfection, cells can be exposed to any kind of DNA-damaging treatment. For UVC irradiation, we use a germicidal lamp at a fluence rate of 0.5 J/m^2 per second and doses in the range of 5–20 Jm^{-2}. The fluence rate is measured with an IL1400A radiometer/photometer (International Light). Before irradiation, cells are rinsed twice with phosphate-buffered saline (PBS). After treatment, fresh medium is added to the cells. They are then incubated for various time periods, typically between 2 and 24 h.

Fixation of Cells and Visualization of eGFP Autofluorescence

Materials

PBS (Dulbecco A) tablets from Oxoid

3.7% paraformaldehyde (dissolve 3.7 g of paraformaldehyde [Sigma] in 100 ml warmed water in a fume cupboard. Addition of a few drops of NaOH helps dissolve the paraformaldehyde. Then add a PBS tablet. Aliquot and store at $-20°$)

30% bovine serum albumin (BSA) stock solution 10× (w/v) (dissolve 4.5 g BSA [fraction V, Sigma] in 15 ml PBS. Put on rotator until the BSA has dissolved completely, aliquot, and store at $-20°$). 3% BSA working solution: Dilute 1 ml of stock sol in 9 ml PBS

0.2 mg/ml of 4′-6′-diamidino-2-phenylindole stock solution 1000× (DAPI, Sigma, dissolve 2 mg of DAPI in 10 ml of water, aliquot, and store at $-20°$). Keep DAPI solution in dark

Glycergel (Dako, Ltd)

Methodology

1. After treatment of the cells and further incubation, the coverslips are removed from the dishes into six-well trays containing PBS and left for 5 min.
2. The coverslips are transferred into further six-well trays containing 3.7% paraformaldehyde and incubated at room temperature for 30 min with gentle agitation.
3. During this incubation period, glycergel is melted at 50°.
4. The coverslips are rinsed in PBS for 5 min. After the last wash, cells are counterstained with 0.2 μg/ml DAPI (dilute 1 μl of stock solution in 1 ml PBS) for 5 min to visualize nuclei and washed again for 5 min in PBS.
5. Two separate drops of glycergel are placed on a slide. The coverslip is touched on a tissue to drain off excess liquid and then placed, cells downward, onto one of the drops on the slide (two coverslips per slide). The slide is left for 2 h to dry.

Visualization of Specialized Polymerases in Replication Factories

In eukaryotes, all proteins required for replication of the DNA accumulate at sites in the nucleus at which DNA replication take place, the so-called replication factories or replication foci (Leonhardt *et al.*, 2000). These foci vary in morphology during the S phase. Using indirect immunofluorescence, it is possible to determine if specialized polymerases such as polη are localized in the replication factories (Kannouche *et al.*, 2001, 2003): (i) these sites can be visualized conveniently with an antibody for the replication accessory protein, the sliding clamp PCNA or, alternatively, (ii) replicating DNA can be observed by pulse-labeling DNA for 15 min with bromodeoxyuridine (BrdU). DNA is visualized with an antibody specific for BrdU.

Colocalization of eGFP-polη and PCNA

Materials

Same as described earlier
Buffer A:100 m*M* NaCl, 300 m*M* sucrose, 3 m*M* MgCl$_2$, 10 m*M* PIPES, pH 6.8, 1 m*M* EGTA, 0.2% Triton-X100 and protease inhibitors tablet
Cold methanol
Cold acetone

Note that for indirect immunofluorescence, cells are kept in the dark (aluminium foil) as soon as they are in contact with the secondary antibody coupled with fluorophores.

Proteins such as PCNA are present in growing cells in two different forms: a soluble fraction in the nucleoplasm and a bound fraction in the replication factories. In order to visualize the proteins associated with the replication machinery, it is necessary to remove the soluble form using a nonionic detergent such as Triton X-100 or methanol before fixing the cells.

Methodology. Two different techniques are described to remove the soluble material.

1. Coverslips containing cells expressing eGFP-polη are rinsed twice in cold PBS for 5 min and transferred into a six-well plate.

2. To remove the Triton-soluble fraction, cells are incubated for 5 min on ice with gentle shaking in 2 ml of buffer A and then fixed with paraformaldehyde as described earlier. Alternatively, after rinsing cells in PBS, 2 ml of ice-cold methanol is added to the wells and cells are kept at $-20°$ for 20 min. The cells are then transferred to a further six-well plate containing ice-cold acetone and incubated for 30 s.

3. Cells are transferred again to further wells containing PBS and washed 2–3 min with gentle agitation. It is important to note that if the Triton procedure has been used to remove the soluble fraction, it is necessary at this step to incubate the cells with cold methanol for 5 min. Indeed, for an unknown reason PCNA foci can be detected only if cells are pretreated with methanol before immunostaining.

4. Cells are incubated for 1 h at room temperature with anti-PCNA mouse monoclonal antibody (PC10, Santa Cruz, Biotechnology) diluted 1:500 in 3% BSA solution, and 50 μl of the diluted antibody is added to the coverslips.

5. After three washes with PBS, the fixed cells are incubated for 45 min at room temperature with TRITC-conjugated affiniPure goat antimouse IgG (Jackson Immuno Research Laboratories, Inc) diluted 1:500 in 3% BSA solution. Fifty microliters of the diluted antibody is added to each coverslip, which are then washed three times with PBS for 5 min each.

6. Finally the cells are counterstained with DAPI solution (as described earlier), and coverslips are mounted on slides using glycergel solution as described previously.

Colocalization of polη with Replicating DNA

Materials

Same as described earlier

10% Triton solution (stock solution): Mix 1 ml Triton X-100 in 9 ml PBS. Put on rotator until completely dissolved. Store at 4° for 2 to 3 weeks. Working solution (0.5–1% Triton): Dilute 1 ml of stock solution in appropriate volume of PBS.

Formaldehyde solution 2%: Dilute 2 ml 37% formaldehyde (Sigma F8775) in 35 ml PBS. Prepare just before using.

2 M HCl

0.1 M borate buffer, pH 8.5

Methodology. Cells are transfected with pCDNA-polη and incubated for 20 h to allow expression of exogenous polη. DNA is then pulse labeled for 15 min with 10 μM BrdU. Indirect immunofluorescence can be then processed.

1. Cells are fixed in paraformaldehyde as described earlier.

2. Cells are washed with PBS and permeabilized by adding 0.5% Triton solution for 10–15 min at room temperature. Coverslips are then transferred into new six-well plates and three washes with PBS are carried out before incubating cells with 50 μl of the primary antibody (for polη detection, use rabbit polyclonal antibody raised against the full-length polymerase diluted 1:300 in PBS/3% BSA) for 1 h at room temperature.

3. After three washes with PBS, the cells are incubated for 45 min at room temperature with TRITC-conjugated affiniPure goat antirabbit IgG (Jackson Immuno Research Laboratories, Inc) diluted 1:500 in 3% BSA solution. The antibody is added and washed off as described previously.

4. After the final wash, antigen–antibody complexes are fixed with 2% formaldehyde solution for 10 min. Cells are treated with 2 M HCl at 37° for 40 min and then neutralized in 0.1 M borate buffer, pH 8.5. After washing in PBS, cells are incubated for 1 h at room temperature with the FITC-conjugated, anti-BrdU monoclonal antibody (Roche). The coverslips are exposed to antibody, washed, and mounted onto slides using the glycergel solution as described earlier.

Microscopy

Cells are observed and photographed using a Zeiss Axiophot 2 or equivalent microscope (Carl Zeiss, Germany) equipped with a 100× NA 1.4 objective lens and an Orca ER CCD camera (Hamamatsu, Japan).

Digital images are acquired and colored using Simple PCI software (Compix, USA). Appropriate filter sets are used to visualize DAPI, FITC, GFP, and TRITC staining (Omega XF 02–2, 100–2, 116–2, and 101–2, respectively).

Ubiquitination of PCNA and Interaction with POLη

Hoege *et al.* (2002) first showed in yeast that when the DNA replication machinery encounters DNA damage, PCNA becomes modified post-translationally by mono- and polyubiquitination. These two steps have different genetic requirements, and genetic analyses suggest that monoubiquitination channels lesions into a translesion synthesis pathway (Hoege *et al.*, 2002; Stelter and Ulrich, 2003). In human fibroblasts, we and others have only been able to detect monoubiquitination of PCNA and we have demonstrated that this increases its affinity for polη, thus providing an attractive mechanism for the polymerase switch (Kannouche *et al.*, 2004; Watanabe *et al.*, 2004).

Detection of Ubiquitinated Human PCNA

Materials

1× Laemmli buffer: 5% (w/v) SDS, 4% β-mercaptoethanol, 62.5 mM Tris–HCl, pH 6.8, 5% glycerol, 3 mM EDTA, and 0.02% bromphenol blue

Lysis buffer B: 50 mM Tris, pH 7.5, 20 mM NaCl, 1 mM MgCl$_2$, 0.1% SDS, protease inhibitors tablet (Roche), and 1 μl/ml Benzonase nuclease (Novagen)

Methodology

1. Cells (5×10^5) are seeded into 60-mm dishes and incubated for approximately 24 h.

2. Cells are exposed to UV irradiation as described earlier or other DNA-damaging treatment as required and are then incubated further for different times to enable replication forks to reach sites of DNA damage.

3. Cells are rinsed in PBS and harvested directly into 150 μl of 1× Laemmli buffer. Alternatively, 150 μl of lysis buffer B is added to the dishes, and cells are incubated at room temperature for 5 min with gentle shaking and transferred to a 1.5-ml tube containing 30 μl of 5× Laemmli buffer. In both cases, samples are boiled for 10 min. For analysis, 20 μl of each sample is loaded onto 8% SDS–PAGE gels and electrophoresed for 1 h 30 min at 120 V. Proteins are transferred onto PVDF filters using standard procedures and then probed with anti-PCNA mouse monoclonal

antibody (PC10, Santa Cruz, Biotechnology) diluted 1:5000 in PBS/0.05% Tween 20.

Interaction of PCNA with polη on Chromatin

Materials

Buffer A: 100 mM NaCl, 300 mM sucrose, 3 mM MgCl$_2$, 10 mM PIPES, pH 6.8, 1 mM EGTA, 0.2% Triton X-100, and protease inhibitors

Lysis buffer C: 50 mM Tris, pH 7.5, 150 mM NaCl, 0.1% SDS

Dilution buffer: 50 mM Tris, pH 7.5, 150 mM NaCl, 5 mM EDTA, 0.2% Triton

Formaldehyde solution 1%: dilute 1 ml formaldehyde 37% solution (Sigma F8775) in 36 ml PBS. Prepare just before using.

1 M glycine solution

Because the levels of polη in most cell lines are close to the detection limit of available antibodies, it is necessary to use either a stable cell line with modest overexpression levels of polη or cells transiently transfected with a polη containing expression plasmid.

Methodology

1. Cells are plated in 100-mm dishes (1–2 10^{6}cells/dish).

2. On the following day, the medium is removed, and the cells are rinsed once with PBS and then irradiated (20 J/m^2) as described earlier.

3. Six hours after irradiation, the medium is removed and the cells are rinsed twice in cold PBS. To isolate the Triton-insoluble fraction, cells (on plates) are incubated for 5 min on ice with gentle shaking in 3 ml buffer A and then buffer is removed (as much as possible).

4. To cross-link proteins, the adhering cellular material is incubated on ice for 10 min in 3 ml 1% formaldehyde. The reaction is stopped by a 5-min incubation with 300 μl in 1 M glycine.

5. As much of the solution as possible is removed and the cells are rinsed twice with cold PBS, scraped off the dishes, centrifuged, and dissolved in 0.2 ml of lysis buffer C for 10 min at room temperature. The samples are sonicated (optimize conditions for shearing cross-linked DNA to 200–1000 bp, depending on sonicator and cells). Our conditions are two to three sets of 10 s, 25–30% maximum power.

6. Samples are centrifuged for 5 min at 13,000 rpm at 4° and the supernatant is transferred into a new 2-ml tube and diluted 1 in 10 with dilution buffer. Samples are then ready for immunoprecipitation.

7. For immunoprecipitation, 2–3 μl anti-PCNA mouse monoclonal antibody (PC-10 Santa Cruz) is added to each sample, which are then incubated overnight in a cold room with gentle mixing.

8. Meanwhile, protein A beads (protein A-Sepharose, Pharmacia, Amersham) are prepared as follows. Beads from the stock suspension are centrifuged for 2 min at 2000 rpm at 4° and the ethanol is removed. The beads are resuspended in 10 volumes of dilution buffer and centrifuged for 2 min at 2000 rpm at 4° and the supernatant is removed. This step is repeated with final resuspension in 1 volume of dilution buffer.

9. Fifty microliters of protein A–agarose–50% slurry is added to each sample and incubated for 1–2 h in a cold room with gentle mixing. Then the beads are washed with dilution buffer four times, resuspended in 60 μl Laemmli buffer, and boiled for 10 min. Fifteen to 20 μl of the bound proteins is analyzed by electrophoresis and immunoblotting as described in the previous section.

References

Hoege, C., Pfander, B., Moldovan, G.-L., Pyrolowakis, G., and Jentsch, S. (2002). RAD6-dependent DNA repair is linked to modification of PCNA by ubiquitin and SUMO. *Nature* **419**, 135–141.

Kannouche, P. L., Wing, J., and Lehmann, A. R. (2004). Interaction of human DNA polymerase η with monoubiquitinated PCNA; a possible mechanism for the polymerase switch in response to DNA damage. *Mol. Cell* **14**, 491–500.

Kannouche, P., Broughton, B. C., Volker, M., Hanaoka, F., Mullenders, L. H. F., and Lehmann, A. R. (2001). Domain structure, localization and function of DNA polymerase η, defective in xeroderma pigmentosum variant cells. *Genes Dev.* **15**, 158–172.

Kannouche, P., Fernandez de Henestrosa, A. R., Coull, B., Vidal, A. E., Gray, C., Zicha, D., Woodgate, R., and Lehmann, A. R. (2003). Localization of DNA polymerases η and ι to the replication machinery is tightly coordinated in human cells. *EMBO J.* **22**, 1223–1233.

Lehmann, A. R. (2002). Replication of damaged DNA in mammalian cells: New solutions to an old problem. *Mutat. Res.* **509**, 23–34.

Lehmann, A. R. (2005). Replication of damaged DNA by translesion synthesis in human cells. *FEBS Lett.* **579**, 873–876.

Leonhardt, H., Rahn, H. P., Weinzierl, P., Sporbert, A., Cremer, T., Zink, D., and Cardoso, M. C. (2000). Dynamics of DNA replication factories in living cells. *J. Cell Biol.* **149**, 271–280.

Stelter, P., and Ulrich, H. D. (2003). Control of spontaneous and damage-induced mutagenesis by SUMO and ubiquitin conjugation. *Nature* **425**, 188–191.

Watanabe, K., Tateishi, S., Kawasuji, M., Tsurimoto, T., Inoue, H., and Yamaizumi, M. (2004). Rad18 guides poleta to replication stalling sites through physical interaction and PCNA monoubiquitination. *EMBO J.* **23**, 3886–3896.

[26] Repair of DNA Double Strand Breaks: *In Vivo* Biochemistry

By NEAL SUGAWARA and JAMES E. HABER

Abstract

Double strand breaks (DSBs) can cause damage to the genomic integrity of a cell as well as initiate genetic recombination processes. The HO and I-SceI endonucleases from budding yeast have provided a way to study these events by inducing a unique DSB *in vivo* under the control of a galactose-inducible promoter. The *GAL::HO* construct has been used extensively to study processes such as nonhomologous end joining, intra- and interchromosomal gene conversion, single strand annealing and break-induced recombination. Synchronously induced DSBs have also been important in the study of the DNA damage checkpoint, adaptation, and recovery pathways of yeast. This chapter describes methods of using *GAL::HO* to physically monitor the progression of events following a DSB, specifically the events leading to the switching of mating type by gene conversion of *MAT* using the silent donors at *HML* and *HMR*. Southern blot analysis can be used to follow the overall events in this process such as the formation of the DSB and product. Denaturing alkaline gels and slot blot techniques can be employed to follow the 5′ to 3′ resection of DNA starting at the DSB. After resection, the 3′ tail initiates a homology search and then strand invades its homologous sequence at the donor cassette. Polymerase chain reaction is an important means to assay strand invasion and the priming of new DNA synthesis as well as the completion of gene conversion. Methods such as chromatin immunoprecipitation have provided a means to study many proteins that associate with a DSB, including not only recombination proteins, but also proteins involved in nonhomologous end joining, cell cycle arrest, chromatin remodeling, cohesin function, and mismatch repair.

Introduction

DNA double strand breaks (DSBs) pose a hazard to the genomic integrity of a cell, possibly leading to cell death or hereditary changes. If left unrepaired, DSBs may lead to chromosome loss, whereas DSBs repaired with poor fidelity may lead to point mutations, deletions, truncations by new telomere formation, or chromosomal translocations. DSBs

METHODS IN ENZYMOLOGY, VOL. 408   0076-6879/06 $35.00
DOI: 10.1016/S0076-6879(06)08026-8

can also initiate important "programmed" genetic recombination processes such as meiotic recombination, immunoglobulin V(D)J joinings and class switching, or, as discussed here, mating type switching in budding yeast (Aylon and Kupiec, 2004; Bassing and Alt, 2004; Haber, 1998; Pierce *et al.*, 2001).

Normally these events and processes occur at a low frequency, so the development of an inducible endonuclease system, where a single DSB can be induced synchronously in a large number of cells, has proven to be useful. Two meganucleases, HO and I-SceI endonucleases, with recognition sites of ≥18 bp, have been used to examine repair at specific sites in budding yeast and in other organisms (Haber, 2000; Johnson and Jasin, 2001). In yeast the *GAL::HO* construction has been used extensively, where the UAS from the *GAL1-GAL10* genes controls the expression of the HO endonuclease gene. The HO endonuclease creates a unique DSB at the *MAT* locus. Because the induction of HO is very synchronous, it is possible to monitor the progression of DNA changes by Southern and polymerase chain reaction (PCR) analyses. Chromatin immunoprecipitation (ChIP), immunostaining and the use of fluorescent protein-tagged fusion proteins have allowed the investigation of what proteins associate with DSBs. This physical monitoring or "*in vivo* biochemistry" permits the examination of processes occurring under physiological conditions in both wild-type and mutant strains.

These methods have proven valuable in the analysis of specific processes such as mating type switching, inter- and intrachromosomal gene conversions, single strand annealing, break-induced recombination, and nonhomologous end joining (Haber, 2000; Malkova *et al.*, 2005). Mating type switching has been examined in our laboratory and is used as an example here. A model outlining the essential steps of mating type switching from *MAT*a to *MAT*α is shown in Fig. 1. This process is initiated when HO cleaves the *MAT* locus at the Y/Z junction. The W, X, and Z sequences of *MAT*a and *HML*α are homologous, whereas the **Ya** and Yα sequences are unique. After HO cuts, 5′ to 3′ resection creates a 3′ tail that can strand invade the homologous Z sequence at *HML*. The invading strand can then be extended by new DNA synthesis to begin copying of the opposite mating type sequence. The newly synthesized DNA can be displaced from *HML* to *MAT* to create gene conversion products that are mostly noncrossover in structure (G. Ira and J. E. Haber, manuscript in preparation).

Physical monitoring of mating type switching begins with an examination by Southern hybridization (Fig. 2A). Initially, at 0 h a Southern blot shows two bands produced by *Sty*I digestion representing the substrate with the HO-cut site and the adjacent restriction fragment, *MAT* distal

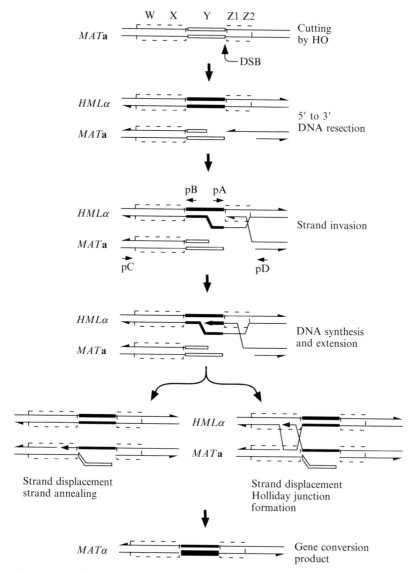

FIG. 1. A model of DSB-induced gene conversion in mating type switching. HO endonuclease induces a DSB that initiates a gene conversion event between *MAT* and *HML*. The individual steps in this process are described in the text.

(Fig. 2B). After galactose induction, DNA cleavage can be assessed by the decrease in the intensity of the substrate band and by the appearance of the HO-cut fragment. The HO-cut fragment is a transient intermediate that disappears as it is processed into the gene conversion product. DNA resection can be observed when denaturing alkaline gels are used, as most restriction enzymes cannot cut single-stranded DNA (ssDNA), which results in intermediate bands of higher molecular weight (Fig. 2C). This effect is more apparent in donorless strains or certain *rad* mutants where the DSB cannot be repaired and more DNA becomes single stranded. The use of denaturing gels and enzymes such as *Sty*I allow us to see discrete bands derived from a population of heterogeneously sized DNA intermediates.

Examination of Southern blots using denaturing gels sometimes reveals a light background smearing in between the bands comprising the ladder (Fig. 2C), suggesting that the ssDNA may be heterogeneous in length, possibly due to mechanical shearing during isolation or due to the presence of native nucleases. Because of this we have also used slot blots for quantitation, as this assay determines the presence of single-stranded intermediates regardless of their length. Furthermore, this method has proven effective when trying to detect smaller amounts of single-stranded DNA produced during the process of DSB repair (Sugawara and Haber, 1992). The method takes advantage of ssDNA binding to positively charged nylon membranes, whereas double-stranded DNA mostly does not. When probed with single strand-specific probes, only the single-stranded DNA is detected (Fig. 2D). These methods and others (Booth *et al.*, 2001; Dionne and Wellinger, 1996; Frankfurt *et al.*, 1996; Garvik *et al.*, 1995; Raderschall *et al.*, 1999; Tapiero *et al.*, 1976) have been used to detect single-stranded DNA generated *in vivo*. Additionally, it is possible to assay the loss of double-stranded DNA in donorless strains by quantitating the loss of the HO-cut restriction fragment or by quantitating the loss of signal from denatured DNA on slot blots (Fig. 2D) (Lee *et al.*, 1998).

After DNA resection starts, proteins such as the RPA complex and Rad51 rapidly associate with the ssDNA. In the example shown (Fig. 2E), the anti-Rad51 antibody was used to precipitate formaldehyde cross-linked chromatin associated with Rad51. DNA was released from the cross-links and PCR-amplified using primers specific to *MAT* and *HML*. Association of Rad51 with the DSB is consistent with the formation of a Rad51–DNA filament. We also detected Rad51 localization at the *HML* donor, which provides an assay for the synapsis of the donor and the Rad51–DNA filament (Sugawara *et al.*, 2003).

Once strand invasion has occurred, the 3′ tail can be used to prime the synthesis of DNA using the *HML* Y sequence as a template. This step can be

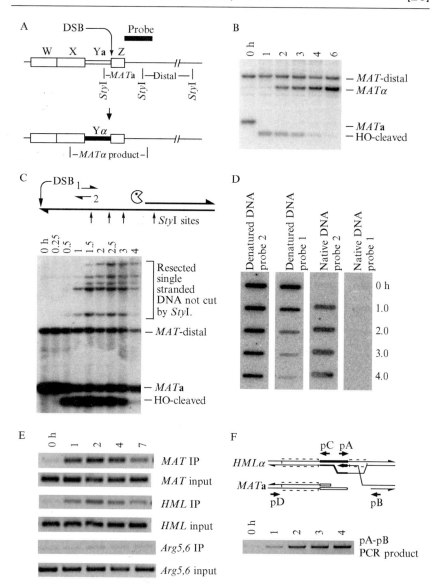

FIG. 2. Monitoring of a DSB-induced gene conversion by Southern blot analysis. (A) The diagram shows *MAT*a, the location of the DSB made by HO endonuclease, the probe (black box), and the location of *Sty*I sites. (B) The *MAT*a (0.93 kb) band is cleaved by HO creating a 0.72-kb band. The larger *MAT*α band (1.88 kb) arises due to the absence of *Sty*I site in the Yα sequence. (C) As DNA is resected 5′ to 3′, *Sty*I sites are progressively destroyed as the sites become single stranded. This results in a ladder of what appear to be partial digestion products of increasing size created during the time course. DNA from a donorless strain was

assessed by PCR analysis by selecting primers specific to the *HML*-Yα σεθνενχε and to the *MAT* (*MAT*-distal) sequence (pA and pB, respectively, in Fig. 2F). In this example, at least 29 nucleotides of *HML*-Yα σεθενχε must be synthesized so that a continuous DNA template can be formed for the pA and pB primers. Completion of gene conversion can be observed by formation of the product band on the Southern blot or by use of a different pair of PCR primers, pC and pD (Fig. 2F), which determine when the centromere proximal part of the DSB is joined covalently to donor sequences.

The following sections provide detailed information on setting up yeast strains for *GAL::HO* inductions and on carrying out experiments described earlier.

Methods

Strains

The *GAL::HO* gene on a *URA3* plasmid can be integrated at *ade3* by homologous recombination (Sandell and Zakian, 1993). The plasmid can be excised using 5-fluoro-orotic acid (Boeke *et al.*, 1987), leaving ade 3::*GAL::HO*. The plasmid bears an autonomous replication sequence and care should be taken to select for stable integrants. *GAL::HO* is also available on centromeric plasmids with *URA3* (pGAL-HO) (Herskowitz and Jensen, 1991; Jensen and Herskowitz, 1984), *LEU2* (pJH727) (A. Plessis), and *TRP1* (pFH800) (Nickoloff *et al.*, 1989). For the recognition sequence, we use either the *MAT* locus or a 117-bp restriction fragment (*Bgl*II to *Hinc*II) derived from *MAT*a. A 36-bp cleavage site, AGTTTCAGCTTTCCGCAACAGTA TAATTTTATAAAC, has also been in used in a colony assay (Pâques and Haber, 1997). Strains are tested to avoid respiratory-deficient (petite) strains (Evans and Wilkie, 1976; Mahler and Wilkie, 1978) which do not grow on YEP-EG, and *gal3* mutants

---

used. (D) When native genomic DNA is loaded onto a slot blot, the single-stranded tail binds to a positively charged nylon membrane and can be detected by strand-specific RNA probes (probe 1 and 2) that do not hybridize with any double-stranded DNA present. (E) Chromatin cross-linked with formaldehyde was isolated and precipitated with the anti-Rad51 antibody. DNA from the precipitates was released from the cross-linking, purified, and amplified by PCR. The PCR product was run on a 2% agarose gel stained with ethidium bromide (reverse images are shown). (F) When a strand invasion structure is extended for a short distance (29 nucleotides) by DNA synthesis, it creates a template that can be amplified by PCR by using sequences unique to *HML* (pA) and *MAT* (pB). Agarose gels were stained with ethidium bromide (reverse image is shown).

(Douglas and Pelroy, 1963) which can be identified by their failure to grow on YEP-galactose under anaerobic conditions.

Media

YEP-lactate is prepared by adding 37 g of 85% lactic acid to 0.9 liters of distilled water and adjusting the pH to 5.5 with NaOH. Yeast extract (10 g) and 20 g peptone are added for a final volume of 1 liter. YEPD is 2% peptone, 1% yeast extract, and 2% glucose. YEP-raffinose is 2% peptone, 1% yeast extract, and 2% raffinose (added after autoclaving). YEP-EG is 2% (w/v) peptone, 1% yeast extract, 2.6% glycerol, 1% succinic acid, adjusted to pH 5.5 with NaOH, 2.5% agar, and 2.5% (v/v) ethanol (added after autoclaving). Adenine can be added to the aforementioned media at 40 mg/liter. Galactose (20%) is made at room temperature and filter sterilized. We use galactose that contains less than 0.01% glucose.

Galactose Induction

Grow a 2-ml culture overnight starting from a single colony in YEPD or selective medium. After the culture has grown, spin down the cells, discard the supernatant and resuspend the pellet in an equal or larger volume of YEP-lactate medium. Repeat this wash with an equal volume of YEP-lactate. Shake the cells at 30° for 3 to 10 hr. Strains undergo a lag phase upon the shift to lactate medium and this step makes cell growth more predictable. Inoculate YEP-lactate medium with enough cells to give a cell density between 3×10^6 and 1×10^7 cells/ml the next day. Shake the culture overnight at 30° providing sufficient aeration. Most strains have a doubling time of about 3 h in YEP-lactate or about 5 h in YEP-raffinose medium, which can be substituted for YEP-lactate.

When the cells have grown to the desired cell density, collect a 50-ml sample for the uninduced control and add 1/10th volume of 20% galactose to the remaining culture. For DNA extractions centrifuge the sample and discard the supernatant. Resuspend the cells in 400 μl extraction buffer (100 mM Tris, pH 7.5, 50 mM EDTA, 2% SDS) and add to a 1.5-ml microcentrifuge tube(s) containing 0.5 ml of acid-washed glass beads (425 to 600 μm in diameter). The cells can be stored at $-70°$ until the DNA can be extracted.

For experiments studying mating type switching, add 1/10th volume of 20% glucose to downregulate the HO induction after 30 or 60 min to prevent recutting of the HO cut site. During long time courses, recutting of the HO recognition site can be seen, presumably because the glucose has been metabolized and the remaining galactose induces cutting. To avoid this, add a second aliquot of glucose (1/10th volume of 20% glucose) after 6 h. The addition of glucose is unnecessary when the product cannot be recut, such as when the product contains a *MAT*-inc mutation.

At the initial time point and at the desired time points, dilute a sample of the cultures 1×10^{-4} and spread 100 μl (or more, if viability is low) on YEPD plates. Examination of the colonies lets one determine the viability of the cells after HO induction, the fraction of colonies with recombined substrates, and the stability of the *GAL::HO* plasmid or other plasmids if present.

DNA Extraction

Suspend the cells in a mixture of 400 μl of extraction buffer and 500 μl glass beads as described earlier plus 400 μl of phenol. Vortex tubes for 2 min. Leave the samples on ice for 1 min and then spin the tubes in a microcentrifuge for 10 min. Remove the top aqueous layer to a new tube and repeat the phenol extraction (optional).

Remove the aqueous layer to a new tube and add 40 μl of 3 M sodium acetate (pH 5.5) and 600 μl isopropanol. Mix and spin in a microcentrifuge for 10 min. Discard the supernatant and add 300 μl TE containing 10 μg of RNase A to each tube. Incubate at 37° for 60 min. Add 30 μl 3 M sodium acetate and 600 μl of isopropanol and centrifuge the DNA for 10 min. Discard the supernatant, carefully rinse the pellet with 80% ethanol, and dry the pellet. Dissolve the DNA in TE.

The DNA is ready for restriction digestion or PCR amplification. Sometimes restriction enzymes will not cut the DNA completely, presumably due to the carryover of reagents used in the extraction. This can sometimes be overcome by increasing the digestion volume (e.g., 300 μl), ethanol precipitating the DNA again, or by further purification. Filtering with Microcon filters according to the manufacturer's (Millipore) directions is useful.

Southern Analysis Using Native and Denaturing Gels

When using native gels, conventional procedures can be used for electrophoresis and to carry out Southern hybridizations of DNA. When designing a Southern analysis it is desirable to keep the restriction fragments in a comparable size range for quantitation purposes as smaller fragments are recovered less efficiently.

Denaturing gels are used to detect and follow the formation of ssDNA after a DSB is formed. Denaturing gels can be prepared as described by Sambrook *et al.* (1989). Prepare an agarose electrophoresis gel in 50 mM

NaCl and 1 mM EDTA, as agarose will decompose if boiled under alkaline conditions. After it has solidified, equilibrate it in 50 mM NaOH and 1 mM EDTA for at least 30 min. Alternatively, melt the agarose in water and allow it to cool to 60° at which point the solution can be adjusted to 50 mM NaOH and 1 mM EDTA.

Precipitate the DNA samples by adjusting the solution to 0.3 M sodium acetate (pH 5.5) and 5 mM EDTA followed by the addition of 2 volumes of ethanol. EDTA is included to prevent the magnesium from precipitating under alkaline conditions. After chilling at −70° and centrifuging, discard the supernatant and rinse the pellet in 80% ethanol. Resuspend the pellet in alkaline gel-loading buffer [1× buffer: 50 mM NaOH, 1 mM EDTA, 2.5% Ficoll (type 400), and 0.025% bromocresol green, stored at 4°]. If desired, size markers can be prepared by adding 1× alkaline gel-loading buffer to 1/10th volume of plasmid or phage DNA digested by restriction enzymes.

During electrophoresis, a glass plate optionally can be placed on the gel to prevent the dye from diffusing from the agarose. Because of the large currents that can be generated with denaturing gels, gels are usually run slowly at lower voltages with circulation. The gel can be stained with ethidium bromide (0.5 μg/ml) in 1× TAE electrophoresis buffer (two washes of 15 min). ssDNA will be faint when viewed under UV light.

Gently shake the gel in 0.25 N HCl for 6–7 min, rinse with water, and soak the gel in 0.5 N NaOH and 1.5 M NaCl for 30 min with gentle agitation. Rinse briefly with water and apply the gel to the DNA transfer apparatus. Conventional Southern blotting methods can be used to transfer the DNA to nylon membranes followed by UV cross-linking and Southern hybridization (Church and Gilbert, 1984).

Analysis of Single-Stranded Tail Formation Using Slot Blots

DNA samples isolated from *GAL::HO* induction experiments can be analyzed by slot blots to quantitate the amount of ssDNA that is formed. Three sets of samples are loaded in a slot blot experiment. For the native samples, load 1 μg/well of DNA isolated from an HO induction experiment, after adjusting the concentration to 10× SSC. For the denatured samples, prepare 0.1 μg of genomic DNA in 0.4 N NaOH. Neutralize the samples with an equal volume of 0.8 M Tris and 0.6 M sodium acetate prepared with stock solutions at pH 8 and pH 5.5, respectively. For controls, include a sample with nondenatured plasmid DNA, samples of unlabeled RNA transcripts complementary to the RNA probes that will be used, and, if possible, denatured genomic DNA from a mutant strain lacking hybridizable sequences as a negative control.

Samples are loaded into the wells of a slot blot apparatus, transferred to nylon membranes, and the wells rinsed with 10× SSC. The DNA is covalently cross-linked to the nylon membrane with UV light (Church and Gilbert, 1984). The blots, one for each strand-specific probe, are hybridized using a suitable hybridization protocol (Church and Gilbert, 1984).

Polymerase Chain Reaction Analysis

PCR is well suited for the detection of recombinant products for two reasons. First, it is capable of detecting very small quantities of product, which enables us to see the earliest formation of recombinant structures. Second, it produces a DNA product of defined length from a population of DNAs that may be heterogeneous in length. The primers pA (Yα) and pB (*MAT*-distal) in Fig. 2F will generate a PCR product when the single-stranded DNA tail from the *MATZ* strand invades *HML* and is used to prime DNA synthesis. The strand invasion/extension intermediate detected by the primers pA (GCAGCACGGAATATGGGACT) and pB (ATGT-GAACCGCATGGGCAGT) arises 30 min before the product appears on Southern blots. The end of the gene conversion process can be assayed with the Yα and *MAT*-proximal primers pC (AGAT GAGTTTAAATCCAG-CA) and pD (TGTTGTCTCACTATCTTGCC), respectively (White and Haber, 1990).

Conventional PCR methods can be used to generate the strand invasion/extension PCR product. A typical reaction is 1 min at 94°, 25 cycles of 15 s at 94°, 20 s at 55°, 20 s at 72°, and 5 min at 72°. The pA–pB PCR product is 520 bp and can be quantitated on a 2% agarose gel. During preinduction growth, a small population of cells (about 1%) may arise that contain recombinant product. Consequently, the number of cycles should be limited so that saturating amounts of the PCR product are not generated. For a mating type-switching experiment, PCR amplification of the product can be diluted to create a standardization curve to ensure that the PCR signal lies in the linear range.

Chromatin Immunoprecipitation Procedure

Cross-Linking Chromatin

1. Prepare a culture as described in the galactose induction protocol. When the cells have grown to 3×10^6 to 1×10^7 cells/ml, collect a 45-ml sample.

2. Dilute an aliquot of cells by 10^{-4} and plate 100 μl on YEPD (before induction and at 4 h) to assess viability. Add formaldehyde to 1% final

concentration and incubate for 10 min at room temperature. The optimal exposure to formaldehyde depends on the protein.

3. Quench the cross-linking by adding 2.5 ml of 2.5 M glycine and incubate samples for 5 min at room temperature.

4. Centrifuge the cells for 5 min, discard the supernatant, and resuspend the cells in 1 ml of ice cold Tris-buffered saline (TBS, 20 mM Tris, pH 7.6, 150 mM NaCl) in a microcentrifuge tube. Wash two additional times with 1 ml of TBS. Freeze the cell pellet at $-70°$ after the final spin.

Cell Lysis and Shearing DNA

1. Resuspend the cell pellets in 0.55 ml of lysis buffer (50 mM HEPES, pH 7.5, 1 mM EDTA, 140 mM NaCl, 1% Triton X-100, 0.1% sodium deoxycholate, 1 mg/ml Bacitracin, 1 mM benzamidine, 1 μg/ml leupeptin, 1 μg/ml pepstatin A, 1 mM phenylmethylsulfonyl fluoride).

2. Transfer the suspension to 1.5-ml centrifuge tubes containing 0.5 ml of glass beads (425–600 μm diameter, acid washed) on ice. Vortex the cells at 4° until 90% of the cells have lysed.

3. Puncture the bottom of the tubes with a red hot syringe needle heated with a Bunsen burner and place the tube on top of an ice-cold 15-ml conical tube. Separate the whole cell lysate from the glass beads by centrifuging the 15-ml tubes to collect the lysates.

4. Transfer the supernatant and pellet to 2-ml tubes. Sonicate the samples until the DNA has an average size of 0.5 kb.

5. Centrifuge the lysate for 10 min at 4°. Remove the supernatant to a new tube and repeat. Freeze the lysate at $-70°$.

Chromatin Immunoprecipitation

1. For the immunoprecipitated (IP) samples, add 350 μl of the lysate to chilled tubes containing antibody and incubate at 4° with gentle agitation for 1 h. For the nonimmunoprecipitated sample (Non-IP or input), remove 50 μl of the lysate to a new tube. Keep on ice until the IP samples are ready for the reversal of cross-links.

2. Wash the protein-A or -G agarose beads twice with 1 ml of lysis buffer by spinning at 3000 rpm for 1 min. Resuspend the beads in an equal volume of lysis buffer. Add 30 μl of the protein-A or -G agarose beads to each IP sample. Incubate the tubes at 4° with slow rotating for 1 h.

3. Wash the beads as described earlier with 1 ml of lysis buffer (twice) and once with high salt lysis buffer, wash buffer, and TE for 5 min each at room temperature. High salt lysis buffer is lysis buffer made with 0.5 M NaCl.

4. Elute the precipitate from the beads by incubating at 65° for 15 min in 100 μl of elution buffer (50 mM Tris, pH 7.5, 1 mM EDTA, 1% SDS).

5. Centrifuge the beads, mix the supernatant in a new tube with 150 μl of TE plus 0.67% SDS, and incubate at 65° for 10 min.

6. Centrifuge the beads and combine the supernatant with the eluate just described. Add 200 μl of TE plus 1% SDS to the Non-IP samples and incubate the eluates and Non-IP samples at 65° for at least 6 h.

7. Add 250 μl TE containing proteinase K (0.4 mg/ml) and glycogen (0.1 mg/ml). Incubate the samples at least 2 h at 37°.

Purification of DNA and PCR Analysis

1. Add 55 μl of 4 M LiCl to the samples and extract with 500 μl of phenol. Centrifuge and save the supernatant. Precipitate the DNA with ethanol for 15 min at room temperature and centrifugation. Wash the pellet with 80% ethanol. Centrifuge for 10 min and remove the supernatant. Allow the pellets to dry and resuspend the DNA in TE.

2. The DNA can be analyzed using conventional or real-time PCR. A typical program is 1 min 94°, 26 cycles of 15 s 94°, 20 s 55°, 20 s 72°, and 5 min at 72°. A hot-start *Taq* polymerase is recommended to minimize competing PCR products. PCR products are generally chosen to be about 300 bp long and can be visualized on a 2% agarose gel. Adjust the conditions to obtain signals in the linear range of a calibration curve prepared with dilutions of the Non-IP 0-h sample. For controls, Non-IP and IP samples from an independent locus should be amplified and quantitated. A sample not treated with formaldehyde provides a control to indicate whether protein binding may have occurred after lysis. While this is a useful control, there may be proteins, such as the RPA complex, that bind DNA *in vivo* and remain bound during the washes even in the absence of formaldehyde cross-linking (Wang and Haber, 2004).

Comments

Harvesting Cells

There is a noticeable increase in the size of the cell pellet in time courses lasting 4 h or longer, especially in strains where the DSB does not induce the DNA damage checkpoint. To compensate for this, smaller samples are taken and diluted with YEP-lactate, based on OD_{600} measurements.

Normalization of PCR Signals

We use two methods to normalize the ChIP signals. The first method is to normalize the IP time point signals to the input signal at 0 h for the locus of interest (e.g., the *MAT* site) and the independent control locus. The ratios of the *MAT* IP to control IP signal are then normalized to the 0-h time point. The second method normalizes the IP signals to the input signal at 0 h for both *MAT* and control loci. The ratio of the *MAT* IP signal to the control input signal can then be calculated. This method is useful if the signals from the 0-h IP and/or control IPs are very low or if the *MAT* and control IP signals are not completely independent.

This procedure is based on the work of Strahl-Bolsinger *et al.* (1997) and Evans *et al.* (2000).

References

Aylon, Y., and Kupiec, M. (2004). DSB repair: The yeast paradigm. *DNA Repair (Amst.)* **3,** 797–815.

Bassing, C. H., and Alt, F. W. (2004). The cellular response to general and programmed DNA double strand breaks. *DNA Repair (Amst.)* **3,** 781–796.

Boeke, J. D., Trueheart, J., Natsoulis, G., and Fink, G. R. (1987). 5-Fluororotic acid as a selective agent in yeast molecular genetics. *Methods Enzymol.* **154,** 164–175.

Booth, C., Griffith, E., Brady, G., and Lydall, D. (2001). Quantitative amplification of single-stranded DNA (QAOS) demonstrates that *cdc13–1* mutants generate ssDNA in a telomere to centromere direction. *Nucleic Acids Res.* **29,** 4414–4422.

Church, G. M., and Gilbert, W. (1984). Genomic sequencing. *Proc. Natl. Acad. Sci. USA* **81,** 1991–1995.

Dionne, I., and Wellinger, R. J. (1996). Cell cycle-regulated generation of single-stranded G-rich DNA in the absence of telomerase. *Proc. Natl. Acad. Sci. USA* **93,** 13902–13907.

Douglas, H. C., and Pelroy, G. (1963). A gene controlling the inducibility of galactose pathway enzymes in *Saccharomyces cerevisiae*. *Biochim. Biophys. Acta* **68,** 155–156.

Evans, E., Sugawara, N., Haber, J. E., and Alani, E. (2000). The *Saccharomyces cerevisiae* Msh2 mismatch repair protein localizes to recombination intermediates *in vivo*. *Mol. Cell.* **5,** 789–799.

Evans, I. H., and Wilkie, D. (1976). Mitochondrial role in the induction of sugar utilization in *Saccharomyces cerevisiae*. *In* "Function of Mitochondrial DNA" (C. Saccone and A. M. Kroon, eds.), pp. 209–217. North-Holland, New York.

Frankfurt, O. S., Robb, J. A., Sugarbaker, E. V., and Villa, L. (1996). Monoclonal antibody to single-stranded DNA is a specific and sensitive cellular marker of apoptosis. *Exp. Cell Res.* **226,** 387–397.

Garvik, B., Carson, M., and Hartwell, L. (1995). Single-stranded DNA arising at telomeres in *cdc13* mutants may constitute a specific signal for the *RAD9* checkpoint. *Mol. Cell. Biol.* **15,** 6128–6138.

Haber, J. E. (1998). Mating-type gene switching in *Saccharomyces cerevisiae*. *Annu. Rev. Genet.* **32,** 561–599.

Haber, J. E. (2000). Lucky breaks: Analysis of recombination in *Saccharomyces*. *Mutat. Res.* **451,** 53–69.

Herskowitz, I., and Jensen, R. E. (1991). Putting the HO gene to work: Practical uses for mating-type switching. *Methods Enzymol.* **194,** 132–146.

Jensen, R. E., and Herskowitz, I. (1984). Directionality and regulation of cassette substitution in yeast. *Cold Spring Harb. Symp. Quant. Biol.* **49,** 97–104.

Johnson, R. D., and Jasin, M. (2001). Double-strand-break-induced homologous recombination in mammalian cells. *Biochem. Soc. Trans.* **29,** 196–201.

Lee, S. E., Moore, J. K., Holmes, A., Umezu, K., Kolodner, R., and Haber, J. E. (1998). *Saccharomyces* Ku70, Mre11/Rad50 and RPA proteins regulate adaptation to G2/M arrest DNA damage. *Cell* **94,** 399–409.

Mahler, H. R., and Wilkie, D. (1978). Mitochondrial control of sugar utilization in *Saccharomyces cerevisiae. Plasmid* **1,** 125–133.

Malkova, A., Naylor, M. L., Yamaguchi, M., Ira, G., and Haber, J. E. (2005). *RAD51*-dependent break-induced replication differs in kinetics and checkpoint responses from *RAD51*-mediated gene conversion. *Mol. Cell. Biol.* **25,** 933–944.

Nickoloff, J. A., Singer, J. D., Hoekstra, M. F., and Heffron, F. (1989). Double-strand breaks stimulate alternative mechanisms of recombination repair. *J. Mol. Biol.* **207,** 527–541.

Paques, F., and Haber, J. E. (1997). Two pathways for removal of nonhomologous DNA ends during double-strand break repair in *Saccharomyces cerevisiae. Mol. Cell. Biol.* **17,** 6765–6771.

Pierce, A. J., Stark, J. M., Araujo, F. D., Moynahan, M. E., Berwick, M., and Jasin, M. (2001). Double-strand breaks and tumorigenesis. *Trends Cell Biol.* **11,** S52–S59.

Raderschall, E., Golub, E. I., and Haaf, T. (1999). Nuclear foci of mammalian recombination proteins are located at single-stranded DNA regions formed after DNA damage. *Proc. Natl. Acad. Sci. USA* **96,** 1921–1926.

Sambrook, J., Fritsch, E. F., and Maniatis, T. (1989). "Molecular Cloning: A Laboratory Manual." Cold Spring Harbor Laboratory Press, Cold Spring Harbor, NY.

Sandell, L. L., and Zakian, V. A. (1993). Loss of a yeast telomere: Arrest, recovery, and chromosome loss. *Cell* **75,** 729–739.

Strahl-Bolsinger, S., Hecht, A., Luo, K., and Grunstein, M. (1997). SIR2 and SIR4 interactions differ in core and extended telomeric heterochromatin in yeast. *Genes Dev.* **11,** 83–93.

Sugawara, N., and Haber, J. E. (1992). Characterization of double-strand break-induced recombination: Homology requirements and single-stranded DNA formation. *Mol. Cell. Biol.* **12,** 563–575.

Sugawara, N., Wang, X., and Haber, J. E. (2003). *In vivo* roles of Rad52, Rad54, and Rad55 proteins in Rad51-mediated recombination. *Mol. Cell* **12,** 209–219.

Tapiero, H., Leibowitch, S. A., Shaool, D., Monier, M. N., and Harel, J. (1976). Isolation of single stranded DNA related to the transcriptional activity of animal cells. *Nucleic Acids Res.* **3,** 953–963.

Wang, X., and Haber, J. E. (2004). Role of *Saccharomyces* single-stranded DNA-binding protein RPA in the strand invasion step of double-strand break repair. *PLoS Biol.* **2,** E21.

White, C. I., and Haber, J. E. (1990). Intermediates of recombination during mating type switching in *Saccharomyces cerevisiae. EMBO J.* **9,** 663–673.

[27] Assays for Nonhomologous End Joining in Extracts

By Joe Budman and Gilbert Chu

Abstract

In mammalian cells, nonhomologous end–joining (NHEJ) repairs DNA double strand breaks created by ionizing radiation and V(D)J recombination. Using human whole cell extracts prepared by the method of Baumann and West (1998), we have described a cell-free system for NHEJ that joins both compatible and noncompatible DNA ends (Budman and Chu, 2005). To measure joining efficiency and assess the processing of DNA ends, we developed a quantitative polymerase chain reaction assay for the joining of two specific DNA ends. The *in vitro* NHEJ reaction recapitulates key features of NHEJ observed *in vivo*: end joining is dependent on DNA-PK and XRCC4/Ligase4, and noncompatible ends are processed by polymerase and nuclease activities that often stabilize the alignment of opposing ends by base pairing. This chapter describes methods for preparing whole cell extracts and for studying the NHEJ reaction *in vitro*.

Introduction

The nonhomologous end-joining (NHEJ) pathway is important for repairing DNA double strand breaks produced by ionizing radiation and anticancer drugs such as adriamycin, etoposide, and bleomycin (reviewed in Lees-Miller and Meek, 2003; Lieber *et al.*, 2003). NHEJ is also required for repairing double strand breaks created during V(D)J recombination (Gellert, 2002). Because the broken DNA ends may not be directly ligatable, processing often prepares the ends for NHEJ. Processing also contributes to the immunological diversity generated by V(D)J recombination.

A cell-free system is essential for the study of NHEJ in biochemical detail. To be physiologically relevant, a cell-free system must recapitulate the key characteristics of NHEJ observed *in vivo*. NHEJ *in vivo* requires several proteins, including Ku, DNA-PKcs, and XRCC4/Ligase4. The reaction also includes factors that process noncompatible DNA ends by templated nucleotide addition or by deletions that often extend back to regions of microhomology. Previous investigators established a cell-free system for NHEJ that was dependent on Ku, DNA-PKcs, and XRCC4/Ligase4, but observed the joining of only compatible ends or 5′-dephosphorylated ends

METHODS IN ENZYMOLOGY, VOL. 408   0076-6879/06 $35.00
        DOI: 10.1016/S0076-6879(06)08027-X

(Baumann and West, 1998; Chappell *et al.*, 2002). We extended this system to include the joining of noncompatible DNA ends (Budman and Chu, 2005). Our *in vitro* joining reactions recapitulated the characteristics observed for NHEJ and V(D)J recombination *in vivo*. They were strongly dependent on Ku, DNA-PKcs, and XRCC4/Ligase4, and were characterized by nucleotide addition, deletion, and frequent use of microhomology.

The method of extract preparation is critically important. To obtain robust NHEJ activity, we found that extracts must be prepared on a large scale and under specific buffer conditions. Using our method of extract preparation, joining of compatible and noncompatible ends was inhibited at least 100-fold by wortmannin, an inhibitor of DNA-PKcs and related kinases, or by immunodepletion with anti-XRCC4 antibodies. We found that extracts prepared by alternative protocols might catalyze DNA end joining, but the reactions did not require DNA-PKcs or XRCC4/Ligase4, and therefore did not reflect bona fide NHEJ.

NHEJ in our cell-free system is an intermolecular reaction that can be measured by quantitative polymerase chain reaction (qPCR). As shown in Fig. 1, incubation of linear plasmid DNA with T4 DNA ligase generates products that are largely circular monomers. However, extract joins the same linear plasmid DNA into linear dimers, trimers, and higher order multimers, but not circular monomers. The failure to observe circular monomers in the NHEJ reaction suggests that the DNA is highly bound to proteins in the extract. Therefore, our assay for NHEJ focuses on intermolecular joining of DNA substrates with different ends. To be confident of the sequence at the DNA ends, we prepare DNA substrates free of unwanted contaminants, as described later.

To measure the joining efficiency of specific DNA ends, we use a quantitative PCR assay. Although the PCR primers amplify DNA fragments joined in a specific orientation, restriction enzyme analysis shows that the DNA fragments are joined in all possible orientations. Head-to-head joining of the same fragment to itself could potentially make a spurious contribution to our measurement of joining efficiency. However, no product was detected when PCR is performed with a single primer, consistent with the known failure of PCR to amplify long inverted repeats. Therefore, the quantitative PCR assay is specific for the joining of two DNA ends in a defined orientation.

To characterize junctions formed by the NHEJ reaction, the DNA sequence spanning the junction is amplified by PCR, subcloned, and sequenced from individual clones. Sequencing reveals that compatible ends are usually joined precisely, indicating that processing is suppressed when not required. Noncompatible ends are joined after processing by different enzymatic activities. In the presence of deoxynucleotides (dNTPs), processing is dominated

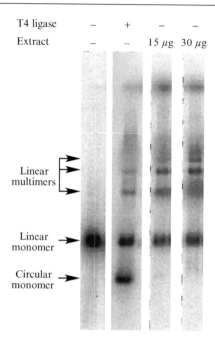

Fig. 1. NHEJ in extract is an intermolecular reaction. Radiolabeled pUC18 (10 ng) linearized with *Eco*RI was incubated with T4 DNA ligase or lymphoblastoid whole cell extract. The DNA was deproteinized, and the reaction products were analyzed by agarose gel electrophoresis followed by autoradiography. The T4 ligase generated circular monomers and linear multimers, but the extract generated only linear multimers.

by polymerase activity that can synthesize DNA across a discontinuous template strand (Budman and Chu, 2005). In the absence of dNTPs, junctions are often processed by a regulated nuclease activity that generates deletions generally limited to less than 20 bp, often back to regions of microhomology. Because it contains all the known NHEJ activities, this cell-free system can be used as the basis for purification of additional NHEJ factors.

Preparation of Human Whole Cell Extracts That Support NHEJ

Cell Lines

We have prepared extracts that support NHEJ from both lymphoblastoid cells and HeLa S3 cells. The lymphoblastoid cell line GM00558C (from Coriell Cell Repositories) was cultured in our laboratory. For HeLa

S3 cells, we use a frozen pellet from a 10-liter culture provided by the National Cell Culture Center.

Protocol for the Growth and Harvest of Lymphoblastoid Cells

The whole cell extract is prepared by the method of Baumann and West (1998). Grow lymphoblastoid cell line GM00558C (Coriell Cell Repositories) in suspension to a density of 8×10^5 cells/ml in 10 liters of medium consisting of RPMI 1640, 15% heat-inactivated fetal bovine serum, 2 mM L-glutamine, and 1% penicillin/streptomycin solution (Invitrogen). Spin down the cells at 900g for 10 min and resuspend the cells in ice-cold phosphate-buffered saline (PBS). Because the cells are grown in large-scale culture, the cells must first be spun down in multiple containers. The resuspended cells should subsequently be pooled into a single container and then spun down to form a single pellet.

Tips

1. The growing lymphoblastoid cells should be split when they reach a density of $\sim 1.2 \times 10^6$ cells/ml. Do not split the cells to a density of less than 2.5×10^5 cells/ml.

2. Grow the cells with adequate aeration (we use 1900-ml tissue culture flasks with 400 ml of cells per flask). The cells should be harvested when they are healthy and still actively doubling. After spinning down the PBS-washed cells, verify that they are still viable. Cells undergoing apoptosis produce poor extracts, probably because DNA-PKcs is cleaved by apoptotic caspases (Song et al., 1996).

3. We have found that large-scale cultures are necessary for robust NHEJ activity. XRCC4/Ligase4 may be limiting in extracts, since other groups have reported the need to supplement their extracts with recombinant XRCC4/Ligase4 to detect NHEJ activity (Lee et al., 2000).

Protocol for Preparing Extract

These steps should be performed at 4°. Resuspend the cell pellet in 5 volumes of hypotonic lysis buffer (10 mM Tris–Cl, pH 8.0, 1 mM EDTA, 5 mM dithiothreitol [DTT]). Centrifuge the cells at 900g for 10 min, and remove as much of the supernatant as possible by pipetting. Add 1 swollen cell volume of hypotonic lysis buffer plus a mixture of protease inhibitors consisting of phenylmethylsulfonyl fluoride to a final concentration of 1 mM and aprotinin, pepstatin, chymostatin, and leupeptin to final concentrations of 1 μg/ml. Incubate the cells on ice for 20 min and then lyse the cells by 20 strokes of douncing with an A pestle. Verify that the cells have been lysed by microscopic inspection, incubate on ice for another 20 min, and transfer the lysed cells to centrifuge tubes. Add 0.5 volume of high salt buffer (50 mM Tris–Cl, pH 7.5, 1 M KCl, 2 mM EDTA, 2 mM DTT), and incubate on ice for 30 min, swirling occasionally. Centrifuge the lysate for 3 h at 54,000 rpm in a Beckmann Ti60 rotor. Carefully harvest the supernatant, avoiding the upper lipid layer, and transfer into a dialysis membrane with a molecular weight cutoff of 8000 Da. Dialyze against several liters of 20 mM Tris–Cl, pH 8.0, 100 mM KOAc, 0.5 mM EDTA, 1 mM DTT, and 25% glycerol. After 1.5 h, replace the dialysis buffer and dialyze for an additional 2 h. Remove the extract from the dialysis bags and centrifuge at 13,000g for 2 min. Remove the supernatant, fast-freeze aliquots in liquid nitrogen, and store at $-80°$. The final protein concentration should be 8–12 mg/ml. Prior to use, centrifuge the extract for 2 min at 13,000g and use only the supernatant.

Tips

1. Some of the protease inhibitors are very labile. Be sure the protease inhibitors have been stored appropriately and add them just before lysis.

2. The presence of Ku, DNA-PKcs, XRCC4/Ligase4, and Artemis in the extract can be checked by Western blot. The DNA-binding activity of Ku and DNA-PKcs can be verified by an electrophoretic mobility shift assay (Hwang *et al.*, 1999), and the enzymatic activity of DNA-PKcs can be verified by a DNA-dependent protein kinase assay (DeFazio *et al.*, 2002).

Extract Variability. We have observed some variability in joining efficiencies and processing activities between different extracts. While most of our extracts join compatible ends with high efficiency, some extracts are more efficient than others in joining certain combinations of noncompatible ends. We have also found that extracts prepared from HeLa cells join noncompatible ends 2- to 10-fold more efficiently than extracts prepared from lymphoblastoid cells. Our HeLa extracts also process noncompatible ends predominantly by polymerase activity, whereas lymphoblastoid extracts contain a more equal mixture of polymerase and nuclease activities. Thus, differences in processing may arise from variations among different cell types or variations among different extract preparations. Despite these differences, all of our extracts have catalyzed DNA-PKcs and XRCC4/Ligase4-dependent NHEJ of compatible and noncompatible DNA ends.

Gel-Based Assay for NHEJ of Compatible DNA Ends

Protocol for the Preparation of End-Labeled DNA Substrate

Digest plasmid DNA with a restriction enzyme, and gel purify the linearized DNA. We use the Qiagen gel extraction kit. Treat the DNA with calf intestinal phosphatase (New England Biolabs) and purify the

DNA (we use a Qiagen PCR purification kit). Incubate dephosphorylated DNA with T4 polynucleotide kinase (New England Biolabs) plus γ-labeled ^{32}P–ATP (3,000 Ci/mmol; PerkinElmer) at threefold molar excess relative to DNA and purify the DNA.

Protocol for the In Vitro *NHEJ Reaction*

Incubate 10 ng of end-labeled linearized plasmid with 4 μl of whole cell extract in a total volume of 20 μl containing end-joining buffer [50 mM Tris–Cl, pH 8.0, 60 mM KOAc, 0.5 mM MgCl$_2$, 1 mM DTT, 0.08 mg/ml bovine serum albumin (BSA), and 1 mM ATP] at 37° for 90 min. Reactions are started by the addition of protein extract. Mix the end-joining reaction gently but thoroughly by pipetting. For control reactions with T4 DNA ligase, incubate 10 ng of end-labeled DNA with 1 μl of T4 DNA ligase (400 U/μl; New England Biolabs) in the manufacturer's T4 DNA ligase buffer at room temperature for 2 h. Stop the reactions by adding 2 μl of 0.5 M EDTA, pH 8.0. To deproteinize the reactions, use a Qiagen PCR purification kit or add 3 μl of deproteinization solution (10 mg/ml proteinase K, 3% SDS, 50 mM EDTA, 100 mM Tris–Cl, pH 7.5) and incubate at 37° for 1 h. Resolve the DNA by agarose gel electrophoresis. Dry the gel onto a Hybond N+ membrane (Amersham) and analyze the reaction products with a phosphorimager.

Tips. The extract and stock solution of end-joining buffer are viscous and therefore should be pipetted carefully.

Results

Under the conditions of this cell-free system, NHEJ is an intermolecular reaction. When 10 ng of linearized, end-labeled pUC18 (New England Biolabs) was incubated with extract under the conditions described earlier, gel electrophoresis revealed that the extract joined the DNA into linear dimers, trimers, and higher order multimers, but not into circular monomers (Fig. 1). This result has also been observed by others (Baumann and West, 1998: Chen *et al.*, 2001). In contrast, incubation of the 2686-bp DNA with T4 ligase generated products that were approximately 75% circular monomers, which is consistent with calculations of the expected ratio of intermolecular to intramolecular joining. For the intermolecular joining of DNA ends, 10 ng of linear pUC18 in a reaction volume of 20 μl corresponds to a concentration for DNA ends of 0.56 nM. For intramolecular joining, the concentration of one end of pUC18 relative to the other end is 0.5 nM (assuming that the DNA is sampling the volume of a perfect sphere). Therefore, we attribute the failure to observe circular monomers in the NHEJ reaction to extensive binding of the DNA to proteins, which prevent circularization.

Comments

This simple gel-based assay has also been used to examine the joining of 5′-dephosphorylated ends (Chappell *et al.*, 2002). The substrates for those experiments were internally radiolabeled DNA fragments that had been treated with calf intestinal phosphatase (CIP).

Quantitative PCR Assay for NHEJ of Compatible or Noncompatible DNA Ends

After observing that NHEJ in this cell-free system is an intermolecular reaction, we developed a quantitative PCR assay for the joining of two specific DNA ends in a defined orientation. To be confident of the sequence at the DNA ends, we take care to prepare substrates free of contaminating sequence.

Protocol for the Preparation of DNA Substrates (Fig. 2)

We use plasmid pRL-null (Promega) as the template for PCR amplification of two distinct DNA fragments with various restriction enzyme sites at the ends. To avoid the repetitive and palindromic sequences found in the cloning sites, we use bases 303–1121 (from the Renilla luciferase gene) as the template for "DNA1" and bases 1636–2504 (from the ampicillin resistance gene) as the template for "DNA2." PCR amplify the DNA fragments using a PCR polymerase such as Pfu to avoid introducing mutations. The 5′ ends of the primers for amplifying DNA1 and DNA2 contain cleavage sites for restriction enzymes that leave 5′ overhangs, 3′ overhangs, or blunt DNA. The sites include *Bam*HI, *Eco*RI, *Kpn*I, *Sac*I, *Eco*RV, *Stu*I, *Swa*I, *Ssp*I, and *Pml*I. The sequences of these primers can be found in Budman and Chu (2005).

Direct digestion of the PCR products would leave small amounts of residual uncut DNA of an almost identical size. To eliminate contamination from unwanted DNA, the PCR products should be subcloned. We use the pCR-BluntII-Topo vector (Invitrogen). Transform the subcloned DNA products into bacteria and plate on the appropriate selective medium. Grow bacteria from a single clone and prepare plasmid DNA (we use the Qiagen maxiprep kit). Digest the plasmid with the appropriate restriction enzyme, gel purify the desired DNA fragment, and determine the DNA concentration.

Protocol for the In Vitro NHEJ Reaction (Fig. 3)

Incubate 18 fmol of DNA1 and DNA2 (with the desired DNA ends) with 4 μl of whole cell extract in a total volume of 20 μl containing end-joining buffer (50 mM Tris–Cl, pH 8.0, 60 mM KOAc, 0.5 mM MgCl$_2$,

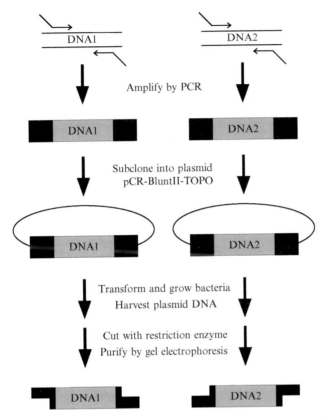

Fig. 2. Synthesis of DNA substrates without contaminating end sequences. Two distinct DNA fragments, DNA1 and DNA2, were amplified by PCR with primers containing various restriction sites at the 5' ends of the primers. To eliminate contamination from residual uncut DNA, the PCR products were subcloned into a plasmid, released by cleavage with the appropriate restriction endonuclease, and gel purified. DNA substrates with various DNA ends attached to the same internal DNA sequences can be created in this way.

1 mM DTT, 0.08 mg/ml BSA, and 1 mM ATP) at 37° for 90 min. To stimulate processing by polymerase activity, 25 μM dNTPs can be added to the reactions. Start the reactions by adding protein extract, and mix by pipetting. Stop the reactions by adding 2 μl 0.5 M EDTA, pH 8.0. Purify the DNA with a Qiagen PCR purification kit.

To inhibit the kinase activity of DNA-PKcs, preincubate the extract on ice for 30 min with wortmannin. Mix thoroughly but gently by pipetting. Dissolve the wortmannin stock solution in dimethyl sulfoxide (DMSO), and the final DMSO concentration in the wortmannin-inhibited

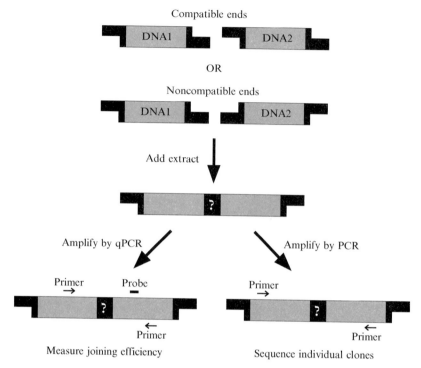

FIG. 3. Assay for NHEJ *in vitro*. Compatible or noncompatible ends were incubated with a whole cell extract from mammalian cells. The joining efficiency was measured with quantitative PCR. Processing of the ends was analyzed by PCR amplification of the DNA junctions and sequencing of individual clones.

extract should be less than 1%. Perform a parallel incubation of extract with DMSO alone to confirm that DMSO is not inhibiting NHEJ.

To immunodeplete XRCC4/Ligase4, preincubate the extract with 1 μl of anti-XRCC4 antibody (Serotec) for every 15 μl of extract for 90 min on a rotary wheel at 4°. Mix thoroughly but gently by pipetting. Wash protein A-Sepharose beads (Santa Cruz Biotech) several times with extract dialysis buffer. Centrifuge the beads for 5 min at 3000g, remove the supernatant, and add the extract/antibody mixture. Incubate for an additional 90 min on a rotary wheel at 4°. Centrifuge at 3000g for 5 min and remove the supernatant carefully. If necessary, centrifuge again to remove any remaining beads.

Tips

1. For consistent results, it is desirable to have a high and uniform yield of DNA from the Qiagen DNA purification column. Instead of using the recommended elution volume of 50 μl, we add 125 μl of elution buffer (10 mM Tris–Cl, pH 8.5) to the column, incubate for 10 min, and then elute the DNA by centrifugation.

2. If dNTPs are present in the end-joining reactions, they should be added to the reaction tubes before the addition of extract. This is because preincubation of the extract with dNTPs gives inconsistent results depending on the duration and the temperature of preincubation.

3. We perform all critical experiments in triplicate.

Protocol for the Measurement of Joining Efficiency by Quantitative PCR

We use the ABI Prism 7900 machine (Applied Biosystems) for qPCR. Some of the following steps may be unique to TaqMan probes and reagents, but the same general principles will apply for other qPCR protocols.

Thermal Cycling. Assemble 20-μl reactions containing 1–2 μl purified DNA from the end-joining reaction, 10 μl of 2× TaqMan PCR Mastermix (Applied Biosystems), 200 nM TaqMan probe (5'-FAM-TGTAACC-CACTCGTGCACCCAACTGAT-3'TAMRA), and 900 nM of each PCR primer (5'-AGGTGGTAAACCTGACGTTGTACA and 5'-CGCTGT-TGAGATCCAGTTCG). Initiate thermal cycling by heating the sample at 50° for 2 min, 95° for 10 min, and then performing 50 cycles of denaturation (95° for 15 s) followed by annealing and extension (60° for 1 min).

The primer and probe sequences were chosen using the recommended Primer Express software, and we verified that these primers and conditions supported exponential amplification. The primer and probe termini are located at least 100 bases from either DNA end so that relatively large deletion products could be amplified. However, the distance from the DNA ends should not be too large, as the PCR amplification will be less than exponential.

Calibration of Joining Efficiency. To determine the percentage of DNA ends joined, it is necessary to prepare preformed DNA junctions. These can be DNA junctions amplified by PCR, as described earlier. Alternatively, the DNA junctions can be amplified by PCR and then subcloned to remove DNA contaminants. We confirmed that both methods give identical results.

Measure the concentration of the DNA junction standards. For each standard, 18 fmol of junction DNA defines 100% joining, since that is the amount of each DNA substrate added to the NHEJ reaction. Prepare serial twofold dilutions down to <0.00001% joining. Treat each dilution with the same DNA purification and dilution procedure used for the experimental samples (Qiaquick PCR purification column). Amplify the dilutions by qPCR and prepare a standard curve of percentage of DNA ends joined versus the number of PCR cycles needed to synthesize a threshold amount of DNA. The threshold cycle amount should be carefully chosen to lie in a linearly exponential area of the curve for all samples. Calculate the joining efficiency of the experimental samples from the standard curve.

Results

In this assay, the value for "percentage of DNA ends joined" refers only to those junctions formed between the DNA ends in the single orientation specified by the qPCR assay. The total percentage of joined DNA ends is significantly higher, since the ends are joined in four possible orientations.

The *in vitro* NHEJ reaction is dependent on DNA-PKcs and XRCC4/Ligase4 (Fig. 4A). Preincubation of the extract with 20 μM wortmannin or immunodepletion with anti-XRCC4 antibody inhibits the joining of compatible and noncompatible DNA ends by at least 100-fold in a lymphoblastoid extract.

Figure 4B shows the effect of dNTPs on the joining of compatible or noncompatible DNA ends. When dNTPs were omitted, the joining efficiency was reduced 100-fold for these noncompatible ends. In contrast, the joining of compatible ends was unchanged. This is because compatible ends are joined without processing, while polymerase activity makes a large contribution to the processing of noncompatible ends (Budman and Chu, 2005). Even in the absence of dNTPs, joining is highly dependent on DNA-PKcs and XRCC4/Ligase4 and therefore reflects bona fide NHEJ.

Analysis of Processing Events

Protocol for PCR Amplification and Sequencing of DNA Junctions

DNA end processing is characterized by polymerase and nuclease activities that produce heterogeneous processing events. To analyze the junctions formed by the NHEJ reaction, the DNA junctions are amplified by PCR, subcloned, and sequenced.

Prepare 50 μl PCR reactions containing 1 μl purified DNA from the end-joining reaction, 200 μM dNTPs, 1 μM each of PCR primer (5′-GAACCATTCAAAGAGAAAGGTG and 5′-GGGAAGCTAGAG-TAAGTAG), and 2 units of Pfu Turbo HotStart DNA polymerase

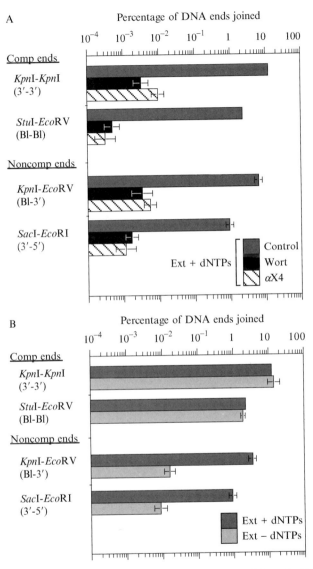

FIG. 4. Extract catalyzes NHEJ of compatible and noncompatible ends. (A) The NHEJ reaction requires DNA-PKcs and XRCC4. DNA substrates with compatible or noncompatible ends were incubated with extract and dNTPs (a) without additional treatment, (b) treated with the DNA-PKcs kinase inhibitor wortmannin (wort), or (c) immunodepleted with anti-XRCC4 (αX4) antibody. Joining efficiency was measured by qPCR. (B) Deoxynucleotides stimulate the joining of noncompatible ends. DNA substrates were incubated with extract (a) in the presence of dNTPs or (b) in the absence of dNTPs. Joining efficiency was measured by qPCR.

(Stratagene) in its appropriate buffer. These primer termini are located 182 and 583 bases from the ends of DNA1 and DNA2, respectively, to permit detection of deletions. After 2 min at 96°, perform 44 cycles of denaturation (94° for 45 s), annealing (55° for 45 s), and extension (72° for 2 min). Finally, perform a final extension at 72° for 10 min. Verify that PCR amplification was successful by resolving 10 μl of the reaction on an agarose gel, and purify the rest of the DNA using a Qiagen PCR purification kit. Subclone the PCR-amplified DNA products into the pCR-BluntII-TOPO vector (Invitrogen). Transform the subcloned DNA products into bacteria and plate on kanamycin continuing medium. Grow bacteria from a single clone, and prepare plasmid DNA for sequencing. Analyze the junctions by DNA sequencing.

Tips

1. To avoid an artifactual bias in the recovered junctions due to unequal PCR amplification, perform several independent PCR reactions for each type of junction to be sequenced.
2. After transforming bacteria with subcloned DNA, prevent outgrowth of duplicate clones by plating the bacteria on selective media as soon as they have recovered from the transformation.
3. It is important to avoid introducing an unwanted selection for junctions with large deletions during the PCR amplification. Be sure that the PCR extension time is two to three times longer than the minimum required extension time for the given polymerase.

Results

Compatible ends are usually joined without processing (Budman and Chu, 2005). Noncompatible ends are processed with polymerase and nuclease activities. Figure 5 shows the DNA junctions formed by lymphoblastoid extract from two types of noncompatible ends. When dNTPs are present in the reactions, the majority of junctions show templated nucleotide addition. As reported previously, 3' overhanging ends were filled in by a polymerase activity that can synthesize DNA across a discontinuity in the template strand. This type of polymerase activity creates base pairing between the opposing ends and may thus stabilize their alignment. When dNTPs are omitted from the reactions, processing is dominated by nuclease activity. The nuclease activity removes single-stranded overhangs or produces small deletions that often extend back to short regions of microhomology. Some polymerase activity is observed even when dNTPs are not added to the joining reactions. We presume that this is due to the generation of nucleotides by nuclease activity in the extract.

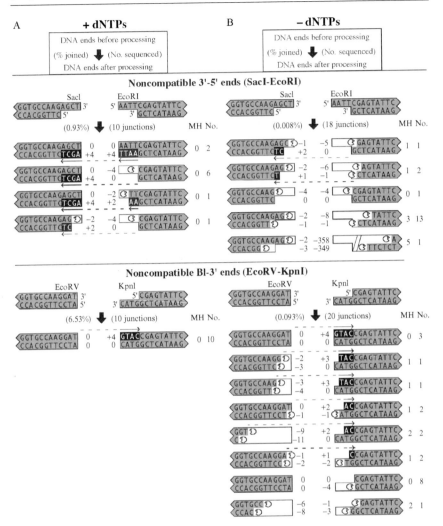

FIG. 5. Noncompatible ends are processed by polymerase and nuclease activities. DNA substrates with noncompatible ends were incubated with (A) extract plus added dNTPs or (B) extract without added dNTPs. Reaction products are shown for two sets of noncompatible ends. Joining efficiency is indicated to the left of the arrow and total number of sequenced junctions to the right of the arrow. Nuclease activity is shown by the open-mouthed icon and white background in the DNA. Polymerase activity is shown by the arrow and black background. Unprocessed DNA is shown by the gray background. The number of bases available for pairing of the two ends via microhomology is shown in the column labeled "MH." The number of times each junction was recovered is shown in the column labeled "No."

Concluding Remarks

This cell-free system is a useful tool for studying the NHEJ pathway in greater detail. Because it contains all the known enzymatic activities for NHEJ, the cell-free system can be used to purify and identify the enzymes that process noncompatible DNA ends. A detailed biochemical understanding of NHEJ will require the reconstitution of the entire reaction using purified proteins.

Acknowledgments

We thank Sunny Kim and Chun Tsai for helpful discussions. This work was supported by a predoctoral fellowship from the Howard Hughes Medical Institute to J.B. and NIH Grant RO1 GM58120 to G.C.

References

Baumann, P., and West, S. C. (1998). DNA end-joining catalyzed by human cell-free extracts. *Proc. Natl. Acad. Sci. USA* **95**, 14066–14070.

Budman, J., and Chu, G. (2005). Processing of DNA for nonhomologous end-joining by cell-free extract. *EMBO J.* **24**, 849–860.

Chappell, C., Hanakahi, L. A., Karimi-Busheri, F., Weinfeld, M., and West, S. C. (2002). Involvement of human polynucleotide kinase in double-strand break repair by non-homologous end joining. *EMBO J.* **21**, 2827–2832.

Chen, S., Inamdar, K. V., Pfeiffer, P., Feldmann, E., Hannah, M. F., Yu, Y., Lee, J. W., Zhou, T., Lees-Miller, S. P., and Povirk, L. F. (2001). Accurate *in vitro* end joining of a DNA double strand break with partially cohesive 3′-overhangs and 3′-phosphoglycolate termini: Effect of Ku on repair fidelity. *J. Biol. Chem.* **276**, 24323–24330.

DeFazio, L., Stansel, R., Griffith, J., and Chu, G. (2002). Synapsis of DNA ends by the DNA-dependent protein kinase. *EMBO J.* **21**, 3192–3200.

Gellert, M. (2002). V(D)J recombination: RAG proteins, repair factors, and regulation. *Annu. Rev. Biochem.* **71**, 101–132.

Hwang, B. J., Smider, V., and Chu, G. (1999). The use of electrophoretic mobility shift assays to study DNA repair. *Methods Mol. Biol.* **113**, 103–120.

Lee, K. J., Huang, J., Takeda, Y., and Dynan, W. S. (2000). DNA ligase IV and XRCC4 form a stable mixed tetramer that functions synergistically with other repair factors in a cell-free end-joining system. *J. Biol. Chem.* **275**, 34787–34796.

Lees-Miller, S. P., and Meek, K. (2003). Repair of DNA double strand breaks by non-homologous end joining. *Biochimie* **85**, 1161–1173.

Lieber, M. R., Ma, Y., Pannicke, U., and Schwarz, K. (2003). Mechanism and regulation of human non-homologous DNA end-joining. *Nat. Rev. Mol. Cell. Biol.* **4**, 712–720.

Song, Q., Lees-Miller, S., Kumar, S., Zhang, Z., Chan, D., Smith, G., Jackson, S., Alnemri, E., Litwack, G., and Khanna, K. (1996). DNA-dependent protein kinase catalytic subunit: A target for an ICE-like protease in apoptosis. *EMBO J.* **15**, 3238–3246.

[28]   Purification and Assays of Saccharomyces cerevisiae Homologous Recombination Proteins

By STEPHEN VAN KOMEN, MARGARET MACRIS, MICHAEL G. SEHORN, and PATRICK SUNG

Abstract

Homologous recombination is an important means of eliminating DNA double strand breaks from chromosomes. The homologous recombination reaction is mediated by the Rad51 recombinase, which requires a number of ancillary factors for maximal efficiency. The development of purification procedures and biochemical assays for yeast Rad51 and other yeast recombination proteins has allowed investigators to begin dissecting the hierarchy of physical and functional interactions among these protein factors that govern the integrity of the homologous recombination machinery. The biochemical studies done with yeast recombination factors have helped formulate conceptual frameworks to guide similar endeavors in other eukaryotes, including humans. Continuing efforts with reconstituted systems that comprise yeast factors will undoubtedly continue to provide insights into the mechanistic intricacy of the homologous recombination machinery.

Introduction

DNA double strand breaks (DSBs) that arise due to radiation exposure or through endogenous pathways are extremely cytotoxic and genotoxic to cells. Failure to process DSBs properly can cause cell death, chromosome aberrations, and tumorigenesis. In eukaryotic organisms, homologous recombination (HR) represents a major means of eliminating DSBs (Pierce *et al.*, 2001; Symington, 2002). Genetic studies, first conducted in the budding yeast *Saccharomyces cerevisiae* and subsequently in other eukaryotic organisms, have identified members of the *RAD52* epistasis gene group that mediates HR (Symington, 2002).

During HR, 3' single-stranded DNA tails derived from the nucleolytic processing of the DSB are engaged by the recombination machinery, consisting of Rad51 and associated factors. The nucleoprotein complex that has assembled then conducts a search for a homologous chromatid, followed by the invasion of the latter to from a DNA displacement loop,

METHODS IN ENZYMOLOGY, VOL. 408
0076-6879/06 $35.00
DOI: 10.1016/S0076-6879(06)08028-1

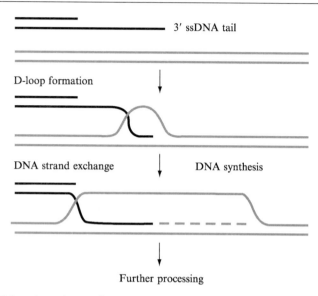

3′ ssDNA tail

D-loop formation

DNA strand exchange   DNA synthesis

Further processing

FIG. 1. D-loop formation. A 3′ ssDNA tail that results from DSB processing is utilized by the HR machinery to invade a chromosomal homolog to form a D-loop. DNA strand exchange and DNA synthesis extend the D-loop. Further processing of the D-loop can occur through distinct pathways (Symington, 2002).

also called D-loop (Fig. 1). The D-loop is extended by DNA strand exchange and by *de novo* DNA synthesis, which is necessary for replacing the DNA sequence eliminated during the initial nucleolytic processing of the DSB. Subsequent steps include resolution of the recombination intermediates and DNA ligation to complete the HR reaction (Sung *et al.*, 2003; Symington, 2002).

Studies employing purified recombinant proteins have provided insights into the biochemical mechanism of HR (Sung *et al.*, 2003; Symington, 2002). Like its *Escherichia coli* ortholog RecA, the Rad51 recombinase polymerizes on single-stranded DNA (ssDNA) to form a right-handed protein filament, often referred to as the presynaptic filament, that comprises 6.2 molecules of the recombinase and 18.6 bases of DNA per helical repeat (Fig. 2). The DNA is held in an extended conformation in the presynaptic filament, with an axial rise of ~5.4 Å per base (Ogawa *et al.*, 1993; Sung and Robberson, 1995; Yu *et al.*, 2001). The reaction steps that lead to heteroduplex DNA joint formation occur within the confines of the presynaptic filament (Ogawa *et al.*, 1993; Sung and Robberson, 1995).

By removing the secondary structure in the ssDNA template, the heterotrimeric single strand DNA-binding factor replication protein A

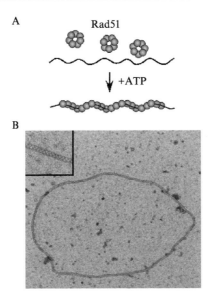

FIG. 2. The presynaptic filament. (A) In the presence of ATP, Rad51 polymerizes onto ssDNA to form a right-handed helical filament called the presynaptic filament. (B) A presynaptic filament of *S. cerevisiae* Rad51 protein as seen under the electron microscope. (Inset) Striations characteristic of the presynaptic filament.

(RPA) facilitates presynaptic filament assembly (Sugiyama *et al.*, 1997; Sung, 1997a; Sung and Robberson, 1995). However, due to its high affinity for ssDNA, an excess of RPA can exclude Rad51 from the DNA (Sugiyama *et al.*, 1997; Sung, 1997a). In *S. cerevisiae*, this inhibitory effect of RPA can be efficiently alleviated by the recombination mediator activity of Rad52 (New *et al.*, 1998; Shinohara and Ogawa, 1998; Sung, 1997a). Rad52 binds ssDNA, Rad51, and also RPA, providing for a mechanism of Rad51 delivery to RPA-coated ssDNA to seed presynaptic filament assembly (Sung *et al.*, 2003). The ability of the presynaptic filament to form a D-loop is enhanced greatly by the Rad54 protein (Petukhova *et al.*, 1998). Rad54 is a member of the Swi2/Snf2 protein family (Eisen *et al.*, 1995; Emery *et al.*, 1991). At the expense of ATP hydrolysis, Rad54 translocates on dsDNA, generating dynamic topological changes that cause transient strand opening in the DNA duplex (Ristic *et al.*, 2001; Sigurdsson *et al.*, 2002; Van Komen *et al.*, 2000). The translocase activity of Rad54 also affects the turnover of Rad51 from dsDNA, a property of Rad54 thought to be important for promoting the intracellular recycling of Rad51 and the release of Rad51 from DNA joints made by this recombinase (Solinger and Heyer, 2001).

This chapter describes the purification of *S. cerevisiae* Rad51, Rad52, Rad54, and RPA. It also details *in vitro* assays that have been developed to study the biochemical functions of these HR factors.

Expression and Purification of Homologous Recombination Proteins

Saccharomyces cerevisiae *Strains*

BJ5464: *MATαura3–52 trp1 leu2Δ1 his3Δ200 pep4::HIS3 prb1Δ1.6R can1 GAL*

LP2749–9B: *MATa his3-Δ1 leu2–3 leu2–112 trp-1 ura3–52 pep4–3*

RDKY1293: *MATα ura3–52 trp1 leu21 his3200 pep4::HIS3 prb11.6R can1 GAL*

RDKY2275: RDKY1293 strain harboring the plasmids pRDK273, pRDK274, and pRDK275 that overexpress the three subunits of RPA (Nakagawa *et al.*, 2001).

Transformation of Yeast Strains with Plasmids Containing the LEU2-d *Marker.* The *LEU2-d* marker is ideally suited to protein overexpression, as yeast cells need to accumulate a high number of copies of plasmids bearing this marker to attain leucine prototrophy. *S. cerevisiae* strains are cotransformed with the *LEU2-d* vector and another vector containing a different selectable marker, such as the *URA3*-containing vector Ycp50 (GenBank X70276), using the lithium acetate procedure described by Ito *et al.* (1983). Transformants are initially plated on synthetic medium lacking uracil. The yeast colonies are grown for 3 to 4 days and are then replica plated onto synthetic medium lacking leucine to select for cells that have taken up the *LEU2-d*-containing plasmid as well. Typically, 15 to 20% of the Ura⁺ transformants are also Leu⁺.

Culture Media

YPD: 1% (w/v) yeast extract, 2% (w/v) peptone, and 2% (w/v) dextrose

Synthetic media: Synthetic media are prepared according to Sherman (1991)

2× Luria broth (LB) with ampicillin: 2% (w/v) Bacto-tryptone, 1% (w/v) Bactoyeast extract, and 1% (w/v) NaCl, pH 7.5, 100 μg/ml ampicillin

YPG: 1% (w/v) yeast extract, 2% peptone (w/v), and 2% galactose (w/v).

Buffers Used in Protein Purification

Cell breakage buffer A: 50 m*M* Tris–HCl (pH 7.5), 10% sucrose, 10 m*M* EDTA, 600 m*M* KCl, and 1 m*M* dithiothreitol

Cell breakage buffer B: 50 mM Tris–HCl (pH 7.5), 10% sucrose, 150 mM KCl, 3 mM EDTA, and 1 mM 2-mercaptoethanol

Cell breakage buffer C: 50 mM Tris–HCl (pH 7.5), 10% sucrose, 10 mM EDTA, 600 mM KCl, and 1 mM 2-mercaptoethanol

Buffer T: 25 mM Tris–HCl (pH 7.4), 10% glycerol, 1 mM EDTA, and 1 mM dithiothreitol

Buffer K1: 20 mM KH$_2$PO$_4$ (pH 7.4), 0.5 mM EDTA, and 1 mM dithiothreitol

Buffer K2: 20 mM KH$_2$PO$_4$ (pH 7.4), 0.5 mM EDTA, and 1 mM 2-mercaptoethanol.

Rad51 Protein

General Comments. Rad51 protein has a DNA-stimulated ATPase activity, forms a nucleoprotein filament on ssDNA, and catalyzes the homologous DNA pairing and strand exchange reaction (Shinohara and Ogawa, 1998; Sugiyama *et al.*, 1997; Sung, 1994). Initially, Rad51 was overexpressed in the *S. cerevisiae* strain LP2749–9B from a 2μ vector containing the *RAD51* gene under the control of the yeast alcohol dehydrogenase 1 (*ADC1*) promoter (Sung, 1994). For improved protein yields, we now employ the constitutive phosphoglycerate kinase (*PGK*) promoter in a *LEU2-d* containing 2μ plasmid for Rad51 overexpression in the same yeast strain (Sung and Stratton, 1996).

Rad51 Expression. The plasmid p51.3 (*PGK-RAD51*) (Sung and Stratton, 1996), containing the *RAD51* gene under the control of the phosphoglycerate kinase (*PGK)* promoter in the vector pMA91 (*2μ, PGK, LEU2-d*) (Kingsman *et al.*, 1990), is used to transform strain LP2947-9B to leucine prototrophy. Two plates (10 cm diameter) of cells derived from at least 10 transformants are used to inoculate 1.6 liter of complete synthetic medium lacking leucine, and the culture is grown overnight (12 to 15 h) with shaking (150 rpm) at 30° to the stationary phase. The culture is diluted eightfold into YPD (4-liter batches in 6-liter flasks) and grown for an additional 8 to 10 h at 30° with shaking (150 rpm). The cells are harvested by centrifugation (5500 rpm in a SLC-6000 Sorvall rotor for 12 min), and the cell pellet is either processed right away or stored at −80°.

Cell Extract Preparation and Ammonium Sulfate Precipitation. All of the steps are carried out at 0 to 4°. For extract preparation, 100 g yeast paste (from 16 liters of culture) is suspended in 150 ml of cell breakage buffer A containing protease inhibitors (phenylmethylsulfonyl fluoride and benzamidine hydrochloride at 1 mM each and aprotinin, chymostatin, leupeptin, and pepstatin A at 5 μg/ml each) and passed once through a French press (SLM Aminco) at 20,000 psi. The extract is clarified by ultracentrifugation in a type Ti45 rotor (Beckman) at 40,000 rpm for 90 min. The supernatant (fraction I, 150 ml) is subject to ammonium sulfate precipitation at a concentration of 0.21 g/ml, which precipitates Rad51 along with ~20% of the total protein. The ammonium sulfate precipitate is pelleted by centrifugation at 13,000 rpm in a Sorvall SLA-1500 rotor for 30 min and is either processed right away for protein purification or stored at −80°.

Rad51 Protein Purification. All of the purification steps are carried out at 0 to 4°. SDS–PAGE followed by Coomassie blue staining is used to screen chromatographic column fractions for their Rad51 content. The ammonium sulfate pellet is dissolved in sufficient T buffer (150 to 200 ml) containing protease inhibitors (aprotinin, leupeptin, chymostatin, and pepstatin A at 5 μg/ml each) to yield a conductivity equivalent to that of 140 mM KCl. The protein solution (fraction II) is then loaded onto a 20-ml Q-Sepharose (Pharmacia Biotech) column (internal diameter of 1.5 cm) equilibrated with buffer T containing 140 mM KCl. The column is washed with 60 ml of the same buffer and is then developed with a linear 300-ml, 150 to 600 mM KCl gradient in buffer T, collecting 5-ml fractions. Rad51 (fraction III, 30 ml) elutes in ~350 mM KCl. The peak fractions are pooled (30 ml), diluted with an equal volume of buffer T containing protease inhibitors (aprotinin, leupeptin, chymostatin, and pepstatin A at 5 μg/ml each), and then applied onto a 6-ml Macro-Prep Hydroxyapatite column (Bio-Rad; packed into a Pharmacia Biotech HR 5/10 column) equilibrated with buffer T containing 50 mM KCl. The column is washed with 20 ml of the same buffer, and the bound proteins are fractionated with a linear 100-ml, 0 to 120 mM KH$_2$PO$_4$ gradient in buffer T containing 50 mM KCl, collecting 2-ml fractions. Rad51 elutes in ~60 mM KH$_2$PO$_4$, and the peak fractions are combined (fraction IV, 18 ml) and loaded onto a Mono Q (HR 5/5) column equilibrated with buffer T containing 100 mM KCl. After washing with 5 ml of the same buffer, the column is developed with a linear 100-ml, 200 to 400 mM KCl gradient in buffer T, collecting 1.5-ml fractions. Rad51 elutes in ~300 mM KCl, and the protein pool (fraction V, 6 ml) is concentrated in a Centricon-30 microconcentrator (Millipore) to 8 mg/ml. The concentrated protein is flash frozen in 10- to 20-μl aliquots in liquid nitrogen and stored at −80°. The concentration of Rad51 is determined by densitometric comparison of multiple loadings of the purified protein against known amounts of bovine serum albumin and ovalbumin in a Coomassie blue-stained SDS–PAGE gel and confirmed by using a molar extinction coefficient of $1.29 \times 10^4 \ M^{-1} \ cm^{-1}$ at 280 nm. The overall yield of nearly homogeneous Rad51 is 5–10 mg.

RPA Protein

General Comments. The efficiency of Rad51-mediated homologous pairing and strand exchange is enhanced by RPA (Sugiyama *et al.*, 1997; Sung, 1994; Sung and Robberson, 1995). Biochemical and electron microscopy data indicate that RPA facilitates filament assembly by removing secondary structure in the ssDNA (Sugiyama *et al.*, 1997; Sung and Robberson, 1995). In addition, RPA sequesters ssDNA generated during DNA strand exchange, thereby preventing reaction reversal (Eggler *et al.*, 2002) and inhibition of Rad51 and Rad54 by the ssDNA substrate (Van Komen *et al.*, 2002). However, an excess of RPA, when added together with Rad51 to the ssDNA substrate, can exclude Rad51 from DNA and attenuate the extent of the recombination reaction (Fig. 3) (Sugiyama *et al.*, 1997; Sung, 1997a). This section describes the purification of RPA from a yeast strain genetically tailored to co-overexpress the three subunits (69, 36, and 13 kDa) using a protocol modified from published methods (Brill and Stillman, 1989; Henricksen *et al.*, 1994; Sung, 1997a).

RPA Expression. S. cerevisiae RPA is overexpressed in strain RDKY2275 containing plasmids that overexpress the three RPA subunits by the use of the *GAL10* promoter (Nakagawa *et al.*, 2001) (a kind gift from Richard Kolodner). Yeast cells are grown at 30° on plates of complete

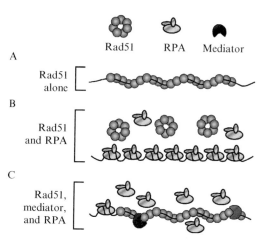

Fig. 3. Function of recombination mediators. (A) Rad51 presynaptic filament. (B) Due to its high affinity for ssDNA, RPA interferes with presynaptic filament assembly by excluding Rad51 from the DNA. (C) Recombination mediator proteins, which bind Rad51, RPA, and ssDNA, promote presynaptic filament assembly by nucleating Rad51 onto RPA-coated DNA.

synthetic medium lacking leucine, tryptophan, and uracil (triple dropout medium), which select for maintenance of the three plasmids. Two plates (10 cm diameter) of cells derived from at least 10 transformants are used to inoculate 1.6 liters (in a 6-liter flask) of the triple dropout medium. The culture is grown with shaking (150 rpm) at 30° overnight to the stationary phase and is then diluted eightfold with YPG (4-liter batches in 6-liter flasks) containing 3% glycerol and 3% lactic acid. After 22 to 24 h of growth at 30° with shaking (150 rpm), cells are harvested by centrifugation (5500 rpm in a SLC-6000 Sorvall rotor for 12 min). The cell pellet is either processed further right away for protein purification or stored at −80°.

Cell Extract Preparation. All the steps are carried out at 0 to 4°. To prepare the cell extract, 150 g of yeast paste (from 40 liters of culture) is suspended in 200 ml of buffer A containing 100 mM KCl and protease inhibitors (phenylmethylsulfonyl fluoride and benzamidine hydrochloride at 1 mM each and aprotinin, chymostatin, leupeptin, and pepstatin A at 5 μg/ml each) and passed once through a French press once at 20,000 psi. The lysate is clarified by centrifugation in a type Ti45 rotor at 40,000 rpm for 90 min.

RPA Purification. All of the purification steps are carried out at 0 to 4°. SDS–PAGE followed by Coomassie blue staining is used to screen chromatographic column fractions for their RPA content. The clarified cell lysate (fraction I, 200 ml) is applied onto a 20-ml Affi-Gel Blue agarose (Bio-Rad) column (internal diameter of 1.5 cm) equilibrated with buffer T containing 100 mM KCl. The column is washed with 80 ml of buffer T containing 800 mM KCl and is eluted with a 120-ml gradient of 0.5 to 2.5 M NaSCN in buffer T, collecting 3-ml fractions. RPA elutes in approximately 1.5 M NaSCN, and the peak fractions are combined (fraction II, 18 ml), dialyzed against 2 liters of buffer T containing 50 mM KCl, 20% sucrose, and protease inhibitors (aprotinin, chymostatin, leupeptin, and pepstatin A at 5 μg/ml each) for 3 h before being loaded onto a Macroprep hydroxyapatite column (6 ml packed into a Pharmacia Biotech HR 5/10 column) equilibrated with buffer T containing 10 mM KH$_2$PO$_4$. The column is washed with 20 ml of the same buffer and is then developed with a linear 100-ml gradient from 10 to 140 mM KH$_2$PO$_4$ in buffer T, collecting 2-ml fractions. RPA elutes in ~55 mM KH$_2$PO$_4$ and the protein pool (fraction III, 10 ml) is applied directly onto a Mono Q column (HR 5/5) equilibrated with buffer T. The column is washed with 5 ml of buffer T and is developed with a linear 75-ml, 50 to 300 mM KCl gradient in buffer T, collecting 1.5-ml fractions. RPA elutes in ~150 mM KCl, and the peak fractions are pooled (fraction IV, 9 ml), concentrated in a Centricon-30 microconcentrator to 10 mg/ml, flash frozen in 10-μl aliquots in liquid nitrogen, and stored at −80°. The concentration of RPA is determined by densitometric comparison of multiple loadings of the purified preparation against known amounts of bovine serum albumin and ovalbumin in a Coomassie blue-stained SDS–PAGE gel and confirmed by using a molar extinction coefficient of 8.8×10^4 M^{-1} cm^{-1} at 280 nm. The overall yield of nearly homogeneous RPA is between 3 and 5 mg.

RAD52 Protein

General Comments. Through its ability to interact simultaneously with ssDNA, Rad51, and RPA, Rad52 helps load Rad51 onto an RPA-coated ssDNA template (Fig. 3) (New *et al.*, 1998; Shinohara and Ogawa, 1998; Sung, 1997b). This "recombination mediator" function of Rad52 is important for overcoming the suppressive effect of RPA on presynaptic filament assembly (New *et al.*, 1998; Shinohara and Ogawa, 1998; Sung, 1997b; Sung *et al.*, 2003). For the purification of Rad52, we take advantage of a hexahistidine (His_6) tag at the carboxyl terminus of this protein that enables it to bind tightly to nickel-NTA agarose.

Rad52 Protein Expression. Rad52 is expressed from the T5 promoter in pQE60-Rad52-His6 (a kind gift from Rodney Rothstein) (Mortensen *et al.*, 1996) in the *Escherichia coli* strain M15(pREP4). A single M15(pREP4) colony that harbors pQE60-Rad52-His6 is used to inoculate 3 ml 2× LB with ampicillin (100 μg/ml), and the culture is incubated for 8 to 10 h at 37° with shaking (185 rpm). This culture is then used to inoculate 200 ml (in a 1-liter flask) of fresh 2× LB with ampicillin (100 μg/ml), followed by an overnight incubation (12 to15 h) at 37° with shaking (185 rpm). The overnight culture is diluted 100-fold into fresh 2× LB with 100 μg/ml ampicillin (2-liter batches in 6-liter flasks) and is incubated at 37° with shaking (185 rpm) until the optical density at 600 nm is between 0.6 and 0.8. Expression of Rad52 is induced by the addition of 1 mM isopropyl-1-D-thiogalactoside, and the culture is grown at 37° for an additional 3 h with shaking (185 rpm). The cells are harvested by centrifugation (5500 rpm in a SLC-6000 Sorvall rotor for 12 min) and either processed right away for protein purification or stored at −80°.

Cell Extract Preparation. All of the steps are carried out at 0 to 4°. An extract is made from 15 g of cell paste (from 8 liters of culture) in 120 ml of cell breakage buffer B containing protease inhibitors (phenylmethylsulfonyl fluoride and benzamidine hydrochloride at 1 mM and aprotinin, chymostatin, leupeptin, and pepstatin A at 5 μg/ml each) by a single pass through a French press at 20,000 psi. The crude lysate is clarified by centrifugation in a type Ti45 rotor at 40,000 rpm for 90 min.

Rad52 Purification. All the purification steps are carried out at 0 to 4°. SDS–PAGE followed by Coomassie blue staining is used to screen chromatographic column fractions for their Rad52 content. The supernatant from the centrifugation step (fraction I, 120 ml) is applied onto a

SP-Sepharose column (internal diameter of 1.5 cm; 20 ml total) equilibrated with buffer K1 containing 150 mM KCl. The column is washed with 60 ml of the same buffer and is then developed with a linear 300-ml gradient of 150 to 650 mM KCl in buffer K1, collecting 5-ml fractions. Rad52 elutes from SP-Sepharose in approximately 360 mM KCl, and the peak fractions are combined (fraction II, 30 ml), diluted with 2 volumes of buffer K2 containing protease inhibitors (aprotinin, chymostatin, leupeptin, and pepstatin A at 5 μg/ml), and applied onto a Q-Sepharose column (internal diameter of 1.5 cm; 8 ml matrix) equilibrated with buffer K2 containing 100 mM KCl. The column is washed with 20 ml of the same buffer, followed by elution with a linear 100-ml gradient of 100 to 450 mM KCl in buffer K2, collecting 2-ml fractions. Rad52 elutes in $\sim$300 mM KCl and the peak fractions are pooled (fraction III, 10 ml) and mixed gently with 1 ml of Ni-NTA-agarose for 3 h at 4°. The matrix is poured into a glass column (internal diameter of 0.5 cm), washed sequentially with 10 ml each of buffer K2 containing 500 mM KCl and the same buffer with 10 mM, 20 mM, and 30 mM imidazole to remove contaminating proteins. Rad52 is eluted with 10 ml of buffer K2 containing 500 mM KCl and 200 mM imidazole, collecting 1-ml fractions. The protein pool (fraction IV; 4 ml) is concentrated to 1 ml using a Centricon-30 microconcentrator and fractionated in a Sepharose 6B gel filtration column (internal diameter of 1.5 cm; 80 ml matrix) with buffer K2 containing 150 mM KCl, collecting 1-ml fractions. The Rad52 pool (fraction V, 8 ml) is loaded directly onto a Mono S column (HR5/5) equilibrated with buffer K2 containing 150 mM KCl. After washing with 3 ml of the same buffer, the Mono S column is developed with a linear 35-ml gradient of 150 to 500 mM KCl in buffer K2, collecting 1-ml fractions. Fractions containing the peak of Rad52 protein, which elutes in $\sim$350 mM KCl, are pooled (fraction VI; 4 ml) and concentrated in a Centricon-30 microconcentrator to 5 mg/ml. The concentrated protein is flash frozen in liquid nitrogen in 5-μl portions and stored at −80°. The concentration of Rad52 is determined by densitometric comparison of multiple loadings of the purified protein against known amounts of bovine serum albumin and ovalbumin in a Coomassie blue-stained SDS–PAGE gel and confirmed by using a molar extinction coefficient of 2.4×10^4 M^{-1}cm^{-1} at 280 nm. The overall yield of nearly homogeneous Rad52 protein is between 3 and 5 mg.

Rad54 Protein

General Comments. Rad54 interacts physically with Rad51 (Clever *et al.*, 1997; Gupta *et al.*, 1997; Jiang *et al.*, 1996; Petukhova *et al.*, 1998), and its ATPase and DNA supercoiling activities are stimulated by Rad51

(Mazin *et al.*, 2000; Van Komen *et al.*, 2000). Rad54 also remodels chromatin and Rad51 stimulates this activity as well (Alexeev *et al.*, 2003; Jaskelioff *et al.*, 2003). Rad54 enables the Rad51 presynaptic filament to efficiently utilize a chromatinized template for D-loop formation (Alexiadis and Kadonaga, 2002; Jaskelioff *et al.*, 2003). We have engineered an amino-terminal hexahistidine (His6) tag into Rad54 to facilitate the purification of this protein by affinity chromatography using nickel-NTA-agarose.

Rad54 Expression. Plasmid pR54.1 containing $(His)_6$-*RAD54* under the control of the *GAL-PGK* promoter in the protein expression vector pPM231 (2μ, *GAL-PGK*, *LEU-2*d) (Petukhova *et al.*, 1999) was introduced into the protease-deficient *S. cerevisiae* strain BJ5464; see earlier discussion for the special procedure for yeast transformation with plasmids that contain the *LEU2-d* marker. Two plates (10 cm diameter) of cells derived from at least 10 transformants are used to inoculate 1.6 liter (in a 6-liter flask) of synthetic medium lacking leucine, and the culture is grown overnight (12 to 15 h) at 30° with shaking (150 rpm) to stationary phase. The overnight yeast culture is diluted eightfold with fresh synthetic medium lacking leucine (4-liter batches in 6-liter flasks) that contains galactose (2% w/v), glycerol (3%), and lactic acid (3%) and is incubated for 22 to 24 h at 30° with shaking (150 rpm). Cells are harvested by centrifugation (5500 rpm in a SLC-6000 Sorvall rotor for 12 min) and either processed right away for protein purification or stored at −80°.

Cell Extract Preparation and Ammonium Sulfate Precipitation. All of the steps are carried out at 0 to 4°. Extract is prepared by suspending 200 g of yeast pellet (from 80 liters of culture) in 300 ml of cell breakage buffer C containing protease inhibitors (phenylmethylsulfonyl fluoride and benzamidine hydrochloride at 1 m*M* each and aprotinin, chymostatin, leupeptin, and pepstatin A at 5 μg/ml each) and passing the cell suspension through a French press once at 20,000 psi. The crude lysate is clarified by centrifugation in a type Ti45 rotor at 40,000 rpm for 90 min. Ammonium sulfate is added to the clarified lysate (fraction I, 300 ml) to 0.28 g/ml to precipitate Rad54 and ~30% of the total protein. The ammonium sulfate precipitate is harvested by centrifugation (13,000 rpm for 30 min in a SLA-1500 rotor) and either processed right away or stored at −80°.

Rad54 Purification. All of the purification steps are carried out at 0 to 4°. SDS–PAGE followed by Coomassie blue staining is used to screen chromatographic fractions for their Rad54 content. The ammonium sulfate pellet is dissolved in sufficient K1 buffer (300 to 400 ml) containing protease inhibitors (aprotinin, leupeptin, chymostatin, and pepstatin A at 5 μg/ml each) to yield a conductivity equivalent to that of 180 m*M* KCl. This protein solution (fraction II, ~500 ml) is passed sequentially through a

Q Sepharose column (internal diameter of 1.5 cm; 25 ml matrix) and a SP-Sepharose column (internal diameter of 1.5 cm; 25 ml matrix) equilibrated with buffer K2 containing 150 mM KCl. The SP column is washed with 60 ml of the same buffer and developed with a linear 120 ml gradient of 150 to 600 mM KCl in buffer K2, collecting 3-ml fractions. Fractions containing the peak of Rad54, which elutes in ~330 mM KCl, are pooled (fraction III; 15 ml) and mixed gently with 1 ml of nickel-NTA-agarose for 3 h at 4°. The mixture is poured into a glass column (internal diameter of 0.5 cm), and the matrix is washed with 10 ml each of buffer K2 containing 350 mM KCl and the same buffer containing 10 mM, 20 mM, and 30 mM imidazole to remove contaminating proteins. Rad54 is eluted from the nickel matrix with 10 ml of K2 buffer containing 350 mM KCl and 200 mM imidazole, collecting 1-ml fractions. Fractions containing the bulk of the Rad54 protein are pooled (fraction IV, 4 ml) and applied directly onto a 1-ml Macro-Prep hydroxyapatite column (packed into a Pharmacia Biotech HR 5/10 column) equilibrated with buffer K1 containing 100 mM KCl. The column is washed with 3 ml of the same buffer and then eluted with a linear 25-ml, 60 to 300 mM KH$_2$PO$_4$ gradient in buffer K1. Rad54 elutes in ~200 mM KH$_2$PO$_4$, and the peak fractions are pooled (fraction V, 3 ml), diluted with an equal volume of buffer K1 containing protease inhibitors (aprotinin, chymostatin, leupeptin, and pepstatin A at 5 μg/ml each), and loaded onto a Mono S column (HR 5/5) equilibrated with buffer K1 containing 150 mM KCl. The column is washed with 2 ml of the same buffer and developed with a linear 30-ml, 150 to 500 mM KCl gradient in buffer K1. Rad54 elutes from Mono S in ~350 mM KCl, and the purified protein (fraction VI, 2 ml) is concentrated to 5 mg/ml using a Centricon-30 microconcentrator, flash frozen in 2-μl aliquots in liquid nitrogen, and stored at −80°. The concentration of Rad54 is determined by densitometric comparison of multiple loadings of this protein against known amounts of bovine serum albumin and ovalbumin in a Coomassie blue-stained SDS–PAGE gel. The overall yield of nearly homogeneous Rad54 protein is 0.5 to 1 mg.

Biochemical Systems for Studying Homologous DNA Pairing and Strand Exchange

In HR reactions, a single-stranded DNA tail is used by the recombination machinery to form a D-loop with a homologous chromatid (Fig. 1) (Sung et al., 2003; Symington, 2002). A variety of in vitro assays, including those depicted in Fig. 4, have been developed to study this central step in HR. These in vitro systems have been used extensively to characterize the recombinase activity of Rad51 and to define the functions of various recombinase ancillary factors, including RPA, Rad52, and Rad54. In these

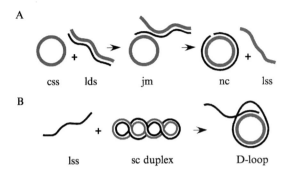

FIG. 4. *In vitro* recombination assays. (A) Pairing between the single-stranded circular (css) and the linear duplex (lds) target yields a joint molecule (jm). Ensuing DNA strand exchange resolves the nascent joint molecule to generate nicked circular duplex (nc) and displaced linear single strand (lss) as products. (B) A D-loop is formed when a linear single-stranded DNA molecule (lss) pairs with a supercoiled duplex (sc duplex).

assays, it is imperative that highly purified proteins are used to avoid artifacts arising through contaminating nuclease and DNA helicase activities. For optimal activity of the purified HR proteins, repeated freeze–thaw cycles are to be avoided. Once thawed from −80°, the HR proteins should be kept in their concentrated storage form on ice, where Rad51, Rad52, and RPA are stable for at least 2 weeks and Rad54 is stable for at least 3 days.

Buffers Used in DNA Substrate Preparation and Homologous DNA Pairing and Strand Exchange Reactions

Reaction buffer A: 35 mM potassium/MOPS (pH 7.2), 40 mM KCl, 2.5 mM ATP, 3 mM MgCl$_2$, and 1 mM dithiothreitol

Reaction buffer B: 35 mM Tris–HCl (pH 7.2), 2.5 mM ATP, 3 mM MgCl$_2$, 100 μg/ml bovine serum albumin, 1 mM dithiothreitol, 15 mM creatine phosphate, and 28 mg/ml creatine kinase

TAE buffer: 40 mM Tris-acetate (pH 7.4), and 0.5 mM EDTA

TE buffer: 10 mM Tris–HCl (pH 7.2), and 0.2 mM EDTA

Gel loading buffer: 30 mM Tris–HCl (pH 7.5), containing 50% glycerol and 0.1% Orange G.

Homologous DNA Pairing and Strand Exchange Reaction

Typically, φ X174 circular ssDNA [viral (+) strand] and linear duplex that contains either a 3′ or a 5′ four base overhang are used as substrates. In this system, pairing between the ssDNA and the duplex yields a DNA joint molecule, which is resolved by DNA strand exchange (also variably referred to as DNA strand transfer or DNA branch migration) to give nicked circular duplex and linear ssDNA as product (Fig. 4A). The substrates and products are separated by agarose gel electrophoresis and are visualized by eithidium bromide staining. Alternatively, one can end label the linear duplex substrate with ^{32}P and visualize the reaction products by phosphorimaging analysis after agarose gel electrophoresis and drying the gel.

DNA Substrates. X174 viral (+) strand and the X174 replicative form I (RFI; 90% supercoiled form and 10% nicked circular form) DNA are from New England Biolabs and Invitrogen, respectively. We find it necessary to purify the X174 viral (+) strand by extraction with phenol/chloroform followed by ethanol precipitation. The replicative form I DNA is linearized by treatment with *ApaL*I or *Pst*I to yield a linear duplex substrate that has either a four base 3' or 5' overhang, respectively. The linearized dsDNA is phenol/chloroform extracted and ethanol precipitated. The DNA substrates are dissolved in TE buffer and stored at 4°, where they are stable for at least 1 year. For longer term storage, $-20°$ or $-80°$ is recommended.

Procedure. The DNA strand exchange reaction (Fig. 4A) is assembled by preincubation of Rad51 with the ssDNA at the ratio of three nucleotides of ssDNA per protein monomer in the presence of ATP, followed by RPA addition and finally the incorporation of linear duplex. The standard reaction has a volume of 12.5 μl, which is sufficient for sampling two time points. The reaction can be scaled up if more time points need to be sampled. The cited reactant concentrations are those in the fully assembled reaction mixture.

1. Rad51 (10 μM) is incubated with X174 viral (+) strand DNA (30 μM nucleotides) in 10 μl reaction buffer A at 37° for 5 min.
2. RPA (1.5 μM) is added in 0.5 μl and the mixture is incubated at 37° for 5 min.
3. The reaction is completed by adding the duplex DNA substrate (15 μM base pairs) in 1 μl and spermidine (4 mM) in 1 μl.
4. The reaction is incubated at 37°, and 5.5-μl portions are withdrawn at the desired times (typically 45 and 90 min) and mixed with an equal volume of 1% SDS and 0.5 μl of proteinase K (0.5 mg/ml). Samples are incubated for 15 min at 37° to allow deproteinization to occur.
5. The deproteinized samples are mixed with 2 μl of gel-loading buffer, and the DNA species are resolved at room temperature in a 0.9% agarose gel in TAE buffer at 100 mA for 3.5 h.

6. The DNA species are stained with ethidium bromide (2 μg/ml in H$_2$O) for 1 h and destained at 4° for 4 to 18 h in a large volume of water.
7. Analysis of the stained gel is carried out in a gel documentation station (Bio-Rad) equipped with a CCD camera. Quantity One software (Bio-Rad) is used for data quantification.

Assays for the Recombination Mediator Activity of Rad52

In the standard reaction described previously, RPA is incorporated after Rad51 presynaptic filament assembly has already occurred. Due to the high affinity of RPA for ssDNA, coincubation of the ssDNA substrate with RPA and Rad51 results in reduced levels of reaction products (Fig. 3) (Sugiyama et al., 1997; Sung, 1997a). The suppressive effect of RPA on homologous DNA pairing and strand exchange can be efficiently overcome by the inclusion of Rad52 protein (Sugiyama et al., 1997; Sung, 1997b). Because ATP hydrolysis by Rad51 is activated by ssDNA and neither RPA nor Rad52 possesses any ATPase activity, the level of ssDNA-dependent ATP hydrolysis represents a reliable means of gauging the extent of Rad51 presynaptic filament formation and for characterizing the recombination mediator activity of Rad52 (New et al., 1998; Shinohara and Ogawa, 1998). This section details the protocol for examining the effect of Rad52 on homologous DNA pairing and strand exchange.

Procedure

1. Rad51 (10 μM) is incubated with RPA (1.5 μM) with or without Rad52 (1 to 2 μM) in 9.5 μl of reaction buffer A for 30 min on ice.
2. X174 viral (+) strand DNA (30 μM nucleotides) is added in 1 μl, followed by a 10-min incubation at 37°.
3. The reaction is completed by adding the duplex DNA substrate (15 μM base pairs) in 1 μl and spermidine (4 mM) in 1 μl.
4. Follow steps 4 to 7 described in the standard protocol given earlier.

The D-Loop Assay to Study Homologous DNA Pairing

The first DNA intermediate made by the HR machinery is a D-loop structure. D-loop formation can be studied using a linear ssDNA molecule (plasmid based or oligonucleotide) and a covalently closed duplex (Fig. 4B) (Petukhova et al., 1998; Sung et al., 2000, 2003; Van Komen et al., 2002). The Rad51-mediated D-loop reaction shows a strong dependence on the Rad54 protein (Petukhova et al., 1998; Van Komen et al., 2000). Because the reaction efficiency is sensitive to the topological state of the duplex (Petukhova *et al.*, 2000; Van Komen *et al.*, 2000), a negatively supercoiled (replicative form I) DNA molecule is most often used.

Specific Comments. The D-loop reaction catalyzed by the combination of Rad51/RPA/Rad54 is optimal when amounts of Rad51 (15 to 8 nucleotides/Rad51 monomer) less than that (i.e., 3 nucleotides/Rad51 monomer) needed to saturate the ssDNA are used (Petukhova *et al.*, 1998, 1999; Van Komen *et al.*, 2002). Because Rad54 becomes inactivated rapidly at 37° (Swagemakers *et al.*, 1998; Van Komen *et al.*, 2002), the D-loop reaction is typically carried out at 25° or below. Rad54 has a robust ATPase activity [$k_{cat} > 2000$/min (Mazin *et al.*, 2000; Van Komen *et al.*, 2000] so an ATP-regenerating system is necessary.

DNA Substrates. To linearize the X174 viral (+) strand DNA, a 10-fold molar excess of a 26-mer oligonucleotide (5'- CGATAAAACT-CTGCAGGTTGGATACG-3') is hybridized to the DNA to create a site for the restriction enzyme *Pst*I. Annealing of the DNA species is achieved by heating to 80° and cooling slowly to 23° over 2 h in TE containing 20 mM KCl and 3 mM MgCl$_2$. Restriction enzyme buffer (from the vendor New England Biolabs) and an excess of *Pst*I are then added to the annealed substrate, and the mixture is incubated at 30° for 3 h. The linearized DNA preparation is phenol/chloroform extracted and then purified as follows: after electrophoresis in a 1% agarose gel in TAE buffer, the linear ssDNA is visualized by brief staining (5 min) with ethidium bromide and then isolated from gel slices using a GeneClean kit using the manufacturer's protocol (Bio 101). The linear ssDNA substrate is stored at 4° in TE buffer, where it is stable for at least 1 year. For longer term storage, −20° or −80° is recommended. The commercially available X174 RF I DNA does not require further purification and is stored as described for the linear ssDNA.

Procedure. The standard reaction has a volume of 12.5 μl, which is sufficient for sampling two time points. The reaction can be scaled up if more time points need to be sampled. The cited reactant concentrations are those in the fully assembled reaction mixture.

1. Rad51 (1.3 to 2.5 μM) is incubated with the linearized X174 viral (+) strand DNA (20 μM nucleotides) in 10 μl of reaction buffer B at 37° for 5 min.
2. RPA (1.5 μM) is added in 0.5 μl and the mixture is incubated at 37° for 5 min.
3. The reaction is completed by adding Rad54 (100 to 200 nM) in 0.4 μl, the duplex DNA substrate (15 μM base pairs) in 0.8 μl, and spermidine (4 mM) in 0.8 μl.
4. The reaction is incubated at 25°, and 5.5-μl portions are withdrawn at the desired times (typically 2 and 5 min) and mixed with an equal

volume of 1% SDS and 0.5 μl of proteinase K (0.5 mg/ml). Samples are incubated for 15 min at 37° to allow deproteinization to occur.

5. Follow steps 5 through 7 in the protocol for standard homologous DNA pairing and strand exchange reaction described earlier.

Acknowledgment

The studies in our laboratory have been supported by research grants from the U.S. National Institutes of Health.

References

Alexeev, A., Mazin, A., and Kowalczykowski, S. C. (2003). Rad54 protein possesses chromatin-remodeling activity stimulated by the Rad51-ssDNA nucleoprotein filament. *Nature Struct. Biol.* **10**, 182–186.

Alexiadis, V., and Kadonaga, J. T. (2002). Strand pairing by Rad54 and Rad51 is enhanced by chromatin. *Genes Dev.* **16**, 2767–2771.

Brill, S. J., and Stillman, B. (1989). Yeast replication factor-A functions in the unwinding of the SV40 origin of DNA replication. *Nature* **342**, 92–95.

Clever, B., Interthal, H., Schmuckli-Maurer, J., King, J., Sigrist, M., and Heyer, W. D. (1997). Recombinational repair in yeast: Functional interactions between Rad51 and Rad54 proteins. *EMBO J.* **16**, 2535–2544.

Eggler, A. L., Inman, R. B., and Cox, M. M. (2002). The Rad51-dependent pairing of long DNA substrates is stabilized by replication protein A. *J. Biol. Chem.* **277**, 39280–39288.

Eisen, J. A., Sweder, K. S., and Hanawalt, P. C. (1995). Evolution of the SNF2 family of proteins: Subfamilies with distinct sequences and functions. *Nucleic Acids Res.* **23**, 2715–2723.

Emery, H. S., Schild, D., Kellogg, D. E., and Mortimer, R. K. (1991). Sequence of RAD54, a *Saccharomyces cerevisiae* gene involved in recombination and repair. *Gene* **104**, 103–106.

Gupta, R. C., Bazemore, L. R., Golub, E. I., and Radding, C. M. (1997). Activities of human recombination protein Rad51. *Proc. Natl. Acad. Sci. USA* **94**, 463–468.

Henricksen, L. A., Umbricht, C. B., and Wold, M. S. (1994). Recombinant replication protein A: Expression, complex formation, and functional characterization. *J. Biol. Chem.* **269**, 11121–11132.

Ito, H., Fukuda, Y., Murata, K., and Kimura, A. (1983). Transformation of intact yeast cells treated with alkali cations. *J. Bacteriol.* **153**, 163–168.

Jaskelioff, M., Van Komen, S., Krebs, J. E., Sung, P., and Peterson, C. L. (2003). Rad54p is a chromatin remodeling enzyme required for heteroduplex DNA joint formation with chromatin. *J. Biol. Chem.* **278**, 9212–9218.

Jiang, H., Xie, Y., Houston, P., Stemke-Hale, K., Mortensen, U. H., Rothstein, R., and Kodadek, T. (1996). Direct association between the yeast Rad51 and Rad54 recombination proteins. *J. Biol. Chem.* **271**, 33181–33186.

Kingsman, S. M., Cousens, D., Stanway, C. A., Chambers, A., Wilson, M., and Kingsman, A. J. (1990). High-efficiency yeast expression vectors based on the promoter of the phosphoglycerate kinase gene. *Methods Enzymol.* **185**, 329–341.

Mazin, A. V., Bornarth, C. J., Solinger, J. A., Heyer, W. D., and Kowalczykowski, S. C. (2000). Rad54 protein is targeted to pairing loci by the Rad51 nucleoprotein filament. *Mol. Cell* **6**, 583–592.

Mortensen, U. H., Bendixen, C., Sunjevaric, I., and Rothstein, R. (1996). DNA strand annealing is promoted by the yeast Rad52 protein. *Proc. Natl. Acad. Sci. USA* **93**, 10729–10734.

Nakagawa, T., Flores-Rozas, H., and Kolodner, R. D. (2001). The MER3 helicase involved in meiotic crossing over is stimulated by single-stranded DNA-binding proteins and unwinds DNA in the 3′ to 5′ direction. *J. Biol. Chem.* **276**, 31487–31493.

New, J. H., Sugiyama, T., Zaitseva, E., and Kowalczykowski, S. C. (1998). Rad52 protein stimulates DNA strand exchange by Rad51 and replication protein A. *Nature* **391**, 407–410.

Ogawa, T., Yu, X., Shinohara, A., and Egelman, E. H. (1993). Similarity of the yeast RAD51 filament to the bacterial RecA filament. *Science* **259**, 1896–1899.

Petukhova, G., Stratton, S., and Sung, P. (1998). Catalysis of homologous DNA pairing by yeast Rad51 and Rad54 proteins. *Nature* **393**, 91–94.

Petukhova, G., Sung, P., and Klein, H. (2000). Promotion of Rad51-dependent D-loop formation by yeast recombination factor Rdh54/Tid1. *Genes Dev.* **14**, 2206–2215.

Petukhova, G., Van Komen, S., Vergano, S., Klein, H., and Sung, P. (1999). Yeast Rad54 promotes Rad51-dependent homologous DNA pairing via ATP hydrolysis-driven change in DNA double helix conformation. *J. Biol. Chem.* **274**, 29453–29462.

Pierce, A. J., Stark, J. M., Araujo, F. D., Moynahan, M. E., Berwick, M., and Jasin, M. (2001). Double-strand breaks and tumorigenesis. *Trends Cell Biol.* **11**, S52–S59.

Ristic, D., Wyman, C., Paulusma, C., and Kanaar, R. (2001). The architecture of the human Rad54–DNA complex provides evidence for protein translocation along DNA. *Proc. Natl. Acad. Sci. USA* **98**, 8454–8460.

Sherman, F. (1991). Getting started with yeast. *Methods Enzymol.* **194**, 3–21.

Shinohara, A., and Ogawa, T. (1998). Stimulation by Rad52 of yeast Rad51-mediated recombination. *Nature* **391**, 404–407.

Sigurdsson, S., Van Komen, S., Petukhova, G., and Sung, P. (2002). Homologous DNA pairing by human recombination factors Rad51 and Rad54. *J. Biol. Chem.* **277**, 42790–42794.

Solinger, J. A., and Heyer, W. D. (2001). Rad54 protein stimulates the postsynaptic phase of Rad51 protein-mediated DNA strand exchange. *Proc. Natl. Acad. Sci. USA* **98**, 8447–8453.

Sugiyama, T., Zaitseva, E. M., and Kowalczykowski, S. C. (1997). A single-stranded DNA-binding protein is needed for efficient presynaptic complex formation by the *Saccharomyces cerevisiae* Rad51 protein. *J. Biol. Chem.* **272**, 7940–7945.

Sung, P. (1994). Catalysis of ATP-dependent homologous DNA pairing and strand exchange by yeast RAD51 protein. *Science* **265**, 1241–1243.

Sung, P. (1997a). Yeast Rad55 and Rad57 proteins form a heterodimer that functions with replication protein A to promote DNA strand exchange by Rad51 recombinase. *Genes Dev.* **11**, 1111–1121.

Sung, P. (1997b). Function of yeast Rad52 protein as a mediator between replication protein A and the Rad51 recombinase. *J. Biol. Chem.* **272**, 28194–28197.

Sung, P., Krejci, L., Van Komen, S., and Sehorn, M. G. (2003). Rad51 recombinase and recombination mediators. *J. Biol. Chem.* **278**, 42729–42732.

Sung, P., and Robberson, D. L. (1995). DNA strand exchange mediated by a RAD51ssDNA nucleoprotein filament with polarity opposite to that of RecA. *Cell* **82**, 453–461.

Sung, P., and Stratton, S. A. (1996). Yeast Rad51 recombinase mediates polar DNA strand exchange in the absence of ATP hydrolysis. *J. Biol. Chem.* **271**, 27983–27986.

Sung, P., Trujillo, K. M., and Van Komen, S. (2000). Recombination factors of *Saccharomyces cerevisiae*. *Mutat. Res.* **451,** 257–275.

Swagemakers, S. M., Essers, J., de Wit, J., Hoeijmakers, J. H., and Kanaar, R. (1998). The human RAD54 recombinational DNA repair protein is a double-stranded DNA-dependent ATPase. *J. Biol. Chem.* **273,** 28292–28297.

Symington, L. S. (2002). Role of RAD52 epistasis group genes in homologous recombination and double-strand break repair. *Microbiol. Mol. Biol. Rev.* **66,** 630–670.

Van Komen, S., Petukhova, G., Sigurdsson, S., Stratton, S., and Sung, P. (2000). Superhelicity-driven homologous DNA pairing by yeast recombination factors Rad51 and Rad54. *Mol. Cell* **6,** 563–572.

Van Komen, S., Petukhova, G., Sigurdsson, S., and Sung, P. (2002). Functional cross-talk among Rad51, Rad54, and replication protein A in heteroduplex DNA joint formation. *J. Biol. Chem.* **277,** 43578–43587.

Yu, X., Jacobs, S. A., West, S. C., Ogawa, T., and Egelman, E. H. (2001). Domain structure and dynamics in the helical filaments formed by RecA and Rad51 on DNA. *Proc. Natl. Acad. Sci. USA* **98,** 8419–8424.

[29] Analysis of DNA Recombination and Repair Proteins in Living Cells by Photobleaching Microscopy

By Jeroen Essers, Adriaan B. Houtsmuller, and Roland Kanaar

Abstract

DNA double strand break repair through homologous recombination has been shown biochemically to require the coordinated action of the *RAD52* group of proteins, including the DNA strand exchange protein Rad51. We have started to develop experimental tools to investigate the close cooperation of homologous recombination proteins in living cells, where proteins operate in the context of chromatin and in the presence of other nuclear processes. This chapter describes in detail methods to establish cell lines stably expressing green fluorescent protein -tagged recombination proteins and photobleaching techniques to investigate the behavior of the proteins with the use of live cell video microscopy. Fluorescence recovery after photobleaching (FRAP), fluorescence loss after photobleaching (FLIP), and their combination in the same cell are useful techniques to gain insights into the dynamic behavior of the recombination proteins. Parameters such as diffusion rates and mobile versus immobile fractions before and after DNA damage induction can be obtained. In addition, residence times of recombination proteins at sites of DNA damage can be determined. Through the application of FRAP and FLIP it is possible to establish whether proteins are present in the same multiprotein

METHODS IN ENZYMOLOGY, VOL. 408   0076-6879/06 $35.00
                      DOI: 10.1016/S0076-6879(06)08029-3

complex, whether this is affected by DNA damage induction, and whether proteins dynamically associate with and dissociate from sites of DNA damage.

Introduction

Discontinuities in double-stranded DNA, particularly double strand breaks (DSBs), are among the most genotoxic DNA lesions. The repair of DSBs is of high importance because the persistence of unrepaired lesions can lead to genetic instability and eventually cancer. Healing of DSBs can occur through a number of mechanistically distinct pathways, including homologous recombination, nonhomologous DNA end joining, single strand annealing, cDNA capture, and *de novo* telomere addition (Kanaar and Hoeijmakers, 1997). This chapter focuses on analysis of the proteins involved in homologous recombination, which is the only pathway assuring accurate repair of DSBs. Homologous recombination exploits extensive regions of DNA sequence homology to repair DSBs precisely using information on the undamaged sister chromatid and therefore operates most efficiently in the S and G2 phases of the cell cycle (Kanaar *et al.*, 1998; Symington, 2002; Takata *et al.*, 1998).

Homologous recombination requires the coordinated action of *RAD52* group proteins, including Rad51, the Rad51 paralogs, Rad52 and Rad54, and the breast cancer susceptibility proteins Brca1 and Brca2 (Wyman *et al.*, 2004). Upon treatment of mammalian cells with ionizing radiation or specific genotoxic agents, such as the interstrand DNA cross-linker mitomycin C, these proteins accumulate at sites where DSBs are induced. The increased local concentration of the proteins at sites of DNA damage can be detected by immunofluorescence as foci (Chen *et al.*, 1999; Choudhary and Li, 2002; Essers *et al.*, 2002; Tashiro *et al.*, 2000). These foci appear to be biologically relevant because cell lines that are defective in homologous recombination generally show an increased sensitivity to ionizing radiation, particularly to mitomycin C. Moreover, mutant cell lines lacking functional BRCA2 or Rad51 paralogs are defective in DNA damage-induced Rad51 and Rad54 foci formation, indicating that BRCA2 and the Rad51 paralogs are essential for the DNA-damaged induced increase in local concentration of these proteins (van Veelen *et al.*, 2005b).

Although the immunofluorescence technique has been very useful in the initial characterization of the response of the homologous recombination proteins to DNA damage, it has its limitations (van Veelen *et al.*, 2005a). First, it requires fixation of the cells and therefore immunofluorescence detection of proteins can only provide snapshots. The dynamics of protein association and dissociation from foci cannot be determined. Second, quantitation of the amount of a certain protein immobilized in foci is not possible. Third, the technique critically depends on the availability of high-quality antibodies. Fourth, subnuclear distribution may not be preserved after fixation. These limitations can be overcome by using homologous recombination proteins that are fused to the green fluorescent protein (GFP) from the jellyfish *Aequorea victoria* (Tsien, 1998). The GFP protein generates a highly visible, efficiently emitting internal fluorophore that can be detected easily in living cells. The availability of spectral variants of GFP and fluorescent proteins from other sources allows the simultaneous detection of different proteins in the same cell. Detection of the proteins in living cells allows the analysis of the response of these proteins to DNA damage by time-lapse video microscopy and to determine their dynamic behavior using fluorescence recovery after photobleaching (FRAP) experiments. The results of these experiments provide insight into the order of addition of different DSB repair proteins into multiprotein complexes formed upon the induction of DNA damage. By extending these studies to cells deficient in certain recombination proteins such as Rad52, Rad54, and BRCA2, insight into the requirements of these proteins in downstream events of DNA repair can be obtained.

To study homologous recombination proteins in living cells we have implemented GFP tagging, live cell confocal microscopy, and FRAP technology (Essers *et al.*, 2002). These studies were the first to probe the dynamic behavior of homologous recombination DNA repair proteins in living cells. We showed that foci containing homologous recombination proteins are dynamic structures that are actively being sampled by the repair proteins. Quantitative analysis of protein diffusion showed that even though the *RAD52* group proteins colocalize in DNA damage-induced foci, the majority of the proteins are not part of the same multiprotein complex in the absence of DNA damage. Our results supported on-site assembly of multiprotein complexes rather than the existence of recombination holocomplexes and provided a mechanism for cross talk between different DNA repair pathways and coupling to other DNA transactions, such as DNA replication.

Fluorescence Recovery after Photobleaching

Fluorescence recovery after photobleaching is a powerful method to determine the mobility of fluorescently tagged molecules. The technique is based on the principle that fluorescent molecules lose their fluorescent capacity, i.e., they are photobleached, when they are irradiated at high-intensity light at their excitation wavelength. The redistribution of fluorescent and bleached molecules after photobleaching in a small area within, for instance, a cell nucleus containing GFP-tagged proteins provides information not only on mobility, but also on immobilization of the molecules under investigation. FRAP was developed in the early 1970s to study the mobility of constituents of the cell membrane. The development of confocal microscopy in the 1980s made it possible to apply FRAP to fluorescently tagged molecules inside the living cell, enhancing the application of FRAP considerably (Houtsmuller and Vermeulen, 2001). Finally, the possibility to label proteins of interest with GFP (or one of its variants) through recombinant DNA technology overcame the biggest difficulty of FRAP experiments, i.e., the necessity to purify the molecules of interest, chemically label, and reintroduce them into the cell (e.g., by microinjection). Because of GFP it is now possible to culture cells that express the fluorescent tag themselves. In addition, the ever-increasing speed and memory of modern computers allow quantitative analysis of complex spatiotemporal FRAP data. Together these developments have enormously boosted the application of FRAP in research of the behavior of proteins in living cells. This chapter gives a detailed protocol for the establishment of cell lines stably expressing GFP-tagged recombination proteins and provides a protocol for the analysis of diffusion coefficients and residence times of GFP-tagged proteins using FRAP technology.

Experimental Approach

Establishing Cell Lines Stably Expressing GFP-Fusion Proteins

There are different approaches for establishing cell lines stably expressing GFP-tagged proteins of interest. This section describes the establishment of cell lines that contain randomly integrated GFP-fusion constructs and constitutively express the encoded GFP-tagged proteins (Fig. 1). The goal is to obtain a cell population in which >80% of the cells are expressing the GFP-fusion protein at physiological levels. This allows functional characterization of the fusion protein and subsequent *in vivo* imaging.

Cloning the cDNA of Interest. A number of experiments using several homologous recombination repair proteins tagged with eGFP and its variants (Clontech) have been published (Essers *et al.*, 2002; Liu and Maizels, 2000; van Veelen *et al.*, 2005b; Yu *et al.*, 2003). It is useful to construct both N- and C-terminal-tagged versions of the protein of interest because location of the tag can influence protein function differentially. Furthermore, constructs are made using standard eGFP vectors (Clontech) that drive

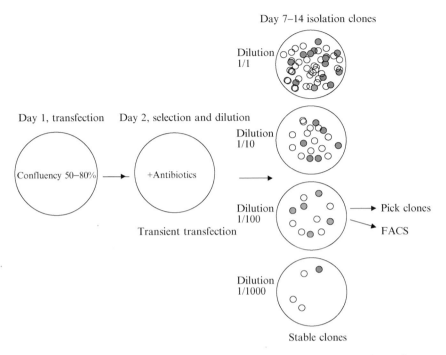

Fig. 1. Scheme for generation of cells expressing GFP-tagged fusion proteins. Adherent cells at a confluency of 50–80% are transfected with the GFP construct at day 1. Transient expression of the GFP-tagged protein at day 2 can be analyzed directly in culture dishes with an inverted microscope equipped with a fluorescence detection setup. Different dilutions are made at day 2 in antibiotics containing medium and colonies will form after 7 to 14 days. After this period the plate containing separate colonies (usually the 1/100 dilution) can be screened for GFP-expressing clones, which can be individually picked or the cells on the plate can be sorted by FACS.

expression of the fusion protein from the CMV promoter. Therefore, it is important to check whether the target cell line is compatible with expression from a CMV promoter. For example, in mouse embryonic stem (ES) cells, we prefer expression from a mouse PGK promoter for more stable expression of the fusion protein. An additional issue to pay attention to is the functionality of the fusion protein. GFP-tagged proteins may not act exactly like their wild-type counterparts, and therefore careful characterization of the fusion protein is crucial before performing detailed microscopic analysis. After finishing the DNA construct, check it carefully by restriction digest analysis and/or DNA sequencing of the junction. At this stage (full-length) expression of the fusion product can be checked using commercially available cell-free *in vitro* transcription- and translation-linked expression systems.

Transfection of the cDNA Encoding the Fusion Protein. Transfect the GFP-cDNA fusion construct using your favorite transfection method that gives the highest transfection efficiency for the target cells. For CHO cells we prefer lipofectamine 2000 (Invitrogen) or Fugene (Roche) because these reagents routinely produce up to 80% transfected cells. One day after transfection the cells should be analyzed by fluorescence microscopy. Verify that the detection method is appropriate in terms of filter sets. In addition to transfecting a GFP cDNA fusion construct, also test the vector alone to confirm that transfection and detection conditions are appropriate. Try using different cell types or vary the conditions of transfection, e.g., amount of plasmid, transfection reagent, and confluency of cells at the moment of transfection when expression problems are experienced at this stage.

The choice of cell line used for transfections is crucial, not only with respect to transfection properties, but also with respect to the possibility to test the functionality of the GFP-fusion protein. To investigate functionality of a GFP-fusion protein, the most unambiguous way is to assay for phenotypic complementation of mutant cells that are missing the endogenous counterpart of the GFP-tagged component. Mutant cells being used are, for example, those derived from random mutagenesis screens using Chinese hamster cells (CHO9, V79, AA8) (Zdzienicka, 1996) or (radio sensitive) patient cell lines. Also, DNA repair defective cell lines can be derived after targeted inactivation of a repair protein in mouse ES cells and mice or chicken DT40 cells. Because all these different cell lines show differences in integration capacity of the GFP-fusion construct, another consideration is the type of cell line that is being used. Cell lines that easily integrate multiple copies of foreign DNA are CHO9, V79, and, to a lesser extent, HeLa cells. Primary fibroblasts, however, either derived from (mutant) mice or patients tend not to incorporate transfected DNA easily.

Stable Cell Line Selection. Before generating stable cell lines prepare a kill curve with G418 or another appropriate antibiotic to be used for selection. Because effective concentrations of antibiotics required for selection vary with each cell line and by lot, empirical determination of an optimal selection concentration for the particular cell line/vector combination is best. For G418, the concentration range suggested for the kill curve is 50–1000 μg/ml. For maintenance of cultures after selection, lowering the antibiotic concentration to between one-half and one-fifth of the selection concentration is possible. However, leaving off selection at all once the cell lines have been established usually works well, is less expensive, and will not suppress potential infections. Selection of stable clones usually starts 1 day after transfection, after checking the efficiency of the transient transfection microscopically. At this point one can also check how many of the cells survived the transfection treatment itself and what the localization of the fusion protein is in the transiently transfected cells. A crucial subsequent step is preparing the optimal dilution of the cells in antibiotics containing media such that individual clones can be isolated after 1–2 weeks of selection. Transfection efficiency and cell survival after transfection are quite variable and depend on the transfection agent, cell line, and the fusion construct used. Therefore, dilutions ranging from 1/1 to 1/5000 should be used. The average time to obtain clones that can be individually picked is 1–2 weeks during which the medium only has to be changed if it contains too many dead cells.

To analyze the spatiotemporal behavior of homologous recombination repair proteins, we described the use of immortalized hamster cell lines, HeLa cells, or other laboratory-used cells stably expressing the GFP-fusion proteins. Alternatively, one can generate GFP-fusion knockin constructs and use them to produce ES cells and mice. In these knockin cells and mice the endogenous locus encoding a protein is altered, such that the fusion protein is expressed from its endogenous promoter and with the use of the endogenous regulatory sequences. There are several reasons to choose a knockin approach. First, because knockouts of several recombination proteins are embryonic lethal, there are no mutants available that can be complemented with GFP-fusion proteins. Therefore, functionality of these fusion proteins cannot be tested directly. By replacing the endogenous gene with GFP-tagged versions, the functionality of these constructs can be tested. Second, targeted integration of the GFP-fusion construct at the endogenous locus will, most of the time, ensure stable and properly regulated expression of the protein. Third, the knockin ES cells can be used to generate mice, from which different cells can be isolated such that tagged proteins can be studied in different cell types. Fourth, these knockin mice can be crossed with defined mutant mice to analyze the behavior of the GFP-fusion protein in defined mutant backgrounds. We do not give a detailed introduction into the gene-targeting technology in mouse ES cells and transgenic mice because this can be found elsewhere (Hogan *et al.*, 2002; Joyner, 2000).

Isolating GFP-Positive Cells. Individual fluorescent cells and colonies can be isolated either by picking individual clones or by fluorescence-activated cell sorting (FACS). Culture dishes containing clones can be screened rapidly for the presence of fluorescence using an inverted fluorescence microscope. After marking the positive clones they can be picked up in the culture hood using a regular microscope. Single clones should be dislodged carefully and taken up with as little medium as possible. After incubation for 5 min at 37° in 100 μl trypsin/EDTA, 1 ml of culture medium is added and clones are mechanically dissociated into a single cell suspension. Finally, they are transferred to 1 well of a 24-well plate. Positive candidates on these 24-well plates are further expanded. Check the fluorescence of these candidate clones. Cells that express low levels of GFP cannot be detected easily in culture dishes on inverted microscopes. An alternative is to randomly pick clones and analyze those by immunoblotting or any other means that allows detection of the protein of interest. When unfamiliar with these methods it is helpful to take a GFP vector alone along as a control. Generating GFP-positive cells from the vector is usually easier than from fusion constructs. Also, in FRAP experiments, cells expressing "free" GFP will serve as very useful controls.

An alternative and usually fast technique to isolate GFP-positive cells is FACS. Instead of picking individual clones, GFP-expressing cells in the mixture can be enriched by sorting. Transiently transfected cells can also be cell sorted, but at this stage the plasmid, after it entered the cell, is episomally maintained. Cellular transcription and translation machinery is used to express the transfected gene of interest and through cell division the plasmids are easily lost. Variations from one cell to another can occur because the number of plasmids that have been taken up differs greatly from cell to cell. In general these cells will have higher expression levels, as the copy number of the plasmid is higher compared to selected clones with stable vector integration. Antibiotic selection results in stable transfectants in which, in general, a lower number of copies is integrated. Therefore, stable transfectants (clones) will usually have lower and more stable expression levels than transient transfectants. Although initially cell sorting can give a homogeneous GFP-positive cell population, be aware that populations like these can lose their fluorescence rapidly because they consist of mixed populations of cells, especially when transiently GFP-expressing cells are sorted.

A common problem with isolated GFP expressing clones is that the fluorescence levels often decrease or become variable after passaging the cells. Although there is no good explanation for this phenomenon, it is probably the case for any stably transfected gene. However, it is likely more apparent with GFP as a detector due to the sensitivity of fluorescence detection. Possibly, the original clone was not pure, but even within a population derived from one clone, variations in GFP expression are often observed. However, also during maintenance of the original clone with cells growing to confluency, with or without selection pressure, cell to cell variation will develop. Generally, depending on how long the cells have been in culture or, more important, how often the cells have been passaged (varying from 1 week to 3 months), a fresh aliquot of the original stock could be thawed to ensure a homogeneous population. Freeze enough aliquots of low-passage number cell populations expressing the fusion protein in a high percentage of the cells (preferably >80%). It is always possible to enrich the population expressing the fusion protein again by FACS. This method allows one to obtain a subpopulation of cells that express GFP at the desired level. This sorting of clones can best be done after the selection procedure has been completed and the construct is integrated in the genome.

Characterization of Cells Expressing the GFP Fusion Protein. As mentioned earlier, the most straightforward technique to show functionality of a GFP-tagged fusion construct is by complementation of mutant cell lines. DSB repair-deficient cell lines stably expressing their missing factor as a GFP-tagged fusion protein can then be analyzed for correction of the mutant phenotype and physiological expression level of the expressed fusion protein. For many homologous recombination repair proteins, however, there are no mutants available that can be used for complementation. As an alternative, wild-type cells can be transfected and biological behavior of the GFP-tagged protein can then be indirectly deduced from various analyses. Regardless of what strategy is followed, the first analysis is to check whether the full-length fusion protein is expressed in the isolated GFP-expressing cells. This can be done by immunoblotting with antibodies against the endogenous protein. To check for breakdown products the blots can also be probed with antibodies against GFP. It is useful to use a number of different GFP antibodies that are commercially available. With the combined results of these immunoblots it is possible to determine whether the full-length fusion protein is expressed at physiological levels. Although GFP antibodies can be quite sensitive, we experienced that while fluorescence can be detectable by confocal microscopy, the fusion protein was undetectable by immunoblotting. For microscopic analysis and colocalization studies using antibodies, eGFP fluorescing cells can best be fixed using 4% paraformaldehyde in phosphate-buffered saline because this fixative does not destroy GFP fluorescence. If antibodies against the protein of interest are available, compare the localization of the GFP fusion protein detected by its fluorescence with the antibody localization in the fixed cells expressing the fusion protein and in untransfected cells.

In case DNA damage-sensitive mutant cell lines are complemented by a GFP fusion protein, cells can now be tested for restoration of DNA damage sensitivity to wild-type levels. Functionality can also be tested indirectly by immunoprecipitation experiments where the interaction of the GFP-tagged protein with a known interaction partner can be verified. If the fusion protein is produced in wild-type cells, these can be tested for possible dominant-negative effects on cell survival or cell cycle progression after DNA damage insult.

DNA Damage Induction Methods

Studying kinetics of GFP-tagged proteins *in vivo* enables the analysis of the spatiotemporal organization and dynamic interplay between different DNA metabolic processes such as cell cycle regulation, DNA replication, and genome maintenance processes. As mentioned earlier, a number of proteins involved in homologous recombination relocalize from a diffuse nuclear distribution to distinct nuclear foci upon treatment of the cells with ionizing radiation, radiomimetic drugs, or DNA cross-linking agents (Chen *et al.*, 1999; Choudhary and Li, 2002; Essers *et al.*, 2002). These treatments result in randomly localized DSBs at which homologous recombination proteins accumulate and can then be detected as foci. In contrast, DNA damage introduced by ultraviolet (UV) light is pan nuclear and nucleotide excision repair (NER) factors involved in the repair of the UV-induced DNA lesions do not assemble into nuclear foci. To study the coordination of the recruitment of NER factors to UV-damaged DNA, different techniques have been developed to induce local UV damage, including UV irradiation through micropores (Mone *et al.*, 2001; Volker *et al.*, 2001), multiphoton damage induction (Meldrum *et al.*, 2003), and laser light-directed 365-nm UVA irradiation (Lan *et al.*, 2004). Probably because the presence of UV-damaged DNA during replication results in an excess number of replication fork arrests and thereby enhanced activation of the DSB repair machinery, these different local damage techniques are also useful to study the coordination of the recruitment and dynamics of DSB repair factors *in vivo*.

Conventional protocols for direct DSB induction, thus not depending on DNA replication, such as treating cells with ionizing radiation, produce numerous DSBs that are stochastically distributed throughout the nucleus and visible as foci that undergo changes in their shape and appearance (van Veelen *et al.*, 2005a). To be able to follow more precisely the order of events that occur after direct DSB induction and kinetics, the accumulation of DSB repair proteins to sites of local DSBs at the cellular level can be visualized in living cells with several methods that produce local DSBs in only a subregion of a nucleus. One of these is α-particle irradiation where the extremely localized energy deposition of one or two traversing charged α particles through a single nucleus produces DSBs along the ion trajectory (Aten *et al.*, 2004; Jakob *et al.*, 2003, 2005). In the standard top-to-bottom irradiation geometry, this technique produces α tracks only in the Z direction. Resolution of these α tracks along the trajectory path is poor due to the flatness of cultured cells and poor resolution of the confocal microscope in the Z direction compared to resolution in X–Y directions. By using a different radiation geometry, where the beam direction was at an angle of $30°$ with respect to the plane of the monolayer of cells, the trajectory of the particle produces a near-horizontal linear DSB track. This way the impact of the ions can be visualized and analyzed in time with a high resolution in X–Y directions in living cells under the microscope. Other techniques being used to introduce DSBs in restricted areas of the nucleus include ultra soft X-rays in combination with partly shielded nuclei (Nelms *et al.*, 1998), focused laser (337 nm) microirradiation of small nuclear volumes in live cells, presensitized by incubation with halogenated thymidine analogues, also known as "laser scissors" (Limoli and Ward, 1993; Lukas *et al.*, 2003; Tashiro *et al.*, 2000), and pulsed neodymium:ytrium laser (532 nm) microbeam irradiation of a small nuclear volume without presensitizing the cells (Kim *et al.*, 2002). Partial irradiation techniques that make use of ultra soft X-rays or focused lasers to induce local DSBs have in common that they produce numerously closely spaced DSBs. In contrast, DSBs produced by local irradiation through α particles produce spatially separated DSBs. This difference in local DNA damage load has an important effect on the biological behavior of the DSB. A high local concentration of DSB appeared to immobilize the chromosomal domains containing the breaks, while the more physiological low concentration of DSBs revealed movement of DSBs containing chromosomal domains (Aten *et al.*, 2004; Nelms *et al.*, 1998). Thus, when using techniques to locally induce DNA damage, DNA damage load is an important issue to consider in interpreting the biological relevance of the results obtained.

Quantitative Photobleaching Experiments

Fluorescence recovery after photobleaching is a powerful method to study the mobility of fluorescently tagged (bio)molecules. In a typical FRAP experiment, for example, using GFP-tagged proteins inside the cell nucleus, fluorescence in a small region is irreversibly photobleached (made nonfluorescent) by brief illumination at high light intensity. Subsequently, the recovery of fluorescence inside that region, due to the influx of mobile fluorescent molecules from outside the bleached region, is recorded at regular time intervals. Quantitative analysis of the resulting fluorescence recovery curves allows determination of several "mobility parameters": diffusion coefficient, immobile fraction, and, in cases where molecules are transiently immobilized, the average time that individual molecules are immobile (Fig. 2). Therefore, FRAP is specifically well suited for the investigation of proteins that interact with DNA, such as homologous recombination repair proteins, as their behavior can be roughly predicted: they have to move through the nucleus to reach their target, either by diffusion or by some form of active transport, and they have to interact with DNA or DNA–protein complexes to exert their function. Because DNA is essentially immobile, at least on the timescale of a typical FRAP experiment, the activity of DNA-transacting enzymes and supportive factors leads to their immobilization. By applying FRAP, one can quantitatively analyze this immobilization by determining both its extent and duration, thereby revealing important information on the cellular behavior of the protein of interest. In addition, information on the rate at which proteins diffuse through the nucleus is also obtained.

FRAP to Study Mobility in the Nucleus

In investigating the behavior of proteins in living cells, one of the first things one may want to know is how the proteins under investigation find their targets. In the case of homologous recombination and DSB repair factors, how do these move through the nucleus and how do they find damaged DNA? Do they move by diffusion or are they efficiently transported in a highly regulated manner? If they diffuse, what is their diffusion coefficient? What does that indicate about the size of the complex the protein is in? For instance, if two DSB repair proteins have different diffusion coefficients, they are in different complexes before binding to DNA. These initial questions can be addressed by straightforward FRAP experiments. In the most simple variant of a FRAP experiment, a small region inside the nucleus is bleached and the recovery of fluorescence in that area is monitored at low laser intensity (Fig. 2A). The most straightforward way to analyze FRAP data is to determine a half-life of recovery as an indication of the mobility of the protein under investigation. Several mathematical formulas have been developed to calculate an effective or apparent diffusion coefficient from this half-life (Feder *et al.*, 1996). A more thorough analysis is to fit the entire experimental curve to analytically derived mathematical models describing diffusion (Blonk *et al.*, 1993; Carrero *et al.*, 2003; Ellenberg *et al.*, 1997).

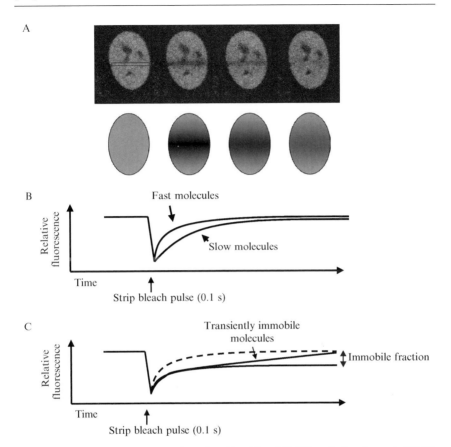

FIG. 2. Fluorescence recovery after photobleaching (FRAP) analysis using strip-FRAP. Cells stably expressing a GFP-tagged protein are subjected to a local bleach pulse, and the kinetics of fluorescence recovery in the bleached area is determined. (A) The upper series represents primary data obtained using a photobleaching protocol on a nucleus expressing a GFP-tagged protein. The fluorescence in a small strip indicated by the rectangle was bleached with a short laser pulse at high intensity and subsequently the recovery of fluorescence in the nucleus was monitored at low laser intensity. Lower series represent a schematic drawing of a nucleus with homogeneously distributed fluorescent molecules subjected to a similar strip-FRAP protocol. (B) The measured fluorescence intensities within the bleached strip are plotted against time. High mobility of GFP-expressing proteins will result in fast recovery of fluorescence in the bleached strip ("fast molecules") and lower mobility in slower recovery ("slow molecules"). (C) Three different scenarios distinguished by strip-FRAP: (1) a situation where all molecules are freely mobile and show recovery rates depending on their molecular weight (represented by the dotted line); (2) a situation where a fraction of the total pool is mobile, with the same diffusion coefficient as in scenario 1, but in addition contains an immobile fraction resulting in decreased total recovery within the time span of the experiment (lower line, "immobile fraction" indicated by the double arrow); and (3) a situation where there is an initial fast recovery and a second slower recovery indication that a fraction of the proteins is transiently immobilized (middle line indicated by the arrow "transiently immobile molecules").

Detailed Protocol

When performing FRAP experiments in cell nuclei it should be realized that due to the small size of the nucleus, differences in shape and size have profound effects on the shape of the fluorescence recovery curves obtained. Therefore, it is highly advisable to pick nuclei for analysis that are similar in shape and size. To determine overall diffusion rates using basic FRAP experiments, we have chosen for a setup where a narrow strip spanning the short axis of an ellipsoid nucleus is bleached (Fig. 2A). The scan area of the confocal microscope is such that pixels are spaced 100 nm apart. The thickness of the strip is 10 pixels. The strip is bleached for 100 ms at the highest laser intensity. Subsequently, fluorescence intensity is monitored at the lowest intensity possible to avoid, as much as possible, bleaching by the monitoring beam. The average fluorescence from the strip can then be plotted against time in several ways, dependent on the normalization chosen. The most obvious way is to normalize data relative to the pre-bleach fluorescence intensity, $I_{norm} = I_t/I_{prebleach}$. However, fitting to most mathematical equations requires the immediate postbleach intensity to be 0 and final recovery to be 1. In that case, data should be normalized as $I_{norm} = (I_t - I_{postbleach})/(I_{final\ recovery} - I_{postbleach})$. In addition, if a comparison is to be made between curves with respect to immobile fraction (see later), i.e., if one wants to compare the degree to which fluorescence has recovered, data should be normalized such that the immediate postbleach intensity is 0 and the prebleach intensity is 1: $I_{norm} = (I_t - I_{postbleach})/(I_{prebleach} - I_{postbleach})$.

Residence Time of Nuclear Proteins in DNA
   Damage-Induced Foci

When an estimate has been obtained for the mobility of the proteins under investigation, in the case of homologous recombination proteins, a second parameter to be obtained is their residence time in the foci. A number of recombination proteins accumulate at sites of DSBs, which is presumably due to interaction with DSBs, specific protein–DNA structures induced by the break, and/or break-induced chromatin modifications. Several methods have been developed to determine such residence times and can be applied to a number of situations, such as proteins associated with nucleoli, nuclear bodies, and telomeres. The most basic experiment is to specifically bleach a site of accumulation and to monitor the recovery of fluorescence similar to the strip-FRAP method. The fluorescence in foci represents the accumulated proteins in the foci as well as the freely diffusing fraction. Therefore it is necessary to obtain reference curves where fluorescence recovery is measured after photobleaching of areas of the same size as the focus in the nucleoplasm. The difference in recovery time between the reference curves and that of a bleached focus is indicative of the residence time at the focus. An alternative method to FRAP is fluorescence loss in photobleaching (FLIP). In a typical FLIP experiment, an area at a distance of a nuclear accumulation of the protein of interest is bleached, either repetitively or with one prolonged bleach pulse, in the latter case at relatively low intensity. If molecules in the focus are permanently bound, fluorescence will not go down in the focus, as unbleached molecules do not exchange with the bleached ones. In contrast, fluorescence in the focus is expected to decrease if molecules in the focus have short residence times compared to the timescale of the FLIP experiment. The rate at which fluorescence is lost is a measure for the turnover of the protein in the focus. The point in time where a new steady state is reached is a good measure for the average residence time in the focus. A variant of FLIP for measuring residence times in multiple foci was introduced in the investigation of foci associated with homologous recombination-mediated DSB repair (Essers et al., 2002) and of telomeric proteins (Mattern et al., 2004) (Fig. 3). In this approach, one-half of the nucleus is bleached, including foci present in that half. Subsequently, the entire image is monitored at regular time intervals. Image analysis is then used to determine the changes in fluorescence level in each of bleached and unbleached foci. The difference in recovery again is indicative for the residence time of the fluorescent protein in the focus. Another alternative method is termed inverse FRAP (iFRAP) (Dundr et al., 2002). It was developed to estimate the rate at which molecules associated with the nucleolus exchange with the surrounding nucleoplasm. The method optimally makes use of the advanced possibilities provided by the operating software of modern confocal microscopes to define bleach regions with complicated shape. In iFRAP the entire nuclear volume is bleached, with exception of the region containing the accumulated proteins under investigation. Immediately after bleaching, the loss of fluorescence in the focus fully represents the releasing molecules, whereas in normal FLIP the loss is the result of unbalance of dissociation and association of fluorescent molecules. Therefore, with this method, determination of the rate of exchange does not require further complicated analytical methods.

When Proteins Do Not Accumulate in Foci

It was described earlier how residence time in foci, i.e., nuclear accumulations, can be studied. However, it is also possible that proteins interact with DNA but do not visibly accumulate in specific region, as discussed earlier. However, in these cases it is still possible that upon DNA damage

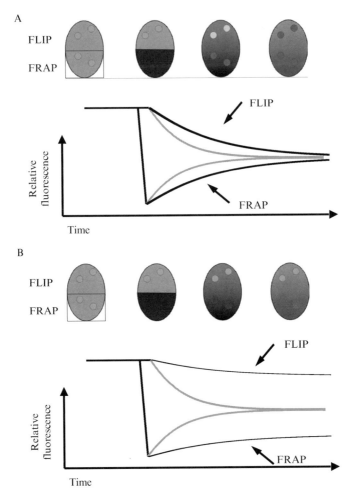

Fɪɢ. 3. Fluorescence recovery after photobleaching (FRAP) and fluorescence loss in photobleaching (FLIP) in DNA damage-induced foci. A region of a cell containing foci, indicated by the rectangle, is bleached by a single laser pulse. Subsequently, the cell is imaged at multiple time points after photobleaching. FLIP is measured in the unbleached half of the cell, and FRAP is measured in the bleached half of the same cell. The measured fluorescence intensities are plotted against time. Two different scenarios are indicated (A and B). FLIP and FRAP of the free diffusing fraction, i.e. the fraction not in foci, are indicated with gray lines and are similar in both situations (A and B). The redistribution is complete when the FLIP and FRAP curves converge and a new equilibrium is reached. (A) A situation is depicted where proteins in foci are mobile, resulting in fast recovery of fluorescence in bleached foci and loss in unbleached foci. (B) A situation is depicted where GFP-tagged proteins are stably associated with foci and are exchanged only marginally with their environment during the experiment. The presence of a stably bound fraction in foci is evident from the fact that the FLIP and FRAP curves of foci do not converge.

induction, a pool of the protein of interest changes behavior, for example, by becoming immobilized, for instance, due to binding to DNA.

In the description of the strip-FRAP method given earlier, it was assumed that proteins are freely mobile. However, if a pool of the protein is immobilized, this immobilization may either be permanent, i.e., much longer than the time scale in which the FRAP experiment is conducted, or transient. If binding is permanent, this will lead to incomplete recovery of the fluorescent signal (Fig. 2C), as the immobile bleached molecules in the strip will not be replaced by fluorescent ones. The percentage of bound molecules can be calculated from the percentage of fluorescence intensity that was not recovered in the strip. Note that correction for the amount of bleached molecules is necessary when it is significant compared to the total amount, which is usually the case in FRAP experiments on the cell nucleus. If immobilization is transient and shorter than the time an experiment takes, fluorescent recovery will be complete, after correction for the fraction bleached, but full recovery will be reached later than when no binding takes place. In this case a primary recovery will be observed due to diffusion and a secondary slower recovery phase due to release, and subsequent diffusion, of the molecules that were immobile at the moment of bleaching (Fig. 2C). The time it takes to fully recover in this secondary process gives an indication of the time individual molecules are immobile. However, a reliable estimate can only be made by more thorough analysis using either mathematically derived equations or computer modeling (Farla et al., 2004).

When transient immobilization/residence time is similar to the redistribution time in the nucleus on the basis of free diffusion, for instance, in the order of 10 to 20 s, it will be difficult to distinguish the resulting FRAP curves from curves representing free diffusion. In such cases additional FRAP and FLIP methods can be applied (Farla et al., 2004). Previous research used a combined FLIP–FRAP approach where the bleach region (FRAP) is monitored simultaneously with an unbleached region (FLIP). In such an experiment, a strip at one pole of the nucleus is bleached for a relatively long period, typically 4 to 8 s, at moderate excitation intensity. Subsequently, the fluorescence is monitored in that region, but also in the area at the other side of the nucleus.

In a scenario where molecules transiently bind to immobile elements in the nucleus, it is often possible to define another scenario where the entire population of molecules is mobile, but moves slower, which results in similar FRAP curves. For example, using simple strip-FRAP methods in a situation where molecules exist in a 30% immobile fraction with an average time of immobilization of 45 s and a 70% mobile fraction with a diffusion rate D of 7 μm^2/s, it will be difficult to distinguish this from a situation where the entire population diffuses at a D close to 1 $\mu m^2/s$. However, a way to cope with this problem is to perform a complementary FLIP–FRAP experiment. The two situations described earlier yield similar curves in the strip-FRAP experiments but are clearly separated in the accompanying FLIP–FRAP experiment. However, scenarios also exist that give similar curves in the FLIP–FRAP experiment, but not in the strip-FRAP: if $D = 4$ $\mu m^2/s$ the FLIP–FRAP curves are similar to the faster but transiently immobile molecules, whereas the strip-FRAP curves are now different. Thus, for a thorough analysis it is useful to apply different FRAP protocols to a particular experimental system.

When Binding Is Very Transient

When binding times are very short ($< \sim 5$ s), FRAP experiments cannot unequivocally distinguish this scenario from a situation where molecules are all freely mobile. However, there are several approaches to cope with this problem. When studying nuclear DNA interacting factors, it is expected that they bind to DNA and become immobilized, at least during the time required for their action. Studying DNA repair processes has the advantage that proteins can be investigated in an inactive state, when no damage is present, and when active in repair after damage induction. Any observed difference in mobility can then assumed to be due to the activity of the protein, either directly or indirectly, in repair. Additional experiments are possible if mutations are known that specifically disable DNA binding. In that case, the mobility of non-DNA-binding mutants can be compared with wild-type mobility (Farla *et al.*, 2004). In addition, it is possible to inhibit the process in which the protein under surveillance is involved.

An alternative powerful set of experiments is to study protein mobility at different temperatures when proteins are active or inhibited (Hoogstraten *et al.*, 2002; Phair and Misteli, 2000; Politz *et al.*, 1999). Because diffusion changes linearly with temperature, a limited effect is expected when temperature is dropped from 310 to 300 K. However, if the molecules are transiently immobilized in a temperature-dependent fashion, e.g., in an enzymatic reactions, their binding time is expected to increase considerably, resulting in a notable drop in mobility.

Potential Problems

In the interpretation and analysis of FRAP experiments, there are some unexpected pitfalls that are discussed briefly.

Low Fluorescence Levels. When the concentration of the tagged protein is low, which is observed frequently when factors are stably expressed at physiological levels, e.g., in transfected cells or knockin mice, fluorescence can only be detected using a laser intensity at which a certain degree of acquisition bleaching will occur. Specifically, when one wants to quantify an observed or expected secondary recovery, this may cause a substantial problem: the secondary recovery may be canceled out or exceeded by monitor bleaching during monitoring fluorescence after bleaching. One obvious method to deal with this problem is to monitor a region similar to the one subjected to FRAP, without applying the bleach pulse, derive the bleach rate, and use that to subsequently correct the FRAP curves. Although this works in the case of a freely mobile population of molecules, the correction method is not applicable when a (transiently) immobile fraction is present. This is due to the fact that in the control experiment performed to derive the bleach rate, the immobile fraction will contribute more to the loss of fluorescence than the mobile fraction, as immobile molecules are constantly in the monitored area, whereas mobile bleached molecules are replaced by mobile fluorescent molecules. This immobile fraction, however, does not contribute to monitor bleaching observed after photobleaching, as it is then largely bleached by the bleach pulse. The result is an overcompensation of the FRAP curve. A frequently used alternative correction for acquisition bleaching is to monitor the entire nucleus and normalize the fluorescence intensity in the bleached area to the entire fluorescence level in the nucleus, minus the bleached region. However, when a confocal microscope is used, this method is also hampered by the same phenomenon as in the example mentioned earlier. If an immobile fraction is present, it will contribute in a different way to acquisition bleaching, outside the bleach area, than the mobile fraction, which is constantly diffusing in and out of the confocal plane. In cases where considerable acquisition bleaching occurs, this way of normalizing data will lead to considerable underestimation of the immobile fraction. The longer the experiment is continued, the smaller the fraction will appear, and eventually, the immobile fraction will be fully obscured, when it is completely bleached by the monitoring beam.

Nucleoplasmic Mobility in the Presence of Abundant Foci. A second problem may occur when proteins under investigation accumulate in multiple foci. In many cases, residence times can very well be determined by FLIP, FRAP, or iFRAP. However, a potential problem may arise if the overall nucleoplasmic mobility of the protein is to be determined whereas the amount of protein in foci is high compared to nucleoplasmic pool. Even if the area that is bleached contains no foci, nucleoli, or local damage, a secondary recovery may be observed due to the exchange of bleached and unbleached molecules with the distant accumulation(s). In the case where focal residence time is in the range of the redistribution time of the protein, this secondary recovery will partly overlap the initial recovery, leading to an apparent slow down of the proteins mobility. A similar effect will take place if the protein shuttles between nucleus and cytoplasm.

Blinking. A third potential threat to proper FRAP analysis is the "blinking" behavior of GFP or its color variants. It has been shown that many fluorescent proteins rapidly switch between a dark nonfluorescent state and a fluorescent state. The period GFPs are in the dark state (off time) is not dependent on laser, whereas the on times are. Because the bleach pulse is at a much higher intensity as the monitoring after bleaching, it can be expected that a part of the recovery of fluorescence is due to a decrease of the pool of molecules in the off state. This is not a great problem if all conditions are kept constant, as it will only introduce a constant bias in the results. However, comparison between different variants of GFP that have different blinking properties is hampered.

Conclusions

GFP tagging of pivotal proteins involved in DNA metabolism combined with live cell microscopy has greatly enhanced our understanding of how these proteins cooperate to regulate and maintain intermingled processes, such as DNA repair and replication. In addition, the combination of GFP-tagged proteins and improved high-resolution labeling of specific DNA and RNA sequences or chromosomal regions *in vivo* also contributed to further insight into the organization of the cell nucleus (Essers *et al.*, 2005; Janicki *et al.*, 2004). Eventually, live cell imaging might help bridge the knowledge of biochemical experiments and cell biology and help translate the biochemical results into understanding mechanisms of DNA transactions *in vivo*. The importance of good molecular and analytical tools for these live cell experiments it is now generally realized. The analyses should be pushed further from cells in culture to cells in tissue context and intact organisms. Using knockin strategies in mouse ES cells for the generation of (low level) GFP-expressing cells, protein localization and dynamics at the level of the whole organism under physiological conditions can be achieved. However, visualization of low expression of GFP in knockin mice and analysis of the kinetics *in vivo* also requires a continuous investment in improving confocal microscopy techniques.

Acknowledgments

The research described in the chapter is supported by grants from the Netherlands Organization for Scientific Research, the Dutch Cancer Society and the European Commission.

References

Aten, J. A., Stap, J., Krawczyk, P. M., van Oven, C. H., Hoebe, R. A., Essers, J., and Kanaar, R. (2004). Dynamics of DNA double-strand breaks revealed by clustering of damaged chromosome domains. *Science* **303**, 92–95.

Blonk, J. C. G., Don, A., Van Aalst, H., and Birmingham, J. J. (1993). Fluorescence photobleaching recovery in the confocal scanning light microscope. *J. Microsc. (Oxf.)* **27**, 363–374.

Carrero, G., McDonald, D., Crawford, E., de Vries, G., and Hendzel, M. J. (2003). Using FRAP and mathematical modeling to determine the *in vivo* kinetics of nuclear proteins. *Methods* **29**, 14–28.

Chen, J. J., Silver, D., Cantor, S., Livingston, D. M., and Scully, R. (1999). BRCA1, BRCA2, and Rad51 operate in a common DNA damage response pathway. *Cancer Res.* **59**, 1752s–1756s.

Choudhary, S. K., and Li, R. (2002). BRCA1 modulates ionizing radiation-induced nuclear focus formation by the replication protein A p34 subunit. *J. Cell Biochem.* **84**, 666–674.

Dundr, M., Hoffmann-Rohrer, U., Hu, Q., Grummt, I., Rothblum, L. I., Phair, R. D., and Misteli, T. (2002). A kinetic framework for a mammalian RNA polymerase *in vivo*. *Science* **298**, 1623–1626.

Ellenberg, J., Siggia, E. D., Moreira, J. E., Smith, C. L., Presley, J. F., Worman, H. J., and Lippincott-Schwartz, J. (1997). Nuclear membrane dynamics and reassembly in living cells: Targeting of an inner nuclear membrane protein in interphase and mitosis. *J. Cell Biol.* **138**, 1193–1206.

Essers, J., Houtsmuller, A. B., van Veelen, L., Paulusma, C., Nigg, A. L., Pastink, A., Vermeulen, W., Hoeijmakers, J. H., and Kanaar, R. (2002). Nuclear dynamics of RAD52 group homologous recombination proteins in response to DNA damage. *EMBO J.* **21**, 2030–2037.

Essers, J., van Cappellen, W. A., Theil, A. F., van Drunen, E., Jaspers, N. G., Hoeijmakers, J. H., Wyman, C., Vermeulen, W., and Kanaar, R. (2005). Dynamics of relative chromosome position during the cell cycle. *Mol. Biol. Cell* **16**, 769–775.

Farla, P., Hersmus, R., Geverts, B., Mari, P. O., Nigg, A. L., Dubbink, H. J., Trapman, J., and Houtsmuller, A. B. (2004). The androgen receptor ligand-binding domain stabilizes DNA binding in living cells. *J. Struct. Biol.* **147**, 50–61.

Feder, T. J., Brust-Mascher, I., Slattery, J. P., Baird, B., and Webb, W. W. (1996). Constrained diffusion or immobile fraction on cell surfaces: A new interpretation. *Biophys. J.* **70**, 2767–2773.

Hogan, B., Beddington, R., Costantini, F., and Lacy, E. (2002). "Manipulating the Mouse Embryo." Cold Spring Harbor Laboratory Press, Cold Spring Harbor, NY.

Hoogstraten, D., Nigg, A. L., Heath, H., Mullenders, L. H., van Driel, R., Hoeijmakers, J. H., Vermeulen, W., and Houtsmuller, A. B. (2002). Rapid switching of TFIIH between RNA polymerase I and II transcription and DNA repair *in vivo*. *Mol. Cell* **10**, 1163–1174.

Houtsmuller, A. B., and Vermeulen, W. (2001). Macromolecular dynamics in living cell nuclei revealed by fluorescence redistribution after photobleaching. *Histochem. Cell Biol.* **115**, 13–21.

Jakob, B., Scholz, M., and Taucher-Scholz, G. (2003). Biological imaging of heavy charged-particle tracks. *Radiat. Res.* **159**, 676–684.

Jakob, B., Rudolph, J. H., Gueven, N., Lavin, M. F., and Taucher-Scholz, G. (2005). Live cell imaging of heavy-ion-induced radiation responses by beamline microscopy. *Radiat. Res.* **163**, 681–690.

Janicki, S. M., Tsukamoto, T., Salghetti, S. E., Tansey, W. P., Sachidanandam, R., Prasanth, K. V., Ried, T., Shav-Tal, Y., Bertrand, E., Singer, R. H., and Spector, D. L. (2004). From silencing to gene expression: Real-time analysis in single cells. *Cell* **116**, 683–698.

Joyner, A. L. (2000). "Gene Targeting, a Practical Approach," 2nd Ed. Oxford University Press, Oxford.

Kanaar, R., and Hoeijmakers, J. H. (1997). Recombination and joining: Different means to the same ends. *Genes Funct.* **1**, 165–174.

Kanaar, R., Hoeijmakers, J. H., and van Gent, D. C. (1998). Molecular mechanisms of DNA double strand break repair. *Trends Cell Biol.* **8**, 483–489.

Kim, J. S., Krasieva, T. B., LaMorte, V., Taylor, A. M., and Yokomori, K. (2002). Specific recruitment of human cohesin to laser-induced DNA damage. *J. Biol. Chem.* **277**, 45149–45153.

Lan, L., Nakajima, S., Oohata, Y., Takao, M., Okano, S., Masutani, M., Wilson, S. H., and Yasui, A. (2004). In situ analysis of repair processes for oxidative DNA damage in mammalian cells. *Proc. Natl. Acad. Sci. USA* **101**, 13738–13743.

Limoli, C. L., and Ward, J. F. (1993). A new method for introducing double-strand breaks into cellular DNA. *Radiat. Res.* **134**, 160–169.

Liu, Y., and Maizels, N. (2000). Coordinated response of mammalian Rad51 and Rad52 to DNA damage. *EMBO Rep.* **1**, 85–90.

Lukas, C., Falck, J., Bartkova, J., Bartek, J., and Lukas, J. (2003). Distinct spatiotemporal dynamics of mammalian checkpoint regulators induced by DNA damage. *Nat. Cell Biol.* **5**, 255–260.

Mattern, K. A., Swiggers, S. J., Nigg, A. L., Lowenberg, B., Houtsmuller, A. B., and Zijlmans, J. M. (2004). Dynamics of protein binding to telomeres in living cells: Implications for telomere structure and function. *Mol. Cell. Biol.* **24**, 5587–5594.

Meldrum, R. A., Botchway, S. W., Wharton, C. W., and Hirst, G. J. (2003). Nanoscale spatial induction of ultraviolet photoproducts in cellular DNA by three-photon near-infrared absorption. *EMBO Rep.* **4**, 1144–1149.

Mone, M. J., Volker, M., Nikaido, O., Mullenders, L. H., van Zeeland, A. A., Verschure, P. J., Manders, E. M., and van Driel, R. (2001). Local UV-induced DNA damage in cell nuclei results in local transcription inhibition. *EMBO Rep.* **2**, 1013–1017.

Nelms, B. E., Maser, R. S., MacKay, J. F., Lagally, M. G., and Petrini, J. H. (1998). In situ visualization of DNA double-strand break repair in human fibroblasts. *Science* **280**, 590–592.

Phair, R. D., and Misteli, T. (2000). High mobility of proteins in the mammalian cell nucleus. *Nature* **404**, 604–609.

Politz, J. C., Tuft, R. A., Pederson, T., and Singer, R. H. (1999). Movement of nuclear poly(A) RNA throughout the interchromatin space in living cells. *Curr. Biol.* **9**, 285–291.

Symington, L. S. (2002). Role of RAD52 epistasis group genes in homologous recombination and double-strand break repair. *Microbiol. Mol. Biol. Rev.* **66**, 630–670.

Takata, M., Sasaki, M. S., Sonoda, E., Morrison, C., Hashimoto, M., Utsumi, H., Yamaguchi-Iwai, Y., Shinohara, A., and Takeda, S. (1998). Homologous recombination and non homologous end-joining pathways of DNA double-strand break repair have overlapping roles in the maintenance of chromosomal integrity in vertebrate cells. *EMBO J.* **17**, 5497–5508.

Tashiro, S., Walter, J., Shinohara, A., Kamada, N., and Cremer, T. (2000). Rad51 accumulation at sites of DNA damage and in postreplicative chromatin. *J. Cell Biol.* **150**, 283–291.

Tsien, R. Y. (1998). The green fluorescent protein. *Annu. Rev. Biochem.* **67**, 509–544.

van Veelen, L. R., Cervelli, T., van de Rakt, M. W., Theil, A. F., Essers, J., and Kanaar, R. (2005a). Analysis of ionizing radiation-induced foci of DNA damage repair proteins. *Mutat. Res.* **574,** 22–33.

van Veelen, L. R., Essers, J., van de Rakt, M. W., Odijk, H., Pastink, A., Zdzienicka, M. Z., Paulusma, C. C., and Kanaar, R. (2005b). Ionizing radiation-induced foci formation of mammalian Rad51 and Rad54 depends on the Rad51 paralogs, but not on Rad52. *Mutat. Res.* **574,** 34–49.

Volker, M., Mone, M. J., Karmakar, P., van Hoffen, A., Schul, W., Vermeulen, W., Hoeijmakers, J. H., van Driel, R., van Zeeland, A. A., and Mullenders, L. H. (2001). Sequential assembly of the nucleotide excision repair factors *in vivo. Mol. Cell* **8,** 213–224.

Wyman, C., Ristic, D., and Kanaar, R. (2004). Homologous recombination-mediated double-strand break repair. *DNA Repair (Amst.)* **3,** 827–833.

Yu, D. S., Sonoda, E., Takeda, S., Huang, C. L., Pellegrini, L., Blundell, T. L., and Venkitaraman, A. R. (2003). Dynamic control of Rad51 recombinase by self-association and interaction with BRCA2. *Mol. Cell* **12,** 1029–1041.

Zdzienicka, M. Z. (1996). Mammalian X-ray sensitive mutants: A tool for the elucidation of the cellular response to ionizing radiation. *Cancer Surv.* **28,** 281–293.

[30] Synthetic Junctions as Tools to Identify and Characterize Holliday Junction Resolvases

By ULRICH RASS and STEPHEN C. WEST

Abstract

Genetic exchanges between chromosomes can lead to the formation of DNA intermediates known as Holliday junctions. The structure of these intermediates has been determined both biochemically and structurally, and their interactions with Holliday junction processing enzymes have been well characterized. A number of proteins, from both prokaryotic and eukaryotic sources, have been identified that promote the nucleolytic resolution of junctions. To facilitate these studies, synthetic DNA substrates that mimic true Holliday junctions have been developed. These now provide an important resource for both the identification and the characterization of novel Holliday junction resolvase activities. This chapter describes methods detailing the preparation and use of synthetic Holliday junctions and how they are best utilized in the study of proteins that might exhibit resolvase activity. Additionally, a method is described that can be used to rapidly screen a TAP-tagged library of proteins for resolvase activity without a need for conventional purification procedures.

METHODS IN ENZYMOLOGY, VOL. 408
                    DOI: 10.1016/S0076-6879(06)08030-X

Introduction

Genetic recombination is a ubiquitous process that creates genetic diversity and facilitates DNA repair. A central intermediate of the process is the Holliday junction (HJ), which has just seen its 40th anniversary as a DNA structure of scientific interest (Liu and West, 2004). Originally proposed to explain the outcome of meiotic recombination in fungi (Holliday, 1964), intermediates containing Holliday structures have been visualized in both prokaryotic and eukaryotic cells (Bell and Byers, 1979; Benbow et al., 1975; Collins and Newlon, 1994; Doniger et al., 1973; Potter and Dressler, 1979; Schwacha and Kleckner, 1994, 1995).

During homologous recombination, Holliday junctions form a covalent link between two DNA double helices and need to be cut to allow the separation of two independent duplex species. To catalyze this cleavage reaction, or resolution, all domains of life feature specialized nucleases. These enzymes, known as Holliday junction resolvases, specifically recognize junctions and introduce symmetrically related nicks at or close to the crossover point to produce nicked duplex products. Because the resolvases introduce cuts with perfect symmetry, the nicks in the products can be sealed by DNA ligase without need for further processing. Resolvases have been isolated from a variety of organisms, including bacteriophage (T4 endonuclease VII, T7 endonuclease I), bacteria (*Escherichia coli* RuvC and RusA, although RusA is encoded by a cryptic lambdoid prophage DLP12), yeast (Cce1, Ydc2), archaea (Hje, Hjc), archaeal viruses (SIRV1, SIRV2), and also from eukaryotic viruses, such as the A22R protein of vaccinia (Birkenbihl et al., 2001; Garcia et al., 2000; Lilley and White, 2001; Sharples, 2001).

The *E. coli* RuvC protein is possibly the best characterized HJ resolvase, both genetically and biochemically, and is known to resolve recombination intermediates that form in the cell during genetic recombination and the recombinational repair of DNA damage (West, 1997). Much less is known about eukaryotic resolvases, as *Saccharomyces cerevisiae* Cce1 and its *S. pombe* equivalent Ydc2 function within mitochondria. In *cce1* mutants, however, residual resolvase activity has been observed indicating the existence of a nuclear resolvase (Kleff et al., 1992). In mammalian cells, resolvase activity has been observed in fractionated extracts and shown to be dependent on the functions of two proteins known as RAD51C and XRCC3 (Constantinou et al., 2001; Liu et al., 2004). However, whether these proteins constitute the resolvase or are associated with the nuclease remains to be determined.

Synthetic Holliday Junctions

Biochemical studies of Holliday junction resolvases have been facilitated by the use of synthetic DNA substrates produced by the annealing of four short oligonucleotides (Kallenbach et al., 1983). These junctions are usually composed of a region of homologous DNA sequences that are flanked by short regions of heterology. Heterologous termini are necessary to block spontaneous branch migration (i.e., movement of the junction point through homologous sequences) that would quickly lead to dissociation of the substrate. The homologous core of synthetic junctions, however, serves as an excellent structural model for the crossover region that exists in naturally occurring Holliday junctions. Junctions that have a homologous core that permits branch migration are often termed "mobile" junctions to distinguish them from "static" junctions where all four arms are composed entirely of heterologous sequences. Given that many HJ resolvases exhibit sequence specificity, or at least a strong preference for cleavage at certain sequences, the use of mobile junctions is usually advantageous because branch migration will permit the resolvase to sample a variety of potential cleavage sequences. In contrast, static HJs have the junction locked at a defined site and conformation. While they can be used in resolution assays with enzymes in which the preferred target sequences have been identified, this limitation needs to be borne in mind.

Structural Features and HJ Resolution

A comprehensive review of the Holliday junction structure is presented elsewhere and the interested reader is recommended to this work (Lilley, 2000). In brief, it is a dynamic DNA structure capable of undergoing conformational changes at the crossover point and, in the case of a mobile junction, branch migration. Synthetic junctions are made by annealing four partially complementary oligonucleotides (usually 40 to 60 nucleotides in length), as shown schematically in Fig. 1A. The junction point links four helical segments that extend toward the corners of a square in its open form (Fig. 1B). Under physiological conditions, however, where divalent metal ions (such as Mg^{2+}) quench electrostatic repulsion of the DNA backbone, the four helical segments stack on top of each other in a pairwise manner to form two continuous right-handed helices. This structure is known as the stacked X structure. Because each helical segment of the open junction can potentially stack on one of two neighboring helices, two stacking conformers of the stacked X exist, which can interconvert (Fig. 1C and D).

In the stacked X structure, pairs of strands take on one of two possible states: they become the continuous strands that follow the course of the helical axis or they become the strands that exchange from one helical axis to the other at the crossover. During branch migration, the crossover point moves by breaking base pairs in helical segments of one plane and rejoining them in the helical segments in the other plane (e.g., imagine pulling out helices 1 and 3 to elongate them, thereby shortening helices 2 and 4). In

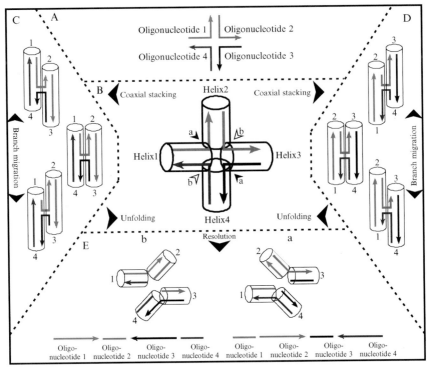

FIG. 1. Schematic representation of a HJ. (A) A synthetic HJ consists of four oligonucleotides (shown in four colors) with their 3′ termini indicated by arrows. (B) The partially complementary oligonucleotides form four helical segments marked as helixes 1–4. Nicks introduced by specialized nucleases at diametrically opposed sites a/a or b/b resolve the HJ. The open square planar form of the junction can undergo pairwise helical stacking of the arms as shown in C and D. In this conformation, pairs of strands are defined as continuous or exchanging. The continuous strands lie in antiparallel orientation. The crossover point, where the exchanging strands pass from one axis to the other, can move (branch migrate) within the homologous core. (E) Resolution of the junction gives rise to two nicked duplex products. Depicted below are the 5′ fragments of the respective oligonucleotides. (See color insert.)

the context of a chromosome, this process can proceed for thousands of base pairs, but in a synthetic HJ will be limited to the region of homology. As a consequence of the combined actions of conformational change and branch migration, a HJ will present itself to a given protein in a variety of different states in solution.

All known HJ resolvases are homodimers that position the two catalytic sites across a junction in order to cleave either the continuous or the exchanging pair of strands. The two mandatory incisions are made within the lifetime of the enzyme–substrate complex to guarantee proper resolution of the bound substrate (Giraud-Panis and Lilley, 1997). As a result, two nicked duplex DNA molecules are released (Fig. 1E) that can be religated to complete a recombination event.

The topological interplay of a resolvase with the characteristically distorted DNA at the crossover point provides for HJ substrate recognition and binding. There are then two possible modes of action: (i) with purely structure-specific resolvases (such as T4 endonuclease VII), binding triggers cleavage, whereas (ii) sequence-specific resolvases (such as RuvC) will only cleave the substrate upon finding their recognition sequence at the crossover point. In this context, it has to be kept in mind that the conformer equilibrium of a given HJ will be biased by the primary sequence, which in turn will favor particular resolution products produced by a purely structure-specific resolvase. However, resolvases that require specific sequences will "wait" until spontaneous dynamics create the appropriate target.

Preparation of Synthetic Holliday Junctions

The following procedures require a workspace suitable for radioactive materials and compliance with radiation safety regulations. The radioactive isotope to be used is ^{32}P, and the maximum single amount to be handled is ~60 μCi (~2.2 MBq).

Sequence Considerations

In principle, HJ substrates can be prepared from any primary DNA sequences. Thus, recognition sequences indicated by experimental data for a given resolvase may be incorporated. It is important, however, to use oligonucleotides that do not self-anneal. Other parameters, such as thermostability, can also be taken into account by increasing the GC content if enzymes from hyperthermophilic archaea are to be assayed at elevated temperatures (Neef *et al.*, 2002).

To assay for an unknown resolution activity, it is important to use HJ substrates with a large homologous core region to provide the greatest variety of crossover point sequences. In such a situation, we suggest the use of the synthetic HJs known as X12 (Benson and West, 1994) or X26 (Constantinou *et al.*, 2001) that have homologous cores of 12 or 26 bp, respectively. These well-established junctions will permit comparison with existing data and eliminate trivial problems that might arise from unexpected structural abnormalities when using untested sequences.

Preparation Procedure

HJ substrates are prepared by annealing four purified oligonucleotides. The resulting junctions are then gel purified to remove any incomplete DNA constructs formed during the process. Using appropriate oligonucleotides, the protocol may also be applied to the production of other DNA substrates, such as control double strands, model replication forks, splayed arm structures, and 3′- and 5′-flap substrates.

Synthetic junctions are normally 5′-^{32}P end labeled on one strand. This enables detection of the native junction and one of the two nicked duplex resolution products. However, if it is necessary to analyze the fate of each DNA strand, four HJ substrates will need to be made in parallel, each individually labeled on a different oligonucleotide.

Holliday junction X26 consists of four oligonucleotides named X26.1–X26.4. Residues in bold constitute the 26-bp branch migratable homologous core.

    X26.1   5′-CCGCTACCAGTGATCAC**CAATGGATTGCTAGGA-CATCTTTGCC**CACCTGCAGGTTCACCC-3′ (60 bases)

    X26.2: 5′-TGGGTGAACCTGCAGGTG**GGCAAAGATGTCCTA-GCAATCCATTG**TCTATGACGTCAAGCT-3′ (60 bases)

    X26.3: 5′-GAGCTTGACGTCATAGA**CAATGGATTGCTAGGA-CATCTTTGCC**GTCTTGTCAATATCGGC-3′ (60 bases)

    X26.4: 5′-TGCCGATATTGACAAGAC**GGCAAAGATGTCCTA-GCAATCCATTG**GTGATCACTGGTAGCGG-3′ (61 bases)

Day 1. Commercially available synthetic oligonucleotides are usually contaminated with shortened oligonucleotide species. If incorporated in the HJ substrate without purification, the precise analysis of cleavage sites will be very difficult. It is therefore necessary to purify all oligonucleotides by denaturing polyacrylamide gel electrophoresis (PAGE) prior to substrate preparation. To do this, run ~250 μg of each oligonucleotide on a 15% acrylamide/bis-acrylamide (19:1) gel containing 7 M urea and TBE buffer (use protean II xi cell system with 20 cm gel length, 1.5 mm thick, Bio-Rad, or similar). Prerun the gel at 15 W for 90 min at room temperature in TBE. Mix the oligonucleotide solution with an equal volume of formamide-loading buffer [90% formamide, 0.1% bromphenol blue (BPB), 0.1% xylene cyanol (XC) in TBE] and boil for 5 min. Before loading the samples, flush the gel wells to remove settled urea. PAGE is carried out for ~2.5 h at 15 W. After electrophoresis, submerge the gel in stains-all (Sigma-Aldrich: 0.1% in formamide, mixed with 2 volumes of isopropanol and 7 volumes H$_2$O) for 10 min in a glass dish and then wash with H_2O for 15 min. The oligonucleotide bands will appear well above the XC dye front. Excise only the upper third of each band to avoid any shortened DNA products. Elute each oligonucleotide in an Eppendorf tube containing 1 ml TE on a rotor overnight at room temperature. *Note:* TBE: 89 mM Tris base, 89 mM boric acid, 2.5 mM EDTA (use a 10× stock solution to prepare gels and buffers); TE: 10 mM Tris–HCl, 1 mM EDTA, pH 8.0.

Day 2. Take the supernatant and ethanol precipitate the eluted oligonucleotides. Dissolve each pellet in 60 μl TE, quantitate the purified DNA by spectrophotometry using 1 $OD_{260\ nm} = 33$ μg/ml, and adjust the volume to make ~10-pmol/μl stock solutions, which equals ~200 ng/μl for X26.1 to X26.4.

For each HJ substrate to be prepared, one of the four oligonucleotides will be 5'-^{32}P end labeled using T4 polynucleotide kinase and [γ-^{32}P]ATP (10 mCi/ml; Redivue, Amersham Biosciences). Typical labeling reactions (10 μl) contain 10 pmol oligonucleotide, 20 pmol (60 μCi) [γ-^{32}P]ATP, and 20 units of T4 polynucleotide kinase (New England Biolabs) in the PNK reaction buffer provided with the enzyme. After incubation for 45 min at 37°, the reaction is stopped by the addition of EDTA to a final concentration of 25 mM, and the volume is increased to 50 μl using 1× PNK buffer. Unincorporated [γ-^{32}P]ATP is then removed using a MicroSpin G-25 column (Amersham Biosciences) following the manufacturer's instructions.

The labeled oligonucleotide is then supplemented with a threefold excess (30 pmol) of each of the three other partner oligonucleotides. The mixture is boiled in a beaker for 2 min on a hot plate, which is then switched off to allow slow annealing as the temperature cools to room temperature overnight. *Note*: If the amount of DNA in a labeling reaction is increased, the amount of [γ-^{32}P]ATP and T4 polynucleotide kinase should be increased accordingly. Always keep the [γ-^{32}P]ATP concentration in excess of the DNA concentration and add sufficient amounts of kinase, as low concentrations of [γ-^{32}P]ATP and DNA will limit incorporation. As both labeled and unlabeled oligonucleotides from the labeling reaction will be incorporated in the HJ substrate alike, it is particularly important to keep the radiolabeled proportion high when trying to detect a (dilute) resolvase activity in the early stages of a protein fractionation or when working with limited enzyme supply. *Note*: The unlabeled strands are added in a threefold excess to ensure that all of the labeled oligonucleotide will be incorporated into a complete HJ product. Excess unlabeled strands will subsequently be removed by gel purification.

Day 3. To purify the Holliday junctions from the annealing mixture, add 15 μl native gel loading buffer (50% glycerol, 0.1% BPB, 0.1% XC in

TBE) and apply to a 12% neutral acrylamide/bis-acrylamide (37.5:1) TBE gel (use a Bio-Rad protean II xi cell system with 20 cm gel length, 1.5 mm thick, or similar). PAGE is carried out for 4–5 h at 200 V at 4° in TBE. At the end of the run, take off one glass plate and cover the wet gel with plastic wrap. Mark the dye fronts with phosphorescent TrackerTape (Amersham Biosciences) and visualize the labeled DNA products by autoradiography (Fig. 2). Typical exposure times are 2–3 min on Kodak BioMax MR film. With the help of the marks left by the TrackerTape, the autoradiograph may be aligned precisely with the gel, allowing excision of the desired band corresponding to the Holliday junctions. We find it helpful to cut the band out of the autoradiograph and then place it over the gel to excise the radioactive band through the window in the autoradiograph. To confirm that the band has been excised properly, reexpose the gel for the same time as the initial autoradiograph and make sure the substrate band has disappeared.

The gel slice is then transferred to a screw-capped 1.5-ml Eppendorf tube and 400 μl TMgN buffer (10 mM Tris–HCl, pH 8.0, 1 mM MgCl$_2$, 50 mM NaCl) is added. The HJ is then eluted overnight by diffusion at 4° with gentle shaking. Quantitate the eluted DNA by spectrophotomety using 1 OD$_{260\ nm}$ = 50 μg/ml. Typically, the DNA concentration will be around 1 ng/μl with only a fraction of the substrate being labeled. The activity of the substrate preparation can be determined by spotting 1 μl in triplicate onto DE-81 paper discs for counting in scintillation fluid in the ^{32}P channel of a liquid scintillation analyzer (Tri-Carb 1500, Perkin-Elmer). Typically, 1 μl will give a reading of around 10,000 cpm. *Notes:* HJ substrates can be kept at 4° and can be used for about 2 weeks. For longer storage, aliquots should be frozen at −20°. In general, each aliquot should be used only once as repeated freeze–thaw cycles tend to destabilize the substrate. The DNA can also be electroeluted from the gel slice using a Biotrap/Elutrap device (Schleicher and Schuell) and a horizontal gel electrophoresis tank (e.g., BRL Horizon 20.25) to increase elution efficiency. Proceed according to the manufacturer's instructions. In brief, the gel slice is placed in 0.5 ml TBE in a compartment between two semipermeable membranes and the electroelution chamber is placed in the electrophoresis tank containing TBE. Upon application of an electric field, the DNA migrates out of the gel slice through one of the semipermeable membranes into the trap section of the device where it is retained by a nonpermeable membrane. Electroelution is carried out for 1 h at 150 V at 4°. Before collecting the DNA from the trap section, the polarity of the current is reversed for 1 min to detach DNA molecules stuck to the nonpermeable membrane at the far end of the trap section. Finally, the DNA is dialyzed

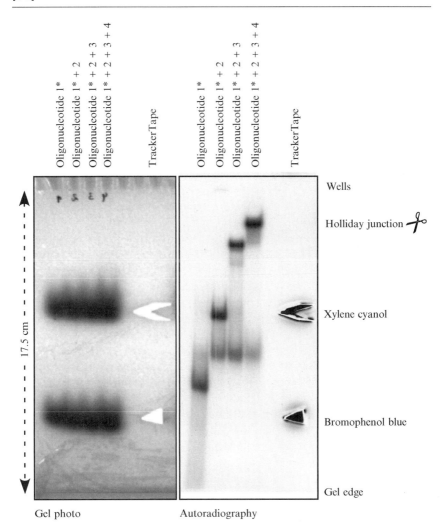

FIG. 2. Purification of synthetic HJ X26 by preparative PAGE. Combinations of oligonucleotides were annealed as indicated and products were separated on a 12% neutral polyacrylamide gel. All reactions contained $5'$-^{32}P-end labeled oligonucleotide 1 (indicated with an asterisk). Reactions lacking one or more oligonucleotides demonstrate the relative running positions of DNA species that may arise from incomplete HJ assembly. The gel photo and the autoradiograph are shown for comparison. In the photograph, dye fronts are marked with white TrackerTape, which is used to align gel and autoradiograph. The band to be excised is indicated with scissors.

for 1 h against 2 liters of TMgN buffer using a Slide-A-Lyser dialysis cassette (Pierce) with 0.5–3 ml capacity and a molecular weight cutoff of 10 kDa.

Holliday Junction Resolution Assays

Depending on the experimental aims, synthetic HJs can be utilized in different ways. The substrates may be incubated with the protein of interest in either purified or crude form, and the nicked duplex products detected by native PAGE. Alternatively, the precise sites of cleavage may be determined by denaturing PAGE.

As an example, we demonstrate a typical analysis designed to check candidate proteins for resolution activity. The candidate proteins are Rad55 and Rad57 from *S. cerevisiae*, which are known to form a heterodimer (Sung, 1997). They were chosen because related proteins in mammalian cells, RAD51C and XRCC3, have been shown to be involved in HJ resolution (Liu *et al.*, 2004). As a positive control, we use the yeast mitochondrial HJ resolvase Cce1. Haploid yeast strains expressing C-terminal fusions of Rad55 or Cce1 with a HisMyc-tag [18 repetitions of c-Myc and 8 histidine residues separated by a double tobacco etch virus (TEV) protease-cleavage site] from their endogenous promoters are constructed using standard methods. The strain expressing Rad55 and a control strain are made to express Rad57 with a 3xHA tag at the C terminus. Extracts from logarithmically growing cells (2 liters) are prepared using a Model 6850 freezer mill (Spex Certiprep), and the Myc-tagged complexes are purified with 9E10-agarose beads (Santa Cruz). Bound proteins are released by digestion with TEV protease and further purified using nickel-NTA agarose (Qiagen). The presence of the target proteins Rad55, Rad57, and Cce1 is confirmed by Western blotting using anti-Myc (9E10) or anti-HA (3F10, Roche) antibodies and by silver staining (not shown). As expected, Rad55-HA copurifies with Rad57-HisMyc in stoichiometric amounts. The final Ni-NTA elution fractions contain ~0.5 ng/μl of each of the target proteins.

Assay Procedure

To assay for resolvase activity, use 1 ng (12.5 fmol) of synthetic HJ DNA (~10,000 cpm ^{32}P) in a total volume of 10 μl (although assays of crude protein fractions may necessitate an increase in DNA substrate and/or reaction volume). If nonspecific nuclease activity is a problem, it is possible to add 250–500 ng of a competitor such as poly[dI-dC] (Roche). This will not affect resolvase activities due to their preference for branched over linear DNA structures. In our example, 1–5 nM of purified Cce1 and Rad55-Rad57 heterodimer are incubated in standard reactions in a buffer containing 100 mM NaCl, 50 mM Tris–HCl, pH 7.5, 10 mM MgCl$_2$, and 2 mM ATP (NaCl is omitted for reactions containing RuvC) for 1 h at 30°. The reactions are terminated and the DNA products deproteinized by the addition of proteinase K and SDS to a final concentration of 2 mg/ml and 0.8%, respectively, followed by further incubation for 30 min at 30°. *Notes:* An input of 10,000 cpm for 1 ng or 12.5 fmol of ^{32}P-labeled HJ X26 corresponds to a specific activity of ~8 × 10^5 cpm/pmol. To faithfully detect resolution products the specific activity for any HJ substrate used should not be lower than 4 × 10^5 cpm/pmol. Optimal protein concentrations, salt, buffer, and pH conditions will need to be determined experimentally for each activity. Although this chapter does not deal specifically with enzyme-driven branch migration, it is known that branch migration and resolution can be linked as observed with RuvABC (West, 1997). The assays described earlier are suitable for detecting branch migration activities as they contain ATP as an energy cofactor. The branch migration of X26 is described elsewhere (Constantinou *et al.*, 2001).

Denaturing PAGE and Analysis

The cleavage products may be separated by the addition of 0.5× volume of formamide-loading buffer (90% formamide, 0.1% BPB, 0.1% XC in TBE) followed by electrophoresis through a 10% acrylamide/bisacrylamide (19:1) gel containing 7 M urea and TBE buffer. The samples are applied to a Sequi-Gen GT nucleic acid sequencing cell (Bio-Rad: 50 cm gel length, 0.4 mm thick), which is run for 2–3 h at 90 W at room temperature, after preheating the gel at 90 W for 1 h to ~50°. The gel is transferred onto DE81-paper (Whatman) backed by a sheet of 3 MM Chr paper (Whatman), covered with Saran wrap (Dow), and dried at 90° for 2 h using the SDS–PAGE drying cycle of a Bio-Rad Model 583 gel dryer. Dried gels are analyzed by autoradiography using Kodak BioMax MR film and a TransScreen LE intensifying screen in an overnight exposure at −80°.

Figure 3A shows results of incubation of junction X26 (labeled in oligonucleotide 1, 2, 3, or 4) with Rad55-Rad57 or Cce1. Cce1 cleaves the HJ symmetrically at sites in strands 1 and 3, and 2 and 4. To map these sites precisely, it is possible to compare the fragments with those produced by a crude yeast extract (e.g., for a junction labeled in oligonucleotide 1, see lane a) and a known resolvase such as RuvC, as shown in lane b (Dunderdale *et al.*, 1994). The major (5'-ATT$^\downarrow$G-3') and minor (5'-TTT$^\downarrow$G-3') cleavage

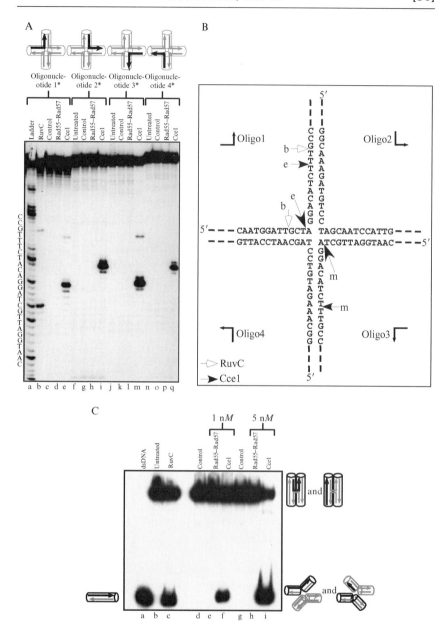

FIG. 3. Analysis of HJ resolution assays. (A) HJ substrate X26 (1 nM), individually 5′-^{32}P labeled on each oligonucleotide as indicated with an asterisk and schematically (bold strands), was incubated with preparations of Rad55–Rad57 (~1 nM), Cce1 (~1 nM), the equivalent volume of a control sample, or left untreated. Lane a: ladder produced by incubating X26 with

sites produced in strand 1 by RuvC (Shah *et al.*, 1994) are used to annotate the digestion ladder. It follows that Cce1 cuts strands 1 and 3 at one major (5'-CT$^{\downarrow}$A-3') and one minor (5'-CT$^{\downarrow}$T-3') site as shown in Fig. 3B. Under the conditions of this experiment, the test protein, Rad55-Rad57, fails to cut the junction. *Notes*: Sequence ladders can alternatively be produced by chemical cleavage of the appropriate 5'-^{32}P-labeled strand (Maxam and Gilbert, 1980). However, due to the nature of chemical cleavage, the product bands run as if they are approximately one nucleotide shorter than fragments of equivalent length produced by enzymatic cleavage. When resolution assays are carried out in larger volumes or to obtain sharper bands, samples can be extracted with phenol, followed by ethanol precipitation, before loading them onto the denaturing gel. When crude protein samples are assayed for resolvase activity it can be difficult to distinguish resolution from nonspecific degradation. In this case, we recommend analysis of a duplex DNA control substrate in parallel with X26. This should contain the ^{32}P-labeled strand of X26 annealed with its complement.

Native PAGE and Analysis

When analyzed by neutral PAGE, the reaction products are mixed with 3 μl native loading buffer (50% glycerol, 0.1% BPB, 0.1% XC in TBE) and applied to a 10% neutral acrylamide/bis-acrylamide (37.5:1) TBE gel. Using a 10-cm-long, 0.5-mm-thick gel (Engineering & Design Plastics Ltd), electrophoresis is carried out for 2 – 2.5 h at 120 V in TBE (the BPB front should reach the bottom of the gel). Gels are dried and exposed to X-ray film as described earlier.

As shown in Fig. 3C, RuvC and the positive control Cce1 partially transform the HJ into a fast-migrating nicked duplex product, which

an appropriate dilution of a crude yeast lysate to induce nonspecific degradation with the respective DNA sequence to the left. Lane b: X26 incubated with RuvC. Products were resolved by denaturing PAGE and the DNA visualized by autoradiography. Cce1 cleaves X26 symmetrically (compare lanes e/m and i/q). (B) Schematic of the homologous core region of X26 indicating the cleavage sites produced by Cce1 (black arrows) and RuvC (white arrows) as mapped in A. Labeled arrows correspond to the lane numbers in A. Note (i) that the RuvC sites were only mapped in oligonucleotide 1 and that equivalent cuts would occur in oligonucleotide 3 and (ii) that the junction point has the ability to move within the homologous core for which the DNA sequence is given. (C) Resolution of X26 by the proteins indicated, analyzed by neutral PAGE and autoradiography. Oligonucleotide 1 was 5'-^{32}P labeled as indicated in bold. Detectable DNA species containing the labeled stand are outlined in bold. Positions of a duplex DNA maker (X26 oligonucleotide 1 with its complement), the intact HJ (in both conformations), and nicked duplex products are indicated.

behaves like the duplex control consisting of oligonucleotide 1 with its complement (lane a). The Rad55–Rad57 heterodimer exhibits no activity under the assays conditions used here. *Note:* Cce1 protein is a true RuvC-like resolvase as indicated by (i) the introduction of symmetric nicks in substrate X26 and (ii) the coordination of these nicks in a way that allows resolution of the Holliday junction. The need to combine denaturing and native analyses to define a resolvase is highlighted by structure-specific nucleases that can give false positive results in one of the two assays. For example, it has been reported that under specific conditions fission yeast Slx1–Slx4 produces identical fragments from oligonucleotides 1 and 3 of HJ substrate X12 but fails to produce resolution products on native PAGE (Coulon *et al.*, 2004). This indicates that Slx1–Slx4 is capable of nicking the HJ at specific sites but is unable to promote the resolving nick and counternick combination. Similarly, Mus81-Eme1 can transform HJ substrates into nicked duplex products on native PAGE (Boddy *et al.*, 2001), but denaturing analysis reveals that the incisions are not symmetrical.

Adaptation to Analyze Immobilized Proteins

Resolution assays are traditionally carried out with both the protein of interest (in purified or crude form) and the substrate free in solution. However, we show here that the assay can be adapted to analyze proteins bound to an IgG matrix via a tandem affinity purification (TAP) tag (Rigaut *et al.*, 1999) fused to its C terminus. This opens the door for two applications: (1) TAP-tagged candidate proteins can be analyzed rapidly for resolution activity without the need for elaborate purification schemes and (2) a TAP fusion library can be screened for resolution activity in a high-throughput approach to identify novel resolvases.

As an example, we present a method to screen a yeast TAP fusion library (Ghaemmaghami *et al.*, 2003) for resolvase activity. The advantage of the TAP system is that the tag has been introduced in the yeast genome to ensure near-natural levels of the tagged target protein, which in turn favors the formation of normal protein complexes.

Cell cultures (100 ml) are grown in YPD to a density of $OD_{600\,nm} = 3$, collected by centrifugation, and disrupted using glass beads in 700 μl binding buffer containing 40 mM Tris–HCl, pH 7.5, 100 mM potassium acetate, 4% glycerol, 5 mM DTT, 0.1% NP-40, 5 mM NaF, 5 mM Na$_4$P$_2$O$_7$, and protease inhibitors (complete EDTA-free inhibitor cocktail tablets, Roche). Cell debris is removed by centrifugation, and supernatants (typically containing 5 mg of total protein) are mixed with 15 μl IgG Sepharose 6 fast flow (Amerhsam Biosciences) preequilibrated in the same buffer.

Binding is carried out for 1 h at 4° on a rotor, before the Sepharose-bound proteins are spun down and then washed for 15 min using the binding buffer. The Sepharose is collected again by centrifugation, the supernatant is removed, and proteins bound to the matrix are assayed for HJ resolution under standard assay conditions. The reactions need to contain 250 ng poly [dI-dC] competitor DNA to minimize nonspecific degradation. Finally, the DNA products are analyzed by neutral PAGE as described earlier, although in this case it is important to take care to load as little of the Sepharose as possible onto the gel.

The result of screening 17 samples from the yeast TAP fusion library is shown in Fig. 4. Cce1-TAP is seen to provide a strong signal (and may be used to calibrate the assay), whereas candidate proteins (lanes c–r) only produce background signals. Mms4-TAP (lane s) gives a positive signal as expected of the structure-specific nuclease Mus81-Mms4, demonstrating the functionality and sensitivity of the assay. Furthermore, as Mms4 can only function as part of the Mus81–Mms4 heterodimer, this result indicates the ability of the assay to analyze protein complexes rather than just the TAP-tagged proteins. Currently, we use this protocol for high-throughput screening in a 96-deep-well format for yeast proteins. However, TAP

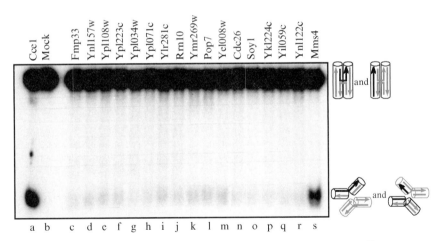

Fig. 4. Analysis of TAP-fusion proteins for HJ resolvase activity. X26 (5′-^{32}P labeled on oligonucleotide 1 shown in bold) was incubated with the indicated yeast TAP-tagged proteins immobilized on IgG Sepharose (see text). The products were analyzed by native PAGE and autoradiography. Detectable DNA species containing the labeled strand are outlined in bold. Positions of the intact HJ (in both conformations) and nicked duplex products are indicated. Cce1 and Mms4 give signals in excess of background.

methods are increasingly applied and improved in mammalian systems (Drakas *et al.*, 2005), and tools are being developed for the large-scale TAP tagging of mammalian genes (Zhou *et al.*, 2004). The assay should be applicable to permit the rapid analysis of any TAP-tagged protein.

Acknowledgments

We thank former post-docs and students who played important roles in the development of resolution assays. This work was supported by Cancer Research UK.

References

Bell, L., and Byers, B. (1979). Occurrence of crossed strand-exchange forms in yeast during meiosis. *Proc. Natl. Acad. Sci. USA* **76**, 3445–3449.

Benbow, R. M., Zuccarelli, A. J., and Sinsheimer, R. L. (1975). Recombinant DNA molecules of ØX174. *Proc. Natl. Acad. Sci. USA* **72**, 235–239.

Benson, F. E., and West, S. C. (1994). Substrate specificity of the *Escherichia coli* RuvC protein: Resolution of 3- and 4-stranded recombination intermediates. *J. Biol. Chem.* **269**, 5195–5201.

Birkenbihl, R. P., Neef, K., Prangishvili, D., and Kemper, B. (2001). Holliday junction resolving enzymes of archaeal viruses SIRV1 and SIRV2. *J. Mol. Biol.* **309**, 1067–1076.

Boddy, M. N., Gaillard, P. H. L., McDonald, W. H., Shanahan, P., Yates, J. R., and Russell, P. (2001). Mus81-Eme1 are essential components of a Holliday junction resolvase. *Cell* **107**, 537–548.

Collins, I., and Newlon, C. S. (1994). Meiosis-specific formation of joint DNA-molecules containing sequences from homologous chromosomes. *Cell* **76**, 65–75.

Constantinou, A., Davies, A. A., and West, S. C. (2001). Branch migration and Holliday junction resolution catalyzed by activities from mammalian cells. *Cell* **104**, 259–268.

Coulon, S., Gaillard, P. H. L., Chahwan, C., McDonald, W. H., Yates, J. R., and Russell, P. (2004). Slx1–Slx4 are subunits of a structure-specific endonuclease that maintains ribosomal DNA in fission yeast. *Mol. Biol. Cell* **15**, 71–80.

Doniger, J., Warner, R. C., and Tessman, I. (1973). Role of circular dimer DNA in the primary recombination mechanism of bacteriophage S13. *Nat. New Biol.* **242**, 9–12.

Drakas, R., Prisco, M., and Baserga, R. (2005). A modified tandem affinity purification tag technique for the purification of protein complexes in mammalian cells. *Proteomics* **5**, 132–137.

Dunderdale, H. J., Sharples, G. J., Lloyd, R. G., and West, S. C. (1994). Cloning, over-expression, purification and characterization of the *Escherichia coli* RuvC Holliday junction resolvase. *J. Biol. Chem.* **269**, 5187–5194.

Garcia, A. D., Aravind, L., Koonin, E. V., and Moss, B. (2000). Bacterial-type DNA Holliday junction resolvases in eukaryotic viruses. *Proc. Natl. Acad. Sci. USA* **97**, 8926–8931.

Ghaemmaghami, S., Huh, W.-K., Bower, K., Howson, R. W., Belle, A., Dephoure, N., O'Shea, E. K., and Weissman, J. S. (2003). Global analysis of protein expression in yeast. *Nature* **425**, 737–741.

Giraud-Panis, M. J. E., and Lilley, D. M. J. (1997). Near simultaneous DNA cleavage by the subunits of the junction-resolving enzyme T4 endonuclease VII. *EMBO J.* **16**, 2528–2534.

Holliday, R. (1964). A mechanism for gene conversion in fungi. *Genet. Res. Camb.* **5,** 282–304.

Kallenbach, N. R., Ma, R. I., and Seeman, N. C. (1983). An immobile nucleic acid junction constructed from oligonucleotides. *Nature* **305,** 829–831.

Kleff, S., Kemper, B., and Sternglanz, R. (1992). Identification and characterization of yeast mutants and the gene for a cruciform cutting endonuclease. *EMBO J.* **11,** 699–704.

Lilley, D. M. J. (2000). Structures of helical junctions in nucleic acids. *Q. Rev. Biophys.* **33,** 109–159.

Lilley, D. M. J., and White, M. F. (2001). The junction-resolving enzymes. *Nat. Revs. Mol. Cell. Biol.* **2,** 433–443.

Liu, Y., Masson, J.-Y., Shah, R., O'Regan, P., and West, S. C. (2004). RAD51C is required for Holliday junction processing in mammalian cells. *Science* **303,** 243–246.

Liu, Y., and West, S. C. (2004). Happy Hollidays: 40th anniversary of the Holliday junction. *Nat. Rev. Mol. Cell Biol.* **5,** 937–944.

Maxam, A. M., and Gilbert, W. (1980). Sequencing end-labeled DNA with base specific chemical cleavages. *Methods Enzymol.* **65,** 499–560.

Neef, K., Birkenbihl, R. P., and Kemper, B. (2002). Holliday junction-resolving enzymes from eight hyperthermophilic archaea differ in reactions with cruciform DNA. *Extremophiles* **6,** 359–367.

Potter, H., and Dressler, D. (1979). DNA recombination: *In vivo* and *in vitro* studies. *Cold Spr. Harb. Symp. Quant Biol.* **XLIII,** 969–985.

Rigaut, G., Shevchenko, A., Rutz, B., Wilm, M., Mann, M., and Seraphin, B. (1999). A generic protein purification method for protein complex characterization and proteome exploration. *Nat. Biotech.* **17,** 1030–1032.

Schwacha, A., and Kleckner, N. (1994). Identification of joint molecules that form frequently between homologs but rarely between sister chromatids during yeast meiosis. *Cell* **76,** 51–63.

Schwacha, A., and Kleckner, N. (1995). Identification of double Holliday junctions as intermediates in meiotic recombination. *Cell* **83,** 783–791.

Shah, R., Bennett, R. J., and West, S. C. (1994). Genetic recombination in *E. coli:* RuvC protein cleaves Holliday junctions at resolution hotspots *in vitro*. *Cell* **79,** 853–864.

Sharples, G. J. (2001). The X philes: Structure-specific endonucleases that resolve Holliday junctions. *Mol. Microbiol.* **39,** 823–834.

Sung, P. (1997). Yeast Rad55 and Rad57 proteins form a heterodimer that functions with replication protein-A to promote DNA strand exchange by Rad51 recombinase. *Genes Dev.* **11,** 1111–1121.

West, S. C. (1997). Processing of recombination intermediates by the RuvABC proteins. *Annu. Rev. Genet.* **31,** 213–244.

Zhou, D., Ren, J.-X., Ryan, T. M., Higgins, N. P., and Townes, T. M. (2004). Rapid tagging of endogenous mouse genes by recombineering and ES cell complementation of tetraploid blastocysts. *Nucleic Acids Res.* **32,** e128.

[31]   *In Vitro* Nonhomologous DNA End Joining System

By YUNMEI MA and MICHAEL R. LIEBER

Abstract

The nonhomologous end joining (NHEJ) pathway is the major pathway
that repairs DNA double strand breaks in multicellular eukaryotic organisms. Unlike homologous recombination, the NHEJ pathway utilizes minimal or no homology between the ends that need to be joined. Although the
resulting NHEJ-repaired junctions can be diverse in sequence, they share a
few common features, including frequent nucleolytic resection of the
ends, near-random junctional additions, and utilization of microhomology.
The *in vitro* NHEJ assay was developed in an attempt to recapitulate the
joining of incompatible ends with purified core proteins and some additional factors. This *in vitro* system allows further understanding of the
biochemical features of the pathway and evaluation of the functions of
other proteins in NHEJ.

Introduction

Double strand (ds) DNA breaks are one of the most deleterious forms
of damage to cells. Eukaryotes have developed two major DNA repair
pathways to minimize the damage from double strand breaks (Lieber *et al.*,
2003). The nonhomologous end joining (NHEJ) pathway functions throughout the cell cycle, and hence is the dominant pathway in multicellular
eukaryotes.

The NHEJ pathway in vertebrates employs at least six gene products:
Ku70, Ku86, DNA-PKcs, Artemis, DNA ligase IV, and XRCC4 (Lieber
et al., 2004). The first protein discovered in this group, Ku (consisting of
Ku70 and Ku86), was found to be present as an autoantigen in patients with
scleroderma-polymyositis overlap syndrome in 1981 (Mimori *et al.*, 1981).
DNA-PKcs was purified based on the fact that it is the only DNA-
dependent protein kinase in the cells (Lees-Miller and Anderson, 1989).
The critical roles of Ku and DNA-PKcs to NHEJ were revealed when it
was reported that they are the defective factors in the X-ray cross complementation (XRCC) cell lines (Finnie *et al.*, 1995; Gu *et al.*, 1997; Smider
et al., 1994; Taccioli *et al.*, 1994). Specifically, Ku70 and Ku 86 are encoded
by the XRCC6 and XRCC5 genes, respectively, and DNA-PKcs is encoded
by the XRCC7 gene. Subsequently, studies showed that Ku and DNA-
PKcs defective cell lines are also deficient in V(D)J recombination, which

METHODS IN ENZYMOLOGY, VOL. 408   0076-6879/06 $35.00

linked the NHEJ pathway to V(D)J recombination (Blunt *et al.*, 1995; Taccioli *et al.*, 1993, 1994). Biochemically, Ku is a sequence-independent, dsDNA end-binding protein and increases the affinity of the DNA-PKcs to the same DNA end. Once activated by DNA ends, DNA-PKcs regulates downstream NHEJ proteins by phosphorylation (Lieber *et al.*, 2003). Subsequently, the finding that DNA ligase IV is an *in vivo*-binding partner of the NHEJ protein XRCC4 revealed the function of XRCC4 as a DNA-binding and activity-stimulating factor of DNA ligase IV, and this protein complex is the only DNA ligase that can function in NHEJ (Grawunder *et al.*, 1997; Teo and Jackson, 1997; Wilson *et al.*, 1997). Cloning of the Artemis gene and its enzymatic characterization determined that this protein (along with DNA-PKcs) is most likely the nuclease involved in V(D)J recombination and the NHEJ pathway (Ma *et al.*, 2002; Moshous *et al.*, 2001; Rooney *et al.*, 2002, 2003). These studies also indicated that Artemis is one of the *in vivo* targets of DNA-PKcs.

Although there have been many biochemical studies of NHEJ proteins, most of these studies have focused on the properties of a single protein/protein complex or the interaction between two protein/protein complexes. Conversely, most of the *in vitro* end joining assays have utilized crude cell extracts and compatible ends (such as ends generated by restriction digestion). Biochemical reconstitution is one of the major ways to establish the functionality of a pathway, and this chapter describes methods for *in vitro* NHEJ of incompatible DNA ends with purified protein factors.

Procedures

Protein Purification

All resins and prepacked columns are from GE Healthcare (Piscataway, NJ) unless otherwise specified. All protein purification steps are carried out at 4° once the cells are collected.

Ku. C-terminal His-tagged Ku70 and nontagged Ku 86 are coexpressed in the baculovirus system and purified as described (Yaneva *et al.*, 1997). Briefly, the cell lysate from Hi5 cells that have been infected with the Ku70- and Ku86-expressing viruses is fractionated through Ni-nitrilotriacetic acid (Ni-NTA, Qiagen, Valencia, CA) batch purification, a double strand DNA Sepharose column (made with double strand oligonucleotide and CNBr-activated Sepharose 4B), and the Mono Q column.

DNA-PKcs. Native DNA-PKcs is purified as described (Chan *et al.*, 1996) except that HeLa cells were used as the source for purification in our laboratory. The nuclear lysate is prepared according to Dignam *et al.* (1983). Subsequently, the clarified lysate is fractionated using a DEAE

Sepharose column, an SP Sepharose column, a DEAE-Mg^{2+} Sepharose column, a double strand oligonucleotide Sepharose column, a Mono Q column, and a Mono S column.

Artemis. An Artemis-myc-His expressing plasmid was generated by inserting the Artemis coding sequence into the pcDNA6/myc-His vector (Version A, Invitrogen) (Ma *et al.*, 2002). For the purification of Artemis-myc-His, transfected cells are collected, washed in 1× phosphate-buffered saline, and resuspended in buffer L (lysis buffer, 25 mM HEPES, pH 7.9, 300 mM NaCl, 10 mM MgCl$_2$, 0.1% NP-40) with protease inhibitors (0.1 mM phenylmethylsulfonyl fluoride, 2 μg/ml leupeptin, 1 μg/ml pepstatin A, 2 μg/ml aprotinin) (Ma *et al.*, 2002). Then, the cell suspension is sonicated and centrifuged at 25,000g for 30 min at 4°, and then the supernatant is mixed with an anti-myc antibody (we routinely use clone 1–9E10.2 [National Cell Culture Center, Minneapolis, MN]). The antibody/cell lysate mixture is incubated for 15–30 min with gentle agitation (e.g., on a roller wheel) and then protein G-Sepharose is added. The antibody–protein G-Sepharose beads are incubated with protein fractions overnight and washed with buffer W (25 mM HEPES [pH 7.9], 650 mM KCl, 10 mM MgCl$_2$, 0.1% NP-40) thoroughly. The Artemis-bound protein G beads are finally washed with buffer LS (25 mM HEPES [pH 7.9], 10 mM MgCl$_2$, 2 mM dithiothreitol [DTT]) and frozen at −80°. These immunobeads can be used as the Artemis enzyme.

Because this is one-step purification, the purity of Artemis is dependent on how careful the washing step is carried out. We usually use at least 10 volumes of the settled beads for each wash, incubate the beads with the washing buffer for 5–10 min with gentle agitation, spin down the beads for 2–3 minutes at low speed,and then carefully pipette out the supernatant. The aforementioned process is repeated at least four times. Alternatively, the beads can be packed into a column for more efficient washing.

Starting with one 145-mm dish of 293T cells, we usually transfect the cells with 20–30 μg of plasmid, and we carry out immunoprecipitation with 20 μg of anti-myc antibody and 40 μl of 50% protein G-Sepharose for the lysate from these cells. Under these conditions, our yield is approximately 4 μg of Artemis-myc-His protein per dish of cells.

XRCC4:DNA Ligase IV. C-terminal His-tagged DNA ligase IV and nontagged XRCC4 are coexpressed in a baculovirus system and purified as described previously (McElhinny *et al.*, 2000). Briefly, the cell lysate from Hi 5 cells coinfected with DNA ligase IV- and XRCC4-expressing baculoviruses is fractionated by Ni-NTA (Qiagen) batch purification and a Mono Q column.

DNA Polymerase (pol) μ*, DNA Polymerase, and Terminal Deoxynucleotidyl Transferase (TdT).* The DNA polymerases used in this assay were obtained from Drs. M. Goodman (pol μ), O. Koiwai (pol λ), and F. Bollum (TdT) (Chang *et al.*, 1988; Dominguez *et al.*, 2000; Shimazaki *et al.*, 2002; Tippin *et al.*, 2004).

The concentration of purified proteins is estimated by comparing to bovine serum albumin (BSA) standards on a Coomassie blue-stained SDS–PAGE gel. The antibodies used to confirm the identity of the proteins by Western blotting are Ku [D6D8, 2D9, anti-p70 (Yaneva *et al.*, 1997)], DNA-PKcs (42–27, 25–4, 18–2), and Artemis (anti-myc antibody, National Cell Culture Center). The activities of DNA-PKcs are confirmed by the DNA-PKcs kinase assay (Yaneva *et al.*, 1997), and the activity of XRCC4: DNA ligase IV is evaluated by a ligation assay (Grawunder *et al.*, 1997).

Nonhomologous End Joining Assay

Substrate Design. In principle, any two DNA ends with incompatible sequences (i.e., ends that are unable to anneal in a manner that generates two nicks) are suitable substrates for the NHEJ assay. Each linear dsDNA molecule has two ends, and monitoring the intermolecular joining of one end of this substrate to the other end keeps the complexity of the outcomes at the lowest level (Ma *et al.*, 2004). One can also use more than one species of dsDNA and select for the desired products at the polymerase chain reaction (PCR) step. We chose oligonucleotides as the NHEJ substrate because the overhang sequence can be altered easily when synthetic DNA is used. If linear plasmid DNA that has incompatible overhang sequences is used, care should be taken when gel purifying the DNA to avoid contamination from a plasmid that has not been fully digested by restriction enzyme.

NHEJ Reaction. The NHEJ substrate (0.25 pmol) is incubated with different combinations of proteins in the NHEJ buffer [25 mM Tris (pH 8.0), 50 mM KCl, 2.5 mM MgCl$_2$, 1 mM DTT, 0.05 mg/ml BSA, 0.25 mM ATP, 10 μM dNTPs) in a total volume of 10 μl (Ma *et al.*, 2004). After the protein DNA mixture is incubated at 37° for 2 h, the mixture is treated with or without phenol/chloroform extraction (it did not make a detectable difference in our hands) and 10% of the joined products is amplified by 35 cycles of PCR (94°, 45 s, 55°, 1 min 15 s, 72°, 30 s) with one of the primers end labeled. We use the ratio of 20:1 for the two primers of the PCR reaction (in our case, the labeled primer:the cold primer = 20:1). This is an effective means to improve the signal-to-noise ratio in our system. The PCR products are resolved by 8 or 10% denaturing PAGE. The gel is then dried and exposed in a PhosphorImager cassette, and the screen is scanned in a phosphor scanner. Alternatively, the dried gel can be exposed directly to an X-ray film (such as Kodak BioMax MR X-ray film).

We note that the PCR conditions should be adjusted to satisfy conditions unique to each substrate. Accordingly, the size of PCR products to be analyzed will also vary based on the size of the substrate and the positions of the primers designed. If oligonucleotides are utilized as the end joining substrate, minor PCR products are likely to occur due to joining of the minor degraded substrate. Therefore, it is a good practice to gel purify the oligonucleotides prior to the end joining assay. For our *in vitro* end joining assay, we use oligonucleotides gel purified within the previous few weeks. Additionally, optimizing the PCR conditions will also improve the signal-to-noise ratio.

Sequencing of the Junctions of NHEJ Products

The radioactive PCR products can be cloned in bulk into Topo TA cloning vector pCR2.1 (Invitrogen, Calsbad, CA, Fig. 1). If certain end joining products need to be recovered, the portion of the dried gel containing bands of desired sizes (based on the obtained gel image) is exposed to an X-ray film if it is not done initially. After the exposure, the film and the gel are aligned, and bands of interest are excised from the dried gel and eluted in TE (10 m*M* Tris, 1 m*M* EDTA, pH8.0) by shaking the mixture at 37° for a few hours. In our case, bands ranging from 45 to 100 nucleotides were recovered. The eluted DNA (0.1–0.5%) is then amplified in a second round of PCR with unlabeled primers to enrich the joining products further (Fig. 1). PCR products from the second round are cloned into pCR2.1. Individual clones can be screened by restriction digestion (e.g., *Eco*RI if pCR2.1 is used) and then positive clones are sequenced on a Li-Cor sequencer (Li-Cor, Lincoln, NE) following the manufacturer's instruction.

General Remarks

1. NP-40 is not available commercially any more and can be replaced by Igepal CA-630 (e.g., Sigma).

2. One obvious concern to a system composed of purified proteins (and DNA) is contamination. Therefore, it is critical to obtain proteins purified carefully. In our practice, proteins with >95% purity judged by Coomassie blue staining suffice for the *in vitro* NHEJ assay.

3. In this *in vitro* end joining system, we intentionally used only one double strand oligonucleotide as the substrate. This limits the outcome of end joining to only three possibilities: ends A to A, ends B to B, and ends A to B (A and B stand for the two ends of the oligonucleotide substrate). In our experiments, we only detected the joining of two different ends (A to B) by PCR. We hypothesize that homologous end joining

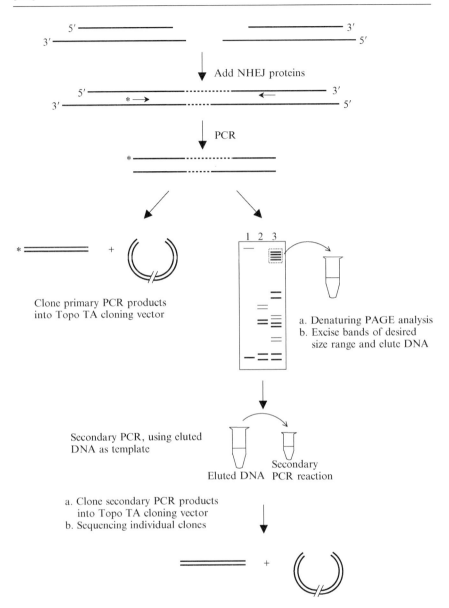

FIG. 1. *In vitro* NHEJ assay between double-stranded DNA molecules with incompatible ends and sequencing of the junctions. Schematic diagram of the *in vitro* NHEJ assay. A dsDNA oligonucleotide was mixed with NHEJ proteins to allow end joining between the incompatible ends of two molecules of the duplex substrate. After the end joining step, the junctions of products were amplified by PCR with one of the primers 5′ end labeled. The resulting PCR products were resolved by denaturing PAGE. For some NHEJ assays, the

(A to A or B to B) may result in hairpin or stem–loop formation during PCR. Therefore, these junctions may not be amplified efficiently. However, joining of homologous chromosomes and sister chromatids has been observed *in vivo* (Celli and de Lange, 2005; Hemann *et al.*, 2001). One can utilize more than one substrate DNA and select for the desired junctions by PCR. However, minor PCR products may arise with increasing numbers of substrate molecules in the assay. Alternatively, a plasmid fragment may be utilized in the end joining assay and the successfully circularized molecules can be selected for by bacterial transformation. However, the joining of plasmid DNA is very inefficient in our hands, raising the possibility that intermolecular end joining is more efficient than intramolecular joining in this *in vitro* end joining assay.

4. In order to estimate the efficiency of the *in vitro* end joining system, we used synthetic double strand oligonucleotides that mimic the joined junctions (Ma *et al.*, 2004). Although a necessary control for our purpose, introducing positive controls in assays that involve PCR may cause contamination problems. Care should be taken if PCR-positive controls are used.

Concluding Remarks

According to the aforementioned procedure, the joining of incompatible DNA ends is qualitatively achievable. Quantitatively, the *in vitro* NHEJ system can join 0.1 to 1% of the input incompatible ends. The *in vivo* NHEJ pathway is more efficient. There are two potential reasons for the difference between the current *in vitro* and the *in vivo* systems. First, some conditions might not be included in the *in vitro* end joining assay. Such conditions include, but are not limited to chromatin structure, nuclear skeleton, and protein concentration. Second, some of the *in vivo* physical and chemical conditions cannot be reproduced faithfully. These conditions may be the concentrations of small molecules (such as ions and ATP), the local concentrations of proteins and DNA ends, and the

---

primary PCR products (radioactive) were cloned directly into the vector (bulk cloning). In other cases, PCR products of desired sizes were excised from the denaturing gel, eluted out of the polyacrylamide gel, reamplified by a second round of PCR, and then cloned. The precise sequence and overhang configuration of the NHEJ substrates have been discussed elsewhere (Ma *et al.*, 2004). However, the sequence and overhang configuration are not critical because NHEJ joins all overhang configurations, regardless of sequence. The phrase incompatible ends is used here simply to refer to ends that require action by a polymerase (for fill-in synthesis) or nuclease (for removal of nucleotides) before both DNA strands of the duplexes can be ligated.

viscosity of the nucleoplasm. However, the ability of purified proteins to join DNA ends with no microhomology will allow optimization of the conditions to improve the efficiency, and the effect of other proteins on NHEJ can be tested with this system.

References

Blunt, T., Finnie, N. J., Taccioli, G. E., Smith, G. C., Demengeot, J., Gottlieb, T. M., Mizuta, R., Varghese, A. J., Alt, F. W., Jeggo, P. A., and Jackson, S. R. (1995). Defective DNA-dependent protein kinase activity is linked to V(D)J recombination and DNA repair defects associated with the murine scid mutation. *Cell* **80,** 813–823.

Celli, G. B., and de Lange, T. (2005). DNA processing is not required for ATM-mediated telomere damage response after TRF2 deletion. *Nature Cell Biol.* **7,** 712–718.

Chan, D. W., Mody, C. H., Ting, N. S., and Lees-Miller, S. P. (1996). Purification and characterization of the double-stranded DNA-activated protein kinase, DNA-PK, from human placenta. *Biochem. Cell Biol.* **74,** 67–73.

Chang, L. M., Rafter, E., Rusquet-Valerius, R., Peterson, R. C., White, S. T., and Bollum, F. J. (1988). Expression and processing of recombinant human terminal transferase in the baculovirus system. *J. Biol. Chem.* **263,** 12509–12513.

Dignam, J. D., Lebovitz, R. M., and Roeder, R. G. (1983). Accurate transcription initiation by RNA polymerase II in a soluble extract from isolated mammalian nuclei. *Nucleic Acids Res.* **11,** 1475–1489.

Dominguez, O., Ruiz, J. F., Lain de Lera, T., Garcia-Diaz, M., Gonzalez, M. A., Kirchhoff, T., Martinez, A. C., Bernad, A., and Blanco, L. (2000). DNA polymerase mu (Pol mu), homologous to TdT, could act as a DNA mutator in eukaryotic cells. *EMBO J.* **19,** 1731–1742.

Finnie, N. J., Gottlieb, T. M., Blunt, T., Jeggo, P. A., and Jackson, S. P. (1995). DNA-dependent protein kinase activity is absent in xrs-6 cells: Implications for site-specific recombination and DNA double-strand break repair. *Proc. Natl. Acad. Sci. USA* **92,** 320–324.

Grawunder, U., Wilm, M., Wu, X., Kulesza, P., Wilson, T. E., Mann, M., and Lieber, M. R. (1997). Activity of DNA ligase IV stimulated by complex formation with XRCC4 protein in mammalian cells. *Nature* **388,** 492–495.

Gu, Y., Jin, S., Gao, Y., Weaver, D. T., and Alt, F. W. (1997). Ku70-deficient embryonic stem cells have increased ionizing radiosensitivity, defective DNA end-binding activity, and inability to support V(D)J recombination. *Proc. Natl. Acad. Sci. USA* **94,** 8076–8081.

Hemann, M. T., Strong, M. A., Hao, L. Y., and Greider, C. W. (2001). The shortest telomere, not average telomere length, is critical for cell viability and chromosome stability. *Cell* **107,** 67–77.

Lees-Miller, S. P., and Anderson, C. W. (1989). The human double-stranded DNA-activated protein kinase phosphorylates the 90-kDa heat-shock protein, hsp90 alpha at two NH2-terminal threonine residues. *J. Biol. Chem.* **264,** 17275–17280.

Lieber, M. R., Ma, Y., Pannicke, U., and Schwarz, K. (2003). Mechanism and regulation of human non-homologous DNA end-joining. *Nature Rev. Mol. Cell. Biol.* **4,** 712–720.

Lieber, M. R., Ma, Y., Pannicke, U., and Schwarz, K. (2004). The mechanism of vertebrate nonhomologous DNA end joining and its role in V(D)J recombination. *DNA Repair (Amst.)* **3,** 817–826.

Ma, Y., Lu, H., Tippin, B., Goodman, M. F., Shimazaki, N., Koiwai, O., Hsieh, C. L., Schwarz, K., and Lieber, M. R. (2004). A biochemically defined system for mammalian nonhomologous DNA end joining. *Mol. Cell* **16**, 701–713.

Ma, Y., Pannicke, U., Schwarz, K., and Lieber, M. R. (2002). Hairpin opening and overhang processing by an Artemis/DNA-dependent protein kinase complex in nonhomologous end joining and V(D)J recombination. *Cell* **108**, 781–794.

Mimori, T., Akizuki, M., Yamagata, H., Inada, S., Yoshida, S., and Homma, M. (1981). Characterization of a high molecular weight acidic nuclear protein recognized by autoantibodies in sera from patients with polymyositis-scleroderma overlap. *J. Clin. Invest.* **68**, 611–620.

Moshous, D., Callebaut, I., de Chasseval, R., Corneo, B., Cavazzana-Calvo, M., Le Deist, F., Tezcan, I., Sanal, O., Bertrand, Y., Philippe, N., Fischer, A., and de Villartay, J. P. (2001). Artemis, a novel DNA double-strand break repair/V(D)J recombination protein, is mutated in human severe combined immune deficiency. *Cell* **105**, 177–186.

McElhinny, S. A., Snowden, C. M., McCarville, J., and Ramsden, D. A. (2000). Ku recruits the XRCC4-ligase IV complex to DNA ends. *Mol. Cell. Biol.* **20**, 2996–3003.

Rooney, S., Alt, F. W., Lombard, D., Whitlow, S., Eckersdorff, M., Fleming, J., Fugmann, S., Ferguson, D. O., Schatz, D. G., and Sekiguchi, J. (2003). Defective DNA repair and increased genomic instability in Artemis-deficient murine cells. *J. Exp. Med.* **197**, 553–565.

Rooney, S., Sekiguchi, J., Zhu, C., Cheng, H. L., Manis, J., Whitlow, S., DeVido, J., Foy, D., Chaudhuri, J., Lombard, D., and Alt, F. W. (2002). Leaky Scid phenotype associated with defective V(D)J coding end processing in Artemis-deficient mice. *Mol. Cell* **10**, 1379–1390.

Shimazaki, N., Yoshida, K., Kobayashi, T., Toji, S., Tamai, K., and Koiwai, O. (2002). Overexpression of human DNA polymerase lambda in *E. coli* and characterization of the recombinant enzyme. *Genes Cells* **7**, 639–651.

Smider, V., Rathmell, W. K., Lieber, M. R., and Chu, G. (1994). Restoration of X-ray resistance and V(D)J recombination in mutant cells by Ku cDNA. *Science* **266**, 288–291.

Taccioli, G. E., Gottlieb, T. M., Blunt, T., Priestley, A., Demengeot, J., Mizuta, R., Lehmann, A. R., Alt, F. W., Jackson, S. P., and Jeggo, P. A. (1994). Ku80: Product of the XRCC5 gene and its role in DNA repair and V(D)J recombination. *Science* **265**, 1442–1445.

Taccioli, G. E., Rathbun, G., Oltz, E., Stamato, T., Jeggo, P. A., and Alt, F. W. (1993). Impairment of V(D)J recombination in double-strand break repair mutants. *Science* **260**, 207–210.

Teo, S. H., and Jackson, S. P. (1997). Identification of *Saccharomyces cerevisiae* DNA ligase IV: Involvement in DNA double-strand break repair. *EMBO J.* **16**, 4788–4795.

Tippin, B., Kobayashi, S., Bertram, J. G., and Goodman, M. F. (2004). To slip or skip, visualizing frameshift mutation dynamics for error-prone DNA polymerases. *J. Biol. Chem.* **279**, 45360–45368.

Wilson, T. E., Grawunder, U., and Lieber, M. R. (1997). Yeast DNA ligase IV mediates nonhomologous DNA end joining. *Nature* **388**, 495–498.

Yaneva, M., Kowalewski, T., and Lieber, M. R. (1997). Interaction of DNA-dependent protein kinase with DNA and with Ku: Biochemical and atomic-force microscopy studies. *EMBO J.* **16**, 5098–5112.

[32] RAG and HMGB1 Proteins: Purification and Biochemical Analysis of Recombination Signal Complexes

By Serge Bergeron, Dirk K. Anderson, and Patrick C. Swanson

Abstract

Two lymphoid cell-specific proteins, called RAG-1 and RAG-2, initiate the process of antigen receptor gene rearrangement, termed V(D)J recombination, by assembling a protein–DNA complex with two recombination signal sequences (RSSs), each of which adjoins a different receptor gene segment, and then introducing a DNA double strand break at the end of each RSS. The study of RAG–RSS complex assembly and activity has been facilitated by the development of methods to purify the RAG proteins and members of the HMG-box family of high mobility group proteins such as HMGB1 that promote RAG binding and cleavage activity *in vitro*. This chapter describes the purification of recombinant truncated and full-length RAG-1 and RAG-2 expressed transiently in mammalian cells, as well as the purification of bacterially expressed full-length HMGB1. In addition, it details several experimental procedures used in our laboratory to study RAG–RSS complex formation and function *in vitro*.

Introduction

The antigen-binding variable domains present in immunoglobulins and T-cell receptors owe much of their structural diversity to a process known as V(D)J recombination, which is responsible for assembling the exons encoding these domains from component variable (V), diversity (D), and joining (J) gene segments (for review, see Bassing *et al.*, 2002). V(D)J recombination is initiated when a multiprotein complex minimally containing two lymphoid cell-specific proteins called RAG-1 and RAG-2 (the "RAG complex") binds a pair of recombination signal sequences (RSSs), each of which adjoins a different gene segment, and then catalyzes a DNA double strand break at each RSS (for reviews, see Fugmann *et al.*, 2000; Gellert, 2002). Each RSS contains a conserved heptamer and nonamer sequence, separated by either 12 or 23 bp of intervening DNA whose composition is more varied (12-RSS and 23-RSS, respectively). Rearrangement preferentially occurs between two RSSs whose spacer length differs (the "12/23 rule"), a restriction that promotes productive exon assembly.

METHODS IN ENZYMOLOGY, VOL. 408
0076-6879/06 $35.00
DOI: 10.1016/S0076-6879(06)08032-3

Four DNA ends are generated by RAG-mediated cleavage: two blunt, 5'-phosphorylated signal ends and two coding ends terminating in DNA hairpin structures. These recombination intermediates arise through sequential strand cleavage and strand transfer steps in which the RAG complex first nicks the RSS at the heptamer and then joins the resulting 3'-OH to the opposing DNA strand via direct transesterification. After DNA cleavage, signal ends are typically ligated heptamer to heptamer to yield a precise signal joint. In contrast, the joining of coding ends is usually imprecise, as hairpin opening, which is most likely catalyzed by a complex containing Artemis and the catalytic subunit of the DNA-dependent protein kinase (DNA-PKcs) (Ma et al., 2002), and subsequent end processing by enzymes that remove nucleotides or add them [terminal deoxynucleotidyltransferase, TdT (Komori et al., 1993)] often lead to gain or loss of nucleotides at the junction. Signal and coding joint formation is mediated by ubiquitously expressed proteins that comprise the nonhomologous end-joining (NHEJ) pathway of DNA double strand break repair, including Ku70, Ku80, XRCC4, and DNA ligase IV (Bassing et al., 2002).

Once RAG-1 and RAG-2 were identified as essential components of the V(D)J recombination apparatus (Oettinger et al., 1990; Schatz et al., 1989), efforts focused on purifying the RAG proteins for biochemical analysis. However, these early efforts were hampered by poor solubility of the full-length RAG proteins. Subsequent studies delineated portions of RAG-1 and RAG-2 essential for supporting V(D)J recombination in cell culture-based assays (Cuomo and Oettinger, 1994; Kirch et al., 1996; Sadofsky et al., 1993, 1994; Silver et al., 1993). These catalytically active "core" forms of RAG-1 and RAG-2 proteins were found to be more soluble than their full-length counterparts, a feature that facilitated their eventual purification. With these proteins in hand, the Gellert laboratory completed a series of elegant biochemical experiments to elucidate the mechanism of RAG-mediated RSS cleavage (McBlane et al., 1995; van Gent et al., 1996). Later studies revealed that binding of the RAG complex to isolated recombination signals (especially a 23-RSS) and RAG-mediated synapsis and cleavage of RSS pairs according to the 12/23 rule is promoted by "architectural" DNA binding proteins of the HMG-box family of high mobility group proteins (Hiom and Gellert, 1998; Sawchuk et al., 1997; van Gent et al., 1997). In this respect, the RAG complex exhibits similarities to several recombinases and transcription factors that interact with, and have their activity stimulated by, HMG-box proteins (Thomas and Travers, 2001). Whether HMG-box proteins facilitate V(D)J recombination in vivo remains unclear, as mice deficient in HMGB1 exhibit no apparent defects in lymphocyte development, although its absence causes lethal hypoglycemia within 24 h of birth (Calogero et al., 1999).

However, it must be noted that *in vitro*, RAG activity is comparably stimulated by either HMGB1 or the related HMGB2 protein (Swanson, 2002a), raising the possibility that the two proteins may play redundant roles in V(D)J recombination *in vivo*, an issue that has not yet been formally addressed. Nevertheless, HMG-box proteins are often included in biochemical assays of the RAG complex to exploit the benefit of enhanced RAG activity imparted by its association with these proteins. Although most biochemical assays have been performed with "core" RAG proteins, the growing recognition that "noncore" portions of RAG-1 and RAG-2 help regulate the activity of the RAG complex *in vivo* Akamatsu *et al.*, 2003; Dudley *et al.*, 2003; Liang *et al.*, 2002; Talukder *et al.*, 2004) naturally necessitated identifying methods to purify the full-length proteins for use in biochemical assays. We and others reported the successful purification of full-length forms of RAG-1 and RAG-2 and their preliminary biochemical characterization (Elkin *et al.*, 2003; Jiang *et al.*, 2004; Swanson *et al.*, 2004; Tsai and Schatz, 2003). This chapter describes methods to purify core and full-length RAG-1 and RAG-2 from transiently transfected 293 cells, as well as bacterially expressed full-length HMGB1, and details several procedures used to characterize the binding and cleavage activity of the RAG complex *in vitro*.

Expression Vectors

Baculovirus expression vectors containing murine core RAG-1 (residues 384–1008) and RAG-2 (residues 1–387), fused at the amino terminus to a maltose-binding protein (MBP) domain and at the carboxy terminus to a polyhistidine sequence and three copies of the myc epitope tag, were obtained from the laboratory of M. Gellert and have been described previously (McBlane *et al.*, 1995). DNA fragments encoding these fusion proteins were subcloned into the mammalian expression vector pcDNA1 (Invitrogen, Carlsbad, CA) using *Bam*HI and *Not*I. Plasmid constructs containing full-length murine RAG-1 or RAG-2 cDNAs were prepared in the laboratory of S. Desiderio and were used for generating DNA fragments encoding noncore portions of RAG-1 and RAG-2. Using this source material for subcloning or as templates for PCR, pcDNA1 expression constructs encoding various forms of wild-type and mutant core and full-length RAG-1 and RAG-2 fusion proteins have been generated (see Table I). The pcDNA1 plasmids are propagated in *Escherichia coli* strain MC1061/P3 (Invitrogen) and are typically purified from overnight 500-ml bacterial cultures (Luria broth supplemented with 30 μg/ml carbenicillin and 7.5 μg/ml tetracyclin) using a Qiagen plasmid maxi kit according to the manufacturer's instructions (Qiagen, Hercules, CA).

TABLE I
FORMS OF RAG-1 AND RAG-2 PURIFIED FROM 293 CELLS AFTER TRANSIENT TRANSFECTION WITH pcDNA1 RAG EXPRESSION PLASMIDS

Designation	Residues[a]	Tags[b]	Mutation(s)	Reference
cMR1hm	384–1008(R1)	MBP, His9, (c-Myc)3	None	Swanson and Desiderio (1998)
cMR1hm384/393	384–1008(R1)	MBP, His9, (c-Myc)3	Ala(384–393)	Swanson and Desiderio (1998)
cMR1hm394/403	384–1008(R1)	MBP, His9, (c-Myc)3	Ala(394–403)	Swanson and Desiderio (1998)
cMR1hm404/413	384–1008(R1)	MBP, His9, (c-Myc)3	Ala(404–413)	Swanson and Desiderio (1998)
cMR1hm414/423	384–1008(R1)	MBP, His9, (c-Myc)3	Ala(414–423)	Swanson and Desiderio (1998)
cMR1hm424/433	384–1008(R1)	MBP, His9, (c-Myc)3	Ala(424–433)	Swanson and Desiderio (1998)
cMR1hm434/443	384–1008(R1)	MBP, His9, (c-Myc)3	Ala(434–443)	Swanson and Desiderio (1998)
cMR1hm594/596	384–1008(R1)	MBP, His9, (c-Myc)3	Ala(594–596)	Swanson and Desiderio (1998)
cMR1hm607/611	384–1008(R1)	MBP, His9, (c-Myc)3	Ala(607–611)	Swanson and Desiderio (1998)
cMR1hm615/618	384–1008(R1)	MBP, His9, (c-Myc)3	Ala(615–618)	Swanson and Desiderio (1998)
cMR1hm995/998	384–1008(R1)	MBP, His9, (c-Myc)3	Ala(995–998)	Swanson and Desiderio (1998)
cMR1hm999/1003	384–1008(R1)	MBP, His9, (c-Myc)3	Ala(999–1003)	Swanson and Desiderio (1998)
cMR1hm1004/1008	384–1008(R1)	MBP, His9, (c-Myc)3	Ala(1004–1008)	Swanson and Desiderio (1998)
cMR1	384–1040(R1)	MBP	None	Swanson and Desiderio (1999)
cM2R1hm	384–1008(R1)	MBP2, His9, (c-Myc)3	None	Swanson and Desiderio (1999)
cM2R1	384–1040(R1)	MBP2	None	Swanson and Desiderio (1999)
cM2R1m	384–1008(R1)	MBP2, (c-Myc)3	None	Swanson (2001)
cMR1hm(wtNBD, mtAS)	384–1008(R1)	MBP, His9, (c-Myc)3	D600A, D708A or E962A	Swanson (2001)
cMR1hm(mtNBD, mtAS)	384–1008(R1)	MBP, His9, (c-Myc)3	Ala(384–393) and D600A, D708A, or E962A	Swanson (2001)
cM2R1m(wtNBD, mtAS)	384–1008(R1)	MBP2, (c-Myc)3	D600A, D708A or E962A	Swanson (2001)
cMR1m	384–1008(R1)	MBP, (c-Myc)3	None	Swanson (2002a)
FLMR1	1–1040(R1)	MBP	None	Swanson et al. (2004)
cMR2hm	1–387(R2)	MBP, His9, (c-Myc)3	None	Swanson and Desiderio (1998)
cMR2h	1–418(R2)	MBP, His9	None	Swanson and Desiderio (1998)
cM2R2hm	1–387(R2)	MBP2, His9, (c-Myc)3	None	Swanson and Desiderio (1999)
cMR2	1–387(R2)	MBP	None	Swanson (2002a)
FLMR2	1–527(R2)	MBP	None	Swanson et al. (2004)

[a] Murine sequence of RAG-1 (R1) or RAG-2 (R2).
[b] MBP, maltose-binding protein; MBP2, tandem MBP domains; c-Myc, SEQKLISEEDLRA epitope tag.

The prokaryotic expression construct pET11d-hHMGB1, encoding full-length human HMGB1 tagged at the amino terminus with polyhistidine, was generated in the laboratory of R. Roeder (Ge and Roeder, 1994). To gain insight into how HMGB1 promotes RAG-mediated RSS synapsis and cleavage *in vitro*, we have prepared and analyzed a large panel of truncated and mutant forms of hHMGB1 in RAG-binding and cleavage assays (Bergeron *et al.*, 2005). Generation of the various HMGB1 expression constructs is detailed extensively in that study and is not described further here.

Protein Expression and Purification

RAG Proteins

The human embryonic kidney fibroblast cell line 293 is used to transiently express the RAG proteins for purification. The cells are maintained under standard humidified conditions (37° and 5% CO_2) in $1\times$ Dulbecco's modified Eagle medium (containing high glucose, L-glutamine, and pyridoxine hydrochloride, but not sodium pyruvate; Invitrogen Life Technologies) supplemented with 50 units/ml penicillin G and 50 μg/ml streptomycin (BioWhittaker, Walkersville, MD) and 10% fetal bovine serum (FBS; v/v). We have used FBS from HyClone (Logan, UT), Invitrogen, and BioMeda (Foster City, CA) with comparable results. The final medium is sterile filtered through a 0.45-μm cellulose acetate membrane before use (Corning 500-ml filter system). Cells are grown to confluence in 10-cm culture dishes, resuspended in the same medium, and seeded into fresh medium (1:5 dilution) the day before transfection. The next morning, the medium is aspirated off and replaced with 10 ml fresh medium and returned to the incubator for 3 h. For transient transfection, instead of using the calcium phosphate precipitation method employed previously (Swanson, 2002b), we now use a polyethylenimine (PEI) transfection procedure adapted from one described by Durocher *et al.* (2002) to reduce media consumption, as a glycerol shock step with a medium change is not necessary with this procedure. An aqueous stock solution of PEI (Polysciences Inc., Warrington PA) is prepared (1 μg/μl, brought to pH 7.0 by the addition of HCl) and stored in aliquots at $-80°$. The stock solution is prepared immediately after opening a new bottle of PEI, as the transfection efficiency declines considerably if solutions are prepared from powdered PEI opened and stored at ambient temperature and atmosphere for longer than 4 weeks (unpublished observations). Plasmid DNA is added to 0.9% NaCl (w/v; sterile filtered) at a final concentration of 10 μg/ml. For coexpression of RAG-1 and RAG-2, 5 μg of each

RAG-expression vector is added; otherwise 10 μg of an individual plasmid is added. To this mixture, PEI is added at a final concentration of 30 μg/ml. After vortexing briefly, samples are incubated at 25° for 10 min, and then 1 ml of the DNA–PEI mixture is added to each dish of 293 cells. Typically, for each RAG protein preparation, two 7-ml solutions of DNA–PEI are assembled for transfection of 14 dishes of 293 cells. After incubation for 48 h, the medium is aspirated, and seven dishes are harvested in 5 ml of sterile phosphate-buffered saline–EDTA (137 mM NaCl, 27 mM KCl, 43 mM KH$_2$PO$_4$, 14 mM Na$_2$HPO$_4$, 2 mM EDTA) per plate and collected in a 50-ml conical tube on ice. The tubes are centrifuged at 274g for 10 min at 4° (Beckman GH3.8 rotor, 1300 rpm) and the supernatant is aspirated. Cell pellets are frozen in a dry ice/ethanol bath and stored at −80° until use.

For protein purification, all steps of the following protocols are performed at 4°. Individually expressed core MBP-RAG-1 and coexpressed MBP-RAG-1 and MBP-RAG-2 are purified using the same procedure. Briefly, each cell pellet is thawed on ice and resuspended in 3.75 ml buffer A [10 mM sodium phosphate (pH 7.4), 0.5 M NaCl, 1 mM dithiothreitol (DTT), 0.25% Tween 20 (v/v)], loaded into a Dounce tissue grinder (Wheaton Science Products, Millville, NJ), and subjected to 20 strokes of a type A pestle. The lysate is clarified by centrifugation at 85,000g (Beckman SW55Ti rotor, 30,000 rpm) for 40 min at 4° and the supernatants collected from two pellets are passed over 1 ml of amylose resin (New England Biolabs, Ipswich, MA) packed in a Poly-Prep chromatography column (Bio-Rad, Hercules, CA) equilibrated in buffer A by gravity flow. The column is washed with 10 ml buffer A (the final five lacking Tween 20), and the MBP-RAG protein(s) is eluted with buffer A containing 10 mM maltose (also lacking Tween 20). Protein-containing samples are dialyzed (Spectra/Por 25,000 MWCO, Spectrum Laboratories Inc., Rancho Dominguez, CA) against buffer R [25 mM Tris–HCl (pH 8.0), 150 mM KCl, 2 mM DTT, and 10% glycerol (v/v)]) for 3 h. Aliquots are snap frozen in liquid nitrogen and stored at −80° until use. For individually expressed MBP-RAG-2, we find that an alternative purification protocol, adapted from Spanopoulou and colleagues (1996), yields protein that is more active in cleavage assays than the method described earlier. Briefly, each thawed cell pellet (two total) is resuspended in 1.5 ml cold RSB buffer [10 mM Tris–HCl (pH 7.4), 10 mM NaCl, 5 mM MgCl$_2$, 1 mM phenylmethylsulfonyl fluoride (PMSF), 0.5% IGEPAL CA-630 (NP-40; v/v)], and allowed to swell for 5 min. Next, 2.25 ml cold buffer LSB [20 mM Tris–HCl (pH 7.4), 1 M NaCl, 0.2 mM MgCl$_2$, 1 mM PMSF, 0.2% IGEPAL CA-630 (NP-40; v/v)] is added and the sample is rocked at 4° for 1 h. The lysate is clarified by centrifugation as described earlier, and the pooled supernatants are applied to a column packed with amylose resin equilibrated in a 1:1.5 ratio of

RSB:LSB (RSB/LSB buffer). The column is washed with 5 volumes of RSB/LSB buffer and then with 4 volumes of buffer WB [20 mM Tris–HCl (pH 7.4), 0.5 M NaCl, 5 mM MgCl$_2$]. The MBP-RAG-2 fusion protein is eluted in buffer WB containing 10 mM maltose and dialyzed as described earlier.

HMGB1

The *E. coli* strain BL21(DE3)pLysS is transformed with a modified version of the pET11d-hHMGB1 expression construct. Bacteria are grown at 37° to an OD$_{600}$ of 0.5 in Luria broth supplemented with 50 μg/ml carbenicillin and 20 μg/ml chloramphenicol, and protein expression is induced by adding isopropyl-ß-D-thiogalactopyranoside to a final concentration of 1 mM. The culture is then incubated at 30° for 4 h, after which the cells are collected by centrifugation and the cell pellet is frozen at −80°. The conditions for the induction phase are derived empirically, as we observe that incubation at 30° yields considerably more protein than at 25 or 37°, and that limiting incubation to 4 h versus overnight also benefits overall expression.

A pellet from a 250-ml culture is resuspended with 7 ml of binding buffer [40 mM Tris–HCl (pH 8.0), 0.5 M KCl, 0.25% Tween 20 (v/v), 50 mM imidazole, 1 mM PMSF, 10 μM leupeptin, and 1 μM pepstatin A] and sonicated on ice using a Fisher Model 500 sonic dismembrator (10 cycles of 10-s pulse, 10-s pause, at 20% power; Fisher Scientific, Pittsburg, PA). The lysates are centrifuged at 46,000g (Beckman SW55Ti rotor, 22,000 rpm) for 40 min at 4°. Supernatants are recovered, passed through 0.45-μm PVDF syringe filters (Millex HV, Millipore, Billerica, MA), and purified by immobilized metal affinity chromatography (IMAC). Yield and purity were compared between the ProBond nickel-chelating resin (Invitrogen) and the chelating Sepharose fast flow (FF) resin (Amersham BioSciences-GE Healthcare, Piscataway, NJ), both of which utilize the iminodiacetic acid-chelating group. In our hands, FF reproducibly yields as much as fivefold more HMGB1 protein than the ProBond resin (with similar purity) and is therefore utilized in the purification protocol described later.

Briefly, FF slurry is loaded into an empty Poly-Prep chromatography column (1 ml bed volume), charged with 5 ml of nickel solution (100 mM NiSO$_4$, 40 mM Tris–HCl, pH 8.0), and washed with 10 ml of binding buffer. After applying the supernatant, the column is washed with 15 ml of binding buffer. To eliminate the void volume, 800 μl of elution buffer [40 mM Tris–HCl (pH 8.0), 0.5 M KCl, 500 mM imidazole, 1 mM PMSF, 10 μM leupeptin, 1 μM pepstatin A] is applied to the column and the flow through is discarded. The proteins are then eluted with 1.2 ml of elution buffer.

Analysis of HMGB1 purified by IMAC reveals the presence of several proteins. The major species are similarly detected by immunoblotting using a rabbit polyclonal anti-HMGB1 antibody raised to residues 166–181 (Pharmingen, San Diego, CA), strongly suggesting that all are derived from HMGB1. Furthermore, the same immunoblotting pattern is observed with unpurified whole bacterial lysates, even if protease inhibitors are added to the purification buffers, indicating that these products are present in bacteria prior to purification. Based on an earlier report suggesting that full-length HMGB1 is difficult to fully denature (Stros, 1998), we initially thought the products represented partially denatured proteins, but then decided to explore further purification using ion-exchange chromatography (IEC). Given the highly acidic nature of the C-terminal tail of HMGB1, we speculated that if it is shortened or eliminated by degradation, full-length HMBG1 could be resolved from degradation products by IEC on a positively charged matrix, as the full-length protein would be expected to bind most tightly to it. This turns out to be true, but two minor high molecular contaminants (~80 kDa) present after IMAC are not removed by this procedure, necessitating incorporation of an additional purification step using a negatively charged matrix. All purification steps involving IEC are performed using a BioLogic LP chromatography system (Bio-Rad). All washes and gradients are prepared using buffer A [40 mM Tris–HCl (pH 8.0), 1 mM DTT] and buffer B [40 mM Tris–HCl (pH 8.0), 1 mM DTT, 1 M KCl]. To purify full-length HMGB1 from bacterial contaminants, the eluate obtained from the Ni^{2+}-FF chromatography step is diluted to 65 mM KCl using buffer A and loaded onto the negatively charged Econo-Pac High S cartridge (Bio-Rad; 1 ml bed volume). Bound proteins are washed with 9 ml of 80 mM KCl and eluted using a KCl gradient of 80 mM to 0.65 M developed with buffer B (25 ml at 1.0 ml/min). Eluted protein fractions centered around 300 mM KCl are pooled, diluted twofold with buffer A, and loaded onto a positively charged Econo-Pac High Q cartridge (Bio-Rad; 1 ml bed volume). Bound proteins are washed with 10.5 ml of 0.22 M KCl (prepared by mixing buffers A and B) and eluted using a KCl gradient of 0.22 to 0.7 M developed with buffer B (40 ml at 1.0 ml/min). Protein fractions containing full-length HMGB1 eluting around 360 mM KCl are pooled, dialyzed overnight at 4° against 1 liter of dialysis buffer [25 mM Tris–HCl (pH 8.0), 150 mM KCl, 2 mM DTT, 10% glycerol], snap frozen in liquid nitrogen, and stored at −80° until use. The purity of the isolated proteins after each step of this purification procedure is shown in Fig. 1. HMGB1 purified using this protocol has the expected molecular weight as determined by mass spectroscopy. The purification of other truncated and mutant forms of HMGB1 is described elsewhere (Bergeron *et al.*, 2005).

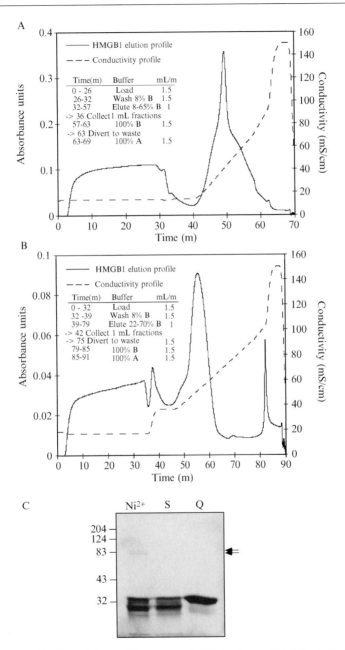

FIG. 1. Purification of bacterially expressed full-length, polyhistidine-tagged human HMGB1. (A) Eluate collected after IMAC was subjected to IEC using an Econo-Pac High S cartridge according to the conditions listed (inset). Conductivity and HMGB1 elution

Oligonucleotide Probe Preparation

Oligonucleotides are synthesized commercially (IDT Inc., Coralville, IA) and those over 35 bases in length are purified further by polyacrylamide gel electrophoresis by the vendor. Standard intact 12-RSS and 23-RSS substrates are assembled from two 50-mer (DAR39 and DAR40; 12-RSS) or two 62-mer (DG61 and DG62; 23-RSS) oligonucleotides (McBlane *et al.*, 1995). The "top" strand (in this case, DAR39 [5′-GATC TGGCCTGTCTTA<u>CACAGTG</u>CTACAGACTGGA<u>ACAAAAACCCT</u> GCAG-3′] or DAR61 [5′-GATCTGGCCTGTCTTA<u>CACAGTG</u>GTAG TACTCCACTGTCTGGCTGT<u>ACAAAAACCC</u>TGCAG-3′]) contains a 16 nucleotide coding sequence followed by consensus heptamer and nonamer sequences (underlined, separated by either 12 or 23 nucleotides), and 6 nucleotides 3′ of the nonamer. To prepare the substrate, top strand DNA (3.33 pmol) is 5′ end labeled in a reaction (20 μl final volume) containing [γ-^{32}P]ATP (Perkin Elmer Life Sciences, Boston, MA; 6000 Ci/mmol, 6.66 pmol) and T4 polynucleotide kinase (10 units; Invitrogen). After incubation at 37° for 40 min, the kinase is heat inactivated by incubation at 65° for 20 min. A small portion of the labeled strand is removed and the remaining DNA is annealed to a fivefold excess of its unlabeled complement. The same procedure can be used to assemble RSS substrates containing a preformed nick introduced at the 5′ end of the heptamer from three oligonucleotides (McBlane *et al.*, 1995). To evaluate the completeness of annealing and to remove excess unannealed DNA, the unannealed and annealed samples are loaded onto a 10% nondenaturing polyacrylamide gel [19:1 acrylamide:methylene(bis)acrylamide, prepared in 1× TBE using a SE400 gel apparatus] (Hoefer Inc., San Francisco, CA; 0.75-mm spacers and a 15-well comb) and fractionated at 250 V for 1.5–2 h at 25°. The gel is wrapped in Saran wrap and exposed briefly to film (BioMAX MR, Kodak).

To elute the DNA from the gel, an electroelution apparatus is used according to the manufacturer's instructions (Hoefer GE200 SixPac Gel Eluter). Using the developed film as a template, the annealed DNA is excised from the gel and placed into an elution tube containing 1× TAE (40 m*M* Tris-acetate, 1 m*M* EDTA; 300 μl]. Typically the gel slice is about

profiles are shown as dashed and solid lines, respectively. Fractions 11–24 were collected. (B) Fractions pooled from A were subjected to IEC using an Econo-Pac High Q cartridge according to the conditions listed (inset). Fractions 14–23 were pooled and dialyzed as described in the text. (C) Protein samples collected after IMAC (Ni^{2+}) and after subsequent S and Q IEC purification steps were fractionated by SDS–PAGE and stained with SYPRO Orange. High molecular weight contaminating proteins removed by IEC using the S cartridge are indicated by arrows.

7 mm long and 5 mm wide, which is then split into two pieces lengthwise. It is difficult to load more than this gel volume into a collection tube and still obtain efficient electroelution. The tube is capped with a blotting paper disk and a porous polyethylene plug, and then the bottom nub is cut off. The tube is placed into a 1.5-ml microcentrifuge tube containing 4× TAE (150 μl). The sample is subjected to electrophoresis at 60 V for 40 min with a 5-s reversed field pulse at the end of the run. The solution remaining around the gel slice is removed carefully with a micropipette and added to the solution in the collection tube. The pooled sample is loaded onto a G25 Sephadex desalting column (NAP-5, Amersham BioSciences) equilibrated in 10 mM Tris–HCl (pH 8.0). Fractions are collected (two or three drops each) and checked for radioactivity using a hand-held survey meter, and the activity of desired fractions is measured using an LSC6500 liquid scintillation counter (Beckman Coulter, Fullerton, CA). Using fresh [γ-^{32}P] ATP, the most radioactive fractions typically contain 50,000–100,000 cpm/μl ($\sim$1 × 10^7 cpm/pmol substrate).

In Vitro Cleavage Assays

The cleavage activity of RAG proteins is evaluated using an *in vitro* assay developed in the Gellert laboratory (McBlane *et al.*, 1995). In our laboratory, cleavage reactions are typically assembled by mixing $\sim$50 ng of each individually expressed RAG protein or $\sim$100 ng of coexpressed RAG protein (in a total of 4 μl dialysis buffer [20 mM Tris–HCl (pH 8.0), 150 mM KCl, 10% glycerol, 2 mM DTT]) and RSS substrate ($\sim$0.02 pmol) in a 10-μl reaction containing sample buffer [25 mM morpholinepropanesulfonic acid (MOPS)-KOH (pH 7.0), 60 mM potassium glutamate, 100 μg/ml bovine serum albumin, and 1 mM MgCl$_2$ or MnCl$_2$). In some reactions shown in Fig. 2, full-length HMGB1 (300 ng) and/or dimethyl sulfoxide (DMSO; 20%, v/v) has been included. Reactions are incubated for 1 h at 37°, quenched by adding 2 volumes of sample loading solution (95% formamide, 10 mM EDTA), and heated to 95° for 2 min. An aliquot of the sample (5 μl) is fractionated on a 15% polyacrylamide [19:1 acrylamide:methylene(bis)acrylamide] sequencing gel containing 7 M urea, and the cleavage products are visualized using a phosphorimager (Storm 860; Molecular Dynamics-GE Healthcare, Piscataway, NJ). Reaction products obtained after incubating an intact or nicked 23-RSS substrate with a wild-type or catalytically inactive (D600A) RAG complex in the presence of Mg^{2+} under various conditions are shown in Fig. 2. Two major products are observed after cleavage of an intact 23-RSS with a wild-type RAG complex in the absence of HMGB1: a small product that comigrates with the 16-mer nicked substrate and a product that comigrates with the sequence predicted

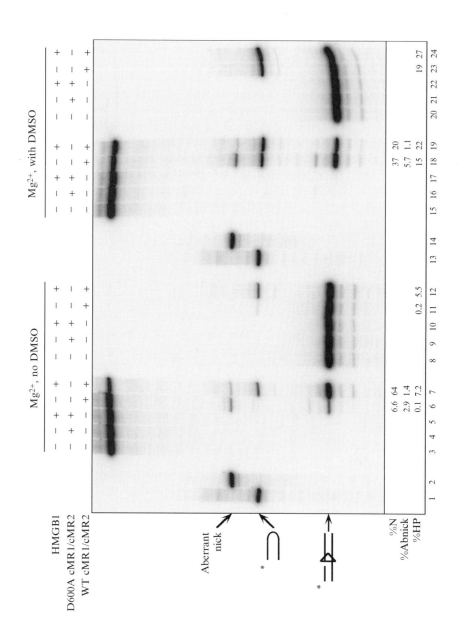

from aberrant nicking at position 27 (equivalent to nicking the substrate as a 12-RSS). Aberrant nicking of a 12-RSS substrate is not usually seen at high levels (Swanson, 2002a). In the presence of HMGB1, the amount of the nicked product is increased, but the aberrantly nicked product is less abundant, and a third product is observed that comigrates with the predicted 32 nucleotide hairpin sequence (which is not denatured on this gel). Interestingly, in cleavage reactions containing Mg^{2+}, RAG-mediated hairpin formation is increased in the presence of DMSO. Under these conditions, the addition of HMGB1 primarily promotes the correct placement of nicks, but does not enhance hairpin formation dramatically. By substituting $MnCl_2$ for $MgCl_2$ in the cleavage reaction, RAG-mediated hairpin formation can be readily visualized in the absence of HMGB1 or DMSO (McBlane et al., 1995).

Electrophoretic Mobility Shift Assays (EMSA)

Binding of the RAG complex to RSS substrates is evaluated by an electrophoretic mobility shift assay using a Hoefer SE400 apparatus. Before starting the experiment, a 4% polyacrylamide gel [29.2:0.8 acrylamide:methylene(bis)acrylamide; stock is filtered through a 0.45-μm cellulose acetate membrane] is poured in 1/2× TBE, allowed to polymerize for about 1 h, and brought to 4° in a cold room. Binding reactions are assembled similarly to cleavage reactions, except that potassium acetate is substituted for potassium glutamate and $MgCl_2$ (or $MnCl_2$) is replaced by $CaCl_2$, which supports RAG-mediated RSS binding, but not cleavage. In addition, DMSO (20%, v/v) is routinely included in the reaction. To assemble complexes on an isolated RSS substrate, reactions are incubated for 10 min at 25°. To form paired RSS complexes, the RAG complex is incubated with HMGB1 (300 mg) and the radiolabeled RSS for 1 min at

FIG. 2. In vitro cleavage assay using purified coexpressed core MBP-RAG proteins and purified full-length HMGB1. Intact (lanes 3–7 and 15–19) or nicked (lanes 8–12 and 20–24) 23-RSS substrates were incubated with a catalytically inactive or a wild-type RAG complex (D600A cMR1/cMR2 or WT cMR1/cMR2, respectively) in standard in vitro cleavage reactions containing Mg^{2+} in the absence (lanes 3–12) or presence (lanes 15–24) of 20% DMSO, with or without added HMGB1 (300 ng) as indicated above the gel. Samples were fractionated on a 15% polyacrylamide sequencing gel containing 7 M urea along with oligonucleotides representing the predicted hairpin (lanes 1 and 13) and aberrantly nicked (lanes 2 and 14) reaction products to serve as markers (positions indicated at left). Reaction products were quantified using a phosphorimager running the ImageQuant software and shown as a percentage of total DNA below the gel. N, 16-mer nicked product; HP, 32 nucleotide hairpin product; Abnick, 27-mer aberrant nicked product.

25°, after which a 50-fold excess of cold partner RSS is added (either intact or nicked, as appropriate). Binding reactions are then incubated at 25° for an additional 10 min. Next, samples are placed on ice for 5 min, and then 4 μl of cold loading solution (25% glycerol, 0.001% bromphenol blue) is added to the side of the tube above the sample. The loading solution is mixed into the sample by gentle flicking and the sample is immediately loaded onto the gel. The mixture is subjected to electrophoresis at 250 V for 1.5 h at 4°. Gels are dried onto Whatman cellulose chromatography paper (0.35 mm; Whatman Inc., Florham Park, NJ) and protein–DNA complexes are visualized using a phosphorimager.

In-Gel Cleavage Assays

We and others have shown that when the RAG proteins are incubated with standard oligonucleotide RSS substrates *in vitro* (either an isolated RSS or RSS pairs), multiple RAG–RSS complexes are detected by EMSA (Hiom and Gellert, 1998; Mundy *et al.*, 2002; Swanson, 2002b). To compare the activity of these discrete protein–DNA complexes on a single nondenaturing polyacrylamide gel directly, an in-gel cleavage assay was developed (Swanson, 2001). In this assay, binding reactions (scaled up fivefold) are assembled in the presence of Ca^{2+} as described in the previous section and fractionated using an EMSA. After electrophoresis is complete, the glass plates are separated so that the gel adheres to one of the plates. Next, a piece of Whatman paper is prewet in cleavage buffer (25 mM MOPS-KOH, 60 mM potassium glutamate, and 5 mM $MgCl_2$ or $MnCl_2$) and placed on top of the gel. The gel is then gently peeled onto the Whatman paper, and both are then submerged in 200 ml cleavage buffer prewarmed to 37° in a Pyrex dish. After incubation for 1 h without shaking, the gel is lifted from the cleavage buffer (using the Whatman paper as a support) and the DNA is transferred to DEAE cellulose chromatography paper (DE81, Whatman) in 1/2× TBE (45 mM Tris-borate, 1 mM EDTA) using a Bio-Rad trans blot apparatus running at 20 V for 18 h at 4°. After transfer, the DEAE cellulose paper is peeled off the gel together with the Whatman paper to avoid tearing it. The DEAE paper is then placed face down on Saran wrap, and the Whatman paper is separated from the DEAE paper. Next, the DEAE paper is wrapped in Saran wrap, taped onto a piece of Whatman paper, and exposed to film (BioMAX MR, Kodak, New Haven, CT) for several hours at −80°.

The desired bands on the film are cut out and the film is used as a template to cut the DEAE paper (which is easier if the paper remains frozen). The cut strips are placed into 1.5-ml microcentrifuge tubes and 300 μl of TES buffer [10 mM Tris–HCl (pH 8.0), 1 M NaCl, 1 mM EDTA] is added. After incubation for 30 min at 65°, supernatant is transferred to a centrifuge tube filter (Spin-X 0.22-μm cellulose acetate filter, Costar) and the DEAE strip is placed on the bottom of the filter insert. The tubes are pulsed for 15 s at high speed in an Eppendorf centrifuge, collecting the eluate in the bottom tube. The DEAE strip is then washed with an additional 100 μl of TES buffer. The pooled eluate is transferred to a 1.5-ml microcentrifuge tube and the DNA is precipitated by adding 3 *M* sodium acetate (pH 5.0) (40 μl), 10 μg tRNA as a carrier, and, after vortexing briefly, 100% ethanol (1 ml). After at least 30 min at −80°, the sample is centrifuged at 16,000g for 20 min at 25° using an Eppendorf microcentrifuge. The supernatant is decanted, and the inside of the tube is wiped dry with a rolled-up half of a Kimwipe, making sure to avoid the pellet. The pellet and side of the tube are washed with 0.5 ml cold 95% ethanol, taking care to retain the pellet, and the inside of the tube is wiped dry. The pellet is dried in a Savant Speed-Vac for at least several minutes to remove residual ethanol, and the pellet is dissolved in 7 μl of 80% formamide containing 0.001% bromphenol blue. The radioactivity in the samples is measured using the Cerenkov method, normalized, loaded onto a 15% polyacrylamide sequencing gel as described earlier, and the reaction products analyzed using a phosphorimager.

Discussion

RAG proteins are essential for the initiation of V(D)J recombination. However, until relatively recently, little was known about the assembly, organization, composition, and activity of the RAG–RSS complexes involved in this process. In addition, how the RAG proteins recognize DNA and what the relative roles of RAG-1 and RAG-2 are in this process remained unclear. With the development of methods to purify the RAG and HMG-box proteins, and the establishment of assays to examine the binding and cleavage activity of the RAG complexes, significant progress has been made in addressing many of these issues. The interested reader will find a fuller discussion of these advances in Swanson (2004). With purified full-length RAG proteins now available for biochemical analysis, the future promises new advances in our understanding of how noncore portions of RAG-1 and RAG-2 regulate aspects of RAG function, including the activity and fidelity of the RAG complex (Steen *et al.*, 1999; Talukder *et al.*, 2004), the ordering of antigen receptor gene rearrangement (Akamatsu *et al.*, 2003; Dudley *et al.*, 2003; Liang *et al.*, 2002), the suppression of alternative reaction outcomes mediated by the RAG complex (Elkin *et al.*, 2003; Sekiguchi *et al.*, 2001; Swanson *et al.*, 2004; Tsai and Schatz, 2003), the sequestration and degradation of RAG-2 (Corneo *et al.*, 2002;

Jiang *et al.*, 2005; Li *et al.*, 1996; Mizuta *et al.*, 2002; Ross *et al.*, 2003), and the ubiquitin ligase activity of RAG-1 (Jones and Gellert, 2003; Yurchenko *et al.*, 2003).

Acknowledgments

This work has been supported by grants to P.C.S. from the American Cancer Society (RSG-01-020-01-CCE) and the National Institutes of Health (R01 AI055599). Laboratory renovation was funded by the Research Facilities Improvement Program of the NIH National Center for Research Resources (C06 RR17417-01).

References

Akamatsu, Y., Monroe, R., Dudley, D. D., Elkin, S. K., Gartner, F., Talukder, S. R., Takahama, Y., Alt, F. W., Bassing, C. H., and Oettinger, M. A. (2003). Deletion of the RAG2 C terminus leads to impaired lymphoid development in mice. *Proc. Natl. Acad. Sci. USA* **100**, 1209–1214.

Bassing, C. H., Swat, W., and Alt, F. W. (2002). The mechanism and regulation of chromosomal v(d)j recombination. *Cell* **109** (Suppl.), S45–S55.

Bergeron, S., Madathiparambil, T., and Swanson, P. C. (2005). Both high mobility group (HMG)-boxes and the acidic tail of HMGB1 regulate recombination-activating gene (RAG)-mediated recombination signal synapsis and cleavage *in vitro*. *J. Biol. Chem.* **280**, 31314–31324.

Calogero, S., Grassi, F., Aguzzi, A., Voigtlander, T., Ferrier, P., Ferrari, S., and Bianchi, M. E. (1999). The lack of chromosomal protein Hmg1 does not disrupt cell growth but causes lethal hypoglycaemia in newborn mice. *Nature Genet.* **22**, 276–280.

Corneo, B., Benmerah, A., and Villartay, J. P. (2002). A short peptide at the C terminus is responsible for the nuclear localization of RAG2. *Eur. J. Immunol.* **32**, 2068–2073.

Cuomo, C. A., and Oettinger, M. A. (1994). Analysis of regions of RAG-2 important for V(D)J recombination. *Nucleic Acids Res.* **22**, 1810–1814.

Dudley, D. D., Sekiguchi, J., Zhu, C., Sadofsky, M. J., Whitlow, S., DeVido, J., Monroe, R. J., Bassing, C. H., and Alt, F. W. (2003). Impaired V(D)J recombination and lymphocyte development in core RAG1-expressing mice. *J. Exp. Med.* **198**, 1439–1450. Epub 2003 Oct 1427.

Durocher, Y., Perret, S., and Kamen, A. (2002). High-level and high-throughput recombinant protein production by transient transfection of suspension-growing human 293-EBNA1 cells. *Nucleic Acids Res.* **30**, E9.

Elkin, S. K., Matthews, A. G., and Oettinger, M. A. (2003). The C-terminal portion of RAG2 protects against transposition *in vitro*. *EMBO J.* **22**, 1931–1938.

Fugmann, S. D., Lee, A. I., Shockett, P. E., Villey, I. J., and Schatz, D. G. (2000). The RAG proteins and V(D)J recombination: Complexes, ends, and transposition. *Annu. Rev. Immunol.* **18**, 495–527.

Ge, H., and Roeder, R. G. (1994). The high mobility group protein HMG1 can reversibly inhibit class II gene transcription by interaction with the TATA-binding protein. *J. Biol. Chem.* **269**, 17136–17140.

Gellert, M. (2002). V(D)J recombination: RAG proteins, repair factors, and regulation. *Annu. Rev. Biochem.* **71**, 101–132.

Hiom, K., and Gellert, M. (1998). Assembly of a 12/23 paired signal complex: A critical control point in V(D)J recombination. *Mol. Cell* **1**, 1011–1019.

Jiang, H., Ross, A. E., and Desiderio, S. (2004). Cell cycle-dependent accumulation *in vivo* of transposition-competent complexes between recombination signal ends and full-length RAG proteins. *J. Biol. Chem.* **279**, 8478–8486. Epub 2003 Dec 8474.

Jiang, H., Chang, F. C., Ross, A. E., Lee, J., Nakayama, K., and Desiderio, S. (2005). Ubiquitylation of RAG-2 by Skp2-SCF links destruction of the V(D)J recombinase to the cell cycle. *Mol Cell* **18**, 699–709.

Jones, J. M., and Gellert, M. (2003). Autoubiquitylation of the V(D)J recombinase protein RAG1. *Proc. Natl. Acad. Sci. USA* **100**, 15446–15451. Epub 12003 Dec 15411.

Kirch, S. A., Sudarsanam, P., and Oettinger, M. A. (1996). Regions of RAG1 protein critical for V(D)J recombination. *Eur. J. Immunol.* **26**, 886–891.

Komori, T., Okada, A., Stewart, V., and Alt, F. W. (1993). Lack of N regions in antigen receptor variable region genes of TdT-deficient lymphocytes. [published erratum appears in *Science* **262**(5142),1957 (1993)]. *Science* **261**, 1171–1175.

Li, Z., Dordai, D. I., Lee, J., and Desiderio, S. (1996). A conserved degradation signal regulates RAG-2 accumulation during cell division and links V(D)J recombination to the cell cycle. *Immunity* **5**, 575–589.

Liang, H. E., Hsu, L. Y., Cado, D., Cowell, L. G., Kelsoe, G., and Schlissel, M. S. (2002). The "dispensable" portion of RAG2 is necessary for efficient V-to-DJ rearrangement during B and T cell development. *Immunity* **17**, 639–651.

Ma, Y., Pannicke, U., Schwarz, K., and Lieber, M. R. (2002). Hairpin opening and overhang processing by an Artemis/DNA-dependent protein kinase complex in nonhomologous end joining and V(D)J recombination. *Cell* **108**, 781–794.

McBlane, J. F., van Gent, D. C., Ramsden, D. A., Romeo, C., Cuomo, C. A., Gellert, M., and Oettinger, M. A. (1995). Cleavage at a V(D)J recombination signal requires only RAG1 and RAG2 proteins and occurs in two steps. *Cell* **83**, 387–395.

Mizuta, R., Mizuta, M., Araki, S., and Kitamura, D. (2002). RAG2 is down-regulated by cytoplasmic sequestration and ubiquitin-dependent degradation. *J. Biol. Chem.* **277**, 41423–41427.

Mundy, C. L., Patenge, N., Matthews, A. G., and Oettinger, M. A. (2002). Assembly of the RAG1/RAG2 synaptic complex. *Mol. Cell. Biol.* **22**, 69–77.

Oettinger, M. A., Schatz, D. G., Gorka, C., and Baltimore, D. (1990). RAG-1 and RAG-2, adjacent genes that synergistically activate V(D)J recombination. *Science* **248**, 1517–1523.

Ross, A. E., Vuica, M., and Desiderio, S. (2003). Overlapping signals for protein degradation and nuclear localization define a role for intrinsic RAG-2 nuclear uptake in dividing cells. *Mol. Cell. Biol.* **23**, 5308–5319.

Sadofsky, M. J., Hesse, J. E., and Gellert, M. (1994). Definition of a core region of RAG-2 that is functional in V(D)J recombination. *Nucleic Acids Res.* **22**, 1805–1809.

Sadofsky, M. J., Hesse, J. E., McBlane, J. F., and Gellert, M. (1993). Expression and V(D)J recombination activity of mutated RAG-1 proteins. [published erratum appears in *Nucleic Acids Res.* **22**(3),550 (1994)]. *Nucleic Acids Res.* **21**, 5644–5650.

Sawchuk, D. J., Weis-Garcia, F., Malik, S., Besmer, E., Bustin, M., Nussenzweig, M. C., and Cortes, P. (1997). V(D)J recombination: Modulation of RAG1 and RAG2 cleavage activity on 12/23 substrates by whole cell extract and DNA-bending proteins. *J. Exp. Med.* **185**, 2025–2032.

Schatz, D. G., Oettinger, M. A., and Baltimore, D. (1989). The V(D)J recombination activating gene, RAG-1. *Cell* **59**, 1035–1048.

Sekiguchi, J. A., Whitlow, S., and Alt, F. W. (2001). Increased accumulation of hybrid V(D)J joins in cells expressing truncated versus full-length RAGs. *Mol. Cell* **8**, 1383–1390.

Silver, D. P., Spanopoulou, E., Mulligan, R. C., and Baltimore, D. (1993). Dispensable sequence motifs in the RAG-1 and RAG-2 genes for plasmid V(D)J recombination. *Proc. Natl. Acad. Sci. USA* **90**, 6100–6104.

Spanopoulou, E., Zaitseva, F., Wang, F. H., Santagata, S., Baltimore, D., and Panayotou, G. (1996). The homeodomain region of Rag-1 reveals the parallel mechanisms of bacterial and V(D)J recombination. *Cell* **87**, 263–276.

Steen, S. B., Han, J. O., Mundy, C., Oettinger, M. A., and Roth, D. B. (1999). Roles of the "dispensable" portions of RAG-1 and RAG-2 in V(D)J recombination. *Mol. Cell. Biol.* **19**, 3010–3017.

Stros, M. (1998). DNA bending by the chromosomal protein HMG1 and its high mobility group box domains: Effect of flanking sequences. *J. Biol. Chem.* **273**, 10355–10361.

Swanson, P. C. (2001). The DDE motif in RAG-1 is contributed in trans to a single active site that catalyzes the nicking and transesterification steps of V(D)J recombination. *Mol. Cell. Biol.* **21**, 449–458.

Swanson, P. C. (2002a). Fine structure and activity of discrete RAG–HMG complexes on V(D)J recombination signals. *Mol. Cell. Biol.* **22**, 1340–1351.

Swanson, P. C. (2002b). A RAG-1/RAG-2 tetramer supports 12/23-regulated synapsis, cleavage, and transposition of V(D)J recombination signals. *Mol. Cell. Biol.* **22**, 7790–7801.

Swanson, P. C. (2004). The bounty of RAGs: Recombination signal complexes and reaction outcomes. *Immunol. Rev.* **200**, 90–114.

Swanson, P. C., Volkmer, D., and Wang, L. (2004). Full-length RAG-2, and not full-length RAG-1, specifically suppresses RAG-mediated transposition but not hybrid joint formation or disintegration. *J. Biol. Chem.* **279**, 4034–4044.

Talukder, S. R., Dudley, D. D., Alt, F. W., Takahama, Y., and Akamatsu, Y. (2004). Increased frequency of aberrant V(D)J recombination products in core RAG-expressing mice. *Nucleic Acids Res.* **32**, 4539–4549.

Thomas, J. O., and Travers, A. A. (2001). HMG1 and 2, and related "architectural" DNA-binding proteins. *Trends Biochem. Sci.* **26**, 167–174.

Tsai, C. L., and Schatz, D. G. (2003). Regulation of RAG1/RAG2-mediated transposition by GTP and the C-terminal region of RAG2. *EMBO J.* **22**, 1922–1930.

Van Gent, D. C., Hiom, K., Paull, T. T., and Gellert, M. (1997). Stimulation of V(D)J cleavage by high mobility group proteins. *EMBO J.* **16**, 2665–2670.

Van Gent, D. C., Mizuuchi, K., and Gellert, M. (1996). Similarities between initiation of V(D)J recombination and retroviral integration. *Science* **271**, 1592–1594.

Yurchenko, V., Xue, Z., and Sadofsky, M. (2003). The RAG1 N-terminal domain is an E3 ubiquitin ligase. *Genes Dev.* **17**, 581–585.

[33] Purification and Biochemical Characterization of Ataxia-Telangiectasia Mutated and Mre11/Rad50/Nbs1

By JI-HOON LEE and TANYA T. PAULL

Abstract

Ataxia-telangiectasia mutated (ATM) is a serine-threonine kinase that is activated by DNA double strand breaks to phosphorylate many cellular proteins involved in cell cycle regulation and DNA repair. We have shown previously that the activation of ATM can be reconstituted in an *in vitro* system using recombinant human ATM. In this system, ATM activity is dependent on the Mre11/Rad50/Nbs1 (MRN) complex and linear DNA, similar to requirements observed in human cells. This chapter describes methods used for the overexpression and purification of human ATM and MRN, as well as a protocol for *in vitro* kinase assays.

Introduction

Many diverse cellular stress responses, including apoptosis, cell cycle arrest, and gene induction, are controlled by phosphorylation and dephosphorylation. Ataxia-telangiectasia mutated (ATM) is a serine-threonine kinase that is activated when cells are exposed to DNA double strand breaks (DSBs) (Bakkenist and Kastan, 2003; Kim *et al.*, 2002). ATM phosphorylates a number of proteins involved in cell cycle checkpoint control, apoptotic responses, and DNA repair, including p53, Chk2, Chk1, Brca1, RPAp34, H2AX, SMC1, Rad17, and Nbs1 (Shiloh, 2003). Phosphorylation of cell cycle-related proteins by ATM initiates a series of events that ultimately halt the growth of damaged cells and facilitate repair (Abraham, 2001; Shiloh, 1997). ATM-dependent phosphorylation events in cells require the Mre11/Rad50/Nbs1 (MRN) complex, an assembly of proteins that plays a critical role in DNA double strand break repair, in addition to its role in ATM signaling (Carson *et al.*, 2003; Uziel *et al.*, 2003).

ATM exists as an inert dimer under normal physiological conditions and dissociates into active monomers upon DSB formation (Bakkenist and Kastan, 2003). We have investigated the kinase activity of recombinant human ATM (both monomeric and dimeric forms) in the presence of recombinant human MRN and DNA (Lee and Paull, 2004, 2005).

METHODS IN ENZYMOLOGY, VOL. 408

0076-6879/06 $35.00
DOI: 10.1016/S0076-6879(06)08033-5

These results were consistent with the observations made in human cells in that MRN and DNA were both required for the activation of ATM dimers by DNA double strand breaks. This chapter describes methods for the transient expression of ATM in human cells, purification of recombinant ATM and MRN by chromatography, and kinase assays of ATM.

Purification of Dimeric ATM

Transfection

Solutions and Materials

293-T cells (ATCC, Manassas, VA)

DMEM medium: 1× DMEM (Invitrogen Corporation, Grand Island, NY) plus 10% fetal bovine serum or FetalPlex animal serum complex (Gemini Bio-Products, Woodland, CA), and 3.7 g/liter sodium bicarbonate

Tissue culture dishes 245 × 245 × 25 (Nunclon, Roskilde, Denmark) pcDNA- FLAG-ATM wt N-terminal FLAG-tagged mammalian expression vector

(Bakkenist and Kastan, 2003)

HA-tagged ATM wt expression vector (Bakkenist and Kastan, 2003)

2× HEBS: 16 g/liter NaCl, 0.74 g/liter KCl, 0.198 g/liter NaHPO$_4$, 2 g/liter dextrose, 10 g/liter HEPES, adjusted to pH 7.2 and filter sterilized

2.5 M CaCl$_2$ in 10 mM HEPES: 0.238 g/100 ml HEPES, 27.75 g/100 ml CaCl$_2$, adjusted to pH 7.2 and filter sterilized

1× TE: 10 ml/liter Tris–HCl, pH 7.5, 2 ml/liter EDTA, pH 8.0, adjusted to pH 7.3 and filter sterilized.

Procedure. 293-T cells are transfected by the calcium phosphate coprecipitation technique, which is carried out as follows (this protocol is to transfect the cells in 100 ml media in a 245 × 245 × 25-mm dish). We use 24 dishes for the purification of dimeric ATM by this method.

1. Split cells to about 40–50% confluency at least 12 h prior to transfection. This gives them enough time to be 70–75% confluent at the time of transfection.
2. In one tube, aliquot 100 μg of pcDNA-FLAG-ATM wt and 200 μg of HA-tagged ATM wt expression vectors, 500 μl of 2.5 M CaCl$_2$ in 10 mM HEPES, and enough TE to bring the total volume to 5 ml. In another tube, aliquot 5 ml of 2× HEBS.

3. Add the CaCl$_2$/DNA mix to the 2× HEBS drop by drop. Do this while bubbling the mixture with a pipette. Fine opalescent precipitate should appear within 10 min.
4. Let the mixture stand at room temperature for 30 min.
5. Add the mixture directly to the surface of the media containing the cells (drop by drop) and swirl the plate gently to mix.
6. Incubate the cells at 37°/5% CO$_2$ overnight.
7. Remove the Ca$_3$(PO$_4$)$_2$ containing medium and replace with fresh DMEM medium.
8. Incubate the cells at 37°/5% CO$_2$ for 36–48 h, harvest the cells by gentle pipetting, and centrifuge (1500g) for 15 min.
9. Remove supernatant, and freeze pellets with liquid nitrogen and store at −80°.

Extraction Preparation

Solution

Lysis buffer: 25 mM Tris, pH 8.0, 250 mM NaCl, 20 mM MgCl$_2$, 0.5 mM phenylmethylsulfonyl fluoride (PMSF), 1 mM dithiothreitol (DTT), 20% glycerol.

Procedure

1. Resuspend the pellets with 50 ml of cold lysis buffer.
2. Homogenize with 50 strokes using a Dounce homogenizer (A type).
3. Centrifuge the homogenate at 10,000g for 15 min at 4°.
4. Transfer the supernatant into a 50-ml tube.

Anti-Flag Affinity Chromatography

Solutions and Materials

Buffer A: 25 mM Tris, pH 8.0, 100 mM NaCl, 1 mM DTT, 10% glycerol
Flag elution buffer: 100 μg/ml Flag peptide (Sigma, St. Louis, MO) in buffer A (5 ml)
Anti-Flag M2 affinity gel (Sigma) (1 ml).

Procedure

1. Mix the prepared cell extracts with 1 ml of anti-Flag M2 affinity gel and incubate at 4° for 1 h on a benchtop rotator.
2. Centrifuge the beads at 500g for 1 min and discard supernatant.
3. Wash the beads with lysis buffer and centrifuge again. Repeat washes two more times.

4. Pack the beads into a small chromatography column.
5. Wash the column at a flow rate of 0.25 ml/min using buffer A until the OD_{280} reaches a constant baseline. Flow rate should not exceed 0.25 ml/min, as the agarose beads are sensitive to high pressure.
6. Elute the flag-tagged ATM with 5ml of flag elution buffer: after running 1 column volume of flag elution buffer, pause the flow for 15 min in order to facilitate the removal of the bound ATM by the flag peptide.
7. Continue to flow and collect the eluted ATM protein.
8. To clean the resin, wash the column with 5 ml of 0.1 M glycine (pH 3.4) and then wash thoroughly with buffer A. This protocol generates approximately 50 mg total protein in 5 ml from 24 dishes (245 × 245 × 25 mm).

Anti-HA Affinity Chromatography

Solutions and Materials

HA elution buffer: 0.4 mg/ml HA peptide (AnaSpec Inc., San Jose, CA) in buffer A (2.5 ml)
Agarose-immobilized goat anti-HA antibody (Bethyl, Montgomery, TX) (200 μl)
Bio-Gel A-1.5m gel (Bio-Rad, Hercules, CA) or other inert resin (800 μl).

Procedure

1. Prepare a chromatography column with 0.2 ml of anti-HA agarose and 0.8 ml of Bio-Gel A-1.5m gel or other inert resin (or use 100% anti-HA agarose in 0.5–1.0 ml).
2. Equilibrate the column with buffer A.
3. Load the pooled fractions from the Flag column onto the HA column at 0.25 ml/min.
4. Wash the column at the same flow rate with buffer A until the OD_{280} reaches a constant baseline.
5. Elute HA-tagged ATM with 2.5 ml of HA elution buffer. After passing 1 column volume of HA elution buffer, pause the flow for 15 min in order to facilitate the removal of the bound ATM by the HA peptide.
6. Continue the flow and collect the eluted ATM protein.
7. To clean the resin, wash the column with 5 ml of 0.1 *M* glycine (pH3.4) and then wash thoroughly with buffer A.
8. Store in aliquots at −80° after flash freezing in liquid nitrogen.

This protocol generates approximately 1.5 to 2.5 μg total protein in 1.5 ml from 32 dishes (245 × 245 × 25 mm).

Purification of Monomeric ATM

The protocol just described is designed specifically to purify dimeric forms of ATM; however, it is also sometimes useful to prepare the monomeric form, particularly if ATM kinase assays are to be performed in the absence of the MRN complex (in the presence of manganese). Monomeric preparations generally require 10-fold smaller cell cultures.

Transfection

1. Prepare 20 to 25 100-mm dishes containing 293-T cells at 70–75% confluency, each containing 10 ml of DMEM medium.
2. Transfect cells with 10 to 20 μg of plasmid DNA (flag-tagged ATM) per dish.
3. Use the same calcium phosphate transfection technique as described earlier, except reduce the amount of reagents accordingly (~10-fold less).

Anti-Flag Affinity Chromatography

Solutions

TGN buffer: 50 mM Tris, pH 7.4, 150 mM NaCl, 1% Tween 20, 0.3% NP-40, 2 mM DTT, 10% glycerol

Lysis buffer: 0.5 mM PMSF and 1 mM NaF plus 3 mM DTT additionally in TGN buffer

Washing buffer: 0.5 M LiCl in TGN buffer

Flag elution buffer: 100 μg/ml Flag peptide in buffer A, 1% Tween 20.

Procedure

1. Resuspend the pellets with 10 ml lysis buffer.
2. Homogenize with 50 strokes using a Dounce homogenizer and then sonicate (three times 20 s using a microconicator tip at setting 2 to 3).
3. Centrifuge the homogenate at 50,000g for 30 min at 4°.
4. Load the supernatant onto a 1-ml column containing anti-Flag M2 antibody conjugated to agarose beads at a flow rate of 0.25 ml/min.
5. Wash the column with at least 5 ml of washing buffer containing 0.5 M LiCl.

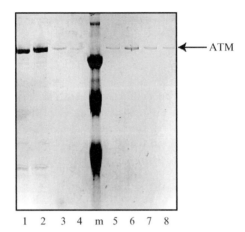

FIG. 1. Coomassie-stained SDS–PAGE gel of recombinant human ATM. Eluate from the anti-Flag column (lane 1), flow through from the anti-HA column (lane 2), and eluted fractions from the anti-HA column (lanes 3–8). Twenty microliters of the fractions were loaded in each lane. Molecular mass markers (m) are 210, 130, and 85 kDa.

6. Wash the column with buffer A until the OD_{280} reaches a constant baseline.
7. Elute the flag-tagged ATM with 5 ml of flag elution buffer containing 1% Tween. After passing 1 column volume of flag elution buffer, pause the flow for 15 min.
8. Continue the flow and collect the eluted protein.
9. To clean the resin, wash the column with 5 ml of 0.1 M glycine (pH 3.4) and then wash thoroughly with buffer A.
10. Pool the concentrated fractions of ATM together and store in aliquots at $-80°$ after flash freezing in liquid nitrogen (Fig. 1).

Purification of the MRN Complex

Transfection and Extraction Preparation

Solutions and Materials

Sf21 (or Sf9) cells
High-titer baculovirus prepared from transfer vectors containing human Rad50, Mre11, and Nbs1. Corresponding vectors from the Paull laboratory are pTP11 (C-terminal His6-tagged Rad50), pTP17 (C-terminal His6-tagged Mre11), and pTP36 (Nbs1), respectively

Nickel A buffer: 0.5 M KCl, 50 mM KH$_2$PO$_4$, pH 7.0, 5 mM imidazole, 20 mM β-mercaptoethanol, 10% glycerol

Lysis buffer: 0.5% Tween 20 and 2 mM PMSF in nickel A buffer.

Procedure

1. Infect 1.4 liter of Sf21 cells in suspension culture (or 5000 cm^2 of Sf21 cells in adherent culture) at an MOI of ~10 with baculovirus expressing Rad50, Mre11, and Nbs1 in combination (use a ratio of 3:1:2 or 3:1:3).
2. After 48 h, harvest the cells, wash the pellet with PBS, and freeze in liquid nitrogen.
3. To prepare the lysate, thaw and resuspend the cells in 100 ml of cold lysis buffer.
4. Resuspend the mixture briefly with a Dounce homogenizer (type A) to remove lumps and then sonicate (three times 20 s using a microconicator tip at setting 2 to 3).
5. Centrifuge the homogenate at 100,000g for 1 h at 4°.
6. Transfer and keep the supernatant into a tube at 4°.

Nickel Chromatography

Solutions and Materials

Nickel A buffer: 0.5 M KCl, 50 mM KH$_2$PO$_4$, pH 7.0, 5 mM imidazole, 20 mM β-mercaptoethanol, 10% glycerol

Nickel B buffer: 0.5 M KCl, 50 mM KH$_2$PO$_4$, pH 7.0, 250 mM imidazole, 20 mM β-mercaptoethanol, 10% glycerol

Buffer A: 25 mM Tris, pH 8.0, 100 mM NaCl, 1 mM DTT, 10% glycerol

Nickel-NTA superflow resin (Qiagen, Valencia, CA).

Procedure

1. Load the prepared cell extracts onto an ~5-ml column of nickel-NTA superflow resin equilibrated in nickel A buffer.
2. Wash the column with nickel A buffer until the OD$_{280}$ reaches a constant baseline.
3. Wash the column with 10% nickel B buffer (~30 mM imidazole) until the OD$_{280}$ reaches a constant baseline.
4. Elute the MRN complex with 50% nickel B buffer (~130 mM imidazole).
5. To clean the resin, wash the column with 100% nickel B buffer (250 mM imidazole).
6. Pool the fractions containing the MRN complex and dialyze into buffer A.

Ion-Exchange Chromatography

Solutions and Materials

Buffer A: 25 mM Tris, pH 8.0, 100 mM NaCl, 1 mM DTT, 10% glycerol

Buffer B: 25 mM Tris, pH 8.0, 1 M NaCl, 1 mM DTT, 10% glycerol

HiTrap Q Sepharose column (Amersham Biosciences, Piscataway, NJ).

Procedure

1. Load the pooled and dialyzed fractions from the nickel column onto the HiTrap Q Sepharose column equilibrated in buffer A.
2. Wash the column with nickel A buffer until the OD_{280} reaches a constant baseline.
3. Elute the MRN complex with 50% buffer B ($\sim$550 mM NaCl).
4. To clean the resin, wash the column with 100% buffer B (1 M NaCl).

Size-Exclusion Chromatography

Solutions and Materials

Buffer A : 25 mM Tris, pH 8.0, 100 mM NaCl, 1 mM DTT,10% glycerol

Superose 6 HR 10/30 column (Amersham Biosciences, Piscataway, NJ).

Procedure

1. Prepare loading sample containing 0.1% Tween 20 with best fraction from Q column.
2. Load the sample onto Superose 6 column equilibrated in buffer A.
3. Collect the eluted protein. The MRN complex is eluted in a sharp peak at $\sim$1.2 MDa in comparison to protein standards. This protocol generates approximately 100 μg total protein in 2 ml from 1.4 liter suspension culture or 5000 cm^2 adherent culture.

ATM Kinase Assay

In vivo, ATM is activated rapidly by DNA DSBs to phosphorylate many target proteins, including p53 and Chk2 (Shiloh, 2003). We have demonstrated that the DNA repair complex MRN stimulates monomeric ATM activity *in vitro* (Lee and Paull, 2004) and also serves as a DSB sensor to activate dimeric ATM (Lee and Paull, 2005). ATM kinase activity has also been shown by other groups to be manganese dependent (Chan *et al.*, 2000), and most experiments investigating ATM kinase activity have been performed using both magnesium and manganese. However, manganese concentrations *in vivo* are negligible, thus we have tested ATM kinase activity in the presence of magnesium alone, which resembles *in vivo* conditions. Our results have shown that the kinase activity of dimeric ATM is stimulated by MRN and DNA in the presence of magnesium alone (Lee and Paull, 2005). This section describes the method for the ATM kinase assay with MRN and DNA using p53 and Chk2 as model ATM substrates.

Kinase Reaction

Solutions and Materials

Purified proteins: ATM, MRN, and ATM substrates (GST-p53, GST-Chk2, or full-length proteins)

$2\times$ kinase buffer: 100 mM HEPES, pH 7.5, 100 mM KCl, 10 mM MgCl2, 2 mM ATP, 1 mM DTT, 10% glycerol.

Procedure

1. Mix 3 ng of dimeric ATM, 250 ng of MRN, 100 ng of p53 or 600 ng of Chk2, 10 ng of DNA for p53 or 2.5 ng of DNA for Chk2, and enough buffer A to bring the total volume to 20 μl.
2. Add 20 μl of $2\times$ kinase buffer.
3. Incubate the mixture at 30° for 1.5 h.

For kinase assays with monomeric ATM, 3 ng of dimeric ATM is substituted with 3 ng of monomeric ATM, keeping the rest of the protein components the same. However, the kinase assay with monomeric ATM is performed in a reaction containing both magnesium and manganese (10 mM each, final concentration). Under these conditions, ATM is active without the MRN complex and DNA, although the MRN complex still stimulates the kinase activity of monomeric ATM in the presence of manganese. Monomeric ATM does not require DNA (Lee and Paull, 2004).

Western Blotting

Solutions and Materials

Immobilon-FL PVDF (Millipore Corporation, Billerica, MA)Odyssey blocking buffer (Li-Cor Biotechnology, Lincoln, NE) Anti-p53, phospho-specific (Ser15) antibody (Ab-3) (Calciochem, San Diego, CA) Phospho-Chk2 (Thr68) antibody (Cell Signaling, Beverly, MA)Alexa Fluor 680-conjugated goat antirabbit IgG (Invitrogen, Grand Island, NY)

1× PBS: 8 g/liter NaCl, 0.2 g/liter KCl, 1.44 g/liter Na_2HPO_4, 0.24 g/liter KH_2PO_4, adjusted to pH 7.4

1× PBST: 0.1% Tween 20 in PBS Odyssey Imaging System (Li-Cor Biotechnology).

Procedure

1. Load kinase reactions onto a 10% SDS–PAGE gel and transfer the proteins to an Immobilon-FL PVDF membrane.
2. Block in Odyssey blocking buffer for 1 h at room temperature. Do not use BSA or milk.
3. Add primary antibody to the blocking buffer (first time use of primary) or use a previously made primary antibody solution. Use primary antibody at 1:2000 dilution.
4. Incubate for 2 h at room temperature or overnight at 4°.
5. Wash with PBST several times.
6. Incubate the membrane with secondary antibody solution in Odyssey blocking buffer that also includes 0.01% SDS. Use Alexa Fluor 680 goat antirabbit IgG at 1:10,000 dilution.
7. Incubate at room temperature for 1 to 2 h and protect from light from this point onward.
8. Wash with PBST several times and perform the final wash with PBS.
9. Dry the membrane before scanning at room temperature and keep it protected from light.
10. Scan the membrane using the Odyssey imaging system.

Alternatively, one can also carry out Western blotting using the traditional chemiluminescence technique. In this case, normal PVDF or nitrocellulose membrane can be used with 5% milk as a blocking solution and 1% BSA containing primary and secondary antibody solutions.

References

Abraham, R. T. (2001). Cell cycle checkpoint signaling through the ATM and ATR kinases. *Genes Dev.* **15,** 2177–2196.

Bakkenist, C. J., and Kastan, M. B. (2003). DNA damage activates ATM through intermolecular autophosphorylation and dimer dissociation. *Nature* **421,** 499–506.

Carson, C. T., Schwartz, R. A., Stracker, T. H., Lilley, C. E., Lee, D. V., and Weitzman, M. D. (2003). The Mre11 complex is required for ATM activation and the G2/M checkpoint. *EMBO J.* **22,** 6610–6620.

Chan, D. W., Son, S. C., Block, W., Ye, R., Khanna, K. K., Wold, M. S., Douglas, P., Goodarzi, A. A., Pelley, J., Taya, Y., Lavin, M. F., and Lees-Miller, S. P. (2000). Purification and characterization of ATM from human placenta: A manganese-dependent, wortmannin-sensitive serine/threonine protein kinase. *J. Biol. Chem.* **275,** 7803–7810.

Kim, S. T., Xu, B., and Kastan, M. B. (2002). Involvement of the cohesin protein, Smc1, in Atm-dependent and independent responses to DNA damage. *Genes Dev.* **16,** 560–570.

Lee, J. H., and Paull, T. T. (2004). Direct activation of the ATM protein kinase by the Mre11/Rad50/Nbs1 complex. *Science* **304,** 93–96.

Lee, J. H., and Paull, T. T. (2005). ATM activation by DNA double-strand breaks through the Mre11-Rad50-Nbs1 complex. *Science* **308,** 551–554.

Shiloh, Y. (1997). Ataxia-telangiectasia and the Nijmegen breakage syndrome: Related disorders but genes apart. *Annu. Rev. Genet.* **31,** 635–662.

Shiloh, Y. (2003). ATM and related protein kinases: Safeguarding genome integrity. *Nat. Rev. Cancer* **3,** 155–168.

Uziel, T., Leventhal, Y., Moyal, L., Andegeko, Y., Mittelman, L., and Shiloh, Y. (2003). Requirement of the MRN complex for ATM activation by DNA damage. *EMBO J.* **22,** 5612–5621.

Author Index

A

B

Subject Index

A

Abasic sites, *see* Apurinic/apyrimidinic lesions

ABH2
  AlkB mutant complementation
    studies, 110
  assays
    carbon-14 aldehyde release, 116
    carbon-14 methylated substrate
      preparation, 113–114
    crude cell extracts, 116–117
    direct reversion of methylated DNA
      base to unmodified form, 115–116
    high-performance liquid
      chromatography of remaining
      methylated bases, 115
    incubation conditions, 114–115
  functional overview, 108
  purification of histidine-tagged proteins
    baculovirus–Sf9 cell system, 113
    Escherichia coli, 112
  substrate specificity, 110
  tissue distribution, 110

ABH3
  AlkB mutant complementation
    studies, 110
  assays
    carbon-14 aldehyde release, 116
    carbon-14 methylated substrate
      preparation, 113–114
    crude cell extracts, 116–117
    direct reversion of methylated DNA
      base to unmodified form, 115–116
    high-performance liquid
      chromatography of remaining
      methylated bases, 115
    incubation conditions, 114–115
  functional overview, 108
  purification of histidine-tagged proteins
    baculovirus–Sf9 cell system, 113
    Escherichia coli, 112

Activation-induced cytidine deaminase

assay *in vitro*
  deaminase preparation, 163–165
  gel electrophoresis, 167–168
  incubation conditions, 166
  oligonucleotide purification, 166–167
  substrate preparation, 165–166
  uracil removal, 167
purification of recombinant human
    enzyme from *Escherichia coli*
  cell growth, 164
  chromatography, 165
  extraction, 164–165
  vectors, 163–164

AID, *see* Activation-induced cytidine
    deaminase

AlkB
  assays
    alkylated phage survival assays in
      Escherichia coli
      M13, 118
      overview, 117
      ϕK, 119
    carbon-14 aldehyde release, 116
    carbon-14 methylated substrate
      preparation, 113–114
    direct reversion of methylated DNA
      base to unmodified form, 115–116
    high-performance liquid
      chromatography of remaining
      methylated bases, 115
    incubation conditions, 114–115
  functional overview, 108–109
  human homologs, *see* ABH2; ABH3
  mechanism, 109
  purification
    histidine-tagged proteins from
      Escherichia coli, 112
    untagged recombinant enzyme, 113

AP sites, *see* Apurinic/apyrimidinic lesions;
    DNA polymerase β

Ape1, 2-deoxyribonolactone
    excision, 49

Apn1, *see* Apurinic/apyrimidinic lesions

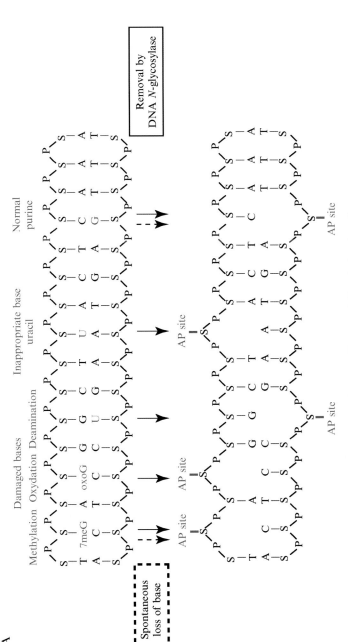

BOITEUX AND GUILLET, CHAPTER 6, FIG. 1. *(continued)*

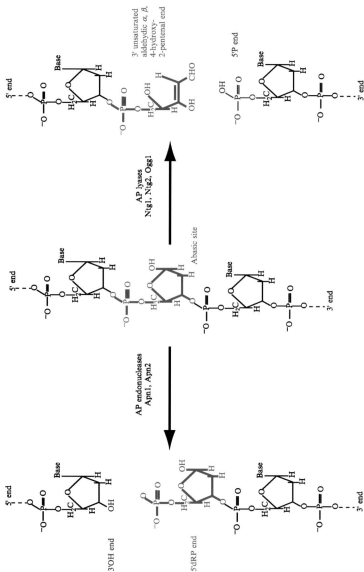

B

5' end

Base

3'OH end

OH

5'dRP end

OH

Base

3' end

AP endonucleases
Apn1, Apn2

Abasic site

OH

Base

5' end

Base

3' end

AP lyases
Ntg1, Ntg2, Ogg1

3' unsaturated
aldehydic α, β,
4-hydroxy-
2-pentenal end

CHO

OH

OH

5' end

Base

5'P end

OH

Base

3' end

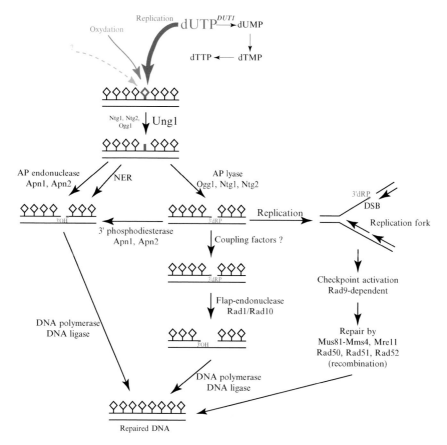

BOITEUX AND GUILLET, CHAPTER 6, FIG. 3. A model in progress. Genetic assays using *apn1 apn2 rad1* and *apn1 apn2 rad14* triple mutants allowed us to investigate the origin and repair of endogenous DNA damage in yeast. This model deals with our present knowledge that is necessarily incomplete and it applies to rapidly growing cells. Cell death is presumably due to a cascade of events that is initiated by the formation of AP sites in DNA and their subsequent conversion into SSBs and ultimately DSBs. These data demonstrate the impact of normal cellular metabolism on the genetic material in living organisms.

BOITEUX AND GUILLET, CHAPTER 6, FIG. 1. Origin and chemical nature of endogenous AP sites and associated 3′- and 5′-blocked SSBs in DNA. (A) AP sites in DNA are generated by spontaneous hydrolysis of the *N*-glycosylic bond of normal purines or damaged bases and by excision of normal, inappropriate and damaged bases by DNA *N*-glycosylases. P, phosphate; S, sugar (deoxyribose). (B) Central: Chemical structure of a regular AP site in DNA. Left: 5′-blocked SSB with a 5′-deoxyribose-phosphate end (5′-dRP). Right: 3′-blocked SSB with a 3′-unsaturated aldehydic (α,β-4-hydroxy-2-pentenal) end (3′-dRP). Because 3′- or 5′-blocked SSBs result primarily from the cleavage of AP sites by AP lyases and AP endonucleases, we suggest that these lesions constitute a single class. Lesions are in double-stranded DNA; because of space constraints, only one DNA strand is represented.

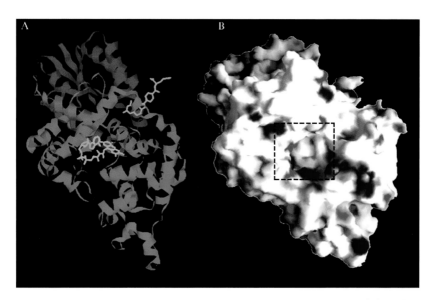

SANCAR AND SANCAR, CHAPTER 9, FIG. 2. Structure of *E. coli* photolyase. (A) A ribbon diagram of the backbone and the two chromophores: MTHF is shown in cyan and FAD is shown in yellow. The α/β nucleotide-binding domain is colored red and the helical FAD binding domain is colored green. (B) Electrostatic surface potential surrounding the substrate-binding site of *E. coli* photolyase. Note the central cavity leading from the surface of the enzyme to the interior and the surrounding band of positive electrostatic potential (blue areas). Areas of high negative potential are colored red.

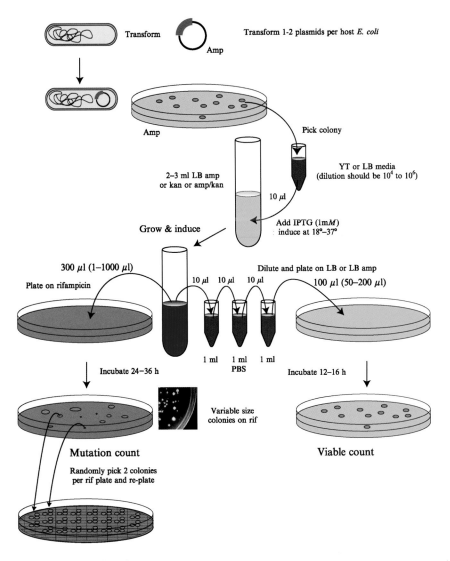

COKER *ET AL.*, CHAPTER 10, FIG. 1. Schematic of the *E. coli* mutation screen. An outline of the *E. coli* mutation assay using a mutable gene (*rpoB*) as a selection marker for deaminase activity (as described in detail in text).

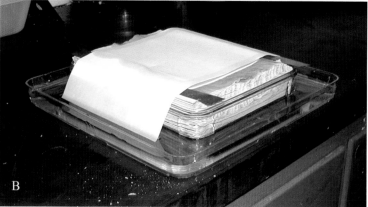

Spivak *et al.*, Chapter 14, Fig. 3. Assembly of blotting sandwich. (A) The agarose gel can be seen sitting on top of the membrane and blotting papers over a thick stack of paper towels fitted inside a plastic box. Strips of X-ray film were placed along the sides of the gel. (B) The wick covers the assembled blotting sandwich, drawing transfer solution through the gel and the membrane and into the paper towels.

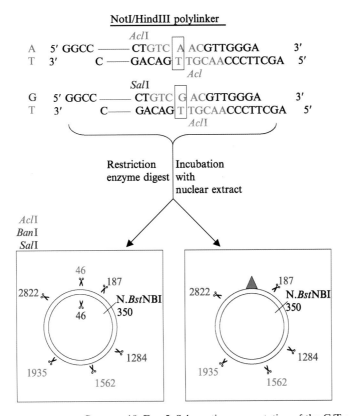

BAERENFALLER ET AL., CHAPTER 18, FIG. 5. Schematic representation of the G/T and G/C substrates. The relevant restriction sites and their nucleotide positions are shown.

STEP 1: Coating of magnetic beads with DNA substrate

STEP 2: Incubation of the coated beads with nuclear extract

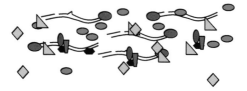

STEP 3: Isolation of the specifically-bound proteins in a MPC

STEP 4: Elution of the specifically bound proteins

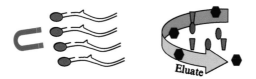

STEP 5: Analysis of the eluted proteins

BAERENFALLER *ET AL.*, CHAPTER 18, FIG. 7. Scheme of the DNA affinity purification of the MMR "repairosome."

A

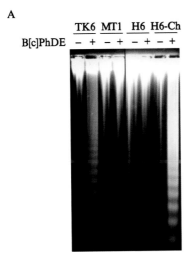

B

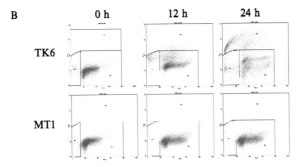

GU AND LI, CHAPTER 19, FIG. 4. MMR-mediated apoptosis in response to DNA damage. (A) DNA fragmentation analysis. Cells were treated with B[c]PhDE (0.25 μM) for 1 h and cultured in fresh medium for 24 h. Genomic DNA was isolated and fractionated through 1.5% agarose gels and visualized under UV light in the presence of ethidium bromide. (B) TUNEL analysis. Cells were treated with 0.25 μM B[c]PhDE at 37° for 1 h, cultured in fresh medium for various time as indicated, and subjected to TUNEL analysis as described in the text. Reproduced from Wu *et al.* (2003) with permission.

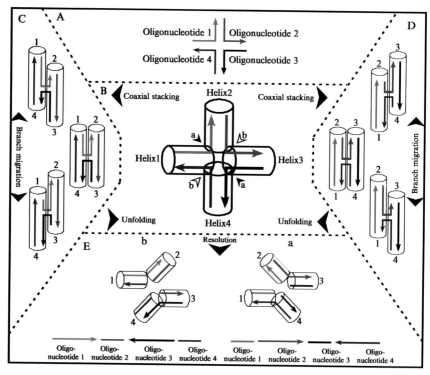

RASS AND WEST, CHAPTER 30, FIG. 1. Schematic representation of a HJ. (A) A synthetic HJ consists of four oligonucleotides (shown in four colors) with their 3′ termini indicated by arrows. (B) The partially complementary oligonucleotides form four helical segments marked as helixes 1–4. Nicks introduced by specialized nucleases at diametrically opposed sites a/a or b/b resolve the HJ. The open square planar form of the junction can undergo pairwise helical stacking of the arms as shown in C and D. In this conformation, pairs of strands are defined as continuous or exchanging. The continuous strands lie in antiparallel orientation. The crossover point, where the exchanging strands pass from one axis to the other, can move (branch migrate) within the homologous core. (E) Resolution of the junction gives rise to two nicked duplex products. Depicted below are the 5′ fragments of the respective oligonucleotides.